2nd International Conference on Materials Science and Technology (ICMST 2016)

IOP Conference Series: Materials Science and Engineering
Volume 360

Kerala, India
5 – 8 June 2016

ISBN: 978-1-5108-7366-7
ISSN: 1757-8981

Printed from e-media with permission by:

Curran Associates, Inc.
57 Morehouse Lane
Red Hook, NY 12571

Some format issues inherent in the e-media version may also appear in this print version.

Printed by Curran Associates, Inc. (2018)

For permission requests, please contact the Institute of Physics
at the address below.

Institute of Physics
Dirac House, Temple Back
Bristol BS1 6BE UK

Phone: 44 1 17 929 7481
Fax: 44 1 17 920 0979

techtracking@iop.org

Additional copies of this publication are available from:

Curran Associates, Inc.
57 Morehouse Lane
Red Hook, NY 12571 USA
Phone: 845-758-0400
Fax: 845-758-2633
Email: curran@proceedings.com
Web: www.proceedings.com

Table of contents

Volume 360

Second International Conference on Materials Science and Technology (ICMST 2016)

5–8 June 2016, Kerala, India

Accepted papers received: 1 May 2018

Preface

Papers

IOP Conf. Series: Materials Science and Engineering **360** (2018) 011001 doi:10.1088/1757-899X/360/1/011001

FROM THE CONVENOR'S DESK

The Department of Physics, St. Thomas College Pala, is highly privileged to organize the Second International Conference on Materials Science and Technology (ICMST 2016) during 5 - 8 June 2016, and as Convenor of the conference it is with legitimate pride and immense gratitude to God that I remember the most enthusiastic responses received for this from scientists all over the world. Following the grand success of our First International Conference ICMST 2012, we have provided once again a really exciting opportunity to consider the tremendous revolutionary changes in Materials Science and Technology and it is quite integrative and challenging to synchronize an extensive dialogue on issues pertinent to the field. That experts of diverse disciplines from across the world were sitting under one roof for four days, exchanging views and sharing findings, was a speciality of this conference. The event has evoked excellent responses from all segments of the Materials Science community worldwide. Around 600 renowned scholars from 30 countries participated in this. We were uniquely honoured to have Prof P. J. Kurien , Honourable Deputy Chairman, Rajya Sabha of India to inaugurate this conference. The presence of illustrious scientists and delegates including the Nobel Laureates Prof. Ada E Yonath from Israel and Prof. Klaus von Klitzing from Germany made the conference rise up to the standards of a distinguished institution like St. Thomas College, Pala.

A conference of this kind provides a nonpareil, a distinctly outstanding platform for the Scholars, Researchers and the Scientists to discuss and share ideas with delegates from all over the world. This had been most fruitful to the participants in identifying new collaborations and strengthening existing relations.

May I take this opportunity to thank all those who have contributed their valuable share, diverse in tone and nature, in the making of this conference.

My whole hearted gratitude is due to the International and National members of the advisory committee especially Dr.Reji Philip of Raman Research Institute, Bangalore, India for their valuable guidance and involvement. I place on record my heartfelt gratitude to our sponsors. I am sure that this conference has contributed to more innovations and inventions in the field and promoted the advancement of Science and Technology in India.

I wish all the success to all who contributed papers to the conference.

Dr. Ginson P. Joseph
Convenor, ICMST 2016

IOP Conf. Series: Materials Science and Engineering **360** (2018) 011001 doi:10.1088/1757-899X/360/1/011001

ICMST 2016: A few snaps

Prof. PJ Kurien, Honourable Deputy Chairman Rajya Sabha of Indian Parliament inaugurating the ICMST 2016

Audience of the Inaugural Session of ICMST 2016

IOP Conf. Series: Materials Science and Engineering **360** (2018) 011001 doi:10.1088/1757-899X/360/1/011001

His Excellency Mar Joseph Kallarangattu, Patron of our college delivering the keynote address

Nobel Laureate Prof. Ada E Yonath giving Plenary talk

IOP Conf. Series: Materials Science and Engineering **360** (2018) 011001 doi:10.1088/1757-899X/360/1/011001

Nobel Laureate Prof. Klaus von Klitzing, Max-Planck-Institut für Festkörperforschung, Germany giving a Plenary talk

Prof. Dr. Katharina Al-Shamery Carl von Ossietzky University Oldenburg, Germany, giving Plenary talk

Prof. Ravi Subrahmanyan, Director, Raman Research Institute, Bangalore India, giving Plenary talk

Dr. Reji Philip, Raman Research Institute, Bangalore, India, giving Plenary talk

A scene from Poster Presentations

Prof. Norbert Kroo, Presidential Advisor & Vice-President, Hungarian Academy of Science, distributing prizes for best poster and oral presentations awards

Peer review statement

All papers published in this volume of *IOP Conference Series: Materials Science and Engineering* have been peer reviewed through processes administered by the proceedings Editors. Reviews were conducted by expert referees to the professional and scientific standards expected of a proceedings journal published by IOP Publishing.

IOP Conf. Series: Materials Science and Engineering **360** (2018) 012001 doi:10.1088/1757-899X/360/1/012001

Synthesis of ZnO Nanoparticle and Evaluation of its Potential to Enhance the Activity of Antibiotics

Reshma S. Kurup[a], Elizabeth Cherian[b], Preema C. Thomas[a]*

[a]*Department of Physics, CMS College, Kottayam, Kerala, India -686001.*

[b]*Department of Botany, CMS College, Kottayam, Kerala, India – 686001.*

**Corresponding author Email:preemacthomas@gmail.com*

Abstract

The properties of zinc oxide (ZnO) vary with its crystallite size or particle and often nanocrystalline ZnO is seen to exhibit superior physical and chemical properties due to their higher surface area and modified electron structure. This paper aims at synthesizing nanosized ZnO particles and the evaluation of its potential to enhance the activity of antibiotics. ZnO is prepared by using both hydrothermal and solvothermal methods. The sample was characterized by X-Ray Diffraction (XRD). The nanosize of particles was further confirmed by using TEM studies. The average crystal size of the prepared ZnO nanoparticles was determined by TEM. The synthesized nanoparticle was used to enhance the antibacterial efficacy of antibiotics such as Ampicillin and Streptomycin. It was found effective in both drugs against the clinically important bacterial strains tested. Drug resistance is an emerging crisis and becoming a menace to the public. As the bacterial strains showing antibiotic resistance are escalating day by day, application of these nanoparticles that enhance the antibacterial property will be promising for the development of new drugs.

Keywords: Hydrothermal, ZnO nanoparticles, TEM, Antibacterial, disc diffusion method

1. Introduction

Nanoparticles usually ranging in dimension from 1-100 nanometers (nm) have properties unique from their bulk equivalent. They possess unique physico-chemical, optical and biological properties which can be manipulated suitably for desired applications [1]. Recent advances in the field of nanosciences has lead to the synthesis of nanosized inorganic and organic particles which are finding increasing applications in industrial, medicine and therapeutics, synthetic textiles and food packaging products [2]. With a wide band gap of 3.37eV Zinc Oxide (ZnO) is one of the most important and widely used metal oxides[3,4]. ZnO can be synthesized into a variety of one-dimensional nano structures like nanorods, nanowires, nanorings, nanosheets, etc., which have a wide range of applications [5-7]. The metallic nanoparticles are thoroughly being explored and

IOP Conf. Series: Materials Science and Engineering **360** (2018) 012001 doi:10.1088/1757-899X/360/1/012001

extensively investigated as potential antimicrobials. Nanoparticles of silver have been studied as a medium for antibiotic delivery, and to synthesize composites for use as disinfecting filters and coating materials [8-10]. The bactericidal effect of silver nanoparticles typically ranging from 2 to 5 nm has been investigated using green fluorescent protein (GFP)-expressing recombinant Escherichia coli [11]. It was reported that the antibacterial activity of gold nanoparticles can be increased by adding antibiotics [12].there are several such reports on other metal nanoparticles such as Magnesium oxides [13], copper oxide [14], Aluminium oxides [15], titanium dioxides [16-18] and zinc oxides [19]. However, reports also reveal that toxic effect of nanoparticles that may affect the biological behavior at the organ, tissue, cellular, subcellular, and protein levels. [20,21]. Hence it is essential to be alert while selecting metal nanoparticle. In the present study, zinc oxide nanoparticle is considered due to its low toxicity.

2. Experimental Details

2.1Synthesis of ZnO nanoparticles

Sample 1 was prepared by hydrothermal method. To prepare the aqueous solutions, 5g of Zinc acetate dihydrate [$Zn(O_2CCH_3)_2(H_2O)_2$] and 2 g of Sodium hydroxide [NaOH] were added in 100ml and 50ml distilled water respectively and stirred well. 5% of Polyethelene glycol (PEG) was added into the Zinc acetate dihydrate solution. The NaOH solution was added drop wise to the Zinc acetate dihydrate $Zn(O_2CCH_3)_2(H_2O)_2$ solution until a pH of 12 was attained. The resulting white colloidal solution was taken in an autoclave and heated to 100ºC for 3 hrs. The residue was washed several times in ethanol and then dried at 50 ºC in hot air oven. Sample 2 was prepared by solvothermal method. Instead of water ethylene glycol was used as the solvent. The rest of the preparation method was same as sample 1.

2.2 Antibacterial assay

The potential of ZnO nanoparticles was evaluated on antibiotics- Ampicillin and Streptomycin and were applied against the bacterial strains viz. *Staphylococcus aureus, Pseudomonas aeruginosa, Escherichia coli* and *Citrobacter freundii.* The experiment was carried out using Kirby-Bauer disc diffusion technique [22]. Antibiotic discs coated with ZnO nanoparticles (test- T), were placed on Muller-Hinton agar medium and kept for overnight incubation. Antibiotic without nanoparticle coating was kept in the same plate as control (C). Agar plates were examined for zone of clearance. The diameter of the clear zone was noted and compared with the control. Sample 1 is used in combination with Ampicillin and sample 2 with Streptomycin.

3. Results and Discussion

3.1 XRD analysis

The XRD pattern of ZnO nanoparticles synthesized by hydrothermal (Sample 1) and solvothermal method (Sample 2) are shown in figures 1 and 2 respectively. The prominent peaks corresponding to the diffraction panes (100), (002), (101), (102), (110), (103) and (112) agree well with JCPDS card 36-1461, confirming the hexagonal wurtzite structure of ZnO nano particles.

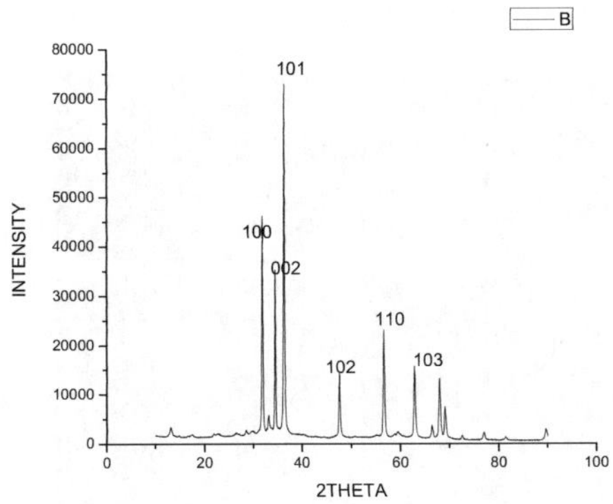

Fig. 1. XRD pattern of Sample 1

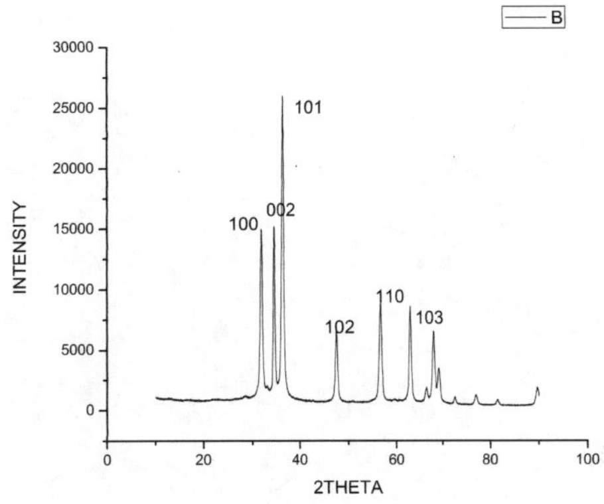

Fig. 2. XRD pattern of Sample 2

3.2 TEM Studies

TEM studies were carried out to find out exact particle size of synthesized ZnO. Figures 3 and 4 show the TEM images of the synthesized ZnO nanoparticles. From the TEM images the average particle size of sample 1 was found to be 62.15nm and that of sample 2 was 9.83nm.

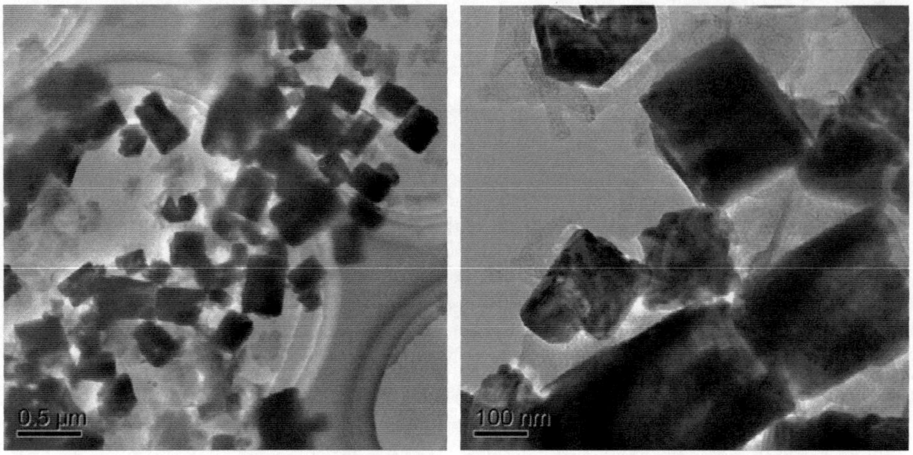

Figure 3. TEM image of synthesized ZnO nanoparticles (sample 1)

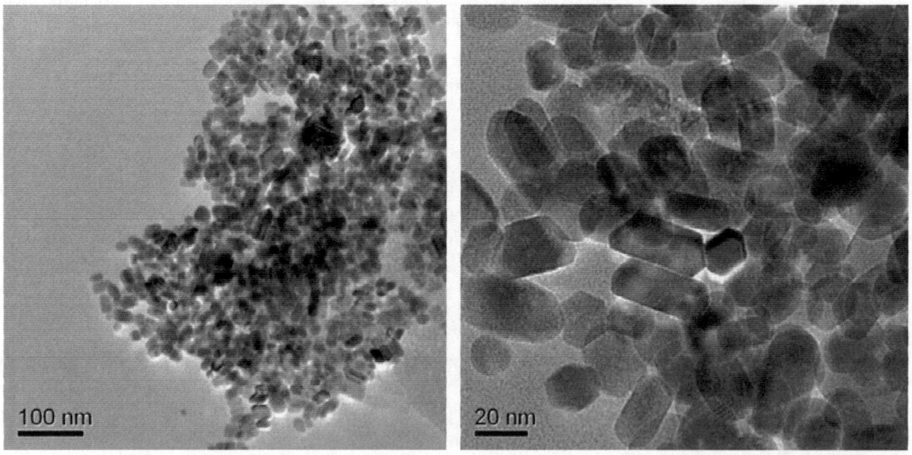

Figure 4. TEM image of synthesized ZnO nanoparticles (sample 2)

3.3 Antibacterial activity

Addition of ZnO nanoparticles to Ampicillin have resulted in slight enhancement of its activity it and maximum inhibition was produced against *C. freundii*. Streptomycin showed better activity against the tested bacterial strains and it was further enhanced by the activity of ZnO nanoparticles. A remarkable enhancement was noticed on *S. aureus* and *Citrobacter freundii* (Table 1). Combination of nanoparticle with Streptomycin was not much effective against *P.aeruginosa* where the diameter of clear zone remained same in both test and control. Earlier reports has showed the susceptibility of *S.aureus* towards silver and gold nanoparticles [23, 24]. There were very few reports on the susceptibility of *C. freundii.* Inhibitory effect is significant when the emergence of drug resistant varieties of these strains is considered.

Table 1. Clear zone obtained for each bacterial stain

S. No.	Bacteria	Ampicillin (clear zone in mm)		Streptomycin (clear zone in mm)	
		T	C	T	C
1	*S. aureus*	8.5	7	26	19
2	*P. areuginosa*	9	7	19	19
3	*E. coli*	10	7	19	16
4	*C. freundii*	14	9	20	8

*T- Test, C- Control

Ampicillin and Streptomycin are broad spectrum antibiotics that are used against both gram positive and negative strains. *S.aureus* is the only gram positive strain used in this study while the other three belongs to gram negative group. *S. aureus* and *C. freundii* were the two bacterial strains that were found to be highly susceptible to the synergic effect of antibiotics with ZnO nanoparticles. Hence the potential of ZnO nanoparticles was found to be equally effective in gram positive and gram negative strains. Comparing the particle size of the nanoparticles used sample 2 with smaller particle size have exhibited better performance than the large sized sample 1. The size of the particle plays a central role in antimicrobial activity [25] The colloidal silver particles, with variable sizes (44, 50, 35, and 25 nm), synthesized by the reduction of $[Ag\ (NH_3)_2]^+$ complexes with carbohydrates were tested for antimicrobial activity [26].

4. Conclusion

ZnO nanoparticles were synthesized using hydrothermal and solvothermal methods. The prepared ZnO nanoparticles were characterized by powder XRD and TEM studies. It was found that the ZnO particles prepared using Ethylene glycol as solvent had an average particle size of 9.83nm and the nanoparticles prepared using water as solvent had an average particle size of 62.15nm. ZnO nanoparticles when combined with antibiotics, an enhancement in the activity of antibiotics was observed and was found effective in all the bacterial strains tested. Further research may help in the formulation of new drugs that may inhibit drug resistant bacterial strains.

Acknowledgments

This work was supported by SRS Major Project by Kerala State Council for Science, Technology and Environment. The authors also thank Toyo University, Japan for the TEM studies.

References and Notes

1. R. Feynman, There's plenty of room at the bottom, *Science*, 254:1300-130, **(1991)**
2. P. Gajjar, B. Pettee, D. W. Britt, W. Huang, W. P. Johnson, J. Anderson, *Journal of Biological Engineering,* 3:9-22, **(2009)**
3. Z. L. Wang, *Materials Today*, 7,26-33, **(2004)**
4. L. E. Greene, B. D. Yuhas, M. Law, D. Zitoun and P. Yang, *Inorganic Chemistry*, Vol. 45, no. 19, pp – 7535-7543, **(2006)**
5. T. Jin, D. Sun, J. Y. Su, H. Zhang and H. J. Sue, *Journal of Food Science*, Vol. 74, No. 1, pp. m46-M52, **(2009)**
6. T. Alammar and A. V. Mudring, *Materials Letters*, Vol. 63, No. 9 – 10, pp. 732 – 735, **(2009)**
7. Z. Fakhroueian, F. M. Harsini, F. Chalabian, F. Katouzian, A. Shafiekhani and P. Esmaeilzadeh, *Advances in Nanoparticles*, Vol. 2, pp 247 – 258, **(2013)**
8. J. S. Kim, E. Kuk, K. N. Yu, J. H. Kim, S. J. Park, H. J. Lee, S. H. Kim, Y. K. Park, Y. H. Park, C. Y. Hwang, Y. K. Kim, Y. S. Lee, D. H. Jeong, M. H. Cho, *Nanomedicine,* 3:95–10, **(2007)**
9. P. Li , J. Li, C. Wu, Q. Wu, J. Li, *Nanotechnology*, 16:1912–1917, **(2005)**
10. J.P. Ruparelia, S.P. Duttagupta, A.K. Chatterjee, S.M. Mukherji., *Proceedings of the 9th Annual Conference of the Indian Environmental Association (Envirovision-2006)*, Goa, India, September 21–23, **(2006)**
11. S. Alexander, K. J. Klabunde, M.R. George, M.S. Christopher, *Langmuir,* 24:7457–7464. **(2008)**
12. G.L. Burygin, *Nanoscale Res. Lett.*, 4:794–801**(2009)**

IOP Conf. Series: Materials Science and Engineering **360** (2018) 012001 doi:10.1088/1757-899X/360/1/012001

13. R. Richards, W. Li, S Decker, C. Davidson, O. Koper, V. Zaikovski, A. Volodin, T. Rieker, K. Klabunde , *J. Am. Chem. Soc.* 122:4921-4925, **(2000)**
14. P.K. Stoimenov, *Langmuir,* 18:6679-86, **(2002)**
15. B. Li, B.E. Logan. , *Colloids Surf B*,36:81-90, **(2004)**
16. T. Matsunaga, R. Tomada, T. Nakajima, H.Wake ,*FEMS Microbiol. Lett.*, 29:211-214. *(1998)*
17. B. Kim, D. Kim, D. Cho, S. Cho, *Chemosphere*, 52:277-281, **(2003)**
18. C. Chawengkijwanich, Y. Hayata, *Int. J. Food Microbiol.* 123:288-292, **(2008)**
19. W. Jiang, H. Mashayekhi, B. Xing. , *Environ.Pollut.* 157:1619–1625, **(2009)**
20. D. Huster, T.D. Purnat, L.L. Burkhead, M. Ralle, O. Fiehn, F. Stuckert, N.E. Olson, D. Teupser, S.J. Lutsekno, *J Biol Chem.*, 282:8343–8355, **(2007)**
21. S. Lanone, F. Rogerieux, F. Geys, A. Dupont, E. Maillot-Marechal, J. Boczkowski, G. Lacroix, P. Hoet, *Part Fibre Toxicol*, 6:14–25, **(2009)**
22. A.W. Bauer, W.M. Kirby, J.C. Sherris, M. Turck. , *American Journal of Clinical Pathol.*,**45**(4):493–496. [PubMed] **(1966)**
23. F. Mirzajania, A. Ghassempoura, , A. Aliahmadib, M.A. Esmaeilib, Vol. 162, (5), P-542–5499, **(2011)**
24. M. E. Abalaka, S. Y. Daniyan, S. O. Adeyemo, D. Damisa , *International Journal of Biological, Biomolecular, Agricultural, Food and Biotechnological Engineering* Vol:8, No:4, **(2014)**
25. J.R. Morones, J.L. Elechiguerra, A. Camacho, K. Holt, J.B. Kouri, M.J. Yacaman, *Nanotechnology*,16:2346–2353, **(2005)**
26. A. Panacek, L. Kvitek, R. Prucek, M. Kolar, R. Vecerova, N. Pizurova, V. K. Sharma, T. Nevecna and R. Zboril , *J Phys Chem*, 110:16248-16253, **(2006)**

IOP Conf. Series: Materials Science and Engineering **360** (2018) 012002 doi:10.1088/1757-899X/360/1/012002

Ferrofluids for Active Shock Absorbers

M D Rao[1,3], P S Goyal[2], Biswajit Panda[2] and R I K Moorthy[1]
[1]Department of Mechanical Engineering, Pillai College of Engineering, New Panvel, Navi Mumbai - 410206, India
[2]Department of Applied Sciences, Pillai College of Engineering, New Panvel, Navi Mumbai - 410206, India

Email: mdurgarao@mes.ac.in

Abstract. This paper deals with synthesis and characterization of a ferrofluid and examining its suitability for developing a damper whose damping behavior can be controlled by applying magnetic field. In particular, ferrofluids, consisting of oleic acid coated Fe_3O_4 nanoparticles dispersed in kerosene, have been synthesized and characterized. Oleic acid coated nanoparticles of Fe_3O_4 have been characterized using XRD, TEM and FTIR. These studies show that sizes of particles are in range of 6 to17 nm and oleic acid is adsorbed on the surface of above particles. The role of damping medium on the damping behavior of a damper has been studied by carrying out studies with several different liquids. The effect of magnetic field on the damping behavior of ferrofluid based damper has also been studied. It is seen that damping ratio depends on the viscosity of damping medium. In particular, it is seen that damping ratio of a ferrofluid based damper increases on application of magnetic field. However, the sensitivity of damping ratio to the magnetic field has been found to be different for different ferrofluids. It seems, magnetic moment or size of the particle plays an important role in deciding the sensitivity of the damper. No attempt has been made to improve the sensitivity of the damper by optimizing various parameters. All the same, present studies provide a proof of concept for smart shock absorber.

1. Introduction
Cars and buses are fitted with suitable shock absorbers so that passengers do not experience bumpy and noisy ride [1]. Shock absorber is a device that controls unwanted oscillatory motion through a process known as damping. Currently available shock absorbers make use of ordinary fluids as damping medium and their damping behavior is decided at the design stage. However, it is of interest to have shock absorbers whose damping behavior could be controlled as per functional requirements. Ideally, one would like to have "*Smart Shock Absorber*", whose damping properties could be controlled by application of suitable electric or magnetic fields. Magnetorheological Dampers [2-4], which use magnetorheological (MR) fluids as the damping medium, have above characteristics but they have not been commercially successful. MR fluids consist of a suspension of large size (~ 10 μm) magnetic particles and these particles aggregate and settle down over a period of time. That is, MR fluids are not stable and their rheological properties change with time. It is felt that for an active damper or shock absorber [5-8], one should use ferrofluid instead of MR fluid as the damping medium.

[3] To whom any correspondence should be addressed.

Ferrofluids are better suited for this purpose. They are stable as sizes of magnetic particles (~ 50 nm) in ferrofluids are much smaller than those in MR fluids [9-11]. It is known that viscosity of ferrofluids also varies with magnetic field [12]. This paper deals with synthesis and characterization of ferrofluids and studying the damping behavior of ferrofluid based damper, both, with and without magnetic field.

Ferrofluid (also referred to as magnetic liquid) is a unique material that has both the liquid and magnetic properties [9,10]. They are colloidal suspensions of magnetic nano-particles (size ~ 10 nm) in a liquid medium. To avoid coagulation of magnetic particles, they are coated with suitable surfactant molecules. The ferrofluid used in present studies was synthesized in our laboratory [13]. It consists of oleic acid coated nanoparticles of magnetite, Fe_3O_4, dispersed in carrier liquid kerosene. The damping behavior of a damper having above ferrofluid as the damping medium, has been studied. The effect of magnetic field on the damping behavior of above damper has also been examined.

The next section gives the details of synthesis and characterization of ferrofluid. Section 3 gives the experimental details of the set-up used for studying the damping behavior of the damper. Results of above studies for several different experimental conditions are given in section 4. In the end, section 5 gives a summary.

2. Synthesis and Characterization of Ferrofluid

The synthesis of ferrofluid involves preparing of nanoparticles, coating them with surfactant molecules and dispersing these particles in suitable carrier liquid [9-11]. We have synthesized ferrofluids using oleic acid coated Fe_3O_4 nanoparticles, which were dispersed in kerosene. Nanoparticles of Fe_3O_4 were synthesized using wet chemical route, which involves co-precipitation of Fe_2^+ and Fe_3^+ in alkali solution. The details of this method of preparing nanoparticles of Fe_3O_4 and coating them with oleic are given in a review paper by Mehta and Upadhyay [9]. We have synthesized two categories of ferrofluids. In case of Category-I, Fe_3O_4 was prepared by reacting ferric chloride ($FeCl_3$) with ferrous sulfate ($FeSO_4$) in an aqueous medium. In case of Category-II, the starting salt for preparing Fe_3O_4 was ferric chloride ($FeCl_3$) and the required ferrous chloride ($FeCl_2$) was obtained by reduction of $FeCl_3$. The pH of the aqueous solution was controlled using ammonia solution. The Iron (III) chloride ($FeCl_3$), ferrous (II) sulfate ($FeSO_4$), oleic acid and 25% ammonia solution used for above studies, were purchased from Sigma-Aldrich, USA. All chemicals were of analytical grade and were used as received. The full details of method of synthesis and characterization of magnetic nanoparticles are given in our earlier paper [13].

The above samples have been characterized using X-ray diffraction (XRD), Transmission electron microscopy (TEM), and Fourier Transform Infra-red (FTIR) spectroscopy. XRD studies provided information about the structure and the composition of the constituents of the nanoparticles. The sizes of above nanoparticles were obtained using TEM. FTIR studies provided information on nature and extent of binding of oleic acid on particle surface. The salient features of these studies for Category-I ($FeSO_4$ based samples) are given below [13].

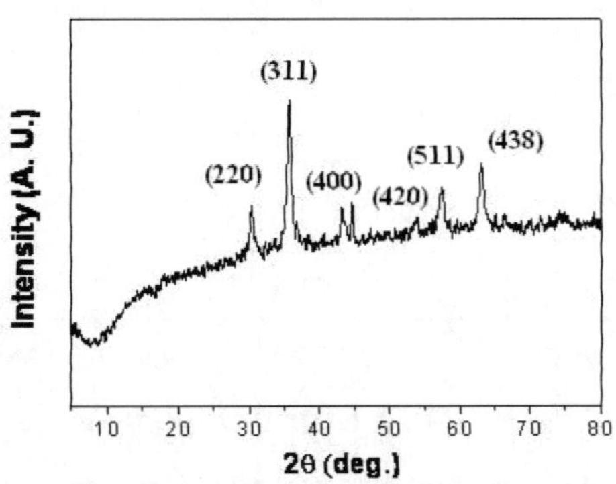

Figure 1. X-ray Diffraction pattern obtained from the magnetic nanoparticles.

Figure 1 shows the XRD pattern obtained from the magnetic nanoparticles. The existence of sharp Bragg peaks suggests that nanoparticles are crystalline in nature. The position of peaks correspond to single phase cubic structure of Fe_3O_4 with a lattice constant a = 8.325 Å, which is in very good agreement with published data [14]. It is noted that there is broadening of the Bragg peaks and this arises because of finite sizes of the particles. The particle size as obtained from the width (β) of (311) Bragg peak using Debye–Scherrer equation $(R = 0.9\lambda/(\beta \cos \theta))$ is about 12.6 nm.

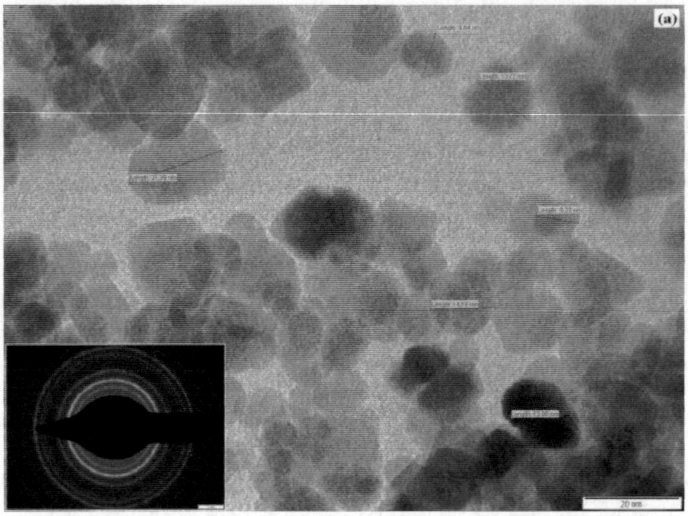

Figure 2. TEM micrograph of Fe_3O_4 nanoparticles.

Figure 2 shows the TEM micrograph of Fe_3O_4 nanoparticles. It is seen that most of the nanoparticles are spheres of varying sizes.

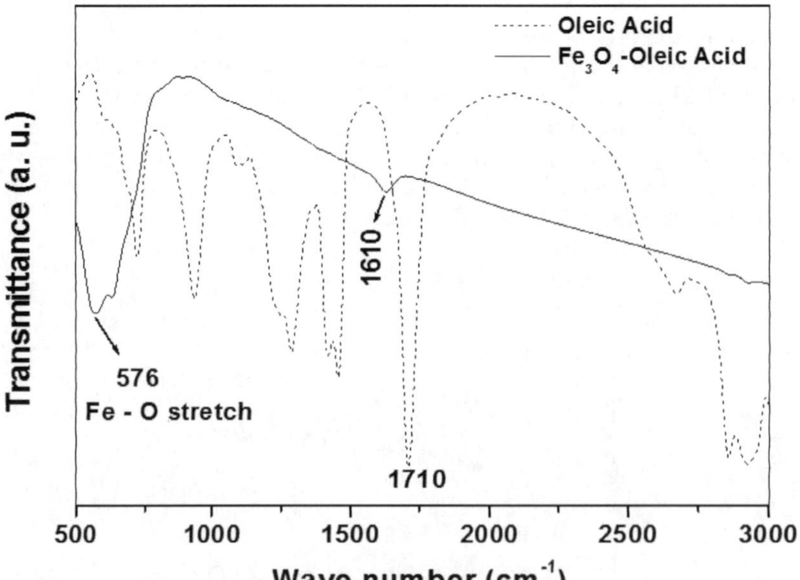

Figure 3. FTIR Spectra for pure oleic acid and for oleic acid coated Fe_3O_4 nanoparticles.

IOP Conf. Series: Materials Science and Engineering **360** (2018) 012002 doi:10.1088/1757-899X/360/1/012002

Figure 3 shows the comparison of FTIR spectra for pure oleic acid and those for oleic acid coated Fe_3O_4 nanoparticles. It is seen that the absorption bands for the pure oleic acid are well resolved, but those of the Fe_3O_4 – oleic acid are rather broad and few. The relevant peaks are in region of 1500-1750 cm^{-1}. C = O vibration from –COOH group of oleic acid gives a peak at 1710 cm^{-1} for pure oleic acid. This peak shifts to 1589 cm^{-1}, when oleic acid is adsorbed on Fe_3O_4 particles. The absence of 1710 cm^{-1} peak in spectra of oleic acid coated nanoparticles suggests that most of the oleic acid in nanoparticle system is adsorbed on Fe_3O_4 nanoparticles. The IR band observed at around 576 cm^{-1} can be ascribed to the Fe–O stretching vibrational mode of Fe_3O_4 nanoparticles [15].

3. Experimental set-up for studying damping behavior of the damper

Figure 4 is a schematic drawing of the experimental set-up that has been used for studying the damping behavior of the damper. The vibrating system is comprised of a beam pivoted at its end and supported by a helical spring at its other end.

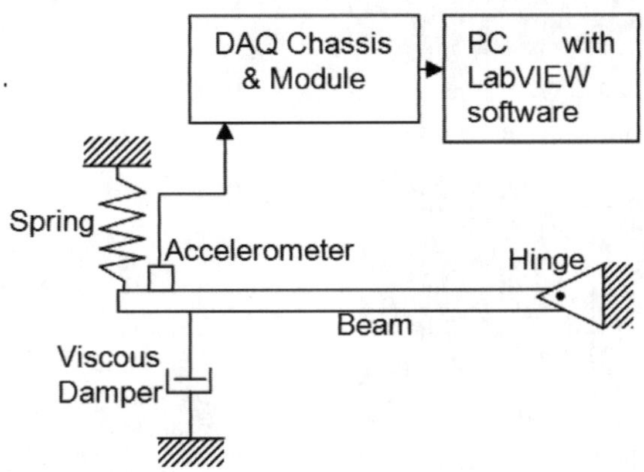

Figure 4. Schematic drawing of the set-up used for measuring damping ratio of a damper

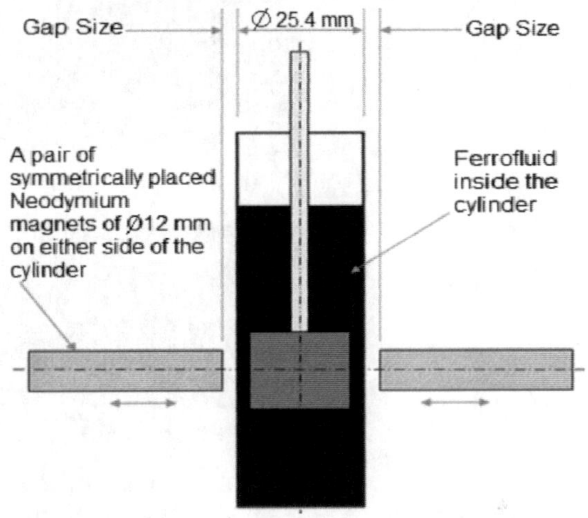

Figure 5. Viscous damper along with the magnets.

The beam is set in free vibration mode by using a small initial angular displacement. Vibrations are sensed by a uni-axial accelerometer which is connected to a suitable data acquisition module (DAQ) through a data cable. The DAQ module is connected to a personal computer in which the National Instruments' LabVIEW programming software (version 12) is installed. The experiment involved measurement of acceleration versus time graphs using the accelerometer for sensing and LabVIEW software for gathering and analyzing the data. Displacement versus time data was obtained by integrating the acceleration data twice. To study the damping behavior of the viscous damper, the damper containing damping fluid is connected at a suitable distance from the spring as shown in figure 4. A piston and cylinder mechanism of the viscous damper is shown in figure 5 with a suitable fluid. It is possible to change the damping fluid of the damper and apply suitable magnetic field to the damping medium using neodymium magnets. The magnetic field can be varied by moving the neodymium magnets (i.e. by varying the gap size) as shown in the figure.

Damping medium slows down and reduces the amplitude of vibratory motion of the piston. The damping ratio ζ is extracted from the displacement versus time plots. It is known that if x_n is the amplitude at the end of nth cycle, then logarithmic decrement δ is given by:

$$\delta = \ln\left(\frac{x_1}{x_{n+1}}\right) \quad (1)$$

IOP Conf. Series: Materials Science and Engineering **360** (2018) 012002 doi:10.1088/1757-899X/360/1/012002

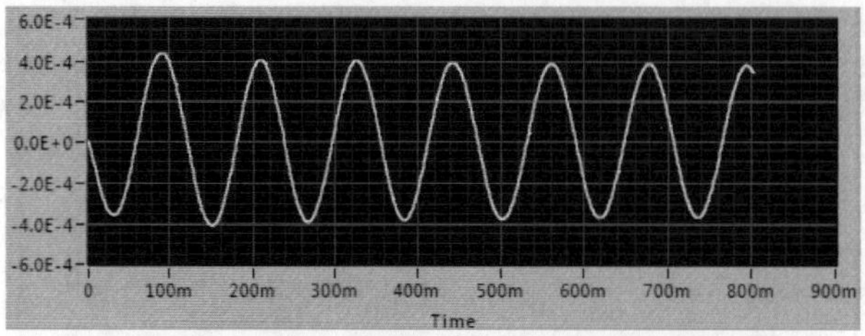

(a) Undamped (no damper connected)

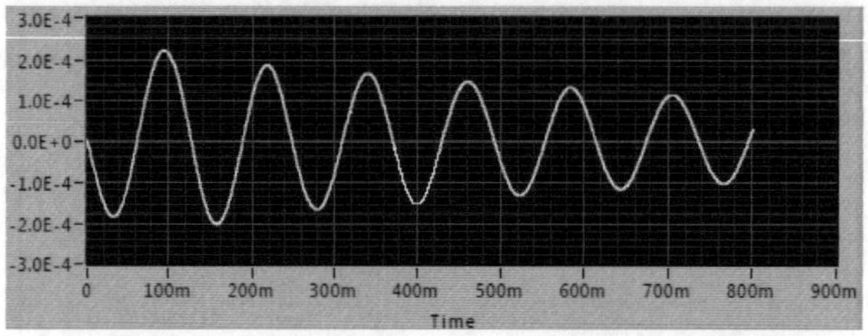

(b) Damper (with water as damping medium)

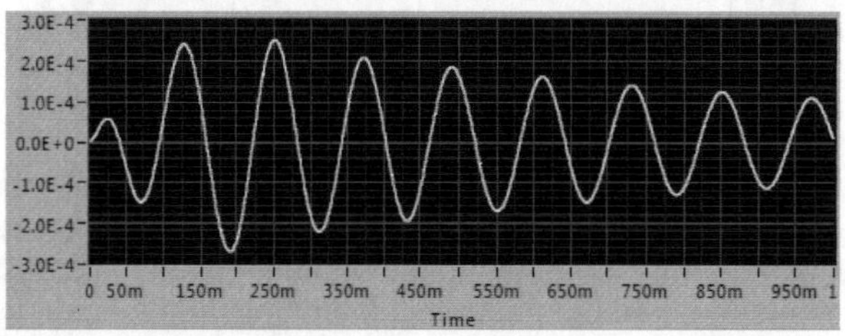

(c) Ferrofluid (FF-Cl_2-II) based damper without magnetic field (H = 0 kG)

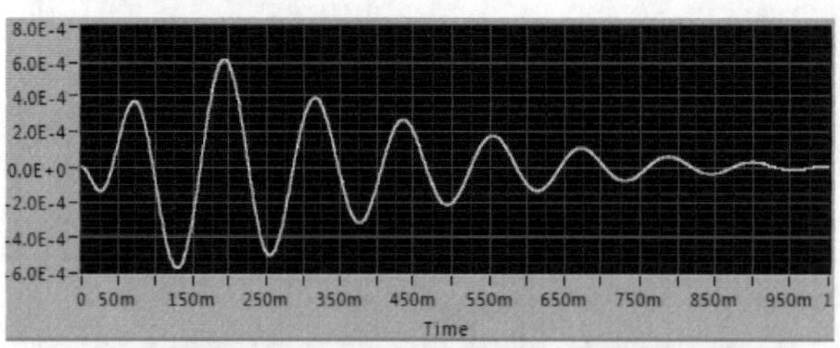

(d) Ferrofluid (FF-Cl_2-II) based damper with magnetic field (H = 3.2 kG)

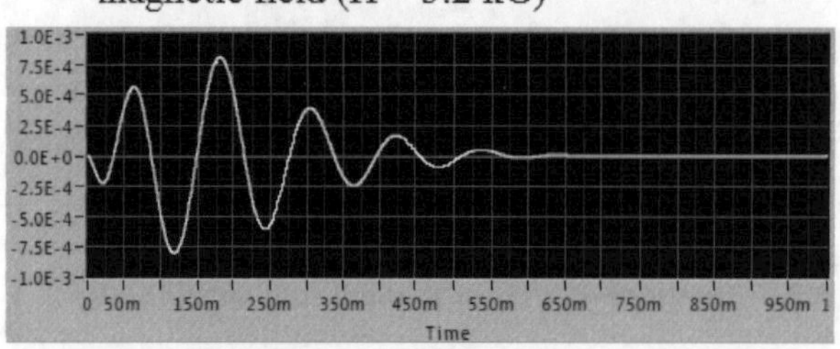

(e) Ferrofluid (FF-Cl_2-II) based damper with magnetic field (H = 4.5 kG)

Figure 6. Displacement versus Time graphs for viscous damper under different experimental conditions.

The damping ratio ζ is given by:

$$\zeta = \sqrt{\frac{\delta^2}{\delta^2 + 4\pi^2}} \qquad (2)$$

We have used the set-up for measuring ζ for several different damping media, namely, viscous liquids (CTAB/ NaSAl system) [16] and three different ferrofluids. The effect of magnetic field on ζ has been examined for ferrofluid based dampers. Typical measured displacement-time graphs are shown in figure 6. Because of integration errors, the first two cycles in the figures have not been considered for computing logarithmic decrement δ.

4. Results and Discussions

4.1. Effect of liquid viscosity on damping ratio

The effect of viscosity (η) of the damping fluid on the damping ratio ζ of the damper has been studied using solution of Cetyltrimethylammonium Bromide (CTAB) and Sodium Salicylate (NaSal) as the damping medium. The viscosity of (CTAB + NaSal) solution depends on the relative concentrations of CTAB and NaSal [16] and thus these studies provided information on the variation of damping ratio ζ with the viscosity of the damping fluid. Measurements have been made for (0.1M CTAB + x NaSal) for x = 0.0, 0.02, 0.05 and 0.08 M. The viscosities of these solutions are 1.5 cP, 30 cP, 50 cP and 100 cP respectively. The measurements were also made for a damper having water (η = 1.0) as the damping fluid. For sake of completeness measurements were also made on an un-damped damper (free vibrations). Results are shown in table 1 and figure 7. As expected, from figure 6, it is seen that the damping ratio increases with increase in the viscosity of damping medium.

Table 1. Effect of viscosity of damping medium on the damping ratio (ζ) of damper.

Damping Medium	Viscosity	$\zeta \times 10^{-3}$
Undamped	-	2.3
Water	1	10.9
0.1M CTAB + 0.0M NaSal	1.5	14.9
0.1M CTAB + 0.02M NaSal	30	123.6
0.1M CTAB + 0.05M NaSal	50	170.4
0.1M CTAB + 0.08M NaSal	100	532.8

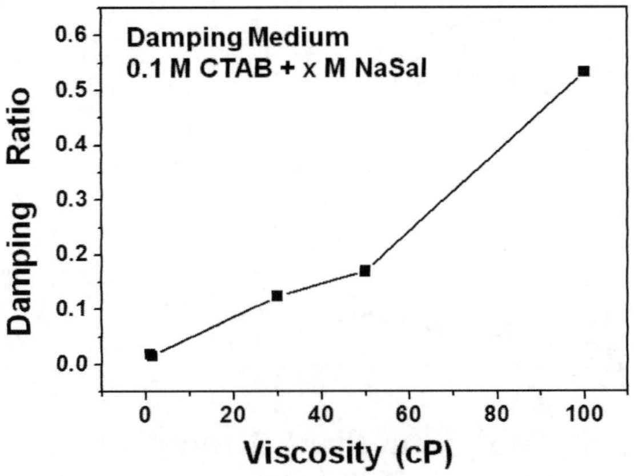

Figure 7. Displacement Damping Ratio vs. Viscosity for (0.1M CTAB + x M NaSal) solutions as damping fluid for x = 0.0, 0.02, 0.05 and 0.08 M.

IOP Conf. Series: Materials Science and Engineering **360** (2018) 012002 doi:10.1088/1757-899X/360/1/012002

4.2. Ferrofluid based dampers: Effect of magnetic field on damping ratio

The damping behavior of ferrofluid based dampers was studied for following three sets of ferrofluids:
(a) Ferrofluids (Category-I) where Fe_3O_4 was prepared using $FeCl_3$ and $FeSO_4$. Measurements were made corresponding to two batches, referre d to as FF-SO_4-I and FF-SO_4-II.
(b) Ferrofluids (Category-II) where Fe_3O_4 was prepared using $FeCl_3$ and $FeCl_2$. Measurements were made corresponding to two batches, referred to as FF-Cl_2-I and FF-Cl_2-II.
(c) Commercially obtained oil based ferrofluid whose composition was not known (referred to as FF-Comm.)

The damping ratio ζ was measured for above samples, both, with and without magnetic field. Measurements were made for magnetic field (H) in the range of 0.0 to 4.5 kG. The magnetic field at piston–cylinder interface was measured using Hall-effect based sensor coupled to a gaussmeter. The results of the above studies are shown in figure 8. It is clear that damping ratio increases with magnetic field for all the ferrofluids. However, the rate of increase of ζ for a given increase in field is different for different ferrofluids. That is, sensitivity of ζ to the magnetic field is not same for all ferrofluids. This is not surprising as zero-field viscosity and particle sizes are expected to be different for different ferrofluids. To get an idea of comparative sensitivities of different ferrofluids, table 2 compares the increase in values of ζ when a fixed field of 4.5 kG is applied to all the five ferrofluids.

Table 2. Variation of damping ratio (ζ) with magnetic field for different ferrofluids, for a ferrofluid based damper.

Damping Medium	Damping Ratio ζ (for H = 0.0 kG)	Damping Ratio ζ_F (for H = 4.5 kG)	ζ_F/ζ (%)
FF-Comm.	0.072	0.085	18
FF-SO_4-I	0.019	0.035	84
FF-SO_4-II	0.023	0.029	26
FF-Cl_2-I	0.017	0.066	388
FF-Cl_2-II	0.020	0.148	740

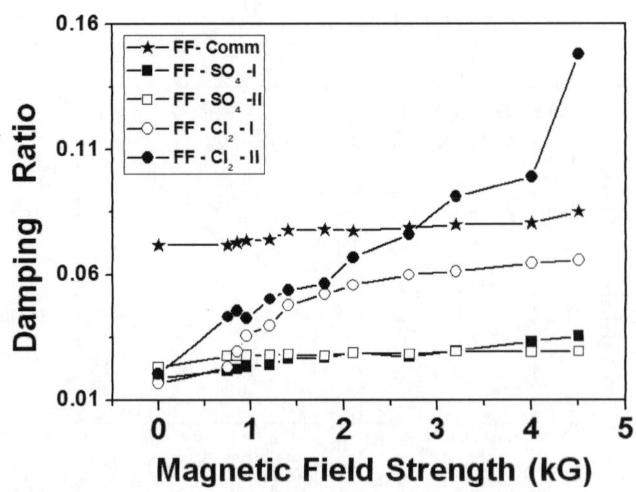

Figure 8. Variation of damping ratio with magnetic field for different ferrofluids.

Table 2 and figure 8 show that zero-field damping ratio (~0.02) is nearly same for all indigenously synthesized ferrofluids. However, this value is much larger (~0.07) for commercial fluid. This may be attributed to the fact that viscosity of commercial ferrofluid is more (perhaps because of larger particle concentration and viscous carrier liquid) as compared to that for indigenously synthesized ferrofluids. It is interesting to note that sensitivity of damper using category-II ferrofluid (Fe_3O_4 prepared using $FeCl_3/FeCl_2$) is much better compared to the one that uses category – I ferrofluids (Fe_3O_4 prepared using $FeCl_3/FeSO_4$) or the one that uses commercial ferrofluid. For example, when magnetic field is increased from 0.0 kG to 4.5 kG, the damping ratio for the former category of dampers increases by about 500% and that for the latter category, increases by about 50%. No effort has been made to understand the reasons for the above. It is known that magnetic nanoparticles form chains in presence of magnetic field and that gives rise to increase in viscosity and the damping ratio. The chains would be more rigid and the damping ratio would be higher if particle size or its magnetic moment is large [17]. It is likely that particle sizes are different in different ferrofluids. The results of such detailed studies will be published in a future paper.

5. Summary

The shock absorbers used in cars and buses make use of ordinary fluids as damping medium and their damping behavior is decided at the design stage. In practice, one needs a shock absorber whose damping could be changed depending on the road condition or the load. It is thus of interest to have "*Smart Shock Absorber*" whose damping properties could be controlled by application of suitable electric or magnetic fields. This paper deals with development of a damper that uses ferrofluid as the damping medium. It is shown that damping ratio of above damper can be varied by applying magnetic field to the damping medium. The effect of magnetic field on damping ratio has been studied using several different ferrofluids.

Ferrofluids, required for above studies, have been synthesized and characterized indigenously. They consist of oleic acid coated nano-particles of magnetite (Fe_3O_4), dispersed in kerosene. Fe_3O_4 is prepared by using two different chemical routes ($FeSO_4$ / $FeCl_3$ and $FeCl_2$ / $FeCl_3$). Nano-particles of Fe_3O_4 prepared by using $FeSO_4$/ $FeCl_3$ were characterized using XRD, TEM and FTIR. It was seen that sizes of magnetic particles are in 6 to 17 nm range, with a dominant population at 12-14 nm.

The damping ratio of a piston–cylinder type damper has been measured using several different damping mediums, such as viscous liquids, commercial ferrofluid and indigenously synthesized ferrofluids. It was seen that there is a direct correlation between the viscosity of the damping medium and the damping ratio. In case of ferrofluid based dampers, it was seen that damping ratio increases with magnetic field. The sensitivity or the rate of increase of damping ratio with magnetic field was, however, found to be different for different ferrofluids. No attempt has been made to understand the reasons for the above. It seems the sensitivity of the damper depends on sizes and concentrations of nanoparticles in the ferrofluid.

The above studies provide a technological demonstration of ferrofluid based active shock absorber. It is, however, important to quantify the role of various parameters on the damping behavior before one designs and manufactures the *Active Shock Absorber* for use in a vehicle.

Acknowledgments

We thank Priam Pillai for encouragement and useful discussions. This research is financially supported by All India Council of Technical Education (AICTE)—New Delhi, under Research Proposal Scheme 8023/RID/RPS-93(Pvt)/2011-12.

6. References

[1] Dixon J C 2007 *The Shock Absorber Handbook* 2nd Edition John Wiley and Sons
[2] Sapinski B 2009 *Mechanics* **28** 18
[3] Kumbhar B K, Patil S R and Sawant S M 2015 *Engineering Science and Technology, an International Journal* **18** 432
[4] Zhu X, Jing X and Li Cheng 2015 *J. Intelligent Material Systems and Structures* **26** 881
[5] Raj K and Moskowitz R 1980 *IEEE Transactions on Magnetics* **16** 358
[6] Nakatsuka K, Yokoyama H and Shimoiizaka J 1987 *J. Mag. Mat.* **65** 359

IOP Conf. Series: Materials Science and Engineering **360** (2018) 012002 doi:10.1088/1757-899X/360/1/012002

[7] Bashtovoi V G, Kabachnikov D N and Kolobov A Y 2002 *J. Mag. Mag. Mat.* **252** 312
[8] Yang W, Li D and Feng Z 2013 *J. Vibrations and Controls* **19** 183
[9] Mehta R V and Upadhyay R V 1999 *Current Science* **76** 305
[10] Odenbach S 2002 *Ferrofluids: Magnetically controlled ferrofluids and their applications* (Ed) Springer Verlag Berlin
[11] Charles S *The preparation of magnetic fluids* 3-18 in [10]
[12] Patel R, Upadhyay R V and Mehta R V 2003 *J. Coll. Int. Sci.* **263** 661
[13] Panda Biswajit and Goyal P S 2015 *AIP Conference Proceedings* **1665** 050020
[14] Lan Qiang, Liu Chao, Yang Fei, Liu Shangying, Xu Jian and Sun Dejun 2007 *J. Coll. Int. Sci.* **310** 260
[15] Nigam Saumya, Barick K C, Bahadur D 2011 *J. Mag. Mag. Mat.* **323** 237
[16] Aswal V K, Goyal P S and Thiyagarajan R 1998 *J. Phys.Chem.B* **102** 2469
[17] Butter K, Bomans P H H, Frederik P M, Vroege G J and Philipse A P 2003 *Nature Materials* **2** 88

IOP Conf. Series: Materials Science and Engineering **360** (2018) 012003 doi:10.1088/1757-899X/360/1/012003

Phase Formation Study of SnS Nanoparticles Synthesized Through PVP Assisted Polyol Method

Benjamin Hudson Baby and D. Bharathi Mohan*

Department of Physics, School of Physical, Chemical and Applied Sciences,

*Corresponding author Email: *d.bharathimohan@gmail.com

Abstract. The optimization study on the preparation of single phase of SnS nanoparticles by varying Sn:S source concentration ratio (1:1, 1:6 and 1:12M) through PVP assisted polyol synthesis method is successfully investigated. The single phase formation, crystal structure, surface morphology and optical properties of SnS nanoparticles have been studied using characterization techniques such as confocal Raman spectrometer, SEM, HR-TEM, UV-VIS-NIR spectrophotometer and spectrofluorometer. Confocal Raman as well as selected area electron diffraction analyses confirmed the single phase formation of SnS NPs with orthorhombic crystal structure at 1:12 ratio whereas SnS_2 and Sn_2S_3 phases formed for lower sulphur concentration. Surface morphology reveals the formation of SnS nanorods predominantly, due to the addition of PVP which act as a capping agent. Band gap calculation from tauc plot showed direct band gap of 1.45 eV for single phase SnS with orthorhombic crystal structure, which can be used as an absorber layer in thin film solar cells. The observed shift in the band gap value is discussed in detail with the support of chemical structure analysis. The PL spectra of samples consists of 3 major peaks; the peak at 411 nm (3.03 eV) corresponds to Sn vacancy (V_{Sn}), the peak at 435 nm (2.86 eV) corresponds to tin interstitial (I_{Sn}) and peak at 459 nm (2.70 eV) corresponds to sulphur vacancy (V_S). Whereas the additional peaks observed by decreasing the sulphur concentration to 0.6 and 0.1 M corresponds to the band edge emission of Sn_2S_3 and SnS_2 respectively.

1. Introduction

Tin (II) sulphide is a suitable material to replace the absorber layer in thin film solar cells (TFSC) due to its excellent physical, optical and electrical properties which is equally good or still better than the existing absorber materials such as CdTe, CIGS, CZTS. SnS has a direct band gap value of 1.38 eV which is close to the optimum value of 1.5 eV for the efficient absorption of electromagnetic radiation in the visible region. It has very high absorption coefficient of 10^{-5} cm^{-1} compared to CdTe [1-3]. It exhibit P-type conductivity, its constituent elements are abundant in nature and these elements are less toxic and environmentally safe to handle. SnS exhibit orthorhombic crystal structure with lattice parameters a=0.4329nm, b=1.1192nm and c=0.3984nm [JCPDS card no 39-0354]. In practice the highest efficiency achieved by SnS solar cell is only 4.4% [4]. This low efficiency may be due to the presence of impurity phases, grain boundaries and also due to the poor device fabrication techniques [5].

In this work, SnS nanoparticles (NPs) are prepared at a reaction temperature of 150 °C under open atmospheric conditions by using simple PVP assisted polyol synthesis method where ethylene glychol is the reaction medium which acts as the reducing agent and PVP is the capping agent which controls the agglomeration and ensures the monodispersity of NPs. It has already confirmed that the reaction temperature and sulphur precursor plays a major role in controlling the surface morphology of SnS NPs [6]. Also it has reported that capping effect enhance the electrical properties of NPs by the rapid separation of electron –hole pair [7]. So here, by varying Sn:S source concentration ratio, optimized the formation of single phase of SnS with an optimum band gap value to use as an absorber layer in thin film solar cells.

2. Experimental Details

SnS NPs were prepared by using ethylene glychol (EG) (Qualigens fine Chemicals) as the reducing agent and polyvinylpyrolidone (PVP) (Spectrochem Pvt. Ltd. Mumbai) as the capping agent to control the particle size of SnS. For the preparation of SnS high pure (reagent grade) $SnCl_2$ and Na_2S salts used as the tin and sulphur precursors obtained from (Merck Specialties Pvt. Ltd, Mumbai.) and Finar (Finar Chemicals Ltd, Mumbai) respectively. Initially 0.1M of $SnCl_2$, 0.1M of Na_2S and 0.15M of PVP prepared in 10 ml of EG separately. Then add Na_2S and PVP simultaneously drop by drop into $SnCl_2$ solution which was already placed in silicon oil bath at 150 °C and after the addition, the solution color changes to dark brown indicating the formation of SnS. Then solution was collected and quenched in a water bath after 20 minutes. This colloidal solution was washed several times using centrifuge at 5000rpm. Then, a thick (few microns) film of SnS was prepared by a simple drop casting technique. Following the above procedure, a set of experiment was carried out by changing Sn:S molarity ratio from 1:1 to 1:6 and 1:12 by fixing the PVP concentration at 0.15M.

The chemical structure of as deposited films was studied using confocal Raman spectrometer (ReinshawinVia Raman Microscope, UK) with an excitation wavelength of 785 nm using Ar+ laser. The surface morphology of the film was studied by using Hitachi (Model: S-3400N) scanning electron microscope (SEM) and high resolution transmission electron microscope (HR-TEM) (JEOL/JEM 2100). The absorbance spectra of colloidal solution were studied using UV-Vis-NIR spectrophotometer (Shimadzu UV-3600 Plus) from the wavelength range of 300 to 1000 nm. The emission spectra was recorded with an excitation wavelength of 350 nm using spectrofluorometer(FLUOROLOG-FL3-11), equipped with Xenon lamp-450.

3. Results and Discussion

Raman spectroscopy is an efficient, sensitive and nondestructive technique to study the phase formation of different tin sulphide phases such as SnS, SnS_2, Sn_2S_3 with their distinct Raman shift. SnS has orthorhombic crystal structure and according to group theoretical predictions it has 21 optical phonon modes and 12 of them are Raman active which are 4 A_g (40, 218±2, 192±2, 95±2 cm^{-1}), 2B_{1g} (70, 208 cm^{-1}), 4B_{2g} (70, 290±4, 160, 85±2 cm^{-1}) and 2B_{3g} (49, 164±2 cm^{-1}). The appearance of different modes depends upon the incident and scattered direction of radiation and direction of electric polarization vector of incident and scattered photons[8,3] 2H-SnS_2 has rhombohedral crystal structure with 6 optical phonon modes and 3 of them are Raman active observing at A_{1g} (315 cm^{-1}) and E_g (140 and 205 cm^{-1}) [9]. For Sn_2S_3, orthorhombic crystal structure with 60 vibrational modes and predominant Raman peaks are observed at 52, 60, 111, 183, 233 and 307 cm^{-1} [3,10]

Figure 1 shows the observed Raman spectra of samples by varying the Sn:S source concentration ratio using an excitation wavelength of 785 nm. Raman analysis confirmed the presence of secondary phases such as SnS_2 (314.5 cm^{-1}) and Sn_2S_3 (59 cm^{-1}) along with SnS phase for Sn:S ratio 1:1 molarity while by increasing to 1:6 molarity ratio, formation of SnS_2 phase is completely suppressed and for 1:12 molarity ratio single phase of SnS is formed. SnS Raman active peaks at 218±2 and 164 ±2 cm^{-1} correspond to the vibration of Sn atom against S and the peak at 290±4 cm^{-1} corresponds to the compressive mode of the layers along c axis [8]. Raman peak at 315 cm^{-1} corresponds to the symmetric interlayer mode of SnS_2 [9].

Table 1 shows the observed Raman modes of SnS, SnS_2 and Sn_2S_3 phases obtained from as prepared samples which is compared with the standard single crystal data [3,8-10]. The RI (%) of different tin sulphide phases from Raman spectra as shown in Table 2 indicating a systematic decrease in the formation of secondary phases such as SnS_2 and Sn_2S_3 by increasing the Sn:S source concentration ratio [3]

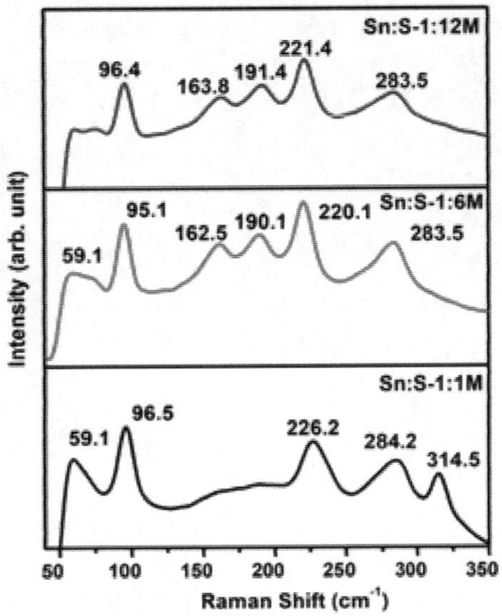

Figure 1. Raman spectra of samples prepared by varying Sn:S source concentration ratio 1:1 to 1:12 molarity using an excitation wavelength of 785 nm.

Table 1. Comparison of Raman modes of SnS, SnS_2 and Sn_2S_3 obtained from as prepared samples compared with the standard single crystal data [3,8-10]

Single Crystal	SnS (cm^{-1})					Sn_2S_3 (cm^{-1})	SnS_2 (cm^{-1})
		A_g		B_{3g}	B_{2g}		A_{1g}
	95±2	**218±2**	**192±2**	**164±2**	**290±4**	**60**	**315**
Sn:S-1:1M	96.5	226.2	--	--	284.2	59.1	314.5
Sn:S-1:6M	95.1	220.1	190.1	162.5	283.5	59.1	--
Sn:S-1:12M	96.4	221.4	191.4	163.8	283.5	--	--

Figure 2 (a) shows SEM image of SnS NPs synthesized with Sn:S source concentration ratio of 1:12 molarity. Figure 2 (b) shows HR-TEM image confirming the predominant phase formation of SnS nanorods with 10-15 nm in width and 30-40 nm in length. The formation of single phase of SnS with orthorhombic crystal structure is confirmed by using selected area electron diffraction (SAED) pattern. Figure 2 (c) reveals the SAED pattern of SnS nanorods which is compared with the standard JCPDS card no 01-071-3679, again confirming pure SnS phase.

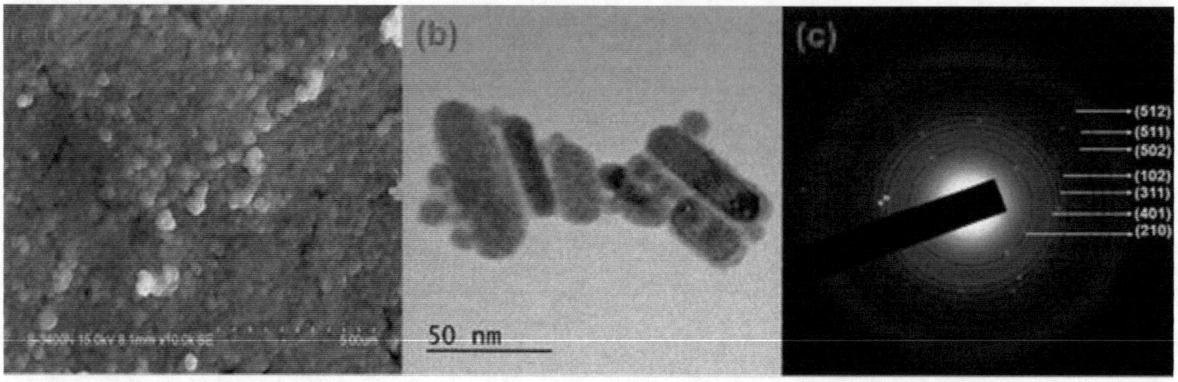

Figure 2. (a) SEM, 2 (b) HR-TEM images and 2 (c) SAED pattern of the sample with Sn:S molarity ratio 1:12.

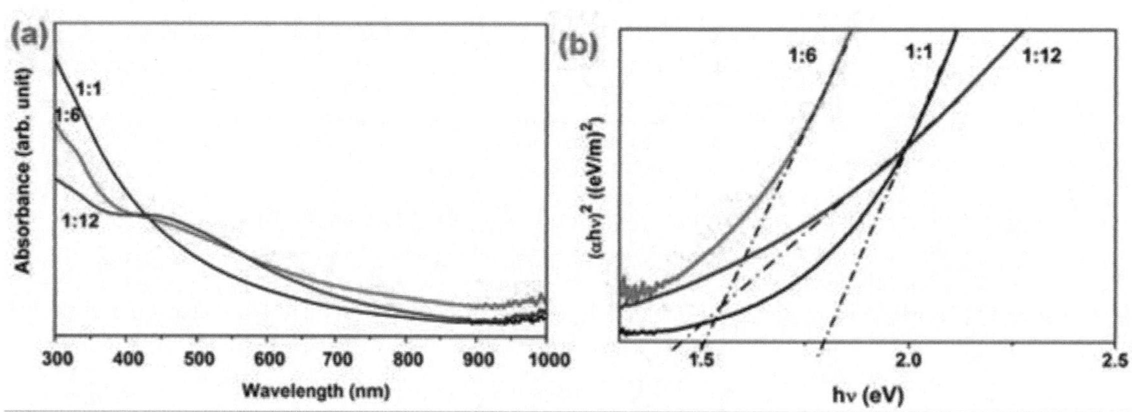

Figure 3. (a) Optical absorbance spectra and (b) tauc plot of samples with Sn:S source concentration ratio 1:1, 1:6 and 1:12 molarity

Table 2. Comparison of band gap value and relative intensity (RI) (%) (obtained from Raman analysis) of different Sn-S phases for the different Sn:S ratios.

Sn:S source concentration ratio (M)	Band gap (eV)	RI (%) of		
		SnS	SnS_2	Sn_2S_3
1:1	1.79	37	31	32
1:6	1.5	64	0	36
1:12	1.45	100	0	0

Figure 3 (a) shows the absorbance spectra of samples dispersed in methanol in the wavelength range of 300 to 1000 nm. Figure 3 (b) shows the direct band gap calculation from x- intercept of tauc plot and the obtained values are shown in table 2 which is compared with the RI (%) of different tin sulphide phases calculated from Raman spectra. The observed shift in band gap value is due to the presence of secondary phases and also due the particle size effect. Initially, by increasing the concentration of S^{2-} ions from 0.1M (1:1) to0.6 M (1:6), the band gap is red shifted by 0.34 eV due to the complete suppression of SnS_2 phase in 1:6M, which has high band gap (Eg=2.3 eV) value among entire tin sulphide phases [5]. Further increase in concentration of S^{2-} to 1.2 M results a red shift of 0.08 eV due to the formation of single phase of SnS (Eg=1.38 eV) with complete suppression of Sn_2S_3 phase (Eg= 1.83 eV) [5]. Sample with single phase of SnS shows a band gap value of 1.45 eV with a blue shift of 0.07 eV (Eg, bulk = 1.38 eV) due to the particle size effect [5]. Optical studies confirmed that the energy band gap of SnS is optimized to 1.45 eV which is matching with the band gap value of an absorber layer in thin film solar cells and it is obtained through controlling the formation of single phase of SnS by varying the concentration of S^{2-} ions.

Figure 4 (a) shows the PL spectra of Sn-S phase (synthesized by different Sn:S molarity ratios) with an excitation wavelength of 350 nm in the range 400 - 680 nm. It mainly consist of three peaks at 411, 435 and 459 nm. In addition to that, sample with Sn:S ratio1:1M shows less intense peaks at 677, 614 and 571 nm, and for sample with ratio 1:6 M shows two peaks at 614 and 677 nm. The emission peak observed at 411 nm (3.03 eV) corresponds to tin vacancy (V_{Sn}) which is responsible for its p type behaviour [11]. The peak at 435 nm (2.86 eV) corresponds to tin interstitial (I_{Sn}) and peak at 459 nm (2.70 eV) corresponds to the sulphur vacancy (V_S) [11] (Fig.4 (b)). The presence of high intense I_{Sn} defect with less intense V_S indicate the formation of Sn rich SnS phase for sample with Sn:S ratio 1:12M. Raman analysis confirmed the formation of single phase of SnS for Sn:S-1:12M and while by decreasing S^{2-} ion concentration secondary phases such as $SnS_2$and Sn_2S_3 are also observed which is once again verified from PL analysis. For 1:6M, two additional peaks observed at 677 nm (1.83 eV) and 614 nm (2.01 eV) corresponds to the band edge emission and Vs of Sn_2S_3 as shown in Fig.4 (b). It is already reported that the direct band gap of Sn_2S_3 is various from 0.96 eV to 2 eV depends due to the quantum confinement effect. The third additional peak at 571 nm (2.17 eV) for 1:1M corresponds to the band edge emission of SnS_2 where the band gap of SnS_2 is 2.3 eV [5]. The observation of broad fluorescence emission spectra with many side emission bands for sample with Sn:S ratio 1:1M and 1:6M indicating enormous delay in recombination rate due to the presence of several shallow trap states or intrinsic near-band edge states slightly below the conduction band involving phonon-exciton and phonon-phonon interactions [12].

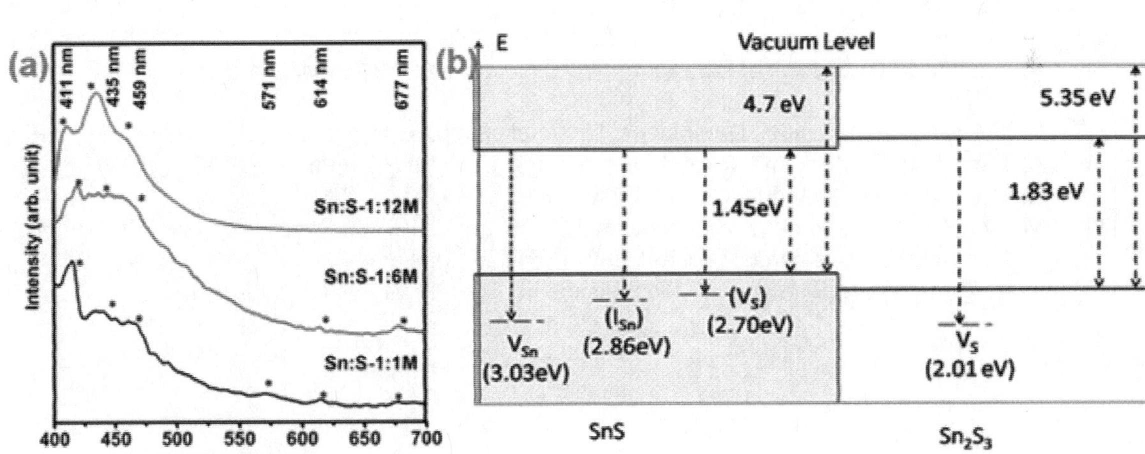

Figure 4. (a) Room temperature PL spectra of Sn-S samples prepared with 1:1, 1:6 and 1:12 molar ratios. (b) Schematic band diagram for the point defects formed in SnS-Sn_2S_3 systems. V_S– sulfur vacancy, V_{Sn}- tin vacancy, I_{Sn}– tin interstitial. The energy levels are drawn from the analysis of absorption and emission spectra.

4. Conclusion

In conclusion, single phase of SnS nanoparticles were successfully synthesized using PVP assisted polyol method for the reaction temperature of 150 °C by varying Sn:S source concentration ratio as 1:1, 1:6 and 1:12M. Chemical structural analysis confirmed the formation of SnS_2 and Sn_2S_3 as the secondary phases for lower sulphur source concentration. While the formation of single phase of SnS nanoparticles with orthorhombic crystal structure was confirmed with source concentration ratio of 1:12M as evident from chemical as well as crystal structural analysis. HR-TEM analysis confirmed the formation of SnS nano rods with an aspect ratio of 0.375. The observed red shift in the band gap value by increasing the source concentration ratio is due to the formation of secondary phases as observed from chemical structural analysis. Single phase of SnS shows a band gap value of 1.45 eV which can be used as an absorber layer in thin film solar cells. Photoluminescence analysis further confirmed the formation of single phase of SnS for 1:12M source concentration ratio and the proposed band diagram shows the presence of three intrinsic defects such as tin interstitial (I_{Sn}), tin vacancy (V_{Sn}) and sulphur vacancy (V_S).

5. Acknowledgments

The work is supported by University Grant Commission (UGC) - Major Project (F. No 4151868/2012 (Sr)). BHB sincerely UGC-BSR for providing Research Fellowship.

6. References

[1] Y. Hirai, Y. Kurokawa, and Akira, Jpn.J. Appl. Phys. Sci. 53, 012301 (2013).
[2] M. A Green, K. Emery, Y. Hishikawa, W. Warta and E. D. Dunlop, Prog. Photovoltaics 23, 1–9 (2015).
[3] B. H. Baby and D. B. Mohan, AIP Conf. Proc. 1665, 050171 1 (2015).
[4] P. Sinsermsuksakul, L. Sun and S. W. Lee, Advanced Energy Materials DOI :10.1002/aenm.201400496 (2014).
[5] L. A. Burton, D. Colombara, R. D. Abellon, F. C. Grozema, L. M Peter, T. J. Sevenije and G. Dennler, Chem. Mater. 25, 4908 (2013).
[6] Z. Niu and Y. Li, Chem. Mater. 26, 72 (2014).
[7] S. Jana, N. Mukherjee, B. Chakraborty, B. C. Mitra and A. Mondal, Appl. Surf. Sci. 300, 154 (2014).
[8] H. R. Chandrasekhar, R. G. Humphreys, U. Zwick and M. Cardona, Phys. Rev. B 15, 2177 (1977).
[9] A.J. Smith, P. E. Meek and W. Y. Liang, J. Phys. C: Solid State Phys.10, 1321 (1977).
[10] H. R. Chandrasekhar and D. G. Mead, Phys. Rev. B 19, 932 (1979).
[11] M. Devika, N. K. Reddy, M. Prashantha, K. Ramesh, S. V. Reddy, Y. B. Hahn and K. R. Gunasekhar, Phys. Status Solidi A 207, 1864 (2010).
[12] D. B. Mohan, V. S. Reddy and C. S. Sunandana, Appl. Phys. A 86, 73 (2007).

Impedance spectroscopy and permittivity investigation of $NdCrO_3$ perovskite ceramic nanoparticles

Jada Shanker[1], G. Narsinga Rao[2], D. Suresh Babu[1,*]

[1]*Department of Physics, Osmania University, Hyderabad, Telangana, India*

[2]*H&S Department, Marri Laxman Reddy Institute of Technology and Management, Hyderabad*

**Corresponding author Email: s_devarasetty1956@yahoo.co.uk*

The $NdCrO_3$ Perovskite Ceramic Compound has prepared by Sol-gel auto-combustion technique. The X-ray diffraction analysis confirmed that single-phase orthorhombic structure. The average size of the grain has estimated and it has found to be ~50nm. The impedance and permittivity investigations of $NdCrO_3$ perovskite ceramic nanoparticles have performed in the frequency range 20Hz-1MHz and temperature range 150K-300K. Impedance spectroscopy studies indicate the presence of grain and grain boundary relaxations in the compound. The polarization mechanism suggests for change in the dielectric constant and loss with temperature. The Impedance analysis supports the typical behaviour NTCR of material. The temperature dependent relaxation time for both grain and grain boundary follows the Arrhenius law.

Keywords: Sol-gel auto-combustion; Impedance; Permittivity.

1. Introduction

ABO_3 perovskites show exciting properties like high temperature superconductivity, colossal magneto resistance and multi-ferrocity [1]. The orthochromaties with a common formula $RCrO_3$ has investigated for their applications in various devices such as solid-oxide fuel cells and catalytic converters [1]. In the recent years, $RCrO_3$ materials exhibited so-called giant dielectric constant (ε') [2]. A literature survey of these materials reveals that no detailed work on NdCrO3 has reported. Therefore, we have studied the frequency and temperature dependence of dielectric and impedance properties of $NdCrO_3$ perovskite ceramics. In this paper, the results of extensive studies on electrical properties of $NdCrO_3$ in the temperature range 150K to 300K have presented.

2. Experimental Details

IOP Conf. Series: Materials Science and Engineering **360** (2018) 012004 doi:10.1088/1757-899X/360/1/012004

Polycrystalline powder of $NdCrO_3$ has prepared by sol-gel auto combustion technique, and detailed procedure has described in previous publication [3]. Resulting powder was annealed at 750^0C for 4 hours. Finally, this powder has used to pelletize under uniaxial pressing. The pellets have sintered separately at 900^0C temperatures for 4hours. These sintered pellets have been used for microscopic and dielectric measurements. The dielectric measurements have carried out using SOLARTRON SI1260 model impedance analyzer.

3. Results and Discussion

The material structural characterization of annealed powders and sintered pellets of $NdCrO_3$ were presented in our previous paper [3]. The annealed powder showed a single phase with orthorhombic symmetry, Pnma space group. The sintered pellets showed around 94% of theoretical density and the average grain size was estimated using SEM micrograph and found to be in the order of 50nm.

Figures 1(a) and 1(b) show the frequency dependence of dielectric constant (ε') and dielectric loss (tanδ) of $NdCrO_3$ at different temperatures respectively. There exist two plateaus in each curve at lower and higher frequency regions. These are shifting towards the higher frequencies with increasing temperature. The Low frequency plateaus may be due to grain boundary response, whereas the high frequency plateau observed may be due to grain effect [2, 4]. The dielectric constant of this material increases with increase in temperature, suggesting the thermally activated relaxation in the material. The dielectric constant and loss decreases rapidly with increasing frequency. It can explain by polarization mechanism. At low frequencies, the dipoles can follow the alternating electric field, giving rise to higher values of dielectric constant and dielectric loss, whereas at high frequencies the dipoles cannot follow the rapidly changing field, hence the reduction in the value of dielectric constant and loss.

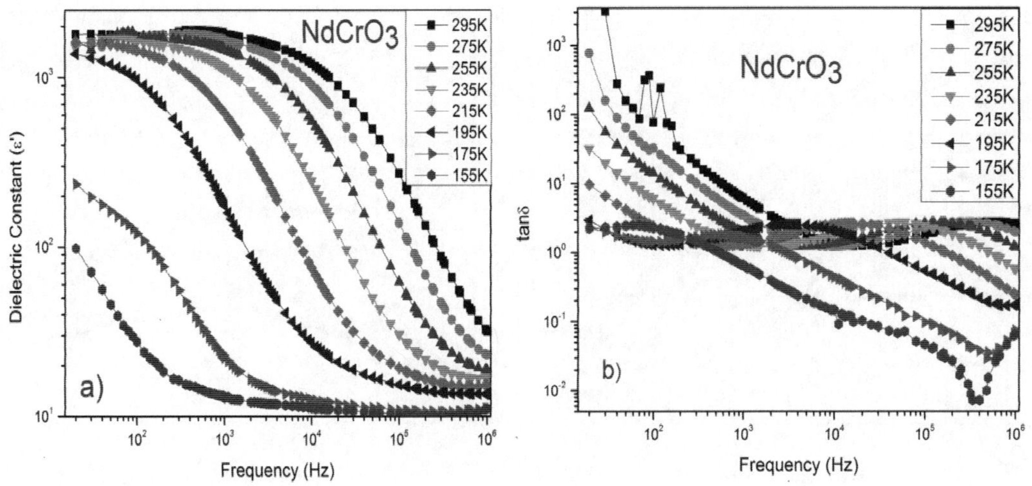

Figure 1. Frequency dependent of (a) dielectric constant (ε') and (b) loss tangent (tanδ) of $NdCrO_3$ at different temperatures.

Figures 2(a) and 2(b) show the frequency dependence of real part of impedance (Z') and imaginary part of impedance (Z'') of $NdCrO_3$ at different temperatures respectively. The Z' is more significant in the low frequency region, and it has decreased with increasing frequency and temperature. The value of Z' is high in the low frequency region, and it has gradually decreased with increasing frequency and attains a constant value at higher frequencies irrespective of temperature. This trend is shifting towards the higher frequency with rise in temperature. The decrease of Z' with increasing temperature indicates a possibility of space charge and subsequent lowering of barrier properties in the material [5, 6].

Figure 2(b) shows two relaxation peaks at lower and higher frequencies for all temperatures. The polycrystalline materials generally show grain and grain boundary properties [7]. Hence, these two relaxation peaks occurred may be due to grain and grain boundary properties and is clear proof of temperature and frequency dependent relaxation. The relaxation time (τ) of grain and grain boundary was estimated from the Z'' V_s frequency curve. Inset Figure 2b showed the variation of grain and grain boundary relaxation time (τ) as a function of temperature. The slope of these plots follows the Arrhenius law. This slop indicating the temperature dependent spread of relaxation in the range of 10^{-2} to 10^{-7} sec suggests the enhancement in the relaxation process of the carrier transport in the material with the rise in temperature [8]. The activation energies of grain and grain boundary relaxations were evaluated, which are found to be $E_{a(g\tau)}$=2.36eV and $E_{a(gb\tau)}$=1.8eV.

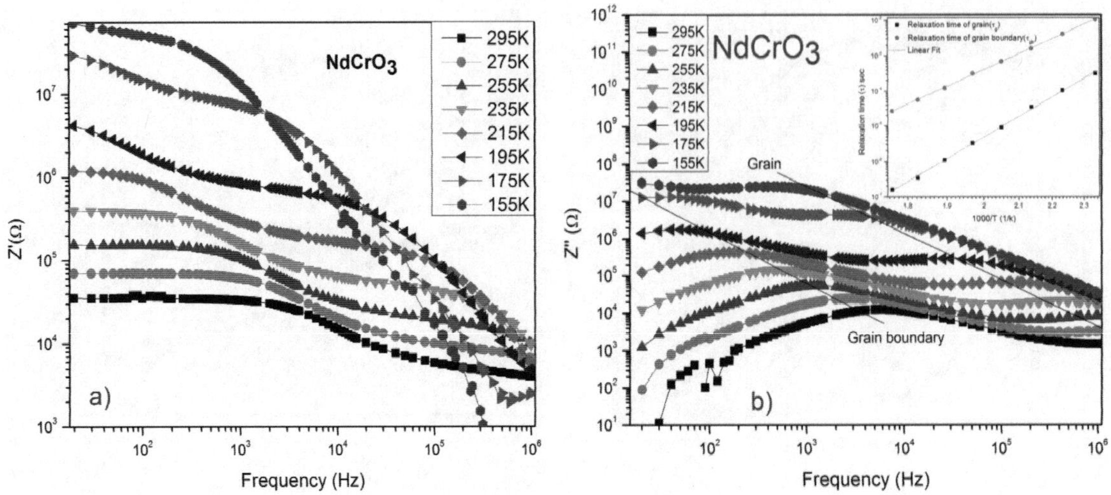

Figure 2. Frequency dependence of (a) Real part impedance (Z') and (b).Imaginary part of impedance (Z") of NdCrO3 at different temperatures. Inset Figure 2(b). Showed the variation of grain and grain boundary relaxation time (τ) as a function of temperature.

Figure 3(a) Complex impedance (Cole-Cole) plots of $NdCrO_3$ at different temperature shows two depressed arcs, which are corresponding to the contributions of two separate polarization relaxation processes. The appearance of two arcs or semicircles at a particular temperature shows the electrical properties of the materials arise mainly due to the contribution of the bulk (grain) and grain boundary effects in the higher and lower frequency regions respectively. The formation of full, partial or no semicircles mainly depends on the strength of relaxation, experimentally available frequency range. The electrical process takes place within the materials may be modelled (RC circuit) based on the bricklayer model. The intercept of the semicircle on Z'-axis gives the value of grain (bulk) resistance (R_g) and grain boundary resistance (R_{gb}) at higher and lower frequencies respectively. Both of these resistances have found to be decreasing with increases in temperature seen in figure 3(b). It suggests that material shows the negative temperature coefficient of resistance (NTCR) type behaviour [9-11]. These plots exhibit depressed semicircles having a centre lying below the real axis confirming the presence of non-Debye type of relaxation phenomenon in the materials.

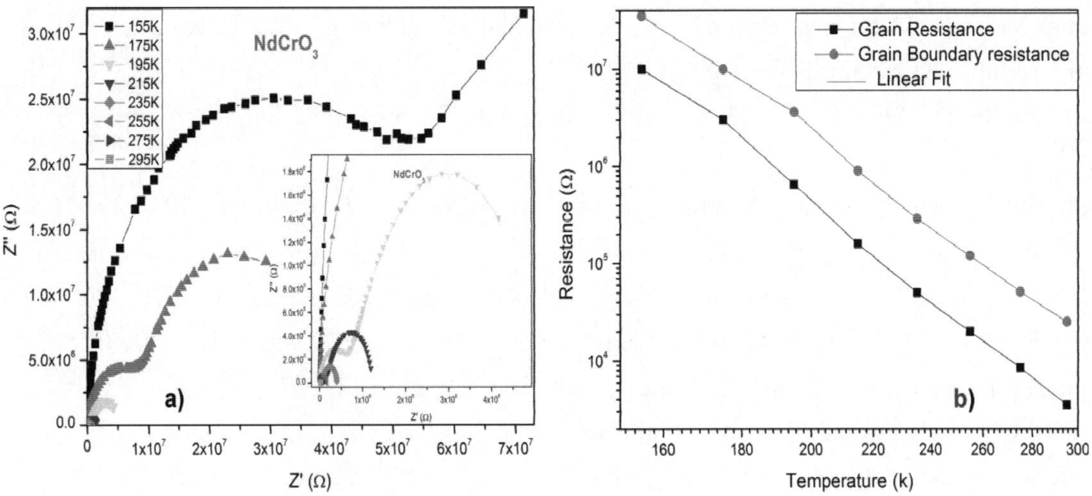

Figure 3. (a) Complex impedance (Cole-Cole) plots of $NdCrO_3$ at different temperature [Inset Fig.3 (a) Zoom view of the same], (b) Temperature dependent resistance of grain and grain boundary of $NdCrO_3$.

4. Conclusion

Polycrystalline perovskite ceramic compound of $NdCrO_3$ has prepared by sol-gel auto-combustion method. Frequency and temperature dependence of dielectric constant analysis suggests that there exist thermally activated relaxation in the material. A polarization mechanism is suggested for change in the dielectric constant and loss with temperature. The Impedance analysis supports the typical behaviour NTCR of material. The temperature dependent relaxation time for both grain and grain boundary follows the Arrhenius law.

Acknowledgments

One of the author Jada Shanker would like to thank UGC-RGNF-SRF for providing financial assistance to carried out this work.

References

[1] Sujoy Saha, Sadhan Chanda, Alo Dutta, T. P. Sinha, J Sol-Gel Sci. Technol (2014) 69:553–563.

[2] Bandi Vittal Prasad, G. Narsinga Rao, J.W. Chen, D. Suresh Babu. Materials Chemistry and Physics 126 (2011) 918–921.

[3] Jada Shanker, M. Buchi Suresh, D. Suresh Babu. Materials Today: Proceedings 3 (2016) 2091–2100.

[4] Bandi Vittal Prasad, B. Venugopal Rao, K. Narsaiah, G. Narsinga Rao, J. W. Chen, D. Suresh Babu. AIP Conf. Proc. 1536, 675 (2013).

[5] Hemant Singh, Amit Kumar, K.L. Yadav, Materials Science and Engineering B 176 (2011) 540-547.

[6] Balgovind Tiwari, R.N.P. Choudhary, Journal of Alloys and Compounds 493 (2010) 1-10.

[7] Subrat K. Barik, R.N.P. Choudhary, A.K. Singh, Adv. Mat. Lett. 201 1, 2(6), 419-424

[8] Jada Shanker, M. Buchi Suresh, D. Suresh Babu, International Journal of Scientific, Engineering and Research Vol. 3, Issue 7, July (2015), page 194-197.

[9] Jada Shanker, B. Vittal Prasad, M. Buchi Suresh, R. Vijaya kumar, D. Suresh Babu, Material Research Bulletin, 94 (2017) 385-398.

[10] Jada Shanker, K. Venkataramana, B. Vittal Prasad, R. Vijaya Kumar, D. Suresh Babu, Journal of Alloys and Compounds Vol. 732, 25 January 2018, Pages 314-327.

[11] Jada Shanker, G. Narsinga Rao, Kasarapu Venkataramana, D. Suresh Babu, Physics Letters A, DOI. https://doi.org/j.physleta.2018.07.002

IOP Conf. Series: Materials Science and Engineering **360** (2018) 012005 doi:10.1088/1757-899X/360/1/012005

SYNTHESIS, CHARACTERIZATION AND PHOTOLUMINESCENCE PROPERTIES OF Ce^{3+} DOPED ZnO NANOPHOSPHOR

Bhagyalekshmi G L*, Neethusha A P and Deepthi N Rajendran
[a]*Department of Physics, Govt. College for Women, Vazhuthacaud, Trivandrum, India*
Email: bhagyalekshmi_gl@yahoo.com

Abstract

Rare earth doped metal oxide semiconducting nanophosphors are relevant because of their unique optical properties and promising applications in optoelectronic devices. The present study involves the synthesis of Ce doped ZnO nano -particles (NPs) by a combustion method. X-ray diffraction results indicated that the synthesized ZnO and ZnO:Ce nanoparticles have a pure wurtzite (hexagonal phase) structure. The average crystallite size was found to be in the nanometer range. The formation of ZnO:Ce3+ nanophosphors was confirmed by Fourier transform infrared and X-ray diffraction spectra. The synthesised nanoparticles exhibit emission peaks in the blue region. The photoluminescence emission spectra of Ce-doped ZnO showed enhanced visible emissions as a result of 5d→4f transition of cerium.

Keywords: Nanophosphor, wurtzite, SEM, Photoluminescence

1. INTRODUCTION

Nanostructured materials play essential role in the fields of energy sources, health and environments. The relevance of nano-sized material is that their optical, electrical and chemical properties can be tuned by changing the size of the particles. Zinc oxide, a unique attractive metal oxide semiconductor, has been widely used in the field of near-UV emission, gas sensors, transparent conductor and piezoelectric applications [1-5]. It is due to its wide band gap 3.37 eV, large bond strength, and large exciton binding energy (60 meV) at room temperature [6]. Luminescent properties of ZnO can be improved by doping with transition metals or lanthanide ions. It provides modification in the electronic structure of ZnO nanoparticles for enhancing optical, magnetic and catalytic properties. Doping with rare earth elements have vast number of applications in lighting, displays, biomedicine, sensing, and security. They exhibit extremely sharp absorption and emission lines due to 4f-4f intra shell transitions and do not photo bleach [7-13].

ZnO doped with rare earth ion such as Eu, Gd, Sm and Ce were synthesised by various methods. The luminescent properties of Ce doped ZnO have been investigated by number of researchers, since cerium has the super-sensitive f–f transitions in some hosts and unusual interaction between the excited states and the host lattice [14]. In order to change the material properties in a controllable fashion suitable temperature and annealing ambient environment have to be chosen [15,16]. The effect of annealing temperature on the structural and optical properties of cerium(Ce) doped ZnO nanoparticles prepared by simple, low cost solution combustion method was discussed in this paper. The prepared nanopowders were characterized by powder X-ray diffraction (XRD), Scanning electron microscopy (SEM), Fourier transform infrared spectroscopy (FTIR), UV-visible spectroscopy and Photoluminescence spectroscopy (PL).

IOP Conf. Series: Materials Science and Engineering **360** (2018) 012005 doi:10.1088/1757-899X/360/1/012005

2.EXPERIMENTAL DETAILS

Ce doped ZnO nanopowders were prepared by combustion method by taking Zinc nitrate hexahydrate, Zn(NO3)2·6H2O and cerium nitrate hexahydrate, Ce(NO3)3·6H2O as oxidant and glycine as fuel. All chemicals are analytical grade. The cylindrical petri dish containing the reaction mixture is placed over a hot plate. Combustion evolves vast amount of heat and non-toxic gases resulting nanocrystalline powders. The prepared samples were annealed at different temperatures. Pure ZnO was also synthesised by combustion method for comparing the properties.

$$Zn(NO_3)_2\cdot 6H_2O + Ce(NO_3)_3\cdot 6H_2O \xrightarrow{\text{Glycine}} Zn_{(1-x)}Ce_xO + \text{gaseous products}$$

For structural analysis, X-ray diffraction measurements of the samples were recorded using Bruker AXS D8 Advance with Cu target radiation (λ =1.5406 A^0).The particle morphology of phosphors has been examined by JEOL, JSM - 6390LV Scanning electron microscopy. FTIR spectroscopic studies have been performed with a IR Prestige-21. The optical absorption studies were recorded on UV-2450 spectrophotometer. Photoluminescence studies carried out by using LS Fluorescence Spectrometer. Sample codes are entitled in Table 1.

Table 1. Sample code description of prepared samples

Sample code	Sample description
Z1	Undoped ZnO calcined at 200^0C
Z2	Ce:ZnO calcined at 500^0C
Z3	Ce:ZnO calcined at 700^0C

3.RESULTS AND DISCUSSION

3.1 X-RAY DIFFRACTION

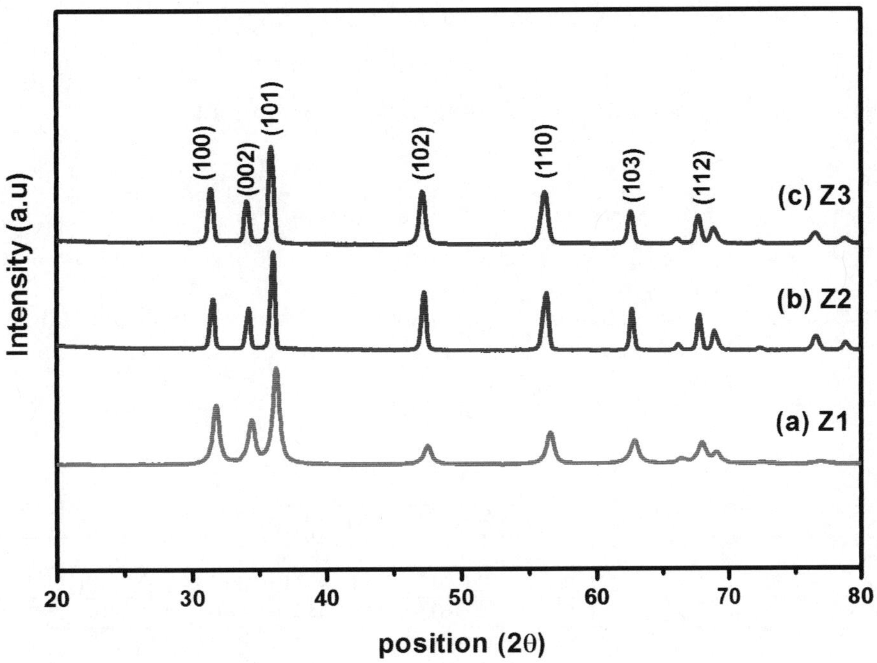

Fig 1. XRD patterns of (a) Z1,(b) Z2,(c) Z3 nanophosphor

Fig 1 shows the XRD results obtained for Z1, Z2 and Z3 nanophosphor. The crystallite size of the nanoparticles was estimated by applying the Scherrer equation [17]:

$$D = \frac{K\lambda}{\beta\cos\theta} \tag{1}$$

where D is the average particle size of the crystallite, K the particle shape factor, λ is the wavelength of X-ray used which is CuKα radiation (λ= 1.5406 Å), β the calibrated the corrected full width at half maximum (FWHM) of the XRD peaks, and θ is the Bragg's angle (half of the peak position angle).

For hexagonal lattice, lattice parameters are obtained by the equation [18]:

$$\left[\frac{1}{d_{hkl}}\right]^2 = \left[\frac{h^2+k^2+l^2}{a^2}\right] + \frac{l^2}{c^2} \tag{2}$$

Unit cell volume (V) of the products is given by the formula,

$$V= (0.866)\ a^2c$$

Bond length of the prepared nanocrystalline hexagonal lattice is obtained by,

$$L = \left\{ \frac{a^2}{3} + c^2 \left[\frac{1}{2} - u \right] \right\}^{0.5} \qquad (3)$$

Where $u = \frac{a^2}{3c^2} + 0.25$, a and c are lattice parameters of the unit cell.

Dislocation density of the nanoparticle is calculated using the formula,

$$\delta = \frac{1}{D^2} \qquad (4)$$

Table 2. Structural parameters of the pure and doped Ce:ZnO nanophosphors

Sample details	'd' spacing (A^0)			Lattice parameters(A^0)			'Cell volume 'V'(A^0)	Bond Length 'L'	Crystallite Size 'D'(nm)	Dislocation density $\delta \times 10^{-3}$ $(nm)^2$
	100	002	101	a	c	c/a				
Z1	2.8134	2.6090	2.4763	3.2486	1.972	1.972	5.2149	1.972	16.7664	3.5572
Z2	2.8464	2.6296	2.5006	3.2867	2.0145	2.0145	5.1770	2.0145	21.5294	2.1574
Z3	2.8122	2.6013	2.4740	3.2472	1.9804	1.9804	5.2064	1.9804	26.2430	1.4520

Crystallographic parameters obtained from XRD are enlisted in Table 2. It is clear that crystalline size increases with increasing annealing temperature. The peak width increases with decrease of annealing temperature. This indicates the growth of crystals and construction of larger clusters. Heating process enhances number of collision between particles leads to the coalesce with one another to form a larger particle [19]. Increase of unit cell volume with temperature may due to the rapid decrease in the density of vacant lattice sites, vacancy clusters, and local lattice disorders and a rapid resumption of lattice parameters [20]. Decrease in dislocation density may be due to the increase of crystal growth.

3.2. SCANNING ELECTRON MICROSCOPY (SEM)

Fig 2. SEM image of (a) Z1,(b) Z2,(c) Z3 nanophosphor

Microstructural charecterization of the synthesized samples were carried out by SEM analysis.SEM images of final products are shown in Fig 2. Pure ZnO appears like tiny sperical particles. In fig 2(b) and 2 (c), crystalline size increases with temperature such that particles were aggregated and also the attractive force between particles will increases and it appears like agglomerate.

3.3. FTIR SPECTROSCOPY (FTIR)

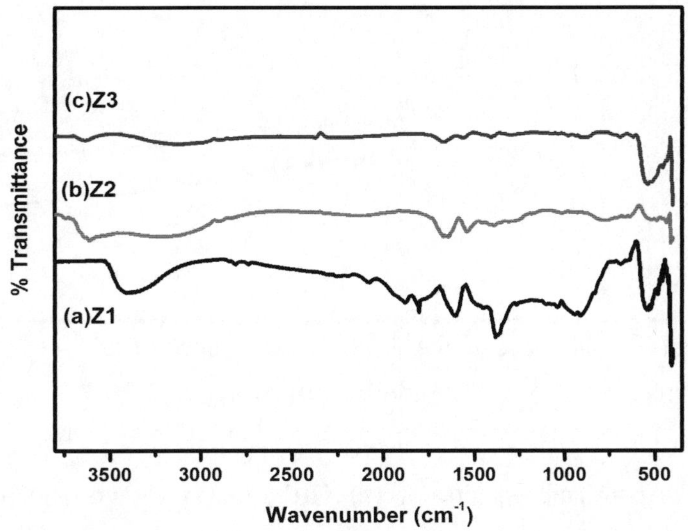

Fig 3. FTIR spectra of (a) Z1,(b) Z2,(c) Z3 nanophosphor

Chemical and structural changes of synthesized nanomaterials can be investigated by FTIR spectroscopic analysis. Fig 3 illustrates FTIR spectra of Z1, Z2 and Z3 nanophosphor, studied in the wave number range 400 - 4000 cm-1. The spectra shows the band located at around 450–490 cm–1 can be attributed to the Zn–O stretching mode in the ZnO lattice [21]. The absorption peak at 526 cm-1 ascribed to stretching frequency of ZnO [22]. Sharpening of metal oxide peak with temperature indicate better crystallinity and at higher temperature peak gets broaden may due to size reduction. On Ce doping samples exhibit a peak centred at 892 cm-1 may be due to the presence of Ce-O stretching vibration [23]. As particle size increases, lower vibrational frequency shift arise, may due to repulsive dipolar interactions and increase of bonding frequency may due to quantum confinement effect. The broad absorption band around 3514 cm-1 attributed to the O-H stretching mode of hydroxyl group. It is observed that the area of hydroxial peak diminishes with rise in temperature. Peak at 1369 cm-1 correspond to C=O bending vibrations and will diminish at higher annealed temperature.

3.4. UV-VISIBLE SPECTROSCOPY

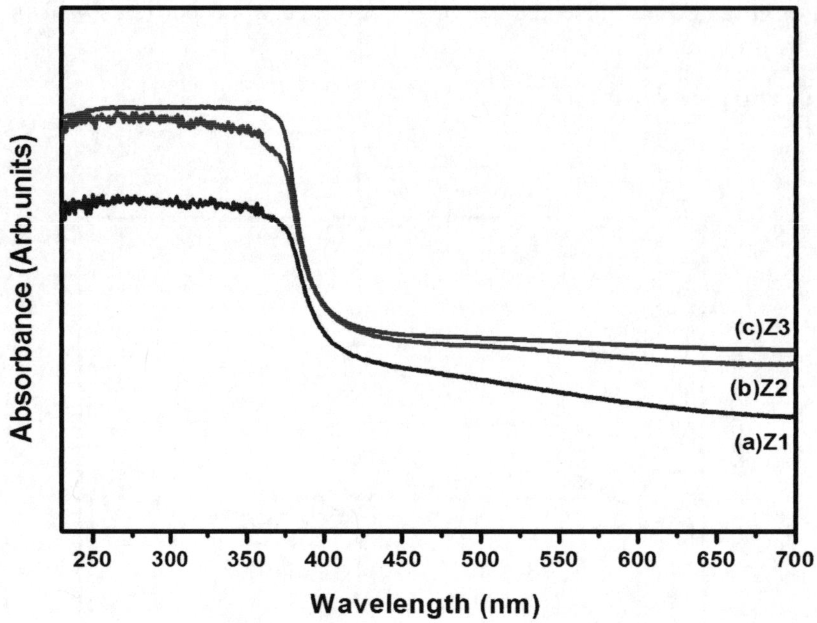

Fig4 . Absorbance spectra of (a) Z1,(b) Z2,(c) Z3 nanophosphor

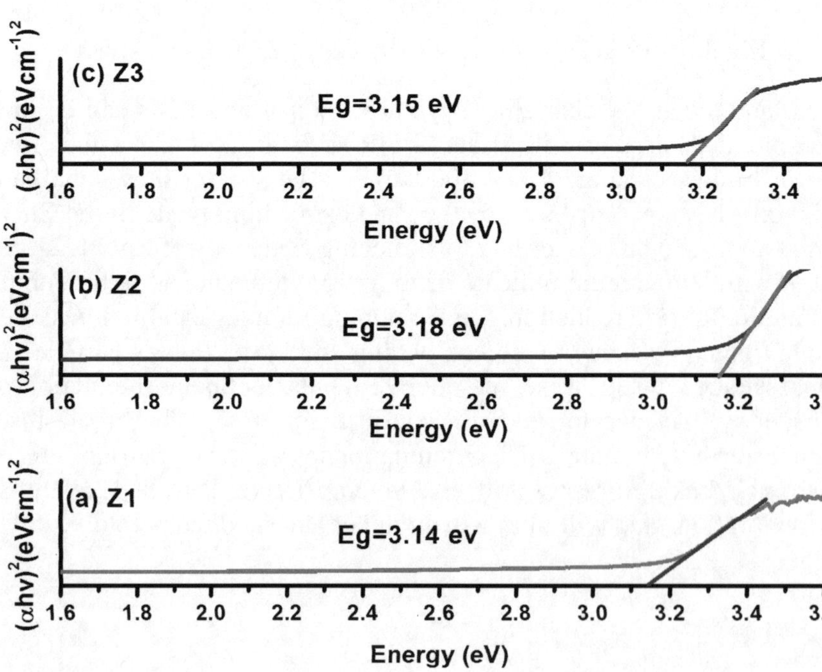

Fig 5. Tauc plot of (a) Z1,(b) Z2,(c) Z3 nanophosphor

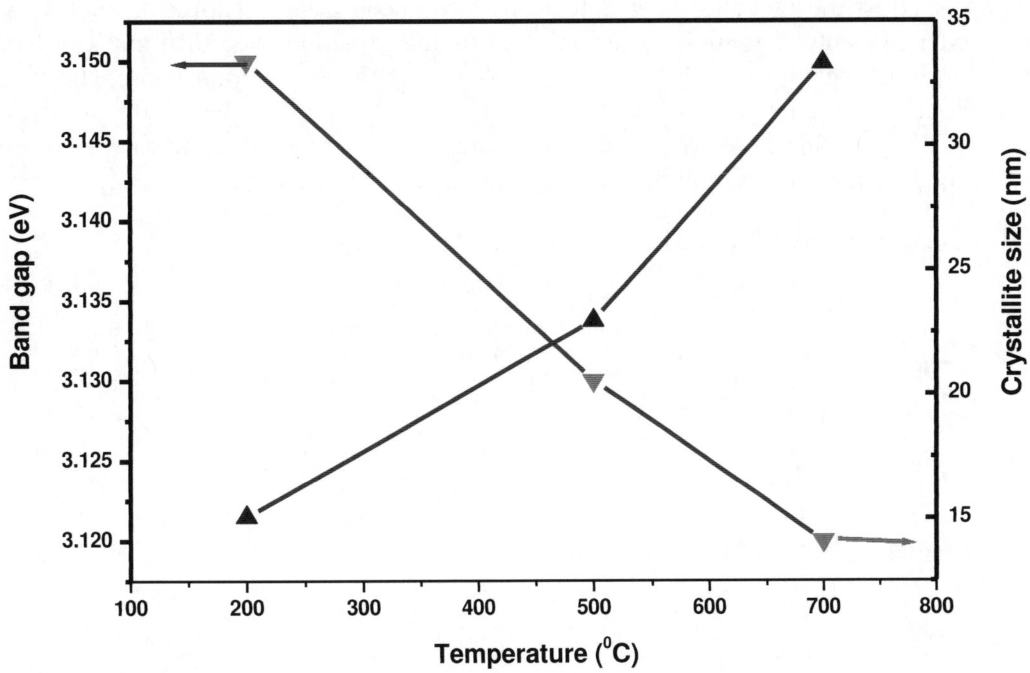

Fig 6. The calcination temperature dependence on the band gap and crystallite size

Table 3. Variation in crystallite size and energy band gap with temperature

Sample details	Crystallite size (nm)	Band gap (eV)
Z1	15.0623	3.15
Z2	22.9682	3.13
Z3	33.2992	3.12

The optical properties of synthesized nanopowders were studied by using optical absorption spectrum. Fig4 exhibits UV-visible spectrum of Z1, Z2, Z3 nanophoshor in the wavelength range 350-550 nm.The fundamental absorption band edge shift towards higher wavelength (red-shift) with calcination temperature. Red shift may be due to the insufficient growth of paricle as a result of solid state diffusion process and blue shift is attributed to the quantum size effect [24]. The optical band gap of the nanopowders were calculated using Tauc relation [25]:

$$\alpha = A(h\nu - E_g)^n \quad (5)$$

Where α is the optical absorption co-efficient, A is a constant, h is Plank's constant, ν is the photon frequency, and Eg is the optical band gap and n is a value that depends on the nature of transition (n=2 for direct band gap, 2/3 for direct forbidden gap and 1/2 for indirect band gap). Eg might be calculated by plotting (αhν)2 on the Y- axis and photon energy (hν) on the X-axis as in Fig 5 and tabulated in table 2 for all samples. Band gap studies reveal that optical band gap undergoes a reduction with increasing temperature. Decrease of band gap may be ascribed to the fact that new defect are introduced after substitution of Ce by Zn atoms in order to electro negativity and ionic radius difference between Zn and Ce. Dopant atoms take up the energy levels located in the bottom of the conduction band. The valence electrons require extra energy to be excited to higher energy states in the conduction band. It weakens the electron and hole wavefunctions, which result from reduction of quantum size effect and enhancement of crystallinity accompanied with increase of particle size.

3.5 PHOTOLUMINESCENCE SPECTROSCOPY

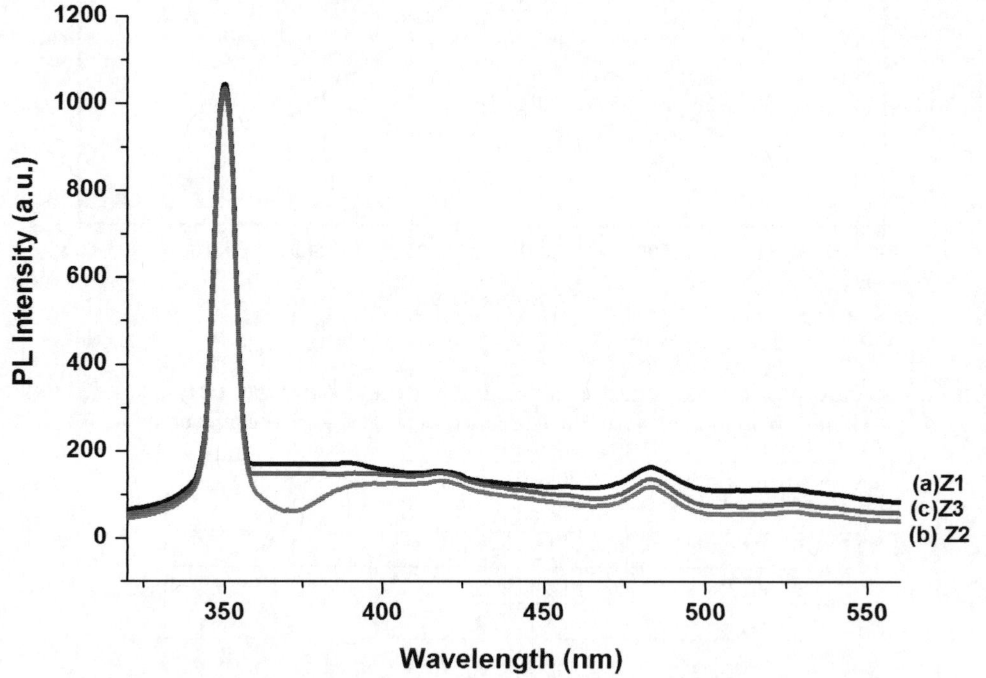

Fig 7. PL spectra of (a) Z1 (b) Z2 (c) Z3 nanophosphor

Photoluminescence investigation confirms the ability of nanoparticle to absorb incident energy and convert it into visible radiation. Fig 7 represents the photoluminescence spectra of Z1, Z2, Z3 nanophosphors. Luminescence peaks of Z1 nanophosphor were observed at 420 nm, 457 nm and 486 nm excited by 280 nm. The most prominent peak located at 486 nm may due to the transitions from 2D3/2–2F7/2 states. Blue emissions correspond to singly occupied oxygen vacancies in deep trapped holes [van Dijken et al., 2000; Li et al., 2003]. Peak at 420 nm may due to Zn vacancies[27].On cerium doping peak positions shifted to higher wavelength region, which may be due to the energy transfer of conduction band electrons of ZnO to the newly created intermediate levels by Ce3+ ions[28].Also, on doping, impurity ions reducing electron-or hole-trapped surface levels on ZnO, leading to increase in luminescence intensity [29].As calcined temperature increases,

IOP Conf. Series: Materials Science and Engineering **360** (2018) 012005 doi:10.1088/1757-899X/360/1/012005

prominent and weak peak positions does not changed but intensity get sharply decreased. Reduction in PL intensity may be due to crystal growth.

4. CONCLUSION

The nanocrystalline powders of ZnO:Ce were prepared by combustion method. The wurtzite structure was confirmed from the diffraction patterns of all prepared samples. It is found that structure of synthesized samples remains unaffected by annealing but there is a change in the crystallite size. FTIR showed that various chemical functional groups are present in the synthesized samples. The SEM results reveals the morphological structure and variation in the grain growth. Optical UV-Vis measurements indicate that band gap shifted to higher wavelength with rise in temperature. It is found that PL intensity increases with rise in calcination temperature. Post growth thermal treatment can drastically modify the luminescence behaviour in the blue region of Ce doped ZnO samples.

ACKNOWLEDGMENT

The authors would like to acknowledge their sincere thanks to NIIST, Trivandrum and STIC, Cochin for providing facilities for the analysis of the prepared samples.

REFERENCE

[1] Jang-Won Kang, Yong-SeokChoi, MinhyeokChoe,Na-YeongKim, Takhee Lee,Bong-JoongKim,CWTu, Seong-JuPark, Nanotechnology 23 (2012)495712–495716.

[2] S.Y. Lee, E.S.Shim, H.S.Kang, S.S.Pang, J.S.Kang, ThinSolidFilms 473 (2005)31–34.

[3]Z.L. Wang, X.Y. Kong, Y. Ding, P.Gao, W.L. Hughes, R. Yang, Y.Zhang,Adv.Funct.Mater.14(2004)943–956.

[4] H. Harima,J.Phys.Condens.Matter16(2004)S5653.

[5] R. Könenkamp,R.C.Word,C.Schlegel,Appl.Phys.Lett.85(24)(2004) 6004–6006.

[6] J.T. Hu, , T.W. Odom, and C.M. Liebe, Accounts of Chemical Research 32 (1999) 435-445

[7] H.X. Shi , T.Y. Zhang , H.L. Wang, J. Rare Earths 29 (2011) 746.

[8] R. Zeng , J.G. Wang , J.Y. Cui , L. Hu , K.G. Mu , J. Rare Earths, 28 (2010) 353.

[9] C.M. Fan , P. Xue , Y.P. Sun, J. Rare Earths, 24(2006) 309.

[10] L. Bian , M.X. Song , T.L. Zhou , X.Y. Zhou , Q.Q. Dai, J. Rare Earths, 27(2009) 461.

[11] M. Zalaj, J. Rare Earths, 32(2014) 487.

[12] X.Z. Li , Z.L. Hu , X.B. Zhao, X.W. Lu, J. Rare Earths, 31(2013) 1157.

[13] J. Du , B.H. Li, J.J. Huang, W.L. Zhang , H.L. Peng , J.G. Zou, J. Rare Earths,

31(2013): 992.

[14] W.M. Yen , Phys. Solid State, 47 (2005) 1393–9.

[15] S. Dutta , S. Chattopadhyay , A. Sarkar , Prog Mater Sci , 54 (2009) 89–136.

[16] Z.B. Fang, Z.J. Yan, Y.S. Tan , Appl Surf Sci , 241(2005) 303–308.

[17] M.-M. B. Mohagheghi, N. Shahtahmasebi, M.R. Alinejad, A. Youssefi and M.S. Saremi, Physica B, 403 (2008) 2431.

[18] M. Z. IqbaL, S. Aliand m. A. Mirza . Coden jnsmac, 48 (2008) 11.

[19] K. Kenevey, F. Valdivieso, M. Soustelle and M. Pijolat, Appl.Catal.B:Environ,29 (2001) 93-101.

[20] M. Soosen Samuel, Lekshmi Bose , K.C. George , Academic Review (2009) 57-65.

[21] Sudheesh K. Shukla, Eric S. Agorku, Hemant Mittal, Ajay K. Mishra, 68 (2014) 217–222.

[22] Shivaji. Rauba. Kulal and Sambhaji. Rau. Bamane, Archives of Applied Science Research, 2 (2010) 205-210.

[23] Harish Kumar, Rani R. Int. Letts. of Chem. Phys,Chem. and Astronomy. 14 (2013)26-36.

[24] W. G. Becker, A. J. Bard, J. Phys. Chem., 87 (1983) 4888.

[25] S. Suwanboon , P. Amornpitoksuk ,A. Haidoux , J.C. Tedenac , J. Allys. Compd., 462 (2008) 335-339.

[26] C.Jayachandraiah, G.Krishnaiah, K. Siva Kumar, International Journal of Chem Tech Research, 6 (2014) 3378-3381.

[27] O.M. Ntwaeaborwa, P.H. Holloway, Nanotechnology 16 (2005) 865.

[28] A. Dutta, S. K. Panda, S. Chaudhuri, J. Solid State Chem. 181 (2008) 2332-2337.

[29] Song Wei Lua , Burtrand I. Leea, Zhong Lin Wang , Wusheng Tong , Brent K. Wagnerc , Wounjhang Parkc , Christopher J. Summersc, Journal of Luminescence 92 (2001) 73–78.

IOP Conf. Series: Materials Science and Engineering **360** (2018) 012006 doi:10.1088/1757-899X/360/1/012006

SYNTHESIS AND CHARACTERIZATION OF P_2O_5 DOPED ZnO

Neethu Sha A P [a, *], Bhagyalekshmi G L, Deepthi N Rajendran
Department of physics,Govt College for Women,Trivandrum,Kerala,India.-695014
[*]*Corresponding author Email*:neethusha15@gmail.com

Abstract

ZnO nanoparticles has drawn great attention due to the presence of ZnO which act as a network former or modifier . It offers wide range of application in optoelectronics with a wide band gap 3.37 eV at room temperature. Now a days ZnO nanoparticles has wide range application in optoelectronic devices. The aim of the present study is to prepare P_2O_5 doped ZnO with different concentrations with good structural. morphological and optical properties. The samples were synthesized using solid state reaction method. The structures of the powdered sample were studied using X-Ray Diffraction technique. The chemical groups of the sample were discussed by using FTIR. UV-visible spectroscopy reveals that as the doping increases the band gap decreases and then increases . Photoluminescence (PL) and SEM studies were also carried out.

Keywords: Nanoparticle, Photoluminescence and Band gap.

1.Introduction

Zinc oxide posses an exciton binding energy of 60meV with a wide band gap which leads to possible light emitting applications well above the room temperature1-2. It can be applied in the field of optical and electrical devices ,piezo electricity, drug delivery ,solar cell ,cosmetics3-7. Lack of stable, conductive p-type epitaxial layers due to its asymmetric doping limits strongly against p doping is the major obstacle to the development of P_2O_5-doping is the major obstacle to the development of ZnO_8.High carrier concentration, reasonable mobility and low resistivity are expressed by P_2O_5 doped ZnO[9].Because of the unique physical properties such as high thermal expansion coefficient, low melting and softening temperature, high electrical conductivities, extensive UV transmissions P_2O_5 have attracted much attention[10].The structural,physical and chemical durability properties of the host can be modified by ZnO[11-13].

In this paper, we report the synthesis of P_2O_5 doped ZnO and the effect of P_2O_5 on the morphological and optical properties. P_2O_5 doped ZnO compounds could be well acheived in future.

2.Experimental details

The P_2O_5 nano materials with three different concentrations 1 wt%,1.5 wt% and 2 wt% were synthesized by solid state reaction method. ZnO and P_2O_5 are weighed accurately and it is placed in an agate motor. The mixture is grinded with substrate, acetone. Then the mixture is placed in an oven for one hour at 80°C.Again grind the mixture and places it in the oven for one hour at 80°C.Then it is calcined to 600°C by using Muffle furnace and analyzed. The pure ZnO samples were present in the white colour, while P_2O_5 doped ZnO for different concentrations becomes light ash and the colour becomes deeper and deeper with the increasing P_2O_5.

Table 1. Sample code of prepared samples

Sample code	Sample description
ZP	Pure ZnO
ZP1	$Zn_{(0.99)}O:P_2O_5\ _{(0.01)}$
ZP2	$Zn_{(0.985)}O:P_2O_5\ _{(0.015)}$
ZP3	$Zn_{(0.980)}O:P_2O_5\ _{(0.02)}$

The scanning electron microscopy is used to find out the morphological study. By using the XRD we can find out the phase analysis and crystal geometry. Fourier transform infrared analysis was carried out by using IR Prestige-21. With the help of Flurescence Spectrophotometer the photoluminescence spectra of P^2O^5 doped ZnO were analysed. UV-Visible spectrophotometer is used to find out the band gap. Sample codes are enlisted in Table 1.

3.Results and Discussion

3.1 XRD Analysis

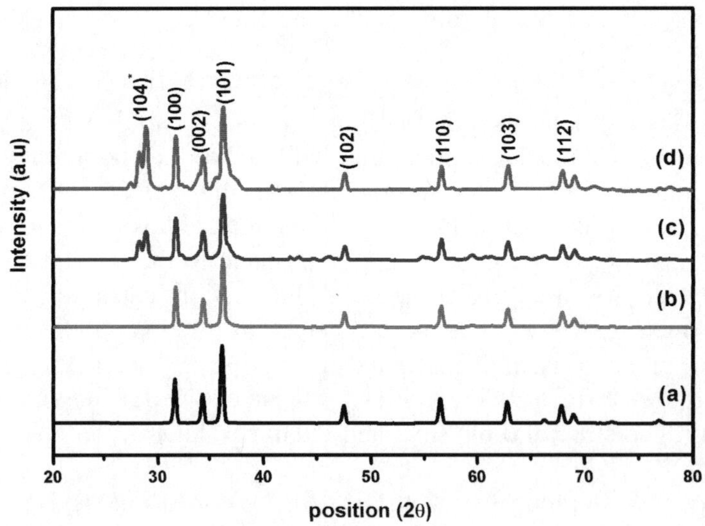

Figure 1. XRD patterns of a) ZP (b) ZP1(c) ZP2 (d) ZP3 nanophosphor

The XRD pattern obtained for P_2O_5 : ZnO NPs are recorded with 2θ ranging from 10^0 to 80^0 and it is represented in the Fig 1.The diffracted peaks at 2θ were assigned to lattice planes respectively in accordance with the JCPDS File No:87-0952 and21-1486 respectively. It is observed that the P_2O_5: ZnO NPs possess hexagonal wurtzite structure. The crystallite size is evaluated using the Scherer Formula[14]:

$$D = \frac{K\lambda}{\beta \cos\theta} \qquad (1)$$

Here K is the Scherer's constant and its value is 0.9. λ is the wavelength of X-Rays, β is the full width of half maximum. From the Scherrer formula P_2O_5:ZnO with three different concentrations having the crystallite size was calculated and is tabulated in Table 2.It can be seen that crystallite size varies with doping.

It is observed that XRD peaks originated from a crystal mainly due to various orientation of crystal planes. In which it comprises of many number of molecules exist in the form of unit cells which plays an important role.

Table 2. Crystallographic parameters of prepared samples

Sample details	'd' spacing (A^0)			Lattice parameters (A^0)			'Cell volume 'V'(A^0)	Crystallite Size'D'(nm)
	100	002	101	a	c	c/a		
ZP	2.82	2.61	2.48	3.26	5.25	1.625	48.3183	29.6375
ZP1	2.83	2.60	2.48	3.28	5.18	1.568	47.4129	40.1298
ZP2	2.82	2.60	2.48	3.26	5.19	1.593	47.6741	44.1407
ZP3	2.82	2.65	2.48	3.26	5.18	1.594	48.2608	43.8129

3.2 FTIR Spectroscopy

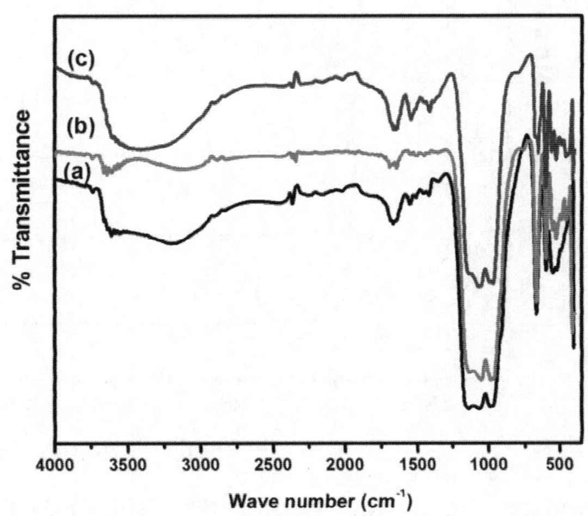

Figure 2. FTIR spectra of a) a) ZP1 (b) ZP2 (c) ZP3 nanophosphor.

By using the FTIR spectroscopy the composition, quality and molecular structure of the samples were carried out. It is observed that the FTIR spectrum of P_2O_5 NPs was recorded in the range of 400-4000cm^{-1} .The characteristic absorption peak is in the region 958.5739 and 1099.51133 respectively and is given in the figure3. The vibrational peaks at 400cm^{-1} to 500cm^{-1} indicates the presence of ZnO[15-16].

IOP Conf. Series: Materials Science and Engineering **360** (2018) 012006 doi:10.1088/1757-899X/360/1/012006

3.3. SEM Morphology

Figure 2. SEM image of a) ZP1, (b) ZP2 (c) ZP3 nanophosphor.

The SEM image of the P_2O_5 doped ZnO was recorded in the SEM operation mode. For electron production the filament voltage applied was 20kV with a magnification of 40000x.It is observed from the figure that it posses spherical shaped structures .It is clear that the distribution of the particles present in the image was computed using image software. It is observed that the morphology is well ordered.

3.4. UV-Visible Spectroscopy

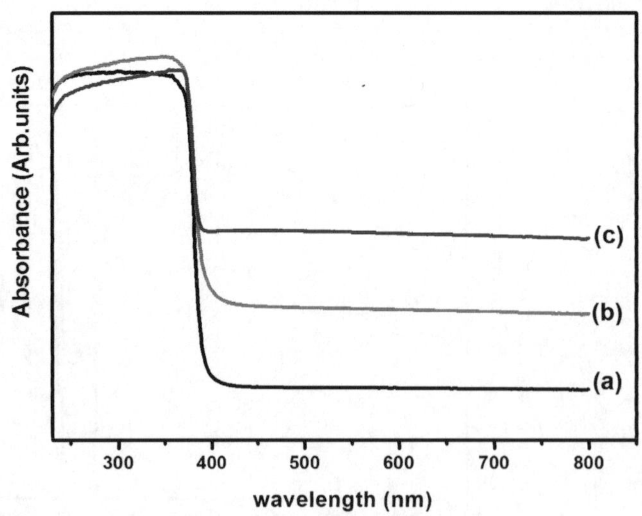

Figure 4. Absorbance spectra of a) ZP1, (b) ZP2 (c) ZP3 nanophosphor

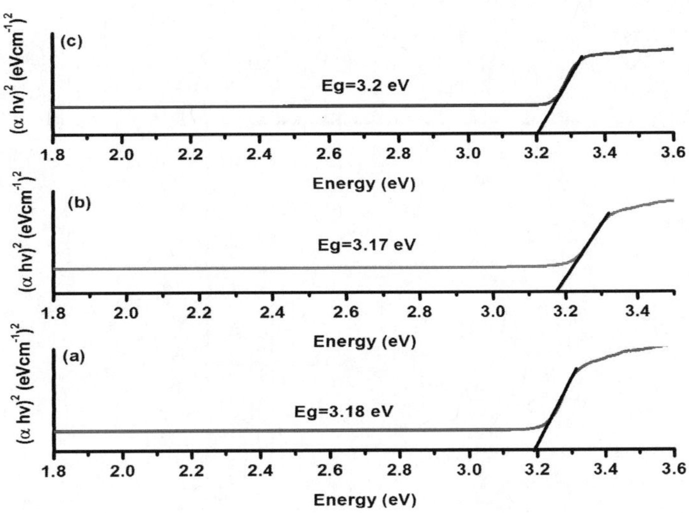

Figure 5. Tauc plot of a) ZP1 (b) ZP2 (c)ZP3 nanophosphor

The UV Visble spectroscopy of P_2O_5: ZnO were taken in the solid state method. It is found that the absorption spectrum is a function of wavelength. The optical band gap values can be calculated by the following relation [17]:

$$\alpha = A(h\nu - E_g)^n \quad (2)$$

where α is the absorption co-effcient, hν the photon energy, A is a constant, and Eg is the optical band gap. n is an index that characterizes the optical absorption and it is equal to 2 and 1/2 for indirect and direct allowed transitions, respectively.

By using tauc plot we can find out the band gap. It is found that the band gap for ZP1, ZP2 and ZP3 are 3.18,3.17 and 3.2 eV respectively. Due to quantum confinement, there is increase in the band gap, with the decrease in particle size and there is an optimum concentration in ZP2 sample.

Table 3.Band gap of prepared samples

Sample details	Bandgap (eV)
ZP1	3.18
ZP2	3.17
ZP3	3.2

3.5. Photoluminescene Spectroscopy

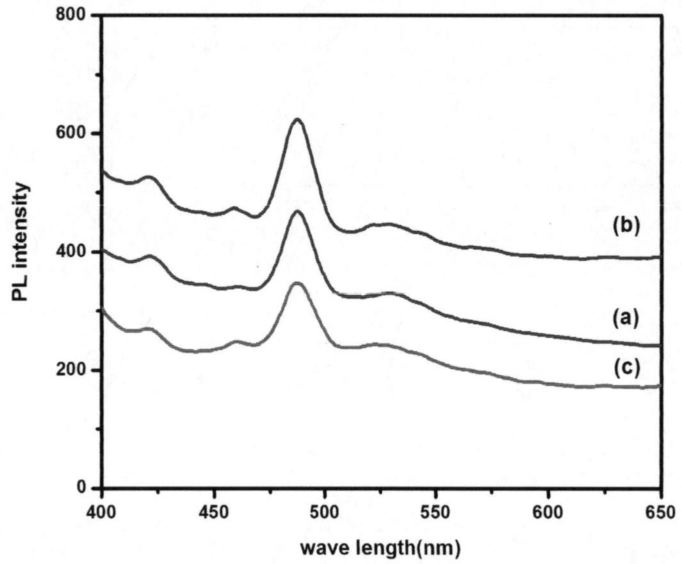

Figure 6. PL spectra of a) ZP1, (b) ZP2 (c)ZP3 nanophosphor

The photoluminescence spectrum of the P_2O_5:ZnO were recorded in the solid state mode. The emission spectra of three different concentrations are shown in the figure with the excitation at 325nm.Most prominent emission of synthesized samples lies in the blue region, lower energy side corresponds to the deep level emission. Also a luminescence quenching effect can be observed in phosphor samples upto certain concentration. It is found that the origin of DLE is due to the oxygen interstitial defect found in the ZnO. The emission at 488 nm is due to the trap –state emission from $^5D_0$ -$^7F_0$[18-19].

4. Conclusion

The structure of the prepared sample was studied using X- ray powder diffraction pattern and FTIR. XRD pattern shows hexagonal wurtzite structure. The chemical groups of the sample were confirmed by the FTIR spectra. The SEM image of P_2O_5 doped ZnO micrograph revealed that the primary particles are uniform, circular in shape and weakly agglomerated. The UV Visible spectroscopic studies of the prepared samples were also carried out and the bandgap of the samples obtained are 3.18eV,3.17eV and 3.2eV respectively. A blue luminescence quenching can be observed in samples upto an optimum concentration.

Acknowledgement

Most of the information is gathered from websites and published literatures. The analysis of the material are carried out from STIC and NIIST, we acknowledge them.

IOP Conf. Series: Materials Science and Engineering **360** (2018) 012006 doi:10.1088/1757-899X/360/1/012006

Reference

1.A. Singh,H.L.Viswakarma, *IOSR Journal of Applied Physics*,6 ,28-32 (2014).

2.G.Gopal,N.A.Spladin,Journal of Electronic Materials,35,538-542(2006).

3 J. B. Baxter , E. S. Aydil, *Appl. Phys. Lett.* 86, 53114 (2005).

*4.*M. H. Huang, S. Mao, H. Feick, H. Q. Yan, Y. Wu, H. Kind, E. Weber, R. Russo, and P. Yang,

Science 292,1897 (2001).

5. J. Song, J. Zhou, and Z. L. Wang, *Nano Lett. 6*, 1656 (2006).

6.. Z. L. Wang, *Annu. Rev. Phys. Chem.* 55, 159 (2004).

7 J. Sawai, H. Igarashi, A. Hashimoto, T. Kokugan, and M. Shimizu, *J. Chem. Eng. Japan* 29, 556 (1996).

8.K.Zhang,M.R. Hao,W. Guo,T.Heeg, D. G. Schlom, W. Z. Shen ,X. Q. Pan, *Physica*

B.407,2825-2828 (2012).

9.Donghua Fan,Rong Zhang,Yang Li ,*Solid state communications*,150,1911-1914(2010).

10.C.R.Kesavulu,Annielle Christine Almeida, Silva,M.R.Dousti,Noelio Olivera Dantas,ASS de

Camargo,Tomaz Catunda,*Journal of Luminescence* 165,77-84(2015).

11.Ngo Thu Huong,Nguyen Viet Tuyen,Nguyen Hoa Hong,Materials Chemistry and

physics,126,54-57(2011).

12.Y.R.Sui,BYao,L.Xiao,L.L.Yang,J.Cao,X.FLi,G.Z.Xing,J.H.Lang,X.Y.Li,SQLv,X.W.Meng,X.YLiu,J.H Yang,*Applied Surface Science* 287,484-489(2013).

13.M.Siti Amalah ,Azmi,M R Sahar,*Journal of Magnetism and Magnetic Materials*, 393,341-346(2015).

14. M.T. Weller, *Inorganic Materials Chemistry*, (1997)

15. C. Li, R. Chen, X. Zhang, S. Shu, J. Xiong, Y. Zheng, W. Dong, Mater. Lett. 65

1327-1330(2011).

16. M. Faisal, S.B. Khan, M.M. Rahman, A. Jamal, K. Akhtar, M.M. Abdullah, J. Mater.

Sci. Technol, 27 , 594-600(2011).

17. J. Tauc, *Amorphous and Liquid Semiconductors*.(1974).

18. Kumar, Nithin,Dorfman,Adam,Hahm,Jong-In,J.Nanosci,Nanotechnol.1915,5(2005)

19.D.M.Bagnall,Y.F.Chen,M.Y.Shen,Z.Zhu.T.Goto,T.Yao,J.Cryst.Growth 605,184 (1998)

Synthesis, Characterization and Anti – Bacterial Activities of SnO_2 Nanoparticles Using Biological Molecule.

Nancy John*, Malu Somaraj[1] and Nisha J Tharayil[2]

[,1] Department of Physics, S.N College, Kollam, Kerala, India*

[2] Department of Physics, S.N College for Women, Kollam, Kerala, India

E-mail: nancypninan@gmail.com

Abstract

Nano – crystalline SnO_2 particles have been synthesized by co–precipitation method using DNA, a biomaterial, as capping agent. The nanoparticles obtained were characterized using XRD, SEM and TEM. The XRD confirms the crystalline nature and formation of SnO_2 nanoparticle. The optical properties were studied using FTIR spectroscopy. The effect of SnO_2 nanoparticles on the growth of some common food and water poisoning pathogens were studied. To understand the mechanism of antibacterial activity and to confirm the role of reactive oxygen species, the samples were annealed at different temperature and the bacterial activity is probed. Antibacterial studies were done against three bacterial strains namely, Staphylococcus aureus (ATCC 9144), Shigella flexineri (ATCC 2908) and Escherichia Coli (ATCC 25922).

1 Introduction

Emerging infectious diseases and the development of drug resistance in the pathogenic bacteria and fungi at an alarming rate is a matter of serious concern. Despite the increased knowledge of microbial pathogenesis and application of modern therapeutics, the morbidity and mortality associated with the microbial infections still remains high [16]. Therefore, there is a pressing demand to discover novel strategies and identify new antimicrobial agents from natural and inorganic substances to develop the next generation of drugs or agents to control microbial infections. Organic compounds used for disinfection have some disadvantages, including toxicity to the human body, therefore, the interest in inorganic disinfectants such as metal oxide nanoparticles is increasing.
As an n – type wide band gap semiconductor ($E_g = 3.6$ eV), SnO_2 is one of the most intensively studied materials owing to its technological applications. The anomalous variation in the size and shape dependent properties of tin oxide nanoparticles makes it a promising candidate in optoelectronics, sensing, laser, solar cell, photocatalytic and biological applications.

Metal oxide nanoparticles are found good inhibitor to bacterial strains. The antimicrobial efficiency of metal oxide nanoparticles depends on the particle size, presence of light, composition of aqueous medium used in assay etc. Electrostatic interactions are responsible for the attachment of nanoparticles to the bacteria. These interactions changes the integrity of cell membrane of bacteria and toxic free radicals are released which induce oxidative stress on bacteria [2, 15]. Using biomolecules as a templating agent to synthesis inorganic nanoparticles is an effective method to fabricate functional materials with well-defined structure and controllable dimensions. Among biological molecules DNA have been extensively used as a bio template because of their physicochemical stability and unique structure [6 – 8].

In this paper, samples were prepared by co – precipitation technique using DNA as capping agent and their activities against various bacterial strains were studied. Antimicrobial studies were done against three bacterial strains namely, Staphylococcus aureus (ATCC 9144), Shigella flexineri (ATCC 2908) and Escherichia Coli (ATCC 25922).

2 Experimental

2.1 Materials

The chemicals used for the synthesis of nanoparticles, namely Stannic Chloride Pentahydrate, and Sodium Carbonate were obtained from MERCK. DNA is obtained from calf thymus. The bacterial cultures were obtained from Institute of microbial technology, IMTECH, Chandigarh.

2.2 Synthesis of SnO_2 nanoparticles

SnO_2 nanoparticles were prepared from Stannic Chloride Pentahydrate and Sodium carbonate using DNA as the capping agent with the help of a magnetic stirrer. The white precipitate obtained was separated from the reaction mixture and washed several times with distilled water to remove impurities and traces of chemicals used. The wet precipitate was allowed to dry naturally and then thoroughly grounded to obtain metal carbonate precursor in the form of fine powder. Upon heating the obtained metal carbonate precursor to 400°C, it decomposes to form metal oxide. The calcinations temperature of the sample were fixed upon the TGA measurements. The white coloured precipitate changed into pale yellow which showed the formation of tin oxide nanoparticles. The nanoparticles annealed at 500˚C and 700 ˚C are denoted by S1 and S2.

2.3 Characterization

The XRD studies were carried out using Bruker AXS D8 Advance model and TEM was recorded using Jeol/ JEM 2100 model from STIC, Cochin. SEM of the samples were taken by Hitachi S- 3400 N and the chemical composition was examined using EDAX attachment Nortan system six attached to the SEM unit. The FTIR studies of the samples were carried out in a Perkin – Elmer FTIR Spectrophotometer between 300 cm^{-1} and 4000 cm^{-1}.

3 Results and Discussions

3.1 EDAX analysis

The chemical composition of SnO_2 nanoparticles prepared at the optimized conditions was extracted from the energy dispersive X – ray spectrum (EDAX) which is shown in Figure 1. The EDAX analysis exhibits clear peaks of only Sn and O elements, whereas no additional peaks were detected, which means the prepared samples are exempted from the starting precursors [4, 8]. The percentage of Sn and O in the sample is shown in table 1.

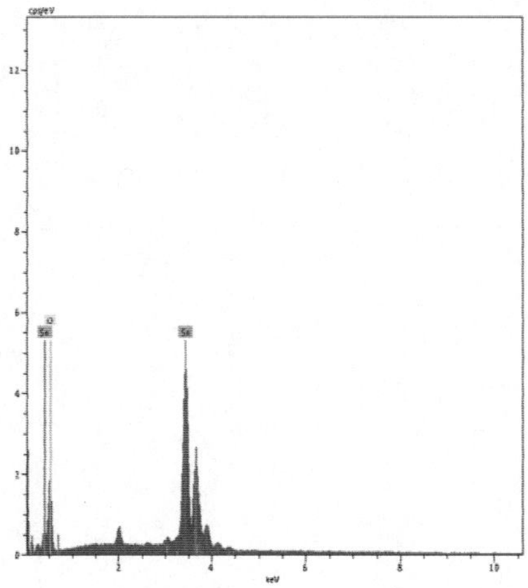

Figure 1. EDAX of S1

Table 1: Result of EDAX analysis showing the percentage of Sn and O in the sample.

Element	Mass %	Atom %
O K	74.12	27.85
Sn L	25.88	72.15
Total	100	100

3.2 XRD analysis

The XRD pattern of the synthesized samples (S1 annealed at 500˚C and S2 annealed at 700 ˚C) are shown in figure 2. The observed and standard (h k l) planes confirmed that the product is of SnO_2 having a tetragonal structure, which is in good agreement with the literature, values (JCPDS No 41 – 1445).

The average crystallite sizes of the samples were calculated using Debye-Scherer equation

$$D = \frac{K\lambda}{\beta cos\theta} \quad \text{...} \quad (1)$$

Here D is the crystallite size and β full width half maximum of the peak with diffracting angle θ.

The microstrain analysis shows that the residual stress and strain of the material is an important mechanical property of the material. The strain present in the lattice also contributes to

IOP Conf. Series: Materials Science and Engineering **360** (2018) 012007 doi:10.1088/1757-899X/360/1/012007

broadening of the XRD peaks. The microstrain, e and dislocation density, *D* can be calculated using the following equations,

$$e = \left(\frac{\lambda}{d\cos\theta} - \beta\right)\left(\frac{1}{\tan\theta}\right) \quad \text{......} \quad (2)$$

$$D = \frac{1}{d^2} \quad \text{......} \quad (3)$$

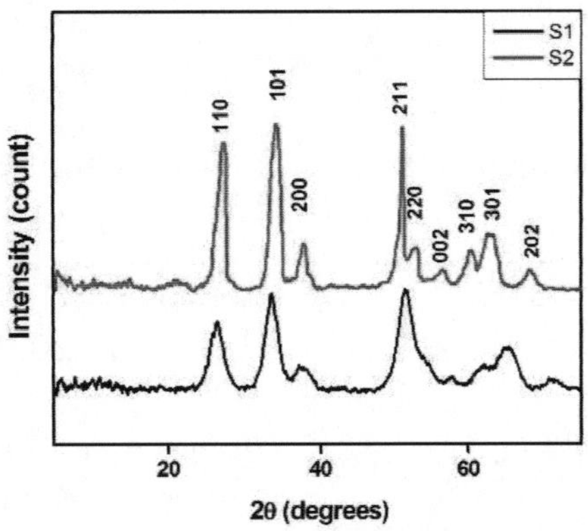

Figure 2. XRD pattern of samples S1 and S2.

Table 2: Parameters obtained from XRD of the three samples with particle size.

Sample	h k l	d observed	d JCPDS	Particle Size (nm)	Microstrain	Dislocation density
S1	110	63.39314	3.3470	5 ± 1	0.2092	4.73
	101	2.66438	2.6427			
	211	1.7721	1.7641			
S2	110	3.39343	3.3470	7 ± 1	0.0079	2.12
	101	2.67045	2.6427			
	211	1.77184	1.7641			

It is clear that from XRD patterns of the samples, the intensity enhanced with annealing temperature. The estimated average particle size, micro strain and dislocation density of the samples are given in table 2. As the annealing temperature increases the diffraction peaks gets sharper. The two possible mechanisms involved in the narrowing of peak, one is particle size effect and the other is lattice strain. These nanocrystals have lesser lattice planes compared to the bulk, which contributes the broadening of peaks in the diffractogram. The non- existence of any other overlapping peak in the standard peak position indicates that the prepared samples have only a single phase.

The microstrain and dislocation density decreases with annealing temperature. This may be due to the fact that when crystallite size become too small, the internal pressure exerted by the surface tension on the nanostructured material will create a stress field which induces lattice strain. When the crystal size is increased, stress reduces and thus lattice strain is reduced. In short, as temperature increases crystalline defects decreases thereby reducing dislocation density and lattice strain.

3.3 Scanning Electron Microscopy (SEM)

Scanning electron microscopy of samples are shown in figure 3. It reveals that the particles are agglomerated. The particles are more dispersed on annealing. The homogeneity in shape and size of the nanoparticles in the sample S2 appears to be better than that of S1 under the resolutions taken.

S1 **S2**

Figure 3. SEM images of samples S1 and S2

3.4 Transmission Electron Microscopy (TEM)

The TEM spectrum of SnO_2 nanoparticles are shown in (Figure 4). The particles are irregular in shape with the size ranges between 5 and 13 nm, which is in reasonable agreement with the results obtained from the XRD patterns.

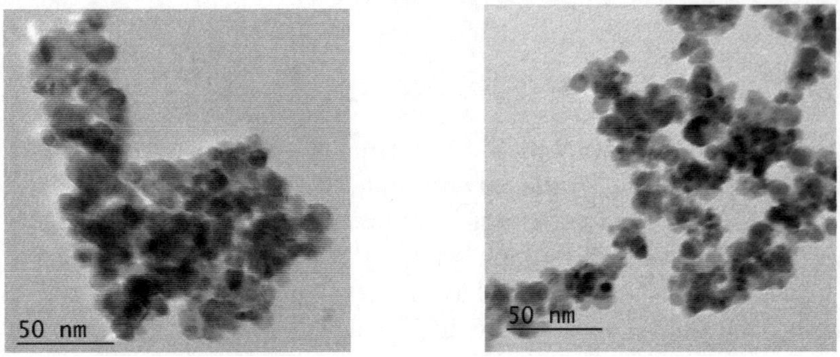

S1 **S2**

Figure 4. TEM images of samples S1 and S2

3.5 Fourier Transform Infrared Spectroscopy (FTIR)

In order to evident the presence of functional groups, FTIR spectroscopic analysis is carried out for synthesized metal oxide nanoparticles. Figure 5 gives the FTIR spectra of samples S1 and S2. The spectrum has been recorded in the range 4000cm^{-1}- 300cm^{-1}

The presence of water molecules adsorbed on the presence of tin oxide during handling or bending and stretching vibrations of OH groups can be seen in 3431 cm^{-1} and 3446 cm^{-1}. The bands at 1612 cm^{-1} and 1636 cm^{-1} have been assigned to lattice vibrations due to decreasing the intensity which leads to overtones and combinations. The bands around 512 and 613 cm^{-1} have been attributed to Sn – 0 stretching modes of Sn – O – Sn, respectively, revealing of the presence of SnO_2. Table 3 shows the bands and assigned vibrations of S1 and S2.

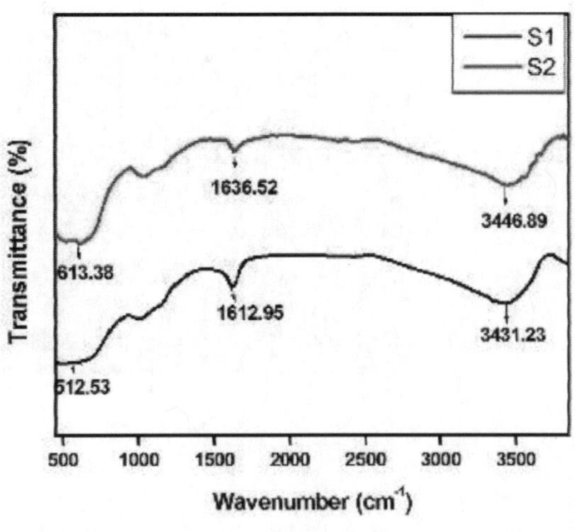

Figure 5. FTIR spectra of samples S1 and S2

IOP Conf. Series: Materials Science and Engineering **360** (2018) 012007 doi:10.1088/1757-899X/360/1/012007

Table 3: The major bands obtained in the FTIR spectrum of S1 and S2

Bands (cm^{-1})	Assignments
512-613	Sn-O stretching vibrations
1612 - 1636	Presence of oxygen radicals
3431-3446	Stretching vibrations of OH attached with surface

3.6 Antibacterial Studies

The antibacterial studies were done in CEPC laboratory and technical division, Kollam, Kerala, India. Test organisms were collected from Institute of Microbial Technology, IMTECH, Chandigarh. The technique used here is filter paper diffusion technique.

The bacterial strains were maintained on their respective medium in slants at 2 - 8°C. Muller Hinton Agar was used for bacterial culture. MHA was prepared and sterilized at 121°C for 15 minutes. After sterilization required volume of the medium (20 ml) was poured in the sterile petri dishes and allowed to solidify. Pure culture was selected as innoculam. 3 – 4 colonies were selected and transferred them into about 5ml of suitable broath such as Tryptone Soya Broath (TSB). It is incubated at 37°C for 2 – 8 hours till light to moderate turbidity develops. A sterile non – toxic swab is dipped on a wooden applicator into the standardized inoculum and the soaked swab is rotated firmly against the upper side wall of the tube to express the excess fluid. The entire agar surface of the plate is streaked with the swab three times, turning the plates at 60 angles between each streaking. The inoculums were allowed to dry for 5 – 15 minutes with lid in place. The disc is allowed to impregnate with the sample, approximately 30 μl, using aseptic technique. This is incubated immediately at 37°C and examined after 16 – 18 hours. The zone showing complete inhibition is measured and the diameters of the zones to the nearest mm is studied.

The antibacterial activity against three bacterial strains namely Staphylococcus aureus, Shigella flexineri and Escherichia Coli is analyzed for the samples. For Staphylococcus aureus and Shigella flexineri, S2 had a higher inhibition zone than S1. It indicates that inhibition zone increases on annealing temperature for the two bacterial strains whereas for Escherichia Coli, S2 has lower inhibition zone than S1. The zone of inhibition of S1 and S2 against bacterial strains are given in table 4. Nano particles tend to absorb on the bacterial surface and dehydrogenation due to respiration process occurs at the cell membrane in bacteria. The high surface to volume ratio of these nanoparticles is the factor for their increased chemical and biological activity. The active oxygen species generated by the presence of SnO_2 nanoparticle interact with the cell membrane of the bacteria and are easily penetrated through it.

Table 4: Results obtained for the anti - bacterial activities of the prepared samples.

Parameters	**Diameter of clear inhibition zone in mm**	
	S1 (mm)	**S2 (mm)**
Shigella flexneri	**17**	**19**
Escherichia Coli	**10**	**6**
Staphylococcus aureus	**7**	**9**

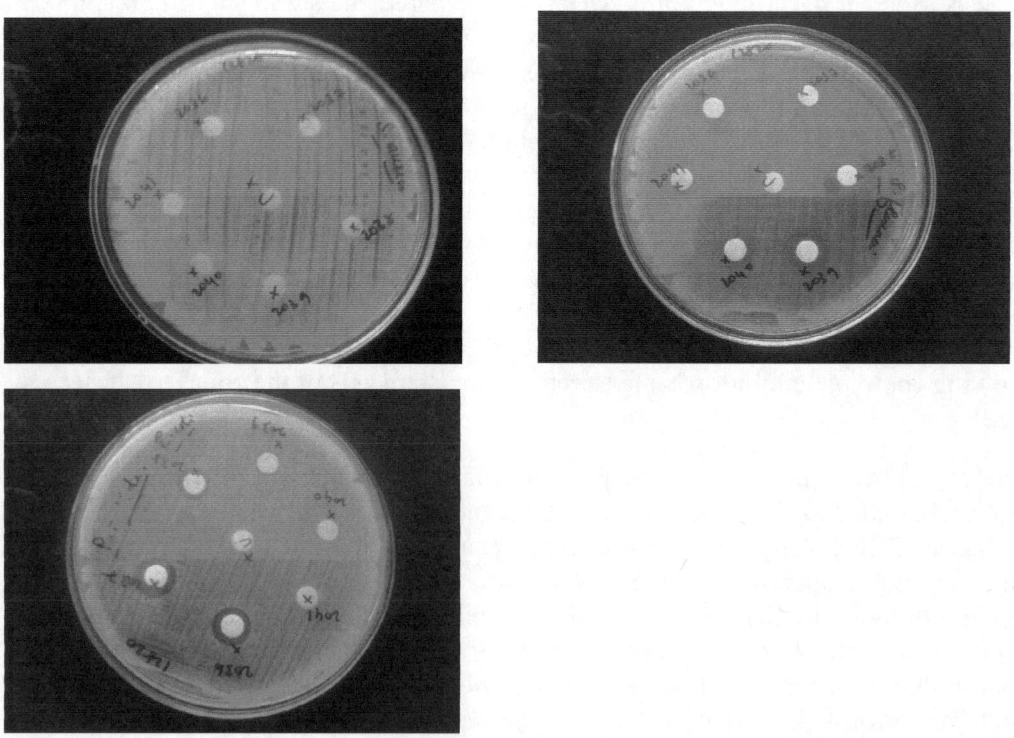

Figure 6. Photograph showing the inhibition zones obtained for different bacterial strains by S1 and S2.

4 Conclusion

An environment friendly method of synthesizing SnO_2 nanoparticles using DNA were successfully done. EDAX shows the presence of Sn and O and their presence in stochiometric proportion. It is clear that from XRD patterns of the samples, the intensity enhanced with annealing temperature. The non- existence of any other overlapping peak in the standard peak position indicates that the prepared samples have only a single phase. The TEM shows particles are of size in between 5 and 13 nm which is in good agreement with XRD results. The morphology and shape of structure improves on annealing. Antibacterial activity increases on annealing temperature. SnO_2 nanoparticles using DNA as capping agents exhibit significant antibacterial activity.

Acknowledgements

The first author is grateful to STIC Cochin, CECRI, Karaikudi, Department of Optoelectronics, Kariavattom and CEPCI, Kollam for providing facilities for analysis.

References

[1] R. Jalal, E.K. Ghorshadi, M. Abareshia, Mmoosavi, A.Yousefi, P. Nancarrow; Materials Chemistry and Physics, 121(2010), 198-201.

[2] P.P Sharmila, S, Sagar, Nisha. J.Tharayil, Nanoscience and Nanotechnology, 8 (2014), 312-314.

[3] Nasrin Talebian,Hoda Sadeghi Haddad Zavvare, Journal of Photochemistry and Photobiology B: Biology, 130 (2014), 132 – 139.

[4] A. Ayeshamarium, R. Perumal Samy, Journal on Photonics and Spinctronics, 2 (2013), (4 – 8).

[5] Ravindra P. Sngh, Vineet K. Shukla, Raghavendra S. Yadav, Prashant K. Sharma, Prashant K. Singh, Avinash C. Pandey. Adv. Matt. Lett., 2 (2011), 313 - 317 .

[6] S. Matin Aminiezhad, Alir eza Rezvani, Mehdi Amouheidari, S. Mohamad Amininejad, Sajjad Rakshani, Zahdan J Res Med Sci,, 16 (2015), 29 – 33.

[7] Sedumadavan Sudhaparimala, Arumugham Gnanamani A. Asit Baran Mandal, American Journal For Nanoscience and Nanotechnology,Bohn P W, Elimelech M, et al, Nature, 452 (2008), 301 – 310.

[8] P. P. Sharmila, Nisha j. Tharayil, International Journal of Material Science and Engineering, 2 (2014), 147 – 151.

[9] Simin Tazikeh, Amir Akbari, Amin Talebi, Emad Talebi, Material Science – Poland, 32 2014, (98 – 101).

[10] J. H. Choi, H. Tabata, T. Kawai; J.Cryst. Growth, 226 (2001), 493.

[11] N. Padmavathy, R. Vijayaraghavan, Sci. Tech. Adv.Matter, 9 (2008).

[12] Vennila Raj, Kamaraj Palanisamy, M. Arthanareeshwari, Chemical Science Review and Letters, 2 (2013), 293 – 299.

[13] H. R. Pouretedal, A. Shafeie, and M. H. Keshavarz, Journal of the Korean Chemical Society, 56 (2012), 484 – 490.

[14] Saba Ahmed, Muhammed Akhyar Farrukh, Madiha Khan, Muhammed Khaleeq–ur-Rahman and Muhammed Ashraf Tahir, Canadian Chemical Transactions, 2 (2014), 122 – 133.

[15] Archita Bhattacharjee, M. Ahmaruzzaman, Journal of Colloid and Interface Science, 448 (2015), 130 – 139.

[16] M. Kolar, K. Urbanek, T. Latal, Antibiotic selective pressure and development of bacterial resistance. Int J Antimicrob Ag, 17 (2001), 357–363.

IOP Conf. Series: Materials Science and Engineering **360** (2018) 012008 doi:10.1088/1757-899X/360/1/012008

Effect microwave sintering in enhancing the infrared transmittance properties of combustion synthesized nanostructured Y_2O_3 ceramics comprising La^{3+} ion in the matrix

Jijimon K.Thomas*, C.T.Mathew

Department of Physics, Mar Ivanios College, Thiruvananthapuram 695015, India

Corresponding author e-mail: jkthomasemrl@yahoo.com ,Ph: +91 471 2530887, Fax: +91 471 2532445

Abstract

Synthesis of nano particles of La^{3+} ion subsituted Y_2O_3 ceramics (~16nm) by a single step auto-igniting combustion technique, followed by microwave sintering to optimum density, and their remarkable infrared transmission characteristics are presented in this paper. The XRD pattern of the as prepared sample clearly indicates that the La^{3+} ions are effectively replacing Y^{3+} ions in the yttria cubic structure and it matches very well with the JCPDS file no-89-5591. As the percentage of La^{3+} increases a slight increase in lattice parameter was observed. The size of the crystallites were calculated using Scherrer formula and they found to be in the size range of 10-20 nm and the average size of the crystallites was 16 nm . The phase purity and nano nature of the crystallites were confirmed using HRTEM. The UV-Visible and FTIR spectroscopy were effectively used to characterise the powder and to obtain the transmittance properties of the sintered samples. The pellets with green density 52% of the theoretical density were sintered using a microwave furnace. The pellets achieved 99% of the theoretical density at 1520°C for a soaking duration of 20 minutes. The pellets were translucent and the SEM analysis revealed that they were well sintered with reduced average grain size of ~430nm. The pellet showed an enhanced transmittance of 77% in the UV-Visible region and 72% in the mid infrared region. The transmittance properties shown by the pellets are superior to those in the case of pure Y_2O_3 and is due to the superior powder quality and the fast uniform densification via microwave sintering. Microwave assisted sintering of the combustion synthesized powder is therefore a promising fabrication technique, especially for infrared transparent ceramics, which considerably reduces the sintering temperature, soaking duration, the grain size and enhances the transmittance properties.

Keywords: Microwave sintering, Infrared transmittance, Combustion synthesis

1. Introduction

Polycrystalline infrared transparent ceramics found to have a large spectrum of applications [1-3]. The research in the field of nanostructured polycrystalline infrared transparent ceramics is in its blooming stage due to its increased demand in the strategic defence and space missions. Among various polycrystalline ceramics yttria(Y_2O_3) ceramics have excellent properties required for good infrared transparent window materials , like high thermal conductivity and absence of birefringence [1], high thermal shock resistance [4], low dielectric loss with temperature [5], moderate refractive index and low thermal expansion coefficient [1], high melting point [6] , greater hardness [7], low emissivity at elevated temperatures [8] and high transparency in the visible and infrared range [4]. The desired mechanical, thermal and electrical properties, in addition to the high infrared cut-off make it ideal for IR windows and protection domes used in strategic applications.

The major challenge in the fabrication of infrared transparent window is the sintering of the sample to high density without compromising the transmittance properties. In most of the cases very high sintering temperature is required for the densification of the nanostructured polycrystalline sample, due to which the grain size increases and adversely affect the transmittance and mechanical properties.

In the present work we use a modified combustion technique to synthesize nano structured $La_{0.15}Y_{1.85}O_3$. The modified combustion technique used in the present study is a relatively simple method viz. the omission of high temperature calcinations for prolonged duration, we could obtain phase pure nano particles of extremely small size (10-20 nm) and further compacted to an optimum density at a much lower temperature with high thermal stability and good transparency to infrared radiation using microwave heating technique. The structure, vibrational spectra and surface morphology of the combustion synthesized powder are also studied and presented. The sintering behaviour of compacted pellets using microwave heating and compared it with that of resistive heating. The transmittance of infrared radiations through the sintered pellets in the UV-Visible and mid infrared ranges are also reported.

2. Experimental Procedure

Nanostructured $La_xY_{2-x}O_3$(where x=0.05,0.10,0.15 0.20 and 0.25) was prepared using a single step auto-igniting combustion synthesis [9-11]. Stoichiometric amount of high purity Y $(NO_3)_3.6H_2O$ (99.99%, Alfa Aesar, USA) dissolved in double distilled water and La_2O_3 (99.99%, Alfa Aesar, USA) dissolved in nitric acid were mixed to make a clear solution. Amount of citric acid, which is used as the complexing agent was calculated based on the total valence of the oxidizing and the reducing agents for maximum release of energy during combustion [12,13]. Nitric acid was used as the oxidising agent and ammonia as fuel and the pH of the solution was monitored till it became ~7. The solution containing the precursor mixture was heated using a hot plate at 250 °C in a ventilated fume hood. The solution boils on heating and undergoes dehydration accompanied by foam. The foam then ignites by itself on persistent heating giving voluminous and fluffy product of combustion.

The phase purity and the crystalline nature of the as prepared $La_xY_{2-x}O_3$ were analysed using different powder characterisation techniques. As-prepared samples were characterized by X-ray diffractometer (X'pert pro, Philips, the Netherlands) with Cu Kα radiation in the range of 20–60° in steps of 0.0840 for the determination of crystalline structure and phase of the nanomaterials. The average crystallite size was calculated using the Scherrer formula [14] $D=0.9\lambda/(\beta \cos\theta)$ where λ is the wavelength of CuK_α radiation, β is the full width at half maximum, and θ is the Bragg's diffraction angle. The nano particulates were examined using high resolution transmission electron microscopy (H600, Hitachi, Japan) operating at 200 kV. The samples for Transmission Electron Microscope were prepared by ultrasonically dispersing the powder in methanol and allowing a drop of this to become dry on a carbon-coated copper grid. The HRTEM image is analysed using imageJ software to obtain the d-spacing and the crystallite size. Additional information regarding the phase purity and the presence of any inorganic impurity was obtained from FTIR spectroscopy taken using Fourier Transform Infrared spectrometer (Spectrum 2, Perkin-Elmer, Singapore) in the range 400-4000 cm^{-1} using the ATR method. The transmission spectrum of as-prepared samples of yttria nanoparticles was recorded using a spectrophotometer (UV-1700, Shimadzu, Singapore). The refractive index of the sample is calculated from the transmission spectrum. The theoretical limit of transmittance in the UV-Visible range for a well sintered pellet using a powder with refractive index n is

IOP Conf. Series: Materials Science and Engineering **360** (2018) 012008 doi:10.1088/1757-899X/360/1/012008

estimated according to the relation $T = 2n/(1+n^2)$, where n is the refractive-index [1] . The as prepared powder was uniaxially compacted into pellets in a 14 mm diameter steel die at 200 MPa using a hydraulic press. The sintering behaviour of the samples were studied using microwave heating and the results were compared with that of resistive heating. The sintering of the disc- shaped pellets were carried out in a microwave furnace with silicon carbide susceptors (VBCC/MF/86,VB Ceramics Consultants, India). In the microwave furnace, microwave heating was realized using a pair of 2.45GHz magnetrons of 1.1KW each .The experimental density of the sintered pellets was calculated using the Archimedes principle. The surface morphological studies of the sintered pellets were performed with a Scanning Electron Microscope (NPEP252, Nova nanosem, USA). The SEM image is analysed using imageJ software to obtain the grain size distribution curve. The pellet used for SEM analysis were further subjected for lapping and polishing using diamond paste of fine grades in a self-designed lapping and polishing machine. These mirror polished 0.5 mm thick pellets were used for the transmittance studies. The transmittances of radiation in the UV-visible and in the IR range were measured using UV-Vis and FTIR spectrometers.

3. Results and Discussion

The phase purity of the crystallites of the as prepared powder was investigated in detail using the X-ray diffraction technique.

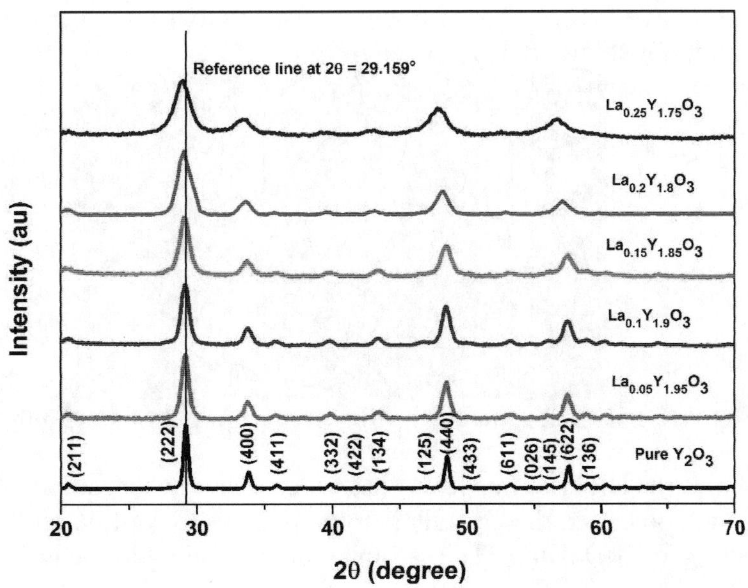

Figure 1. XRD pattern of $La_xY_{2-x}O_3$(where x= 0, 0.05,0.10,0.15,0.20 and 0.25)

Fig. 1 shows the X-ray diffraction patterns of Y_2O_3 and $La_xY_{2-x}O_3$(where x=0.05,0.10,0.15 0.20 and 0.25) . All the peaks corresponds to the cubic phase of yttria and indexed using JCPDS-895591. No other peaks were observed which clearly indicate that La^{3+} ions are effectually replacing Y^{3+} ions in the yttria cubic structure. But the ionic radius of La^{3+} (1.032 Å) is larger than that of Y^{3+} (0.90 Å) ions [15] and resulting increase in d-spacings which is evident from the shift of the diffraction peaks to the lower 2θ values as indicated by the reference line in Fig.1. Table 1 shows the variation in d-spacing and cell parameters for the (222) peak with La^{3+} ion concentration. The shift in peaks towards lower 2θ values were also reported in Y_2O_3/La_2O_3 system with increase in calcination temperature due to the expansion of Y_2O_3 lattice along with the lanthanum ion penetration [16]. The average crystallite size calculated using Scherrer formula is 16 nm for the sample $La_{0.15}Y_{1.85}O_3$.

$La_xY_{2-x}O_3$	d-spacing (Å)	Cell parameter(Å)
x = 0	3.0601	10.604
x = 0.05	3.0639	10.613
x = 0.10	3.0657	10.620
x = 0.15	3.0693	10.632
x = 0.20	3.0722	10.642
x = 0.25	3.0767	10.658

Table 1. Change in d-spacing and cell parameter corresponding to the (222) peak with La^{3+} ion concentration

As the concentration of La^{3+} ions increase the lattice distortion increases[17] and the peaks get broadened due to the lattice strain factor in addition to the size effects. The suitable composition for the application as infrared transparent window is found to be x=0.15 because the presence of La^{3+} above that composition will adversely affect the thermal conductivity of the sample so we limit our studies to $La_{0.15}Y_{1.85}O_3$.

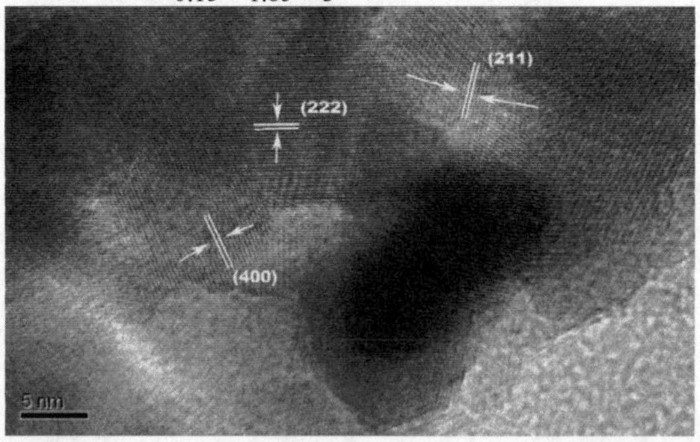

Figure 2 : HRTEM image of different crystallographic planes

Figure 2 shows the HRTEM image of the as-prepared powder sample. Even though the powder is slightly agglomerated the crystallographic planes are well defined and they are in good agreement with the XRD data . The d-spacing corresponding to the *(222)* plane of cubic crystal structure of $La_{0.15}Y_{1.85}O_3$ is 3.0691 Å .

Phase purity of the sample was confirmed using FTIR spectrum shown in figure 3. The transmittance plot confirms that the $La_{0.15}Y_{1.85}O_3$ powder transmits IR in the range 2.5µm-20µm. The peak at 562 cm^{-1} corresponds to Y-O stretching [18-21]. Broad peak around 3400 cm^{-1} and 1661 cm^{-1} correspond to O-H stretching and H-O-H bending modes of water adsorbed by the powder. Peaks at 1510 cm^{-1} and 1410 cm^{-1} are assigned to the asymmetric and symmetric vibrations of carboxylate group formed during the combustion process. No other absorption peaks are observed in this range which shows that no residual nitrate or organic matter is present in the precursor powder.

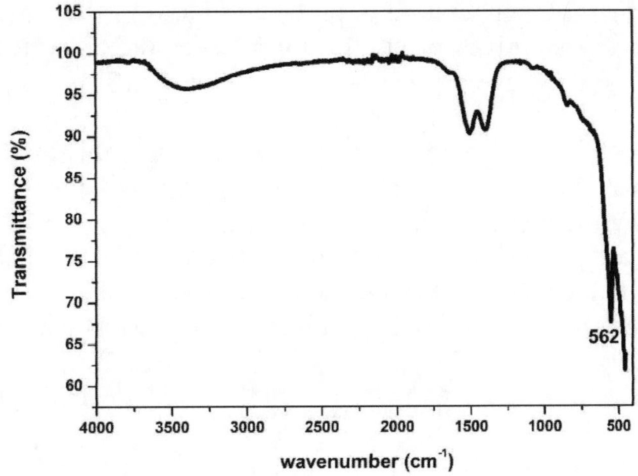

Figure 3 : FTIR transmittance spectrum of as synthesised $La_{0.15}Y_{1.85}O_3$

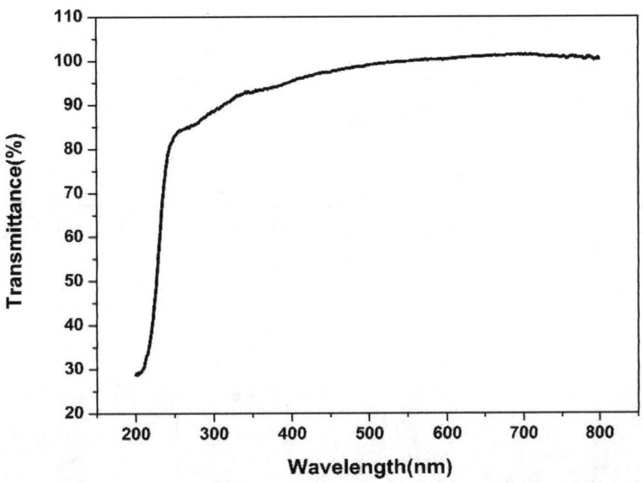

Figure 4 : UV-Visible transmission spectrum of as synthesised $La_{0.15}Y_{1.85}O_3$

The UV-visible spectrum was recorded in the range 200-800 nm and is shown in figure 4 . It is observed that the sample shows maximum transmission in the visible region which decreases towards the UV region. The absorption edge is at 225 nm. The heavy absorption in the high energy end of the UV region makes the sample a potential candidate for UV filters, which filter higher energies of the ultraviolet spectrum from 10 nm to 120 nm which are ionising and it also find application as microwave filters. The optical band gap energy could be determined by extrapolating the wavelength of the onset absorption in the UV region and using the intercept in the wavelength axis in the equation $E=hc/\lambda$ where h is the Plank's constant, c is the velocity of electromagnetic radiation and λ is the wavelength corresponding to the absorption. The corresponding band gap for the wavelength 225 nm is 5.55 eV.

Sintering behaviour was studied on the disc shaped $La_{0.15}Y_{1.85}O_3$ pellets using a conventional furnace having molybdenum heating elements. The relative densities of sintered pellets at different sintering temperatures for a soaking time of two hours were calculated. In conventional heating furnace the pellets were heated at a constant rate of 10 °Cmin^{-1} and were sintered to 98.8% of theoretical density by holding for two hours at 1590°C is found to be the lowest value of sintering temperature of pure nano $La_{0.15}Y_{1.85}O_3$ using conventional sintering methods without any additives or application of pressure, reported so far. To improve the sintering strategy, microwave sintering was effectively utilized to densify the sample pellet. Being a low dielectric loss material susceptor aided microwave sintering was used. The pellets were sintered to 99% of the theoretical density at a considerably low temperature of 1520°C by holding it for just 20 minutes. The relative density achieved by the pellet at different temperature during the course of study is shown in figure 5. It is clear from the figure that at lower temperature there is no much difference in the densification pattern of the sample. But at higher temperature the microwave absorption of the sample increases which is evident from the remarkable enhancement in densification after 1200°C during microwave sintering.

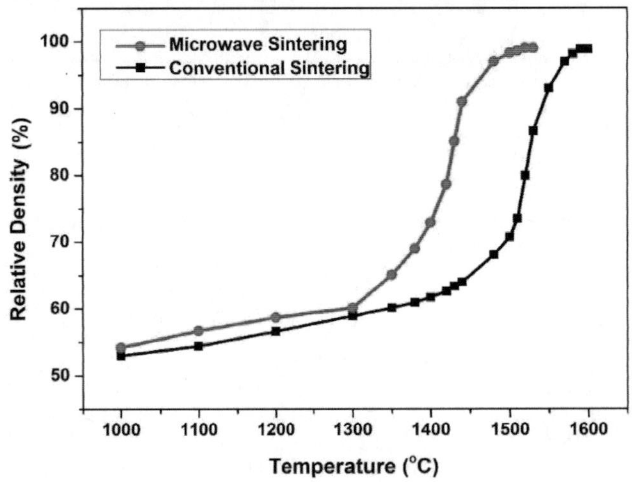

Figure 5 : Variation of relative density of $La_{0.15}Y_{1.85}O_3$ with sintering temperature

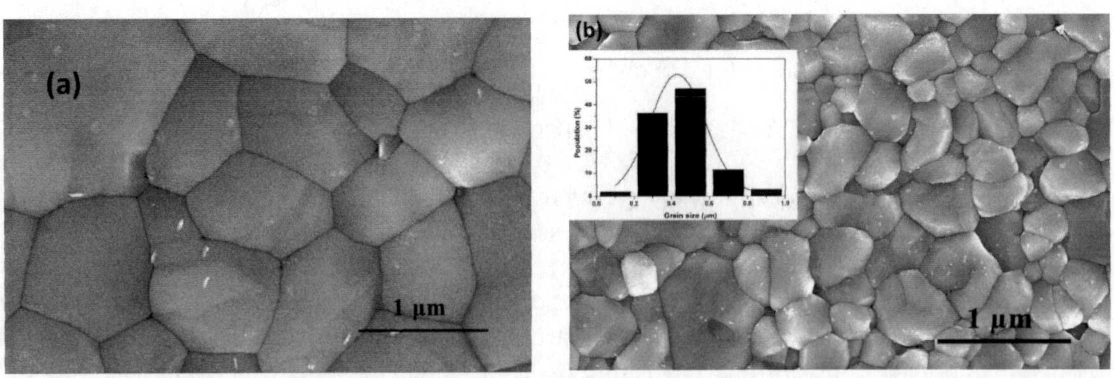

Figure 6: (a) SEM micrograph of conventionally sintered $La_{0.15}Y_{1.85}O_3$ pellet and **(b)** microwave sintered $La_{0.15}Y_{1.85}O_3$, grain size distribution is shown inset.

The sintered pellets were hand polished and thermally etched at 1440°C for 1 hour in air. The surface morphology of the sintered sample was studied using Scanning Electron Microscopy. The SEM micrograph of the $La_{0.15}Y_{1.85}O_3$ ceramics after the initial polishing and etching is shown in Figure 6. The pellet sintered via resistive heating is shown in figure 6(a). The grain boundaries were sharp with minimum porosity without showing any abnormal growth or melting at the grain boundaries and the full-grown grains are in the size range of 0.6 – 1.8 μm. The grain sizes are accurately measured using the Image J software. Figure 6(b) shows SEM micrograph of microwave sintered $La_{0.15}Y_{1.85}O_3$ with 99% of theoretical density from which it is evident that the grain sizes are considerably low compared to that in pellet sintered via resistive heating and they lie in the size range 0.1 – 1 μm. and the average grain size for the entire distribution is 0.43μm In microwave sintering a substantial reduction in the size of the grain is observed. The reduced grain size and reduced porosity may enhance the transmittance properties of the sample to a greater extent [22].

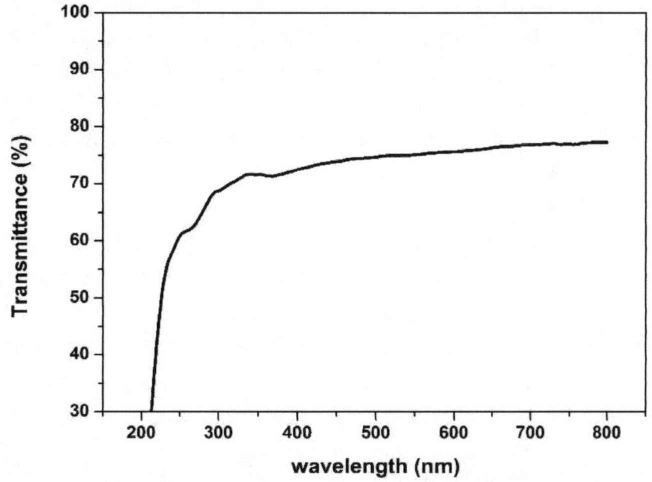

Figure 7: UV-Visible transmittance of the $La_{0.15}Y_{1.85}O_3$ pellet sintered to 99% theoretical density via microwave heating.

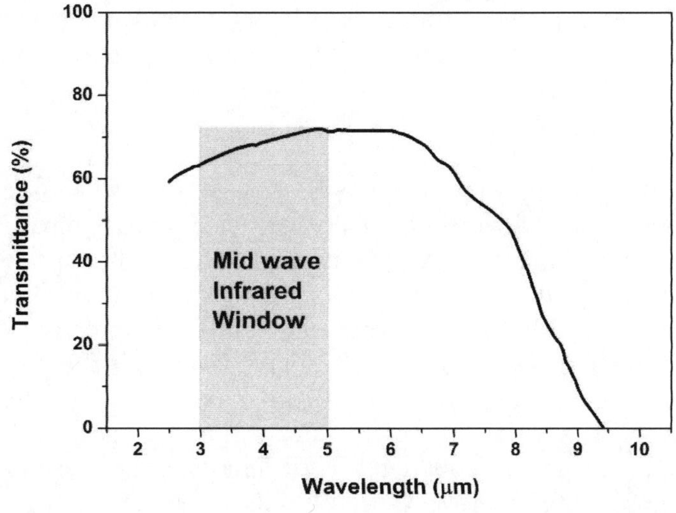

Figure 8: IR spectrum of $La_{0.15}Y_{1.85}O_3$ pellet sintered via microwave heating to 99% of theoretical density

In the absence of absorption or scatter the transmittance of a body is given by the relation,

$$t = \frac{2n}{n^2 + 1}$$

where n is the refractive index. The refractive index is one of the fundamental properties of a material which is closely related to the electronic polarisability and local field inside the material. Reddy et al.[23] formulated an empirical relationship between the refractive index and band gap energy,

$$E_g e^n = 36.3$$

Using the above relation the refractive index of the present sample with band gap energy of 5.81eV is calculated as 1.83. The theoretical limit of transmittance through the sintered pellet according to the relation is calculated as 84.1%. The pellets after the initial hand polishing followed by polishing using diamond pastes of fine grades in a self-designed lapping and polishing machine and the final mirror polished pellets are used for the transmittance studies.

The UV-Visible spectrum of the sintered pellet of density 99% of the theoretical density and thickness 0.5mm is shown in the figure 7. The graph shows a maximum transmittance of 77% at 800 nm and it is found to be translucent.

The IR spectrum of the microwave sintered sample in figure 8 shows that it has a maximum transmittance of 72 % at 5μm . The good transmittance in the Mid IR range makes it ideal for infrared transparent windows and domes. A slight increase in the infrared cut off is also observed. The improvement in the infrared cutoff is due to the fact that the size of the La^{3+} ion is slightly greater than the size of the Y^{3+} ion due to which there is a reduction in vibrational frequency of the system and the absorption edge shifts towards the longer wavelength region according to the relation $\omega = \frac{1}{2\pi}\sqrt{\frac{k}{\mu}}$ where $\mu = \frac{m_1 m_2}{m_1 + m_2}$ is the reduced mass of the system. The greater the masses of the atoms the greater the reduced mass and the smaller the vibrational energies [1] . The increase in the infrared cut-off along with the enhanced transmittance properties makes $La_{0.15}Y_{1.85}O_3$ a better candidate for infrared transparent window.

4. Conclusions

A single step auto-igniting combustion synthesis is used to prepare nano structured $La_{0.15}Y_{1.85}O_3$ X-Ray Diffraction (XRD) analysis of the as prepared powder confirms the formation of phase pure cubic $La_{0.15}Y_{1.85}O_3$ nanoparticles. All the peaks in the XRD were indexed to a body centred cubic structure with a calculated lattice parameter a=10.632Å agreeing very well with the XRD data reported in Joint Committee on Powder Diffraction Standards (JCPDS) file no. 89-5591. The XRD analysis and High Resolution Transmission Electron Microscopic (HR-TEM) studies reveal that the average crystallite size is ~16nm and the lattice planes are well defined. Fourier transform Infrared spectrum of the as prepared sample is recorded in the range 400-4000 cm^{-1} .The vibrational spectroscopic studies confirms the structure of the sample. The absorption spectra of the as prepared $La_{0.15}Y_{1.85}O_3$ is measured in the range 200-800 nm and it shows that the material absorbs heavily in the shorter wavelength region of the ultraviolet spectrum and transmit long wavelength uv and visible light and hence found applications in the filters and sensors of UV radiations. The pure white $La_{0.15}Y_{1.85}O_3$ powder is uniaxially compacted in to pellets in a 14mm diameter steel die and sintered to 99% theoretical density without any sintering additives or applying pressure at 1520°C for 20 minitues in a microwave furnace. The SEM micrograph gives the surface morphology of the well sintered pellet and from the SEM it is found that the average grain size is ~ 430nm. The transmission spectrum of the sintered pellet in the Uv-Vis region reveals that the material is showing 79% of transmittance in the visible region and 72% in the Mid IR region. The presence of La^{3+} ion increases the infrared cut-off. The better transmittance of the pellets attribute to the quality of the nano powder synthesised by the modified combustion technique, high sintered density with minimum porosity and the smaller grain size of the sintered pellet. The results clearly indicate that the ultra-fine $La_{0.15}Y_{1.85}O_3$ nanopowder synthesized by the modified auto-igniting combustion technique synthesised using single step combustion method followed by microwave heating can be used very effectively for the fabrication of improved infrared transparent windows and domes.

Acknowledgments

This work is supported by the Department of Science and Technology- Science and Engineering Research Board, Government of India under Grant No. SB/S2/CMP-0021/2013.

References

1. D.C.Harris, Materials for Infrared Windows and Domes, Properties and Performance. SPIE Press, Bellingham, WA, 1999.
2. M. Grujicic, W.C. Bell, B. Pandurangan, *Mater Design.* 34, 808–819 (2012).
3. D.D. Silva, and A.R. Boccaccini,Rec. *Pat.Mat. Sci.* 1, 56-73(2008).
4. D.C. Harris, *Infrared Phys.Tech.* 39,185-201 (1998).
5. J. Chen, Z.P. Gao, J.M. Wang, D.H. Zhang, *J.Ele. Tech. Chi.* 5 [4] , 320-324 (2007).
6. A.C Pastor, R.C Pastor, *Mater.Res.Bull.* 2 ,555-559(1967).
7. Y. Tsukuda, *Mater. Res. Bull.* 16, 453-459 (1981) .
8. R.M. Sova, M.J. Linevsky, M. E. Thomas, F. F. Mark, *Infrared. Phys. Tech.* 39, 251-261(1998).
9. J. James, R. Jose, Asha M. John, J. Koshy, U.S.Patent 6761866, 2004.
10. J. James, R. Jose, Asha M. John, J. Koshy, U.S Patent 6835367, 2004.
11. J.K. Thomas, H. Padma Kumar, R. Pazhani, S. Solomon, R. Jose ,J. Koshy, *Mater. Lett.* 61, 1592–1595(2007).
12. J.K. Thomas, H. Padma Kumar, S. Solomon, C.N.George, K.Joy, J.Koshy, *phys. status. solidi. (a)* 204 [9], 3102–3107(2007).
13. J.K Thomas, H. Padma Kumar, Annamma John, V.Suni, K.Joy, S.Solomon , J.Koshy, *J.Phys.Chem.Sol.* 70, 703-706(2009).
14. J.I.Langford, A.J.C.Wilson, *J.Appl.Crystallogr,* 11,102-113(1978).
15. R.D.Shannon, *Act.Cryat.A* 32[5]: 751-767 (1976).
16. X. Mao, X. Li, M. Feng, J. Fan, B.Jiang and L.Zhang, *J. Euro. Ceram. Soc.,* 35: 3137-3143(2015).
17. H.Zhang, Q. Yang, B. Zhang and S. Lu,*J. Raman. Spectrosc.,* 42[6]: 1384-1387 (2011).
18. F.Maglia, I.G.Tredici, U.A.Tamburini, *J. Eur. Ceram.Soc.* 33,1045–1066(2013).
19. Z. Huang, X. Sun, Z. Xiu, S. Chen, C.T. Tsai, *Mater. Lett.* 58,2137-2142 (2004).
20. L.Wen, X. Sun , Q. Lu , G. Xu , X. Hu , *Opt. Mater.* 29,239-245 (2006).
21. L. An, A.Ito, T. Goto, *J. Eur. Ceram. Soc.* 32,1035-1040 (2012).
22. R. Apetz and M.P.B van Bruggen, *J.Am.Ceram.Soc.* 86[3],480-486 (2003).
23. R.R.Reddy, S.Anjaneyulu, *Phys.Status.Solidi.(b).* 174,1091-1093(1992).

Synthesis, Structural and Optical Investigation of CdSe semiconductor Quantum Dots

Prajwal S. Kapatkar[1], Rajshekhar B. Shettar[1], S B. Kapatkar[2] and N R. Patil[2*]

[1]*Department of Electronics and communication, BVB College of Engineering and technology, Hubli, India*

[2] *Department of Physics, BVB College of Engineering and technology, Hubli, India*

**Corresponding author Email:* patilnr23@gmail.com

ABSTRACT

In this paper, we present a simple synthetic route for the development of high quality CdSe quantum dots TGA capped in aqueous solutions. CdSe semiconducting Quantum dots were prepared by a chemical method at room temperature. The crystal structure, morphology, spectral and compositional properties were analyzed from powder X-ray Diffraction, High resolution transmission electron microscopy (HRTEM), Scanning electron microscopy (SEM), Transmission electron microscopy (TEM), UV-Vis absorbance and Photoluminescence (PL), Fourier transform infrared (FTIR) and Fourier transform RAMAN (FT-RAMAN) techniques. The XRD results depicted the formation of CdSe quantum dots with the average crystallite size of 3 nm and 4 nm, which is in accordance with observations from HRTEM and TEM micrographs predict an average particle diameter of the QDs range between 2-3 nm. The UV-Vis absorbance peaks indicate the broad absorption behavior of the synthesized CdSe quantum dots. The calculated band gap energy was 2.55 eV and further it is found the particle size estimated from the maximum excitonic peaks for CdSe 2.7 nm. The photoluminescence studies evidenced that the luminesce property of quantum dots can be varied by systematically tuning precursor concentrations. The PL emission spectrum depicts the strong excitation peaks for the synthesized CdSe QDs with peaks centered at 603 nm. The capping of TGA was confirmed from FTIR and FT-RAMAN studies. From the FTIR spectra it is observed that, for the synthesized CdSe QDs almost all the peaks are common with slightly shifting behavior towards larger and shorter wave numbers. FT-Raman analysis was performed to study the sample qualities such as microcrystallinity, homogeneity and surface conditions for the as synthesized CdSe quantum dot. The resulting colloids are very stable, and no precipitate is observed over a period of 6 months.

Keywords: CdSe quantum dots, XRD, photoluminescence, TEM, SEM, FT-IR, FT-Raman.

1. Introduction:

Nanostructured materials have attracted a great transaction of attention in the last few years for their unique characteristics that cannot be obtained from conventional macroscopic materials. Owing to the quantum size effects and surface effects, nanoparticles can display novel optical, electronic, magnetic, chemical and structural properties that might find many important technological applications. In the past decade, colloidal II-VI semiconductor nanocrystals, often referred to as quantum dots or QDs, have attracted scientists' intensive attention in the preparation methods, fundamental studies, because of their size-dependent (and thus tunable) photo- and electro-luminescence properties and promising applications in optoelectronics and biomedical applications [1]. QDs, whose size falls in the range of 1~10 nm, have been made to be the ideal fluorescent labels for the wavelength-luminance multichannel transmission system, because of their unique optical properties. The recognition of strongly size and shape dependent physical/chemical properties of nanostructure materials has stimulated efforts towards the fabrication of nanocrystals in a systematic and controlled way [2]. Chemically synthesized nanostructures and their assembly are of fundamental importance due to their unique dimension dependent properties and their potential applications as building blocks in nanoelectronics, nano-optronics, nanosensors & actuators, and in biology [3-4]. Recently much interest has been aroused in the preparation and assembly of semiconductor QDs due to their narrow and intensive emission spectra, continuous absorption band, high chemical and photo bleaching stability, processability and surface functionality. There is a wide range of very efficient light emitting QDs, which can be synthesized both in organic or as aqueous solutions [5-6].

2. Experimental Details

2.1. Sample preparation:

All the chemicals used in the present work were of analytical grade, and the reactions were carried out under open-air conditions using triply distilled water. Selenium (99.99%), cadmium chloride (CdC12) powder (99.9%), thioglycolic acid (TGA) (98% pure) were obtained from Sigma-Aldrich and hydrazine hydrate (N2H4H2O) (99%) and acetone (AR) were obtained from Sd-Fine Chemicals, India.

0.1M of highly pure $CdCl_2$ powder and 0.2 M of thioglycolic acid (TGA) are added in 250 ml of deionised water and stirred for 10-15 minutes which resulted into a white buffer solution. 0.05M of pure elemental Selenium (99.9%) is added with 0.05 M of Hydrazine hydrate. This dark brown solution is added to the white buffer solution drop by drop at room temperature. Initially the colour of the solution was light orange and after 1-2 hours the solution turns out to yellow colour.

IOP Conf. Series: Materials Science and Engineering **360** (2018) 012009 doi:10.1088/1757-899X/360/1/012009

2.2. Characterization:

The structural properties of the synthesized samples were studied using powder X-ray diffraction on a Rich Seifer with monochromatic nickel filtered Cu Kα radiation (λ= 1.5461 Å) in the 2θ range of 10° to 80°. The size and shape of the synthesized nanocrystals were estimated using the JEOL JEM 2100. High Resolution Transmission Electron Microscopy (HR-TEM) with an accelerating voltage of 200 kV coupled with EDS. Scanning electron microscopy (SEM) was applied to investigate the size and morphology, which was carried out under JEM-3010 with an operating voltage 200 kV. Transition electron microscopy (TEM) was obtained using a JEOL JEM-2100 instrument with accelerating voltage of 200 kV.

UV-Vis spectroscopy analysis was carried out to study the optical properties using Analytica spectrophotometer in the wavelength range of 200 nm to 800 nm. The photoluminescence spectra of the samples were recorded with a Jobin Yvon Fluorolog-3-11 spectrofluorimeter with Xenon lamp 450W as source at room temperature using 400 nm as the excitation wavelength. The FTIR spectra of the samples were obtained by the conventional KBr pellet technique, in a Perkin Elmer spectrometer operating in the range of 4000 to 400 cm^{-1}. The FT-RAMAN spectra of the samples were obtained by using Bruker's FT-Raman Spectrometer 2010.

3. Results and Discussion

3.1. Structural analysis:

X-ray powder patterns were recorded in the powder of nanoparticles, and the pattern is shown in Figure 1.The peaks of CdSe have been obtained due to diffraction corresponding to (1 1 1), (2 2 0), (3 1 1) reflections, which is in a very good agreement with cubic CdSe structure. The information on the particle size (A) of CdSe Qd's has been obtained from the following Scherrer relations,

$$A = 0.94 \lambda / \beta \cos\theta$$

The peak broadening indicated the fine nanocrystalline nature and the reduced size for the synthesized samples.

IOP Conf. Series: Materials Science and Engineering **360** (2018) 012009 doi:10.1088/1757-899X/360/1/012009

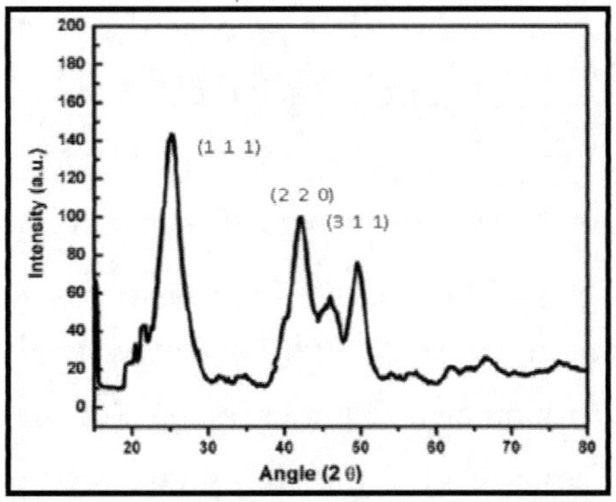

Figure 1. X-ray powder patterns of TGA capped CdSe nanoparticles.

3.2 Spectroscopic Analysis: In order to study the optical properties of the as-synthesized CdSe QDs UV-Vis absorption and photoluminescence studies were carried out and shown in Figure 2. The absorption spectra consisted of an absorption shoulder at 482 nm. The particle size estimated from the maximum excitonic peaks for CdSe, 2.7 nm.

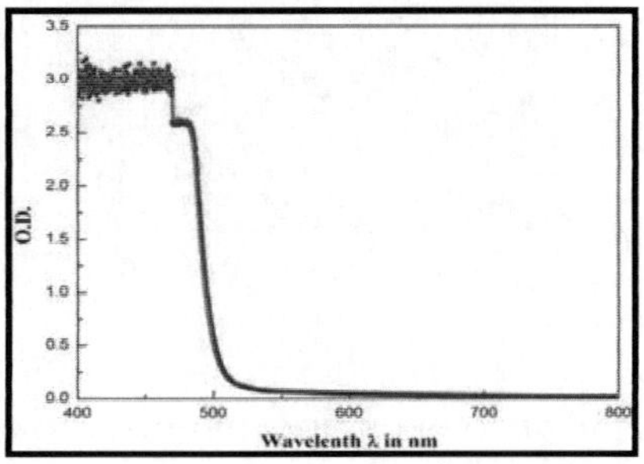

Figure 2. UV-VIS spectrum of TGA capped CdSe QD.

The Tauc Plot of TGA capped CdSe QDs is shown in Figure 3. The standard band gap value for CdSe bulk semiconductor is 2.6 eV. Further, the room temperature photoluminescence spectra of CdSe QDs capped with TGA is recorded using 450 nm excitation wavelength for the sample with the help of a computer controlled rationing luminescence spectrometer and given in Figure 4. The PL emission spectrum depicts the strong excitation peaks for the synthesized CdSe QDs with peaks centered at 603nm.

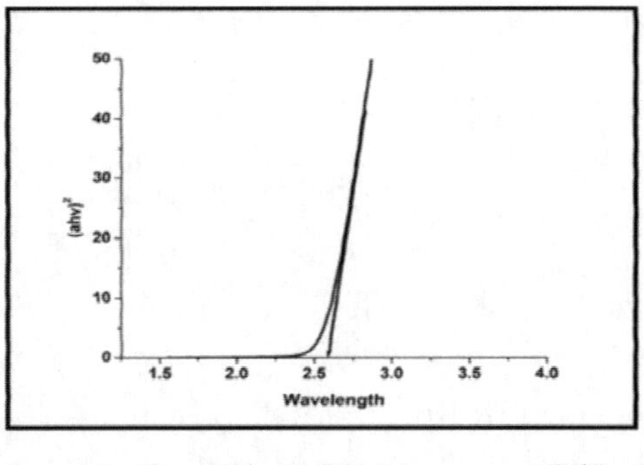

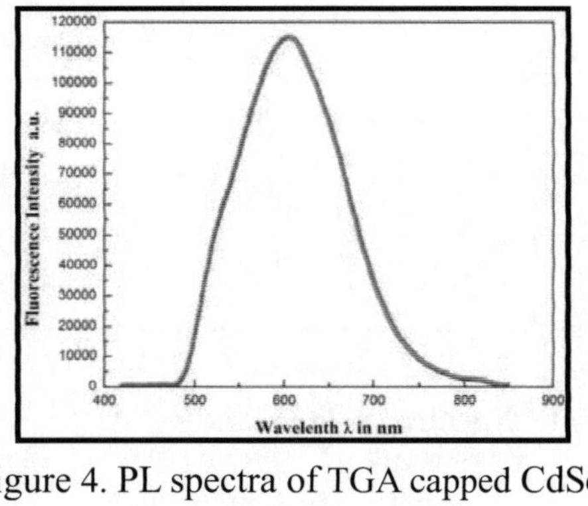

Figure 3. Tauc Plot of TGA capped CdSe QDs Figure 4. PL spectra of TGA capped CdSe QDs

IOP Conf. Series: Materials Science and Engineering **360** (2018) 012009 doi:10.1088/1757-899X/360/1/012009

The surface morphology and the particle size of the synthesized CdSe QDs were analyzed from HRTEM studies. The Figure 5 is a typical HRTEM of this sample. The clear display of lattice fringes in the micrographs reveals the good crystalline nature of the samples. The QDs were found to be well dispersed, spherical and monodisperse without any agglomeration. The average particle diameter of the QDs is found in the range between 2-3 nm. The sizes and shapes of the CdSe quantum dots were analyzed by scanning electron microscopy (SEM) and transmission electron microscopy (TEM) are given in Fig.6 (a) and Fig. 6(b). The SEM image demonstrated that the QDs were spherical morphology. To further elucidate the particle size of QDs, TEM study was performed to record the TEM images of samples. The particle sizes were found in the range of 2-4 nm, spherical in shape and are polycrystalline in nature.

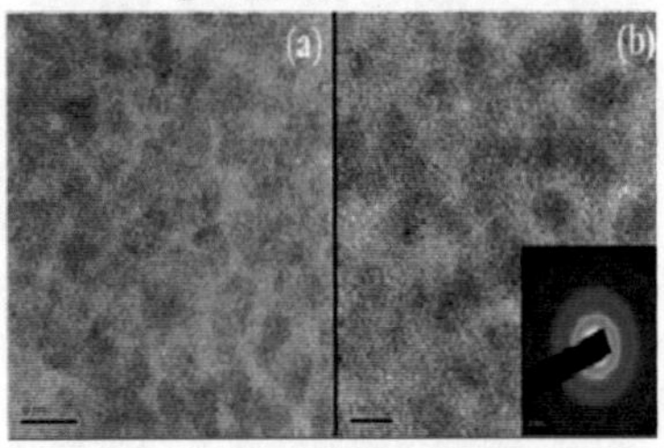

Figure 5. HRTEM pictures of TGA capped CdSe nanoparticles.

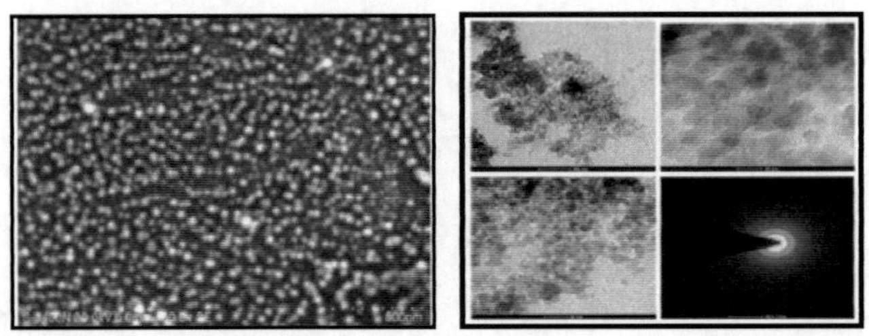

Figure 6. (a) SEM and (b)TEM pictures of TGA capped CdSe QD

In order to determine the chemical composition and to confirm the capping of TGA in the synthesized CdSe QDs the FTIR analysis was performed. Infrared spectroscopy furnished information about the bond that originated between the surface ligand and the CdSe surface. For illustration, Figure 7 contains the infrared spectra of the pure TGA, ligand and of the CdSe QDs stabilized with this ligand. Additionally, the Figure 8 presents the assignment of the main FT-IR band and provides a scheme that represents how the surface ligand binds to the CdSe QDs.

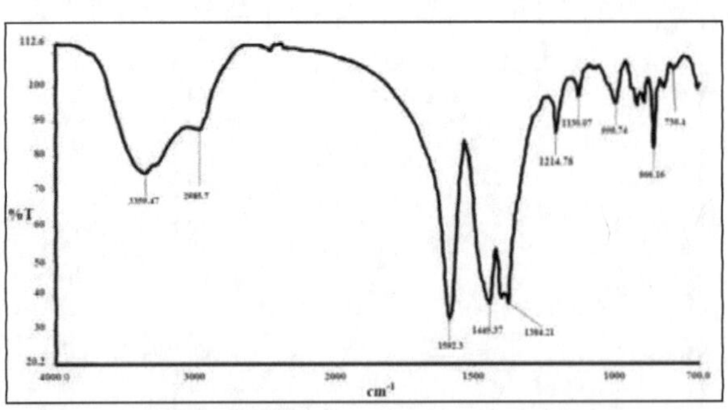

Figure 7. FTIR Spectra of TGA and TGA capped CdSe QD

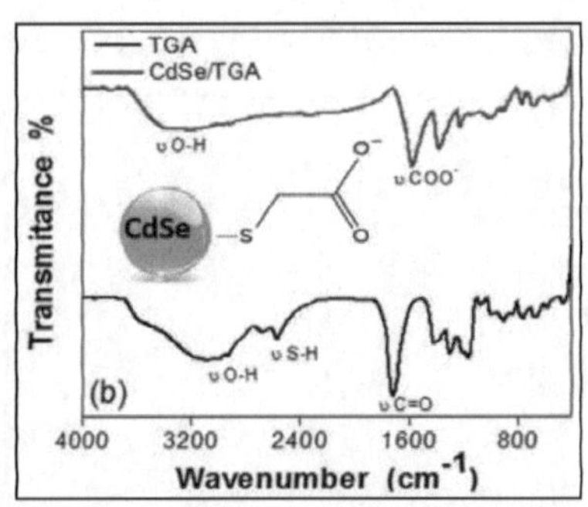

Figure 8. FTIR Spectra of TGA capped CdSe QD

From the FTIR spectra it is observed that, for the synthesized CdSe QDs almost all the peaks are common with slightly shifting behavior towards larger and shorter wave numbers. The peaks at 1380 cm^{-1} are due to shifting of asymmetrical vibration of carboxylic group of in TGA. The peak at 1672 cm^{-1} corresponds to the co stretching of carboxylic group. The peaks around 2969 and 3225 cm^{-1} are due to sp^3 stretching of C-H and due to vibration of O-H group present in TGA. The FT-IR assignments agreed with data recorded for similar systems such as water-soluble CdTe, which means that CdSe and CdTe QDs interact with thiol surface ligands in aqueous medium in much the same way. The Figure 9 depicts that, chemical reaction for the representation of TGA capped.

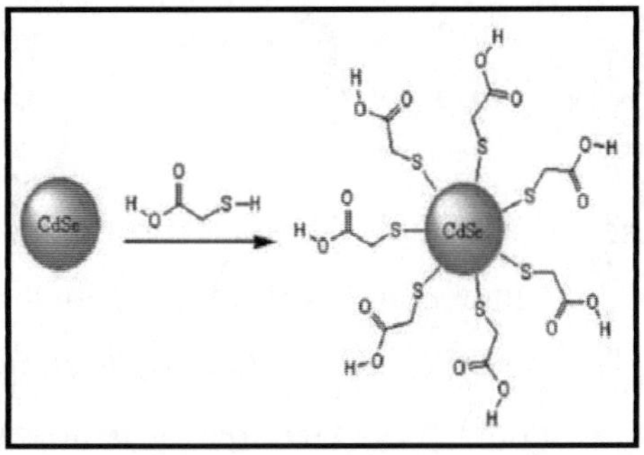

Figure 9. Pictorial representation of chemical reaction for the representation of TGA capped CdSe QD

The Figure 10 depicts FT-Raman analysis was also performed to study the sample qualities such as microcrystallinity, homogeneity and surface conditions for the as-synthesized CdSe quantum dot. The characteristic Raman peaks for CdSe-TGA QDs are observed at 203 cm^{-1} and 279 cm^{-1}. The peaks can be identified as transverse optical (TO) and longitudinal optical (LO) phonon mode, which are blue shifted from the corresponding phonon modes of bulk CdSe (TO=170 cm^{-1}, LO=210 cm^{-1}) system.

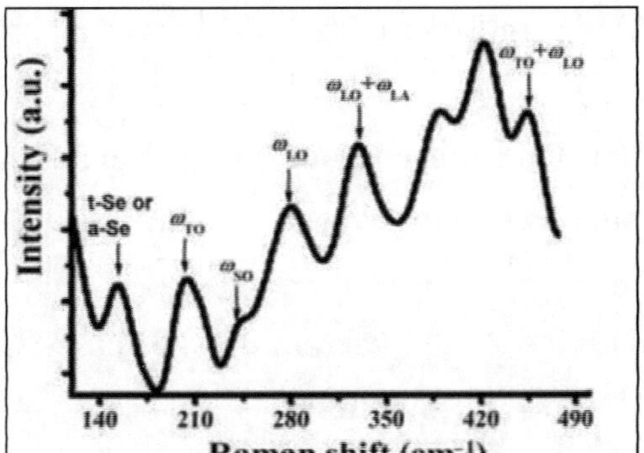

Figure 10. FT Raman spectra of TGA capped CdSe

For nanocrystallites, the confinement effect induces a broadening and downward shift of the Raman peaks. The Raman shifting due to the phonon confinement effect can be described by the confinement model. When the particle size decreases in the nanometer scale, a volume contraction occurs within the nanoparticle due to the size-induced radial pressure and this might lead to a substantial increment in the force constant value which in turn decreases interatomic distances. In vibrational transitions, the wave numbers vary approximately in proportion to $k^{1/2}$, where k is the force constant. Consequently, an enhanced magnitude of the force constant would be characterized by a shifting of the Raman band towards a higher wave number side. The respective LO and TO modes satisfy the well known Lydanne-Sachs- Teller (LST) expression given by

$$\epsilon_\infty / \epsilon_d = \omega_{TO}2 / \omega_{LO}^2$$

4. Conclusions

In summary, stable, fluorescent colloids of CdSe QDs capped with TGA are prepared by simple chemical route under ambient conditions. A new precursor of selenium ions, that is, a hydrazine hydrate-Se complex, is proposed. The method is simple and the most viable solution route to prepare CdSe quantum dots. The size of nanoparticles is conveniently tuned by controlling the concentration of ions, volume of solutions and time of reaction. Colloids of different size are characterized by various experimental techniques such as X-ray powder diffraction, HRTEM, SEM and TEM optical absorption, photoluminescence, FTIR and FT-Raman measurements. It is found that these QDs are homogeneous, spherical, and small. Thus, the present synthesis is a simple and cost-effective in the preparation of high-quality CdSe nanoparticles.

Acknowledgments

The author (PSK) would like to express heartfelt thanks to the Principal, HOD and Management, K.L.E. Society Belgaum for providing financial support.

References and Notes

1. Haizheng Zhong, Zelong Bai, and Bingsuo Zou, J. Phys. Chem. Lett. 3, 3167−3175 **(2012)**
2. M. Bruchez, Jr., M. Moronne, P. Gin, S. Weiss, and A. P. Alivisatos, Science. 281, 2013 **(1998).**
3. Kumar P. and Singh K., Wurtzite ZnSe quantum dots: Synthesis, characterization and PL properties, J. Opt. Biomed. Mat. 1, 59-69 **(2009).**
4. Mostafa A. El-Sayed *Acc. Chem. Res. 37* (5), 326–333 **(2004)**
5. Brus, L. Zero-dimensional IEEE J. Quantum Electron. QE-22 (9), 1909-1914 **(1986)**
6. J.F. Wang, M.S. Gudiksen, X.F. Duan, Y. Cui, C.M. Lieber, Science, 293, 1455-1468 **(2001)**

IOP Conf. Series: Materials Science and Engineering **360** (2018) 012010 doi:10.1088/1757-899X/360/1/012010

Synthesis and Spectroscopic Characterization of Cdse Nanoparticles for Photovoltaic Applications

P Sanjay[1], K Deepa[2], J Madhavan[2] and S Senthil[1]
[1]Department of Physics, Government Arts College For Men (Autonomous), Nandanam, Chennai - 35,Tamil Nadu, India.
[2]Department of Physics, Loyola College, Chennai - 34, Tamil Nadu, India.
Email: ssatoms@yahoo.co.in

Abstract. Cadmium selenide nanoparticle [CdSe] is an important chemical substance gaining great importance and widely used as an additive in the production of various, industrial products like rubber, cosmetics, catalyst, optical materials solar cells etc. CdSe being a wide band gap material and with better lattice matching properties made it suitable option for solar cell applications, the CdSe nano rods are synthesized by a simple solvothermal method at the morphology, phase and the optical properties of Cdse nano particles are studied using powder X-ray diffraction [XRD], High resolution transmission electron microscope [HRTEM], UV-visible absorption spectroscopy, and Photoluminescence [PL] Spectroscopy, the HRTEM images confirm the formation of CdSe nano rods. The presence of elements was confirmed by Fourier transform infrared spectroscopy [FTIR], energy dispersive X-ray spectroscopy [EDAX], SEM images showed the morphology of the samples.

1. Introduction

Nowadays nano structured material research is a rapidly growing field of science, where the efforts of chemists, physicists, materials scientists and recently biologist and engineers have merged. Inorganic semiconductor nano crystals have exhibited interesting and novel size – dependent physical properties, which greatly differ from those of the corresponding bulk materials. For example, semiconductor nano crystals show blue shifts in both optical absorption and emission spectra with the decrease of nano crystal size due to quantum confinement effect [1]. The size and shape of inorganic nano materials are well known to have an important influence on their widely varying electrical and optical properties [2], which are important in various applications such as catalysis, optoelectronics, luminescent materials, lossing materials, solar cells, light emitting diodes and biomedical imaging. As one of the most important II – VI group semiconductors and nanocrystalline wurtzite structured Cadmium selenide (Eg = 1.74 eV) has attracted great interest in their various promising optoelectronic application owing to its excellent optical conductivity, such as photoelectron – chemical cells, photoconductors. One dimensional semiconductor nanocrystals having unique structural, optical and electrical properties have been considered as structural unit for modern electronic devices sensors, photonics materials. Several synthesis routes have been employed for production nanorods and nanowires [3-4].

This include template directed method, vapor phase approach, vapor liquid solid growth, sol-gel technique, solvothermal synthesis, solution phase growth based on capping agents. Sonochemical, radiolytic method among others reported. Therefore, the need to develop different methodologies for synthesis of nano materials is inevitable in synthetic chemistry. Among the II – VI semiconducting nano materials, Cds and CdSe are widely synthesized, because of their high crystallites size dependent features of photoluminescence and absorption [5] and great deal of work has been done to characterize Cds and CdSe nano crystals. In the ongoing research on "bottom – up" approach to produce nano particles, the hydrothermal method is turning out to be the most feasible as well as versatile approach. From the environmental perspective, solvothermal method are more environmentally benign them many other synthesis method [6]. The investigation on the role of hydrazine hydrates as the reducing agent proves that it not only controls the morphology of the nano particles but also provides stability to the CdSe nano rods by preventing the oxidation of nano particles due to its reducing capacity [7], the as - prepared nano particles have characterized structurally and optically using powder X -ray diffraction, UV-visible spectroscopy, photoluminescence, Fourier transform infrared spectroscopy (FTIR), High resolution transmission electron microscopy (HRTEM), etc. studies.

2. Experimental Procedure

All the chemicals used were of analytical grade and were of highest purity, Cadmium nitrate ($Cd(NO_3)_2$ $4H_2O$, Merck 99%) and Sodium selenite (Na_2SeO_3, Merck 90%), no need to undergo any post treatment after the reactions with excess of ($N_2H_4H_2O$) and ammonia (NH_3 H_2O).During the synthesis, the molar ratio of $Cd(NO_3)_2$ $4H_2O$ and Na_2SeO_3 was kept 2:1. Cadmium nitrate $Cd(NO_3)_2$ $4H_2O$ (0.01 mol) was dissolved in 10 ml of Milli Q – water and then NH_3H_2O was slowly added into the solution, which initially led to the formation of white precipitate, however, with further addition of ammonia, a clear solution was formed. This indicates the conversion of Cd^{2+} into Cd $(NH_3)_4^{2+}$. The Se source, Na_2SeO_3 (0.005 mol) was stirred for 5 mins with 15 ml of hydrazine hydrate ($N_2H_4H_2O$) and it was mixed with the previously prepared solution (Cd source), this resulted in colorless and transparent solution. The final solution was transferred into Teflon – coated autoclave and then filled with Milli Q – water up to 70% of filling. The pH of the solution was found to be 11 before heating. The autoclave was selected and heated at 180°C for a reaction time of 4 hr. After the completion of the reaction, the autoclave was allowed to return to room temperature. Finally, the deep dark red product was collected, washed repeatedly with Milli Q – water, ethanol and then dried at 80°C.

The prepared CdSe nanopowder was characterized by X – ray diffraction (XRD). Measurements were carried out at room temperature by using Siemens X – ray diffraction D500 with CuK_{α}. The UV – visible absorption spectrum of the sample was recorded in the range 190 – 1200 nm, Photoluminescence (PL), Scanning Electron Microscope (SEM) was used to examine the surface morphology of samples. The energy dispersive - ray (EDAX) technique was used to determine their elemental compositions. The FTIR spectroscopic measurement is carried out to study the chemical groups present in the sample. The high resolution transmission electron microscopy was done with a JEOL JEM 3010 microscopy. The sample for HRTEM analysis was prepared by dissolving the cdse nanoparticles with few drops of dilute methanol.

3. Results and Discussion

3.1. XRD analysis

The powder XRD result for cdse nanoparticle is shown in figure 1. All the reflection peaks can be indexed to wurizite phase of CdSe. From the XRD pattern, it is clear from the broadening of diffraction peaks of that the particles crystallize at nano scale region. The intensity of the peaks indicates that the CdSe nano particles are of high crystalline and there is no trace of cubic phase. The products are pure in phases with the calculated lattice constant, a = 4.218 A°, b = 4.218 A° , c = 6.887 A°, which are in good agreement with the values given in literature (JCPDS 77 – 2307). The average grain size (D) of CdSe is calculated by the Debye – Scherrer formula, D = 0.89λ/β cosθ, where, λ is the wavelength of CuK_{α} line, β is full width at half maximum and θ is the diffraction angle. From the XRD analysis the average particle size of the nano rods was estimated at 35nm.

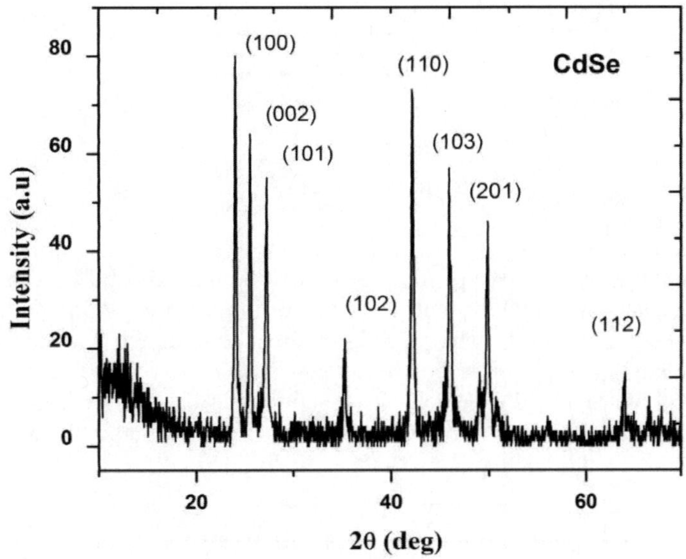

Figure 1. XRD pattern of CdSe nanoparticles.

3.2. UV – Vis Spectroscopy

The optical properties were observed by UV – Vis spectroscopy. The UV – Vis absorption spectroscopy of CdSe nano structural material was measured and shown in figure 2, in the wavelength range 200 – 1200 nm the absorption was measured at room temperature. The absorption edge of CdSe narrowed at 244 nm to be blue shifted the blue shift of the absorption curve results in a reduction of the band gap energy.

IOP Conf. Series: Materials Science and Engineering **360** (2018) 012010 doi:10.1088/1757-899X/360/1/012010

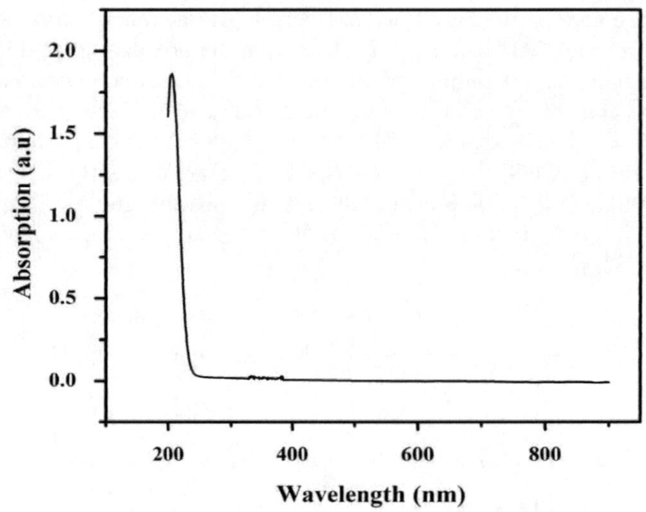

Figure 2. UV-Vis absorption spectra of CdSe nano particles.

3.3. Photoluminescence Spectroscopy (PL)

Figure 3 shows the photoluminescence spectra of the prepared CdSe nano rods. The influence of the width of the CdSe nano rods/nano shows on the shifting behavior in the photoluminescence spectrum has been reported by few researchers [8,9]. The photoluminescence emission spectra of the CdSe nano rods excited at a wavelength of 600 – 720 nm. The sharp peak observed at 642 nm confirms a blue shift compared with bulk wurtzite CdSe which may result which may result from the quantum size effect. The strong PL intensity attains good crystalline quality of the synthesize nano rods.

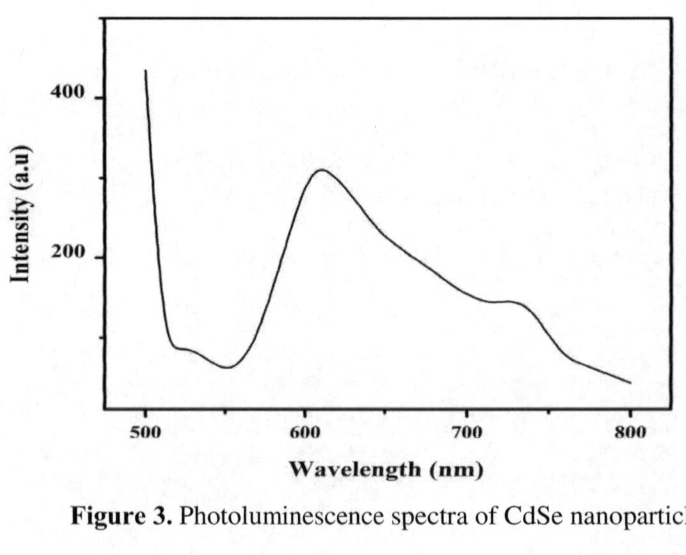

Figure 3. Photoluminescence spectra of CdSe nanoparticles.

3.4. SEM analysis

The surface morphology of the prepared CdSe sample was studied using Scanning Electron Microscope. The SEM micrographs of the CdSe with different magnification at room temperature are shown in Figure 4. The CdSe nano rods were observed at different magnification level at 50 nm. The solvents hydrazine and ammonia ($NH_3.H_2O$) play an important role in the growth of as – formed cadmium selenide nano rods. When the loss of hydrazine is faster, selenide molecules will agglomerate together and condense as rod shaped crystals. The SEM investigation of all the nano CdSe samples reveals that the crystallites are nanometer in size.

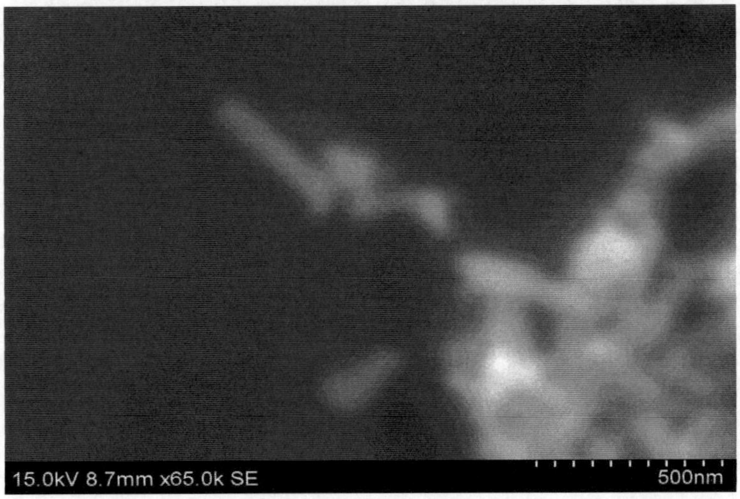

Figure 4. SEM image of the CdSe nanoparticles.

3.5. EDAX analysis

To identify the type of element in the samples Energy Dispersive X- ray Spectroscopy (EDAX) was used. The EDAX spectrum was taken using Jeol 6390LV model Scanning Electron Microscope. The EDAX spectra of the synthesized CdSe nano particles were recorded and they are displayed in figure 5. From the results it is confirmed that the elements such as Cd and Se are present in the samples.

IOP Conf. Series: Materials Science and Engineering **360** (2018) 012010 doi:10.1088/1757-899X/360/1/012010

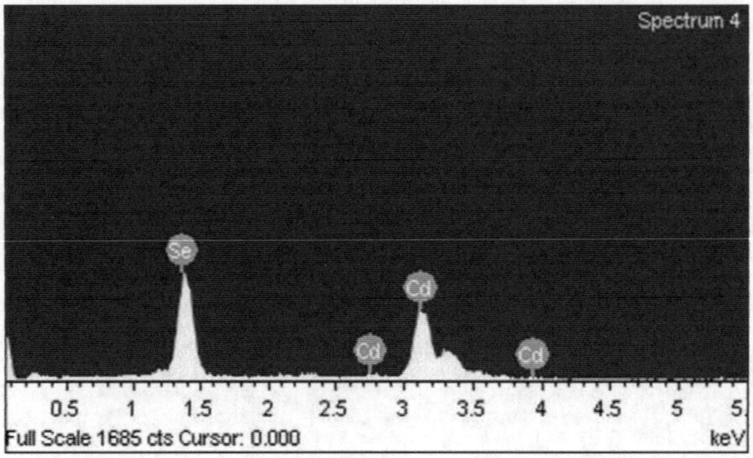

Figure 5. EDAX spectra of the CdSe nanoparticles.

3.6. High Resolution Transmission Electron Microscope (HRTEM)

High – resolution Transmission Electron Microscope is an important research tool for the study of the structure of individual CdSe nano rods. The formation of individual nano rods of CdSe with narrow size distribution is evident from the HRTEM images. Figure 6 shows the HRTEM images of the CdSe nano rods it confirms the uniform size and shape distribution with CdSe nano rods. The influence of hydrazine hydrate to control the morphology of CdSe nano rods has been reported by few research groups. In the present study we used an increased quantity of hydrazine hydrate and still managed to get CdSe nano rods of good crystalline nature. The average particle size of the nano rod is estimated as 35 nm.

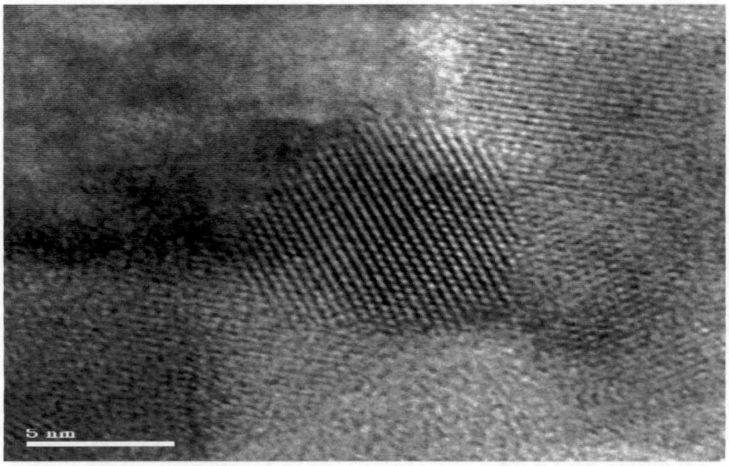

Figure 6. HRTEM image of the CdSe nanoparticles.

3.7. FTIR analysis

FTIR spectroscopy is used to identify and characterize the organic species present in the CdSe nonstructural material. The FTIR spectrum was recorded in the range 400 – 4000 cm^{-1} employing Brukker model IFS 66V FTIR spectrometer. The FTIR spectrum of CdSe nano particles is shown in Fig. 7. The FTIR spectrum of the compound shows two prominent peaks at the weak band at 3730 cm^{-1} and 2926 cm^{-1} arise from C – CH_3 asymmetric stretching. 2510 cm^{-1} and 2614 cm^{-1} are assigned to asymmetric and symmetric stretching vibration of C – CH_2 from the methylene chain respectively. A peak at 738 cm^{-1} is due to Cd - Se band stretching. The absorption band found between 500 cm^{-1} and 700 cm^{-1} are due to the metal oxygen bonding vibrations. Thus the FTIR spectra confirm the presence of functional group and their mode of vibrations.

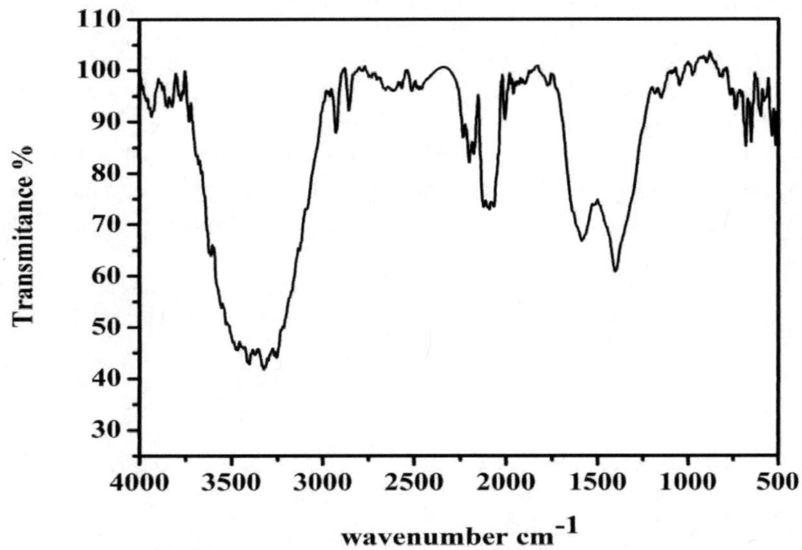

Figure 7. FTIR spectrum of the CdSe nano particles.

4. Conclusion

A CdSe nano rod was successfully synthesized by a simple hydrothermal method and easy process and with better control over the morphology and crystalline quality. The particle size and morphology are verified by powder XRD and HRTEM. The blue shift in the photoluminescence spectrum of CdSe nano rods has been confirmed by the absorption spectrum. UV- Vis spectrum shows blue shift as compared to bulk materials. FTIR analysis shows the presence of surfactant in the final product. This study opens up new avenues for research to find suitable experimental conditions and the possibilities of using different reaction mechanisms to bring out better control over the size/morphology of the semi conducting nano particles.

References

[1] Wang Q, Pan D, Jiang S, Ji X, An L and Jiang B 2006 A solvothermal route to size- and shape controlled CdSe and CdTe nanocrystals *J. Cryst. Growth.* **286** 84-90.

[2] Ma C, Moore D F, Ding Y, Li J and Wang Z L, 2004 Nanobelt and nano saw structures of II-VI semiconductors *Int. J. Nanotechnol.* **1(4)** 431.

[3] Duan X F, Huang Y, Cui Y, Wang J F and Lieber C M **2001** Indium phosphide nanowires as building blocks for nanoscale electronic and optoelectronic devices *Nature.* 409.

[4] Gates B, Wu Y, Yin Y, Yang P and Xia Y, 2001 Single-crystalline nanowires of Ag2Se can be synthesized by templating against nanowires of trigonal Se *Journal of the American Chemical Society.* **123** 11500-11501.

[5] Ramalingam G and Madhavan J, 2011 Investigation on the structural and morphological behaviour of CdSe nanoparticles by hydrothermal method*Archives of Applied Science Research,* **3 (3)** 217-224.

[6] Ramalingam G, Melikechi N, Dennis Christy P, Selvakumar S and Sagayaraj P 2009 Structural and optical property studies of CdSe crystalline nanorods synthesized by a solvothermal method. *Journal of Crystal Growth* **311** 3138-3142.

[7] Riman R E, Suchanek W L and Lencka M M 2002 hydrothermal crystallization of ceramics Cristallisation hydrothermale de ceramiques. Annales de Chimie Science des Matériaux **27** 15-36.

[8] Yang J, Zeng J H, Yu S H, Yang L, Zhang Y H and Qian Y T, Formation process of CdS nanorods via solvothermal route Chemistry of materials. **12** 3259-3263.

[9] Peng Q, Dong Y, Deng Z and Li Y 2001 Low-Temperature Elemental-Direct-Reaction Route to II–VI Semiconductor Nanocrystalline ZnSe and CdSe *Inorg. Chem.* **40 (16)**, pp 3840–3841.

IOP Conf. Series: Materials Science and Engineering **360** (2018) 012011 doi:10.1088/1757-899X/360/1/012011

Synthesis, Structural, Morphological and Optical Characterization of TiO_2 and Nd^{3+} Doped TiO_2 Nanoparticles by Sol Gel Method: A Comparative Study for Photovoltaic Application

P Sanjay[1], E Chinnasamy[1], K Deepa[2], J Madhavan[2] and S Senthil[1,]

[1] Department of Physics, Government Arts College for Men (Autonomous), Nandanam, Chennai - 35, Tamil Nadu, India.
[2] Department of Physics, Loyola College, Chennai – 34, Tamil Nadu, India.
Email: ssatoms@yahoo.co.in

Abstract. Nano structured materials are currently receiving wide attention due their optical, electronic, magnetic, chemical, physical and mechanical properties. Semiconductor nano crystals have been widely studied for their fundamental properties. Pure and Nd^{3+} doped titanium dioxide nano powder was successfully synthesized by sol-gel method. The morphological and structural properties of as-prepared samples were characterized by X-ray diffraction (XRD), High resolution scanning electron microscope (HRSEM). The HRSEM analysis shows the partial crystalline nature of undoped, and doped TiO_2 nanoparticles and it was performed to obtain the particles size. The energy dispersive X - ray (EDAX) technique was used to determine their elemental compositions. The optical properties of doped and undoped TiO_2 nano particles were investigated by UV-Vis spectroscopy and Photoluminescence (PL). The band gap of the doped TiO_2 nanoparticles was found to be less than un-doped TiO_2 nanoparticles.

1. Introduction

Titanium dioxide (TiO_2) has been extensively studied oxide as a pigment and in sunscreens, paints, ointments and toothpaste and as a photocatalytic material for self cleaning coatings. Environmental purifiers, antifogging mirrors and in many other applications. Semiconductor nano crystals have been widely studied for their fundamental properties especially titanium dioxide (TiO_2). Nano sized titanium dioxide materials have been the focus of great interest because they exhibit modified physical-chemical properties in comparison with its bulk. Inexpensiveness, excellent chemical stability, non toxicity, high photo-catalytic property, a wide band gap and high refractive index of TiO_2 make it attractive for practical applications [1]. The use and performance for a given application are strongly influenced by the crystalline structure, the morphology and the size of the particles. The use of TiO_2 nano particles is in great demand because of its properties and application in various fields such as photocatalysis , solar energy conversion and electronic devices. The photocatalytic performance of TiO_2 based devices is largely influenced by the particle size, apparently at the nanometer scale. Following the literature the photocatalytic activity of TiO_2 has been improved by optimizing the nanostructure size using various synthetic methods such as sol-gel, hydrothermal, solvothermal, co-precipitation etc. Furthermore, doping with alkali metals, transition metal ions or rare earth metal ions (lanthanides) or non metal ions has been considered as a promising way for improving the photocatalytic efficiency of nano sized TiO_2.

The optical and electronic properties of nano structured TiO_2 can be tailored by a variety measures including thermal treatments, supported film growth, and metal-ion doping [2]. In particular, doping with lanthanide metal ions has been shown to increase the photocatalytic efficiency for selected reaction. It is critical to assess the effects of lanthanide-ion doping on the structure of titania which will allow greater control over the desired properties. Doping of TiO_2 with neodymium ion would introduce a distortion in TiO_2 lattice [3]. The interstitial neodymium does not affect the charge balance in the anatase lattice, as a substitutional neodymium does.

Sol-gel process is one of the most successful techniques for preparing nanocrystalline metallic oxide materials due to low cost and easy for fabrication. Generally in a typical sol-gel process a colloidal suspension or a sol is formed due to the hydrolysis and polymerization reaction of the precursors, which on complete polymerization and less of solvent leads to the transition from the liquid sol into a solid gel phase, the wet gel can be converted into nano crystals with further drying and hydrothermal treatments. In this paper, we report synthesis and characterization of Nd^{3+} doped TiO_2 nanoparticles obtained by sol-gel method. The prepared samples were characterized by X-ray diffraction (XRD), High resolution scanning electron microscope (HRSEM), the energy dispersive (EDAX) studies, ultraviolet - visible spectroscopy (UV-Vis) and photoluminescence (PL) studies.

2. Experimental Details

Pure titanium dioxide nanocrystals were prepared using the sol-gel method. Titanium isopropoxide (TIP) was used as the precursor for titania sol preparation. The sol corresponds to the overall volume ratio of Ti $[OCH(CH_3)_2]_4$: C_3H_8O: CH_3COOH: Distilled H_2O = 5:30:4.4:30. Ti $[OCH(CH_3)_2]_4$ was first dissolved in isopropanol and distilled water to form titania sol and then stirred for 1h at room temperature. The pH of sol was adjusted to 2-3 by adding 1-2 drops of ammonia with stirring in room temperature for 12h. The prepared sol was left to stand for the formation of gel and dried at 100°C for an hour in a furnace to remove the solvents. The obtained gel was milled into powder and calcined at 400°C for 4h to keep anatase phase. The neodymium (Nd^{3+}) doped TiO_2 was synthesized using the same procedure as the reference sample. The doping rate with Nd^{3+} is equal to 0.5% wt. The prepared sample was characterized by X-ray diffraction [XRD] measurements which were carried out at room temperature by using Siemens X-ray diffraction D500 with CuKα. Scanning electron micrograph images were taken with a JEOL (JSM-840 A), High resolution scanning electron microscope (HRSEM) equipped with an EDAX spectrometer. The absorption spectrum of the sample was measured on a UV-3101 Shimadzu visible spectrometer. Photoluminescence (PL) study was carried out on a Perkin-Elmer LS-55 luminescence spectrometer using a Xe-lamp at room temperature.

3. Results and Discussion

3.1. XRD analysis

The crystalline phase of undoped and doped TiO_2 nanoparticles was analyzed by X-ray diffraction [XRD]. Figure 1a and 1b shows the powder XRD pattern of as-prepared pure and Nd^{3+} doped TiO_2 nano particles. The presence of sharp diffraction peaks in the XRD confirm that products are highly crystalline and the crystalline size was calculated by the Debye Scherrer formula $D = 0.89\lambda/\beta \cos\theta$, where D is the crystalline size, λ is the wavelength of X-ray radiation (0.154 nm), 'β' is the full width half maximum and θ is the diffraction angle [4,5]. It was found that the average crystalline size of pure and Nd^{3+} doped TiO_2 surface were 18 and 20 nm respectively.

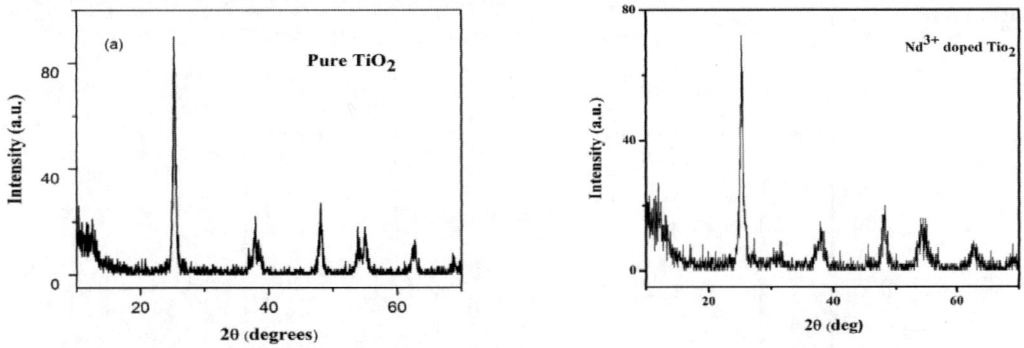

Figure 1. Powder X-ray diffraction analysis of (a) Pure TiO_2 nanoparticles (b) Nd^{3+} doped TiO_2 nanoparticles.

3.2. UV-Vis Absorption spectroscopy

Optical properties were observed by UV–Vis spectroscopy. Figure 2a and 2b demonstrates the optical absorption spectra of undoped and Nd^{3+} doped TiO_2 nano particles. The absorption edge of TiO_2 nano particles at 374 nm moved to a longer wavelength after doping with Nd^{3+} (0.5%) showing the absorption edge at 386 nm. After doping with Nd^{3+} the response of TiO_2 nanoparticles to visible light was increased and showed blue shift (towards increased wavelength). The blue shift of the absorption curve results in a reduction of the band gap energy. The band gap energy of the materials were calculated using the formula, $E_g = h\frac{c}{\lambda} \times 6.2415 \times 10^{18}$ eV. Where, h is Planks constant (6.626 × 10−34J s), C is Speed of light (3.0×10^8m/s) and λ is Cut-off wavelength and its value is 372×10^{-9}m. The calculated band gap energy of pure TiO_2 is 3.3eV and the Nd^{3+} doped TiO_2 is 3.02 eV. The optical band gap energy decrease with the Nd^{3+} doped TiO_2 nanoparticles.

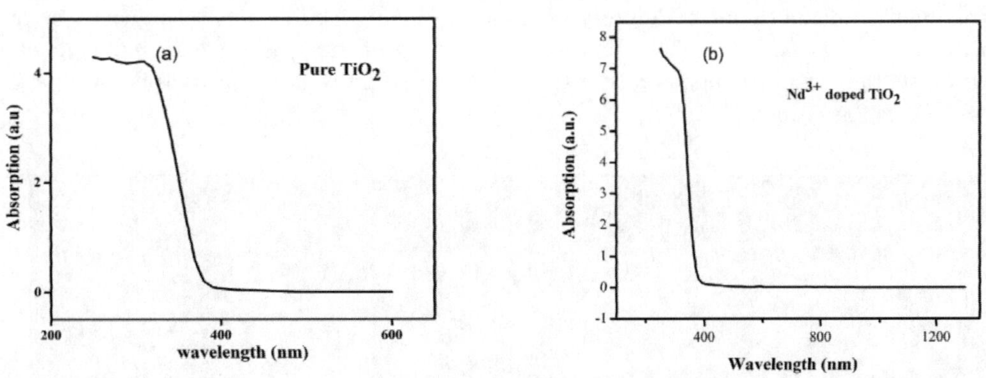

Figure 2 UV-Vis absorption spectra of (a) Pure TiO_2 nanoparticles (b) Nd^{3+} doped TiO_2 nanoparticles.

3.3. Photoluminescence spectroscopy

Photoluminescence spectroscopy (PL) is a practical method for probing the electronic structure of nano materials [6], the PL emission spectra of the TiO_2 and Nd^{3+} doped TiO_2 excited at a wavelength of 285 nm at room temperature is shown in figure 3. The TiO_2 nano particles showed the emission peak in a range of 350-400 nm and another one in range of 400-430 nm, there after doping with Nd^{3+} one peak at 350-390 nm and another one in the range of 420-430 nm corresponds to the radioactive transition of the excited electron from fermi level. The PL intensities of Nd^{3+} doped TiO_2 (0.5%) were

IOP Conf. Series: Materials Science and Engineering **360** (2018) 012011 doi:10.1088/1757-899X/360/1/012011

lower in comparison to those of the TiO_2 nanoparticles because the neodymium cause some changes in the electronic structure of the doped TiO_2 nanoparticles. The PL emission is directly related to the recombination of excited electrons and holes, so the lower PL intensity indicates a delay in recombination rate.

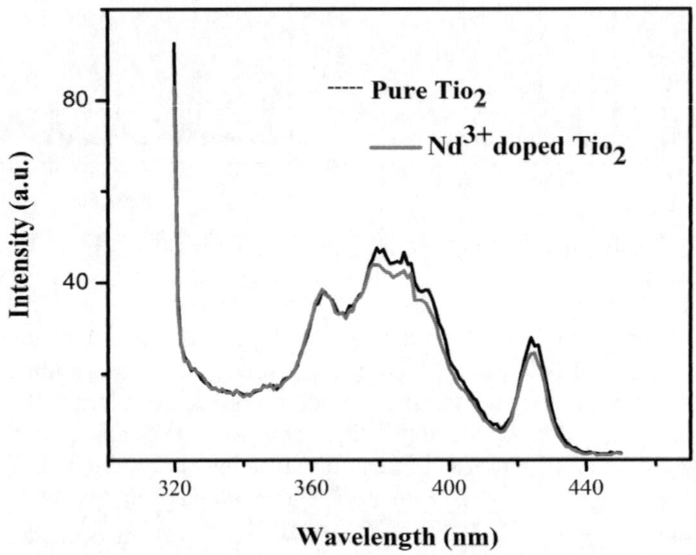

Figure 3. Photoluminescence spectra of (a) Pure TiO_2 nanoparticles (b) Nd^{3+} doped TiO_2 nanoparticles.

3.4. High Resolution Scanning Electron Microscopy (HRSEM)

High resolution scanning electron microscope (HRSEM) was used to examine the surface morphology of prepared samples. Figure 4a and 4b shows HRSEM images of the undoped and Nd^{3+} doped TiO_2. The HRSEM investigation of the synthesized samples reveals that the crystallites are of nanometer size and all samples show uniform morphology in the form of TiO_2 nano clusters.

Figure 4. HRSEM image of (a) Pure Tio_2 nanoparticles (b) Nd^{3+} doped Tio_2 nanoparticles.

3.5. Energy Dispersive X-Ray (EDAX) analysis

To identify the type of elements present in the sample, Energy dispersive X-ray spectroscopy (EDAX) was used. The EDAX spectra of the undoped and Nd^{3+} doped TiO_2 nano particles were recorded and they are shown in figure 5a and 5b. From the results, it is confirmed that Ti, O and Nd^{3+} ions are present in the sample, the presence of Nd^{3+} is confirmed which indicates the doping has entered into the sample.

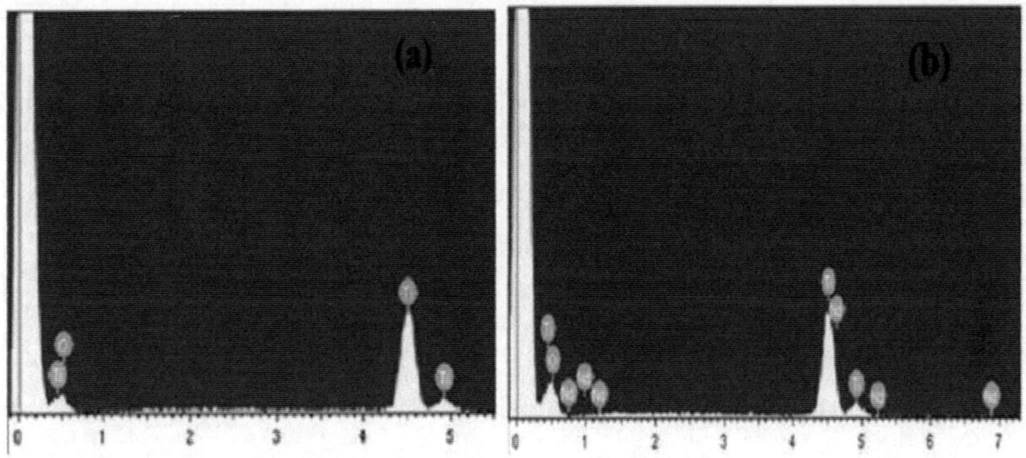

Figure 5. EDAX spectra: (a) Pure Tio_2 nanoparticles. (b) Nd^{3+} doped Tio_2 nanoparticles.

4. Conclusion

The pure and Nd^{3+} doped TiO_2 nanoparticles were synthesized by sol-gel technique. The powder X-ray diffraction analysis confirmed the crystalline phase of the synthesized TiO_2 nanoparticles. From the optical absorption spectra, a blue shift has been observed as compared to the bulk excitonic peak, which clearly shows the effect of quantum confinement and the presence of strong PL intensity which is due to both high crystalline and good surface states of the synthesized nanoparticles. The HRSEM and EDAX studies confirmed the morphological features of the nanoparticles and the composition of the undoped and Nd^{3+} doped TiO_2 nanoparticles.

References

[1] Xue-wei W U, Dajian W U and Xiao L I U, 2009 silver- doping induced lattice distortion in Tio_2 nanoparticles *Chinese Physics. Letters.***26** 7

[2] Li W, Wang Y, Lin H, Shah S I, Huang C P, Doren D J, Rykov S A, Chen J G and Barteau M A, 2003 Band gap tailoring of Nd^3-doped TiO_2 nanoparticles *Appl Phys Lett.* **83** 4143-4145

[3] Xu J, Ao Y, Fu D and Yuan C, 2008 A simple route forth preparation of Eu, N-codoped Tio_2 nanoparticles with enhanced visible light- induced photo catalytic activity *Journal of Colloid and Interfacial Science.* **328** 447–451

[4] Chi C, Choi J, Jeong Y, Lee O Y and Oh H J 2011 Nitrogen and europium doped TiO_2 anodized films with application sin photo- catalysis *Thin Solid Films.* **519** 4676–4680

[5] Ramya S, George R P, Subba Rao R.V, Dayal R K 2010 Effect of biofouling on anodized and sol–gel treated titanium surfaces: a comparative study *Biofouling.* **26** 883–891

[6] Hu S, Wang A, Li X, Wang Y and Lowe H, 2010 Hydrothermal Synthesis of Ionic Liquid [Bmim]OH-Modified TiO2 Nanoparticles with Enhanced Photocatalytic Activity under Visible Light, *Chem. Asian J.* **5** 1171–1177

IOP Conf. Series: Materials Science and Engineering **360** (2018) 012012 doi:10.1088/1757-899X/360/1/012012

Rapid graphene growth using ethanol-CVD

A K Singh and A K Gupta

Department of Physics, Indian Institute of Technology Kanpur, Kanpur 208016, INDIA

skanil@iitk.ac.in

Abstract. We studied fast growth of graphene on copper using ethanol precursor, by low pressure chemical vapor deposition. We have used three different types of copper foils to investigate the effects of copper purity and crystallographic orientation on the nucleation density of graphene for a fixed growth time of 10s. We have used X-ray diffraction (XRD) and Electron back-scattered diffraction (EBSD) to investigate the facets present in the copper foils. In this rapid growth process, the nucleation density of graphene islands is found to be independent of facets in all the three Cu foils but we observed a variation in nucleation density and island size on different copper foils keeping all the growth parameters same.

1. Introduction

Ethanol is a promising liquid precursor for the Chemical vapor deposition (CVD) of graphene on copper due to its low cost and easy handling. It offers different and faster growth kinetics, probably because of oxidizing environment [1] and one can get full graphene coverage on copper in a few seconds [2,3,4]. Therefore, further exploring this efficient growth method and understanding growth mechanism is important. Recently Lisi et. al. [2] reported fast growth of graphene and they could not identify individually separated graphene islands with duration of a few seconds growth. Faggio et. al. [3] studied the fast growth process using unpolished and electro-polished copper foils. Previously, we have reported [4] full coverage of graphene on copper in just 30 sec. In this fast growth process, we have attributed, individual graphene island shapes to an interplay between island growth rate and its relaxation rate.

Wrinkles, grain boundary, surface roughness, step-bunches, chemical impurity and facet orientation in copper foil affects the nucleation density of graphene islands. Liu et al. studied [5] effect of purity of copper substrate on graphene growth using methane and found that graphene growth on the low purity Cu foil was bilayer while on high purity Cu, graphene was single layer. Cu facets Cu(100) and Cu(111) have been reported to be more auspicious for monolayer graphene growth [6,7].

In this paper, we report on the nucleation density and growth of graphene islands on different Cu foils, assessing how the nucleation density depends on facets, chemical impurity and surface defects. Scanning electron microscopy (SEM), X-ray diffraction (XRD) and Electron-backscattered diffraction (EBSD) mapping are presented to correlate the graphene nucleation density with slightly different purity copper foils containing different facets.

2. Experimental Details

We have used three types of commercial Cu foil: two from Alfa Aeser with 99.8% purity (CF-1) and 99.999% purity (CF-2) and 3rd from Sigma Aldrich with 99.98% (CF-3) purity. As received Cu foils

IOP Conf. Series: Materials Science and Engineering **360** (2018) 012012 doi:10.1088/1757-899X/360/1/012012

were cut into pieces of 0.5×0.5 cm2 size and immersed in acetic acid, acetone and 2-propanol sequentially, for 5 min in each liquid. These Cu foils are first put in 8-cm-long quartz boat, which is placed inside a long quartz tube kept inside a tubular furnace. The detailed description of our CVD system can be found elsewhere [4]. It consists of a hot-wall tube furnace where the samples can be rapidly inserted and extracted from the hot zone by sliding Quartz tube. Ethanol vapor was flown inside the quartz tube with Ar/H2 (5% H2 in Argon) as carrier gas and CVD was performed at 980°C for 10s.

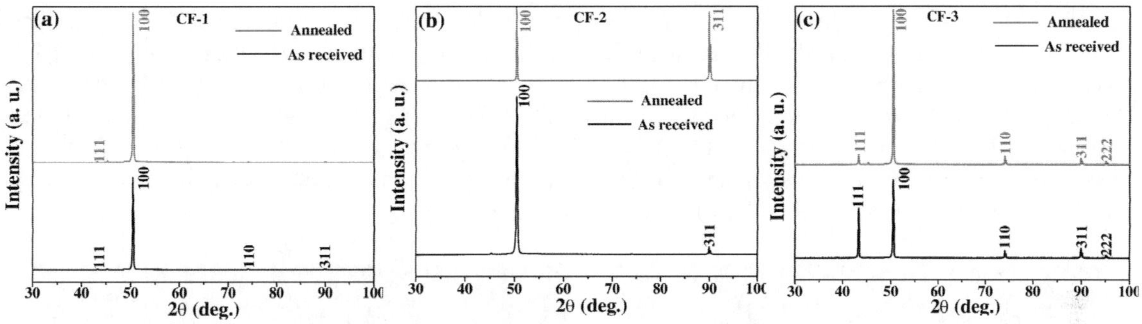

Figure 1. XRD pattern of all three copper foils as received and 20 min Ar/H_2 annealed state (CVD was stopped just before the deposition): (a) Copper foil-1, (b) copper foil-2 and (c) copper foil-3.

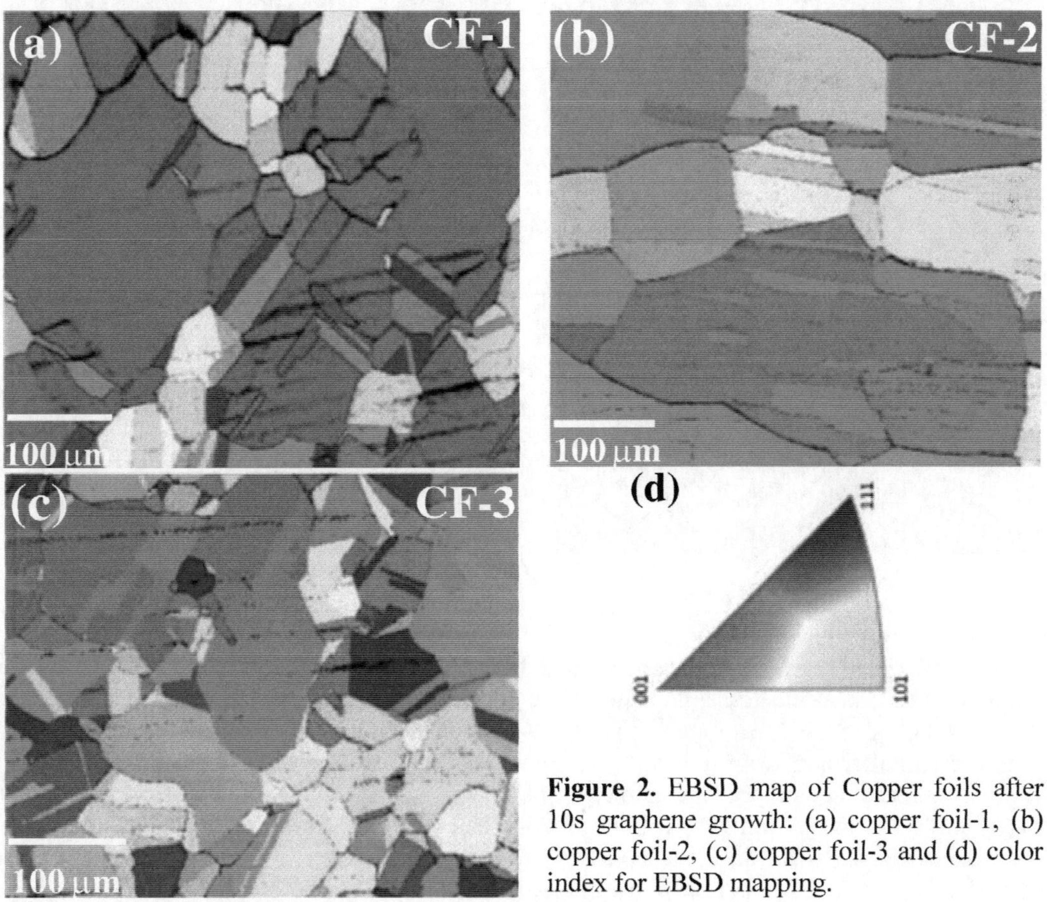

Figure 2. EBSD map of Copper foils after 10s graphene growth: (a) copper foil-1, (b) copper foil-2, (c) copper foil-3 and (d) color index for EBSD mapping.

IOP Conf. Series: Materials Science and Engineering **360** (2018) 012012 doi:10.1088/1757-899X/360/1/012012

We used a EVO-18 Zeiss SEM with low beam voltage in secondary electron detection mode to study the graphene nucleation density on Cu foils. To determine the facets present in the Cu foil, we performed XRD and EBSD orientation mapping on a different SEM (Zeiss, EVO-50) using EBSD detector.

3. Results and Discussion

The crystallographic orientation of the Cu foil affects carbon adsorption. It means facets can affect nucleation density of graphene islands. In order to explore the role of Cu facets on graphene nucleation density, XRD and EBSD mapping of all the three Cu foils were performed. Figure 1 (a), (b) and (c) show the XRD patterns of the CF-1, CF-2 and CF-3 in as received as well of Ar/H2 annealed foils, respectively. Different facets detected from XRD are marked in the figure. Figure 2 (a), (b) and (c) show the EBSD map of the CF-1, CF-2 and CF-3 after graphene deposition. Figure 2 (d) shows inverse pole color index for EBSD map. EBSD measurement gives the crystallographic orientation in x, y and z direction, but here we will only consider z plane of copper surface.

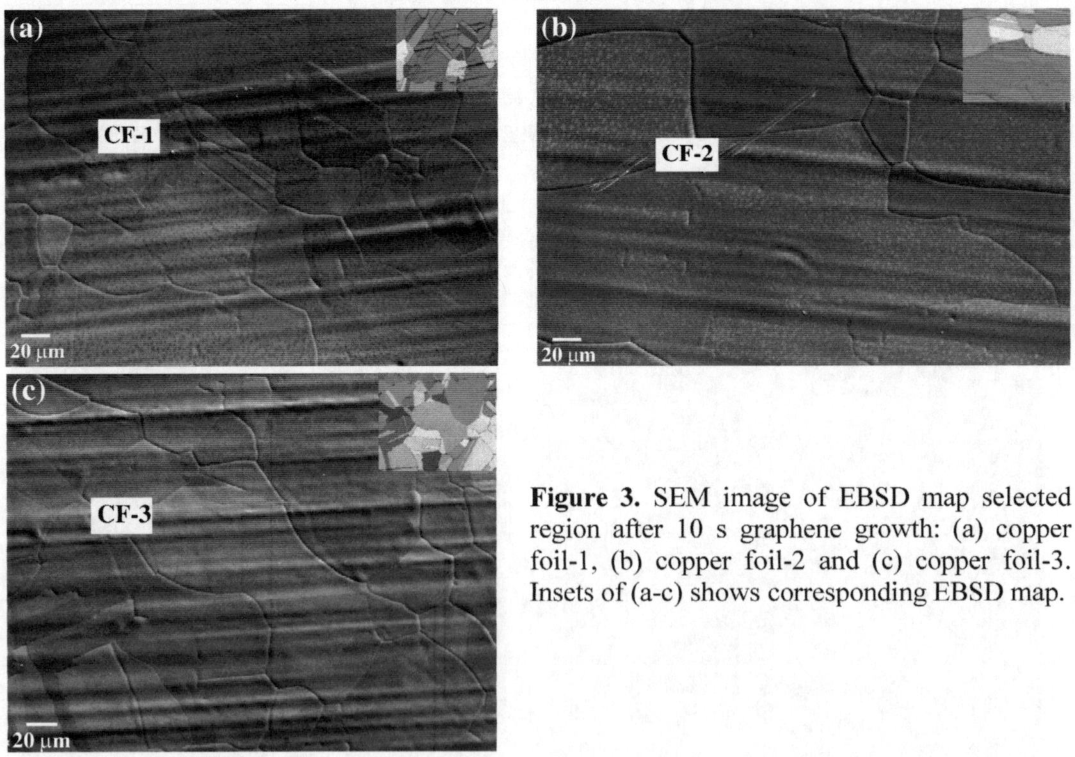

Figure 3. SEM image of EBSD map selected region after 10 s graphene growth: (a) copper foil-1, (b) copper foil-2 and (c) copper foil-3. Insets of (a-c) shows corresponding EBSD map.

Figure 1 (a) shows the XRD with various copper planes as in CF-1. After annealing, the copper foils become more textured along <100> direction as confirmed from the EBSD map [Figure 2 (a)]. CF-2 has (100) and (311) facets as received as well as after annealing [Figure 1 (b)]. After annealing, the height of (100) peak decreases but the height of (311) peak increases indicating the increase in higher index facets in CF-2 with annealing. XRD pattern in Figure 1 (c) shows the facets present in CF-3 before and after annealing indicating its polycrystalline nature. After annealing the intensity of (100) plane increases and other plane, namely (111), (110), (311), and (222), intensity decreases. CF-3 has (100) and (111) facets in higher volume fraction as confirmed from EBSD map [Figure 2 (c)]. If we compare low index facets in the three copper foils – all the foils are polycrystalline and preferred orientation of grains in CF-1 is (100), in CF-2 is (100) and (311) and in CF-3 it is (100), (111) and (110).

IOP Conf. Series: Materials Science and Engineering **360** (2018) 012012 doi:10.1088/1757-899X/360/1/012012

Figure 3 (a), (b) and (c) show the SEM images of selected regions of Figure 2 (a), (b) and (c), respectively, after 10s graphene growth. By correlating these SEM images with corresponding EBSD maps (see inset of figure), it is clear that graphene nucleation density is independent of underlying facets for all the three copper foils. To investigate whether the impurity present in Cu foil affects the nucleation density, we analyzed nucleation density present in all three copper foils. We find nucleation density on CF-3 to be 5 times higher than CF-2 and that on CF-1 to be two times higher than CF-2. The impurities present in CF-1, CF-2 and CF-3 are 0.2%, 0.001% and 0.02%, respectively indicating the nucleation density is not influenced by small change in copper foil purity. We have also altered the position of the copper foil inside the quartz boat, but the nucleation density remains independent of this position. Nucleation density, copper foils' purity and the dominant texture of the foils are summarized in Table 1.

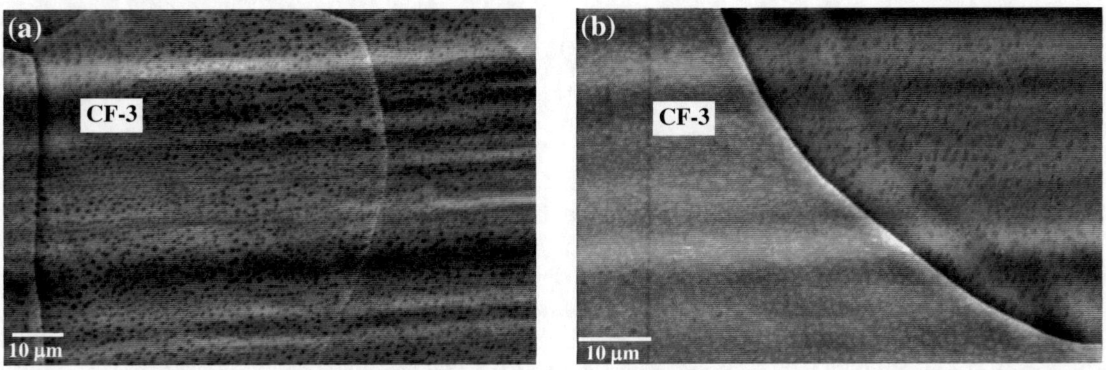

Figure 4. SEM images of copper foil-3 to see the effect of surface defects density on nucleation density: (a) effect of polish line on nucleation density and (b) effect of surface contamination on nucleation density.

From the SEM images of copper foils (Figure 3), it is clear that the density of surface defects, such as polish lines [5] and grain boundaries, in CF-2 is the lowest while in CF-3, it is the highest. This indicates a correlation between nucleation density and surface defects density. Figure 4 further illustrates this point on the same Cu foil but having different defect density in different regions. Nucleation density of graphene islands along the direction of the polish lines is much higher as seen in Figure 4 (a). Surface contaminants present in copper foils also affect the nucleation density [Figure 4 (b)].

Table 1. Summary of nucleation density, copper foils' purity and the dominant texture present in copper foils.

Sample Name	Thickness (μm)	Vendor/Product Number	Purity (%)	Facets	Nucleation density in 500 μm^2
CF-1	25	Alfa Aesar/ #13382	99.8	(100)	82-112
CF-2	25	Alfa Aesar/ #10950	99.999	(100), (311)	46-65
CF-3	25	Sigma Aldrich/ #349208	99.98	(100), (111)	190-280

Graphene island's size is found to depend on copper foils: CF-2 has largest island size while CF-3 has smallest island size among the three copper foils. This anticorrelates with the nucleation density. From this we surmise that the carbon containing species present inside the tube above the copper foils have uniform density but the adsorption of these species depends on the surface defect density. The later does not directly depend on the purity of copper foil and facets. Our previous report showed that the shape of islands is influenced by underlying facets present in the copper foil [4]. We believe that in our experimental conditions the diffusion of carbon containing species depends on the facets but not their adsorption.

4. Conclusion

In summary, we have analyzed graphene nucleation density on three different copper foils for 10 s growth time using ethanol precursor. By correlating electron microscopy images with the EBSD maps we find nucleation density does not depend on underlying facets for all the three copper foils. Further, the nucleation density is not influenced by small change in copper foil purity; it depends strongly on density of surface defects such as like scratch due to polishing.

Acknowledgments

We would like to gratefully acknowledge CSIR of the Government of India for his financial support.

References

[1] Li J, Kazakov A and Dryer F L 2004 *J. Phys. Chem.* A **108** 7671
[2] Lisi N, Buonocore F, Dikonimos T, Leoni E, Faggio G, Messina G, Morandi V, Ortolani L and Capasso A 2014 *Thin Solid Films* **57** 1139
[3] Faggio G, Capasso A, Malara A, Leoni E, Nigro M A, Santangelo S, Messina G, Dikonimos T, Buonocore F and Lisi N 2014 *Nanotech. Mater. Devices Conf. (NMDC)* IEEE 9 pp 69 - 72
[4] Singh A K and Gupta A K 2015 *Bull. Mater. Sci.* **38** 1723
[5] Liu W, Li H, Xu C, Khatami Y and Banerjee K 2011 *Carbon* **49** 4122
[6] Zhao L, Rim K T, Zhou H, He R, Heinz T F, Pinczuk A, Flynn G W and Pasupathy A N 2011 *Solid State Commun.* **151** 509
[7] Wood J D, Schmucker S W, Lyons A S, Pop E and Lyding J W 2011 *Nano Lett.* **11** 4547

IOP Conf. Series: Materials Science and Engineering **360** (2018) 012013 doi:10.1088/1757-899X/360/1/012013

Rietveld refinement and experimental determination of optical and electrical properties of K^+ stabilized α-MnO_2 nanostructures

Reenu Elizabeth John[1], Anoop Chandran[2], Sarath Krishnan V[1], Gijo Jose[1], Ajai Jose[1], Joshy Jose[1], Marykutty Thomas[3*] and K C George [1*]

[1]Department of Physics, St. Berchmans College, Changanassery, 686101, India
[2] School of Pure and Applied Physics, MG University, Kottayam, 686560, India
[3]Department of Physics, BCM College, Kottayam, 686001, India
Email: drkcgeorge@gmail.com
marykutythomas@yahoo.co.in

Abstract. In the midst of strenuous efforts in fabricating materials for supercapacitors, lithium ion batteries, molecular sieves and catalysts, manganese dioxide plays a vital role due to their unique tunnel framework. Alpha phased manganese dioxide (α-MnO_2) nanoparticles are synthesized using a low temperature coprecipitation method. The structural and morphological characterization carried out by XRD, SEM and HRTEM techniques, confirms the formation of α-MnO_2 nanoparticles. The EDS results confirmed the non stoichiometry and the presence of K^+ ions within the crystal lattice. The non stoichiometry is due to the large number of oxygen vacancies within the crystal. Rietveld refinement of XRD pattern was carried out to study in detail the role of K^+ ions in maintaining the stability of the 2x2 tunnel framework of α-MnO_2. The UV-Vis absorption and photoluminescence spectrum were studied for analyzing the energy band structure of MnO_2. A direct band gap of 1.66 eV is obtained from the famous Taucs plot. The photoluminescence exhibits strong emission line in the visible region at 743 nm and defect related emission peaks at 757 and 751 nm. The dielectric studies were carried out using dielectric spectroscopy. The temperature dependence of orientation polarization can be understood from the study of variation of dielectric constant and losses with temperature. Moreover, the phase purity of the sample and the Mn-O vibrational frequencies were analyzed using FTIR spectrum.

1. Introduction

Manganese dioxide (MnO_2) is an important transition metal oxide which shows versatile properties due to their variable oxidation states. The different polymorphic states of MnO_2 arise due to the difference the structural arrangement of the basic MnO_6 octahedra. Among the different polymorphic states alpha phased MnO_2 (α-MnO_2) consists of a 2x2 and 1x1 tunnel structure formed by corner sharing and edge sharing of MnO_6 octahedra. This unique structure is useful in lithium ion batteries and molecular sieves which can be applied to areas like long lasting battery technology and water harvesting technology. Various approaches have been used to fabricate manganese dioxide, such as self-reacting microemulsion [1], precipitation [2], room-temperature solid reaction [3], sonochemical [4], hydrothermal methods [5-8] and co-preciptation [9]. In the present work alpha phased MnO_2 (α-MnO_2) nanaparticles were synthesized using a coprecipitation method. Here in order to study the role of K^+ ions in maintaining the structural framework of the tunnel structure refinement was done with and without K^+ ions.

2. Experimental section

Potassium permanganate (98.5 %, E. Merck), and Manganese Nitrate (99 %, E. Merck) were used as received without further purification. Ultra-pure water was used for the preparation of all reagents solutions. Manganese Nitrate ($Mn(No_3)_2$)and Potassium Permanganate ($KMno_4$) were first dissolved in distilled water the appropriate molar ratio. The solution was heated at 60°C for 5hrs with continuous stirring .The products were washed and dried. The sample were annealed at 70°C for 3hrs.

The X-ray diffraction (XRD) patterns were collected with a Bruker AXS D8 advance X-ray powder diffractometer equipped with a scintillation counter using Cu $K\alpha$ radiation (λ = 1.5406A°). Energy dispersive spectrum (EDS) spectrum was obtained from JEOL Model JED – 2300 instrument. Transmission electron microscopy (TEM) was carried out on a JEM-2100 transmission electron microscope operated at an accelerating voltage of 60-200 KV in 50 V steps. The band gap studies were carried out by a UV- Vis spectrometer (Varian, Cary 5000). The Fourier transform infrared spectroscopy (FTIR) and photoluminescence (PL) studies were done using Thermo Nicolet, Avatar and Perkin Elmer LS 45 spectrometers.

3. Results and discussion

The phase and purity of the products are first examined by XRD. Figure 1 shows the refined XRD pattern along with the crystal structure with and without potassium in the lattice. All the peaks in the XRD pattern can be indexed to a pure body-centered tetragonal α-MnO_2 phase, with lattice constants of a = 9.72 A°, c = 2.86A°, which are in agreement with the standard values of JCPDS 44-0141. With the aid of the famous Scherrer formula the average particle size of the nanoparticles are found to be 24 nm. Absence of additional peaks of other phases indicates the phase purity of the final product. The Rietveld refinement is carried out for a tetragonal unit cell having space group I 4/m using pseudo Voigt function. In MnO_2, Mn is associated with Wykoff position 8h, oxygen at 8h and K+ ions at 2b. Refined unit cell parameters comes out to be a = 9.8165 Å and c = 2.8557Å. A detailed study of the crystal structures are required so as to analyze the importance of K^+ ions in maintaining the structural framework of α-MnO_2 [10]. The structural framework consists of basic MnO_6 octahedra in an edge shared and corner shared fashion resulting in a (2x2) + (1x1) tunnel structure. It is the size of the tunnel that is capable of accommodating a K^+ ion for structural stability. The potassium ions are distributed throughout the crystal structure. The crystal structure obtained incorporating K^+ ions with the refinement parameters matches well with the reported crypomelane architecture [10]. The 2 x 2 tunnel structure is capable of incorporating a K^+ ion, whereas for the refinement procedure done without K^+ ions the tunnel framework is completely disfigured. Also a collapse of the basic MnO_6 octahedra which is the building block of the tunnel framework is observed. Hence the K^+ ions which gets intercalated to the 2 x 2 tunnel space of the crystal structure is capable of holding the tunnel framework. To gain better understanding of the nanoparticles, SEM and HRTEM images (Figure 2) of the sample is necessary. The SEM image clearly reveals that the synthesized nanoparticles have no defect or dislocation and also they are single crystalline. The particles obtained from HRTEM shows that the particle has a size distribution in the range 19nm – 35nm. These results are in good agreement with XRD result.

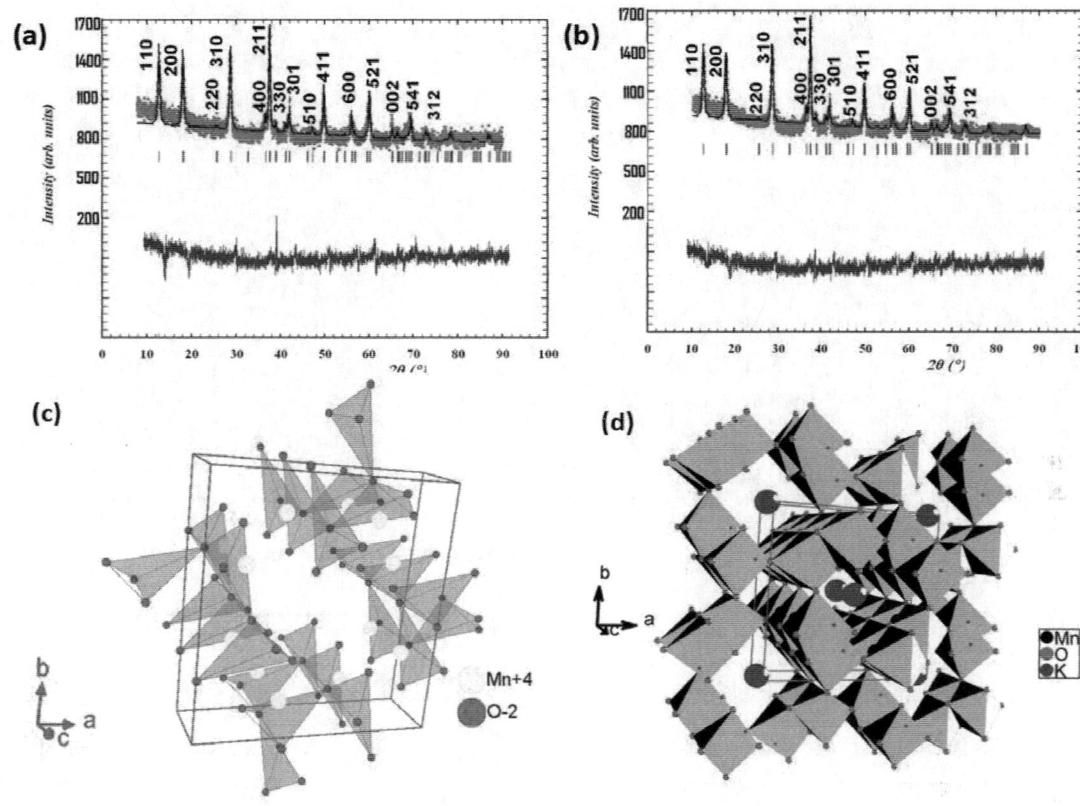

Figure 1. Rietveld refined XRD pattern of (a) without K^+ ions (b) with K^+ ions. Crystal structure of prepared sample (c) without K^+ ions (d) with K^+ ions

The compositional constituency of the prepared sample can be analysed using Energy dispersive spectroscopy (EDS). The EDS spectra (Figure 3 (a)), confirms the presence of Mn, O and K in the sample. The potassium ions stabilize the tunnel framework of the crystal structure. From the analysis it was found that the MnO_2 nanoparticles are nonstoichiometric with a large number of oxygen vacancies. This results in an mixed oxidation state of Mn throughout the sample. Diffuse reflectance spectral studies in the UV- Vis region were used to estimate the optical band gap of the synthesized nanorods. The calculated band gap from the Tauc plot is 1.68 eV and is higher than the reported values for nanostructured MnO_2 [11]. This high value may be attributed to better quantum confinement effect of MnO_2. Figure 3 (c) shows the Pl spectra of MnO_2 nanoparticles with an excitation wavelength of 500 nm. As indicated in the figure the spectrum exhibits prominent emission bands with peaks at 1.66 (743 nm), 1.65 (751 nm) and 1.63 eV (757 nm).

The intense emission peak at 1.66 eV corresponds to the band edge emission. The peak at 1.63 eV may be assigned to the oxygen vacancy related defects. MnO_2 being a highly non stoichiometric material, is rich in oxygen vacancies. The EDS result also confirms the presence of oxygen vacancies. A weak peak at 1.65 eV may be assigned to the surface dangling bonds or surface defects. The phase purity of the sample can be analyzed using FTIR spectrum (Figure 3 (d)). The bands around 3407, 1602 cm^{-1} and 1116 cm^{-1} corresponds to the O-H vibration mode of traces of adsorbed water .The peak at 511 cm^{-1} is due to

IOP Conf. Series: Materials Science and Engineering **360** (2018) 012013 doi:10.1088/1757-899X/360/1/012013

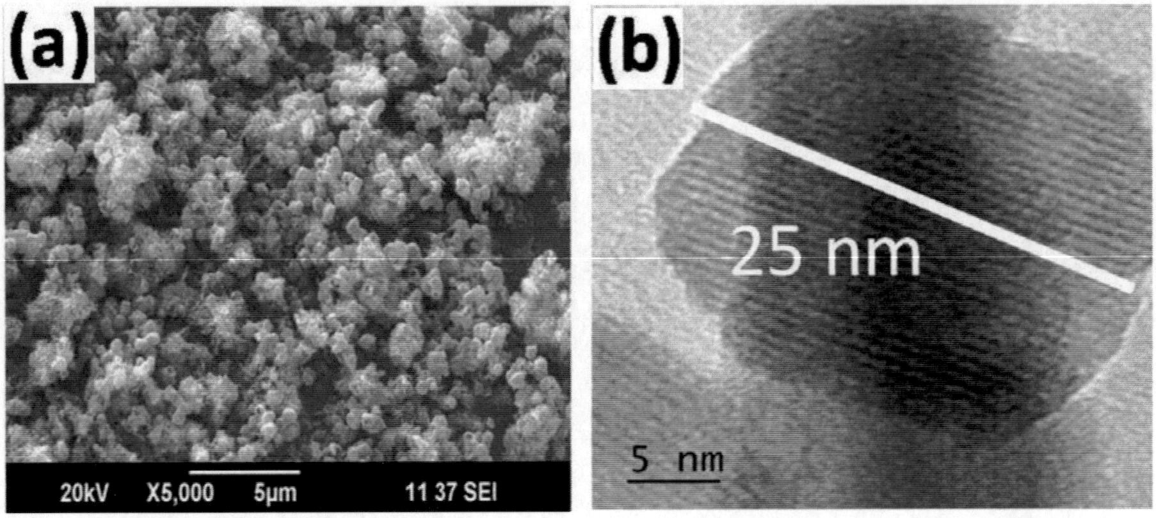

Figure 2. (a) SEM images and (b) HRTEM images of synthesised MnO_2 nanoparticles.

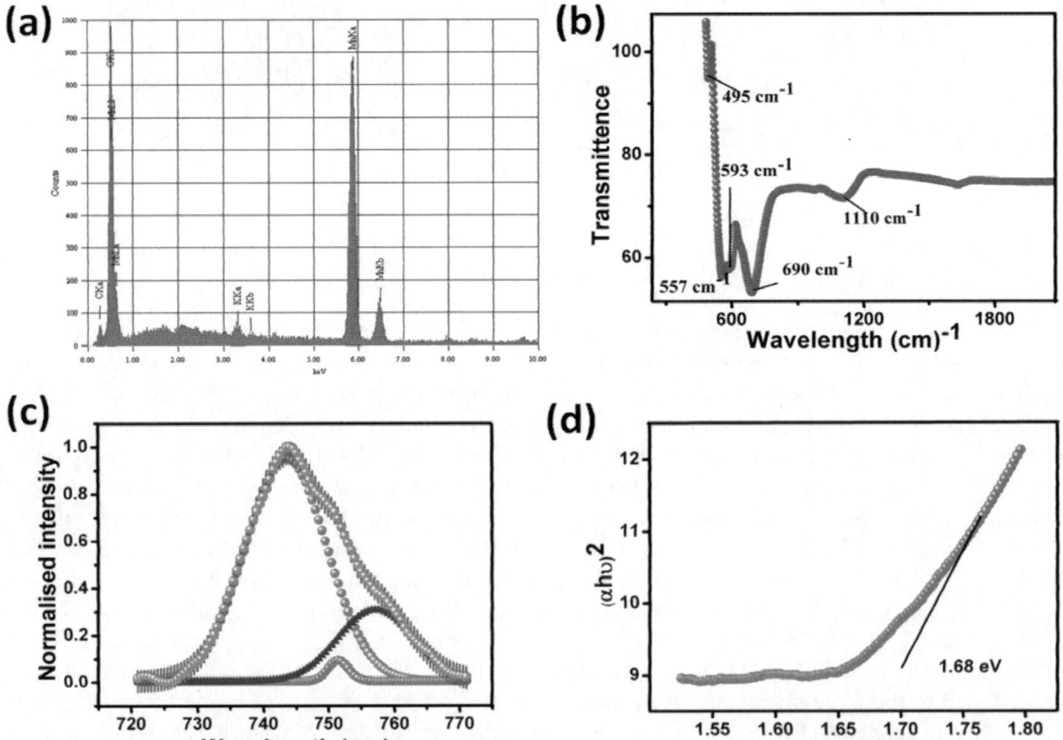

Figure 3. (a) EDS spectrum (b) Tauc plot (c) pl spectrum and (d) FTIR spectrum of α-MnO_2 nanoparticles.

IOP Conf. Series: Materials Science and Engineering **360** (2018) 012013 doi:10.1088/1757-899X/360/1/012013

displacement of oxygen anions relative to Mn cation along the direction of octahedron chain, or we can say it is due to metal-oxygen (Mn-O) bending vibrations of [MnO_6] octahedral in α-MnO_2 nanoparticles.

The net polarisability of a composite depends on several components namely, electronic, atomic, interfacial and orientation polarisability. All these contribute towards the dielectric constant of a given material. Only orientation polarisability depends directly on temperature. Figure 3 shows the variation of dielectric constant and dielectric loss with temperature. As the temperature increases the dielectric constant increases to a very high value at 80^0 C and then the values start to decrease. When an electric field is applied on a polar material, the potential energy of the associated dipoles change. The orientations aligned with the field will have a lower potential energy compared to that of the orientations aligned against the field. Less amount of energy is required to get aligned with the applied electric field and more energy is required to get aligned the applied field.

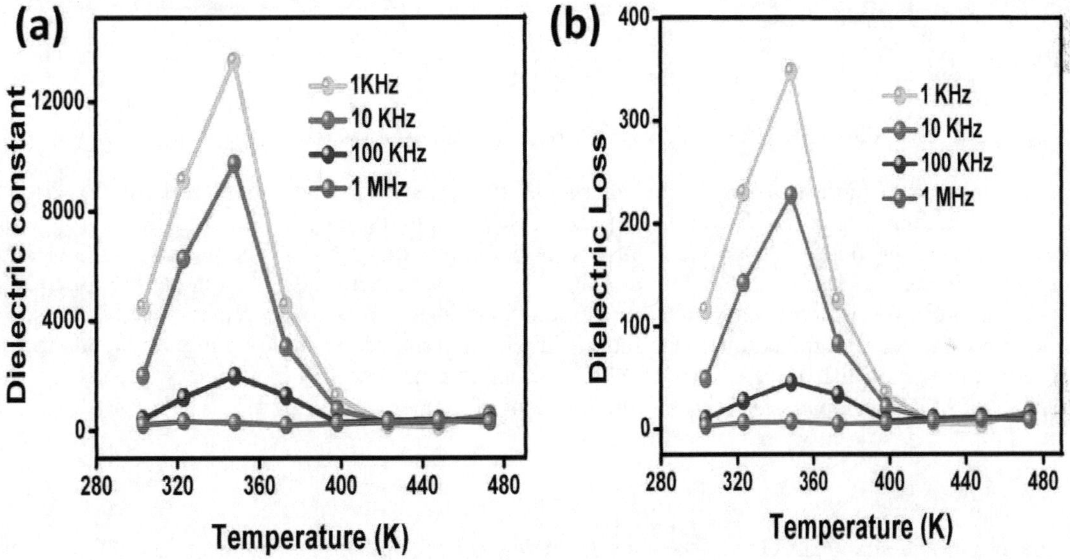

Figure 4. Variation of (a) dielectric constant and (b) dielectric loss with temperature.

At room temperature conditions, only some of the molecules will have the required energy to align themselves with the applied field. On increasing the temperature from 30^0C to 80^0C the energy in the system is enough to increase the potential energy of the remaining orientations to that critical value where they get themselves aligned with the applied electric field resulting in a high net orientation polarization and hence an increase in dielectric constant values is observed.

On increasing temperature further above 80^0C, the dielectric constant values decreases. This is due to the reason that as temperature increases to higher values, the molecular vibrations of the molecules increases along with random thermal motion to such an extent that, the range of deviation from aligning itself to the applied electric field increases resulting in molecules less closely aligned to themselves and applied electric field (Figure 4).

IOP Conf. Series: Materials Science and Engineering **360** (2018) 012013 doi:10.1088/1757-899X/360/1/012013

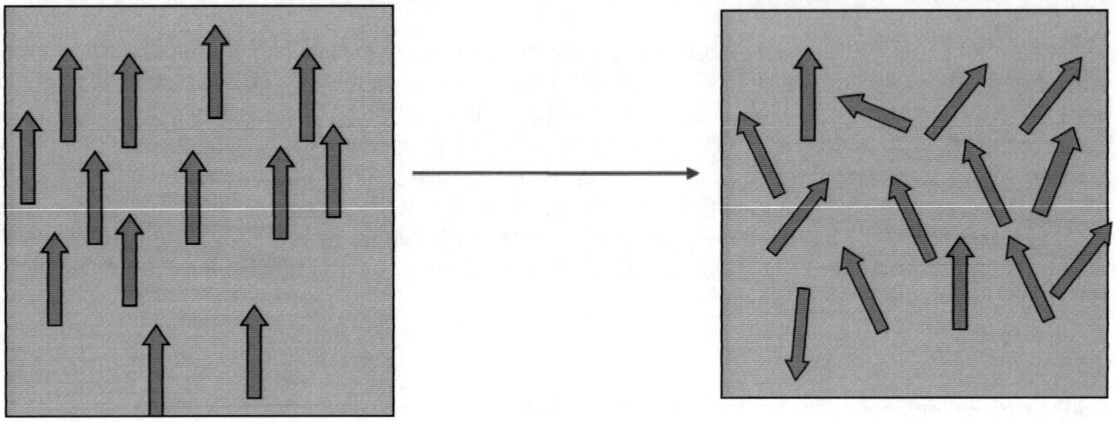

Figure 5. Effect of temperature on orientation polarization

As temperature increases further the energy of the orientations becomes close to the critical potential energy required to align the dipoles or molecules against the applied electric field. Above all mentioned factors contribute to a decrease in net orientation polarisability and hence the dielectric constant values decreases. The same trend is seen in the case of loss factor also. As temperature increases the loss factor increases up to a particular temperature and then starts to decrease. The increase in frequency causes a lag in orientation polarization which in turn results in lowering of the dielectric constant. From the graph it is obvious that at MHz frequency range the value of dielectric constant and dielectric loss of MnO_2 nanoparticles becomes independent of temperature.

4. Conclusion

Manganese dioxide nanoparticles are synthesized using a coprecipitation technique. The XRD pattern and Rietveld refinement confirms the phase and the crystal structure formed. The morphology of the sample and material composition has been identified by HRTEM and EDS, respectively. The role of K^+ ions in maintaining the 2 x 2 tunnel framework is studied in detail. The band gap of the sample is found to be 1.68 eV larger than the reported values, which is due to quantum confinement effects. The photoluminescence spectrum throws light on the band structure of α-MnO_2 nanoparticles. The intense emission peak at 1.66 eV corresponds to the band edge emission. The peak at 1.63 eV may be assigned to the oxygen vacancy related defects. A weak peak at 1.65 eV may be assigned to the surface dangling bonds or surface defects due to Mn interstitials. As temperature increases the dielectric constant increases and then decreases after attaining a particular temperature. The increase is due to the increase in mobility of permanent dipoles which leads to orientation polarization.

REFERENCE

1. Subramanian V, Zhu H and Wei B 2008 *Chem. Phys. Lett.* 453 242

IOP Conf. Series: Materials Science and Engineering **360** (2018) 012013 doi:10.1088/1757-899X/360/1/012013

2. Yuan A, Wang X, Wang Y and Hu J 2009 *Electrochim. Acta* 54 1021
3. Zolfaghari A, Ataherian F, Ghaemi M and Gholami A 2007 *Electrochim. Acta* 52 2806
4. Wang H Q, Yang G F, Li Q Y, Zhong X X, Wang F P, Li Z S and Li Y H 2011 *New J. Chem.* 35 469
5. Wang X and Li Y 2002 *JACS* 124 2880
6. Zhang X, Yu P, Wang D and Ma Y 2010 *J. Nanosci. Nanotechnol.* 10 898
7. Zhang X, Yang W S, Yang J J and Evans D G 2008 *J. Cryst. Growth* 310 716
8. Subramanian V Zhu H Vajtai R Ajayan P M and Wei B 2005 J. *Phys. Chem. B* 109 20207
9. Xing S T, Han R R, Ma Z C, Wu Y S and Zhou Z C 2011*Cryst. Eng. Comm.* 13 6033
10. Cockayne E and Li L 2012 *Chem. Phys. Lett.* 544 53
11. Hao A M and Wang N Q 2014 *Adv. Mater. Res.* 936 591

Effect of cerium doped Baddeleyite on their antibacterial activity by co-precipitation method

R.Mekala[1*], V. Rajendran[2] and B.Deepa[2]
[1,*,2] PG and Research, Department of Physics,
Presidency College, Chennai-05, Tamil Nadu, India.

*Corresponding author Email: kalatamils@gmail.com

Abstract. The present work, antibacterial activity of pure and Ceria doped Baddeleyite (ZrO_2) nanoparticles (Nps) were synthesized by Co-precipitation method using different bacterial strains (gram positive and gram negative). The obtained samples were characterized by using XRD; SEM and antibacterial study. XRD and SEM confirmed the crystalline nature and morphology of the as synthesized products. The antibacterial studies revealed that doped sample exhibits higher activity compared to pure sample, which can be attributed to an increase of defect concentration (oxygen vacancy) in the sample. In addition, the possible formation mechanism of pure and Ce-doped ZrO_2 Nps have also been proposed.

Keywords: Baddeleyite, Ce-doped ZrO_2, Antibacterial Activity, Morphology

1. Introduction

Nanostructured materials have promised several break-through applications in the Field of nano medicine and biomedical sciences due to their physical and chemical properties. Among nanostructured materials, metal oxides have emerged as a class of materials which is increasingly being studied for health-related applications[1]. Aside from various metal oxides, Baddeleyite (ZrO_2) is recognized as an important ceramic material, used in many industrial applications such as high temperature solid oxide fuel cells (SOFC), high temperature pH sensors, oxygen sensor, refractory materials, capacitors, catalyst substrates, and as abrasives
[2-5]. ZrO_2 is also being an important material in the field of medicine because of good mechanical properties, and biocompatibility for aesthetic reasons[6]. When ZrO_2 is doped by rare earth metals, it shows an outstanding structural stability, high thermal conductivity and irradiation stability[7]**.** Many rare earth elements have been tried, in the past, to be implanted to ZrO_2, to improve its optical properties, but few paper of the transition element have been concerned with antibacterial property of Nps implanted in the ZrO_2[8-10]. Amid from many rare earth element Ceria Nps is less toxic and have wide range of engineering and biological application[11-14]. Recently, Ce Nps have emerged as a fascinating material in biomedical science due to their ionic radii and ability to switch oxidation states between III and IV based on environmental conditions[15]. Nano ceria with the ionic radii for Ce^{3+} is (1.14 Å), for Ce^{4+} (0.97 Å) and the switch between oxidation state is comparable to that of biological antioxidants[16]. Newly, researchers have begun to investigate the antibacterial property of Ceria and Ce-doped materials and their use in biological applications[17].To our knowledge, there is no much work has been done on the antibacterial activity of ZrO_2 Nps.

Recently, a variety of methods based on wet chemical routes have been extensively employed to synthesis of ZrO_2 Nps, the Co-Precipitation method is a simple, cost- effective, gives high purity

IOP Conf. Series: Materials Science and Engineering **360** (2018) 012014 doi:10.1088/1757-899X/360/1/012014

products, non-toxic and there is no need for any special instruments[18,19]. The present study aims to investigate the influence of the tetravalent Ceria doped ZrO_2 Nps on its structural, morphology, optical property and antibacterial activity Nps.

2. Experimental

2.1 Synthesis of Ce-doped ZrO_2 Nps

Ceria-doped ZrO_2 Nps were prepared by co-precipitation method using urea (NH_2CONH_2) as the precipitator. The reagents are Zirconyl nitrate hydrate ZrO $(NO_3)_{2.X}H_2O$ and Cerium (III) Nitrate Hexahydrate $CeN_3O_9.6H_2O$. The initial compositions are 0.2 mole of ZrO $(NO_3)_{2.X}H_2O$ and 0.012 mole of $CeN_3O_9.6H_2O$. First, the reagents together were dissolved in methanol (150ml) under stirring for 1 hr. Then urea was added into the above colloidal solution. Then the solution of the mixture was dried at 100°c for 24 hrs in a muffle furnace until the white powders were obtained. Subsequently, the obtained powder was calcined for 600°c for 6hrs respectively.

2.2 Characterization Technique

X-ray diffraction study was carried out for the sample by X-PERT PRO Diffractometer operated at 40kV with a current of 30mA using Cu-Kα (λ=0.1540nm), employing the scanning rate of 0.02°/s in the 2θ ranging from 20-70°. Surface morphology of the ZrO_2 nano samples were examined by SEM on a JOEL JSM 6390 Scanning Electron Microscope operated at an accelerating voltage at 15kV.

2.3 Evaluation of antibacterial activity of Ce-doped ZrO_2 Nps

The antibacterial activity of pure and Ce-doped ZrO_2 Nps were examined under two Gram Positive bacterial pathogens such as (S. aureus, B. subtilis) and two Gram Negative bacterial pathogens such as (K. pneumoniae, P. aeruginosa) by Agar Diffusion Method. These bacterial strains were grown in nutrient broth at 37°c for 24hrs or until visible growth was established. The antibacterial activity was demonstrated by the diameter of the zone of inhibition developed in and around the sample. Zone of inhibition is the area in which the bacterial growth is stopped due to the bacteriostatic effect of the compound and it measures the inhibitory effect of the compound towards particular microorganisms.

3. Result and Discussion

3.1 XRD analysis

The XRD patterns of pure and Ce-doped ZrO_2 precursors were shown in Fig.1. For all the precursors, the broadened humps are observed whether ceria doping or not, indicating their amorphous nature. The result demonstrates ceria doping has no obvious influence on the phase structure of the precursor. In spite of the samples, the crystal planes can be indexed to a zirconia phase with monoclinic (JCPDS No: 37-1484) and tetragonal (JCPDS No: 79-1769) structures. However, from Fig 1-a shows the m-ZrO_2 but in Fig 1-b shows there was shifting takes place from m-ZrO_2 to t-ZrO_2. Moreover, the crystallite size of nano powders was calculated by Debye Scherer's equation,

$$D=K\lambda/\beta \mathrm{Cos}\theta \rightarrow (1)$$

The specific surface area of the as synthesized samples were calculated by using the formula,

$$S= 6/ D^*\rho \rightarrow (2)$$

Where, S – Specific Surface area (m^2/g), D – average particle size, ρ – density of ZrO_2 (g/ cm^{-3}) and both the calculated values are shown in Table 1.

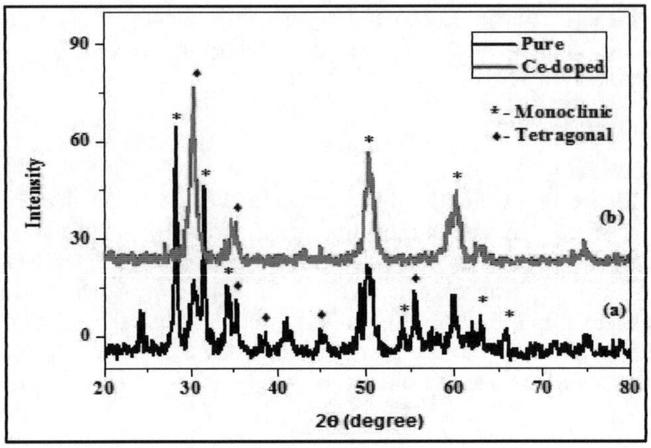

Fig. 1. XRD patterns of a) Pure $ZrO_2$and b) Ce-doped Zirconia Nano crystallites.

Powder	Crystallite Size (nm)	Specific Surface Area (m^2/g)
Pure	12.09	87.37
Ce-doped	8.52	123.98

Table 1 Crystallite Size of nano-sized Zirconia by XRD analysis.

3.2 SEM analysis

SEM analysis is employed to visualize the size and shape of the pure and Ceria-doped ZrO_2 Nps. SEM micrographs of both pure and doped ZrO_2 Nps under different magnifications are shown in Fig. 2 (a-b). From Fig.2. (a-b) the particles observed are clumpy agglomerated and spherical in shape with smooth and fused surface. The unvarying size distribution of the particles may be due to the sintering process which allows the growth by aggregation of particles through their grain boundaries.

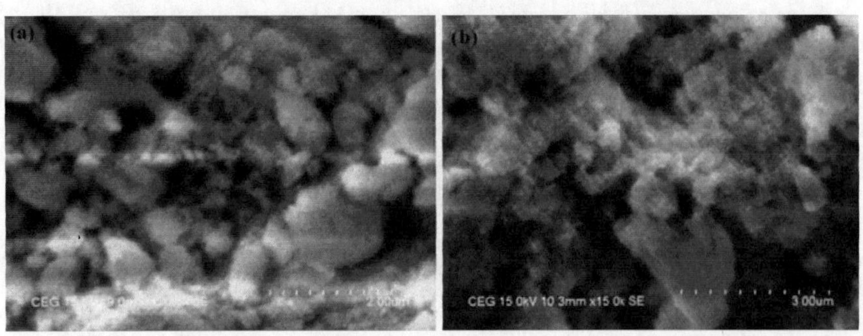

Fig.2. SEM micrograph of a) Pure ZrO_2 b) Ce-doped ZrO_2 by Co-Precipitation Method.

3.3 Antibacterial Assessment of Pure and Ce-doped ZrO_2 Nps

In the present Study, pure and Ce-doped ZrO_2 Nps are studied extensively to explore their utility as a potential antibacterial agent. The antibacterial activity has been studied against (G+) bacteria such as

IOP Conf. Series: Materials Science and Engineering **360** (2018) 012014 doi:10.1088/1757-899X/360/1/012014

(S. aureus, B. subtilis) and (G-) bacteria such as (K. pneumoniae, P.aeruginosa) using the concentration 50µg/ml. Both pure and doped sample pronounced significant growth inhibitory effect against both (G+) and (G-) bacteria due to their large surface area by their nano size which was shown in Fig.3. Zone of inhibition values determined for the pure and doped samples was shown in Table.2. The antibacterial property shows more efficient for Ce-doped ZrO_2 Nps compared to pure ZrO_2 in (G+) bacteria (i.e. S. aureus) than in (G-) bacteria which are clearly visualized in the antibacterial photographs. This difference in antibacterial performance may be due to the following reasons: the Ce-doping may enhance the ROS generation capability of ZrO_2 Nps. In the case of Ce-doped ZrO_2 sample, the probability of electron-hole pair recombination may be reduced due to the trapping of electrons by Ce ions by the following process[20].

$$Ce^{4+} + e^{-} \rightarrow Ce^{3+}$$
$$Ce^{3+} + O_2 \rightarrow O_2^{-} + Ce^{4+}$$

The mechanisms of the antibacterial activity of ions are not fully explained. However, the literature provides information on three hypothetical mechanisms[21] (Figure.4). Firstly, ceria and zirconia ions are penetrating into the bacterial cell; it will lead to affect the production of intracellular ATP. When the ATP production is started to affect, it will tends to disrupt the process of DNA replication. Secondly, amassing of ions in the cell membranes of bacteria, and thus with changes in their permeability. The transportation of protons through the cell membrane is prohibited. Hence, it leads to the destruction of the cell membrane and the death of the bacterial cell. Thirdly, the ion induction of reactive oxygen species (ROSs). Oxygen radicals are able to react with the components of the membrane, cell wall of bacteria, and as well as on other cell components (e.g., mitochondria), causing irreversible changes in their structure and thus the death of the bacterial cell.

Hence, the augmentation in the antibacterial activity of ZrO_2 Nps with Ce doping may be attributed to the electrons trapped in the Ce sites. The trapping electrons will lead to more ROS generation, resulting in higher antibacterial activity of Ce-doped ZrO_2 Nps. From the above probable discussions we clearly came to know about the enhanced antibacterial activity of pure and doped ZrO_2 Nps.

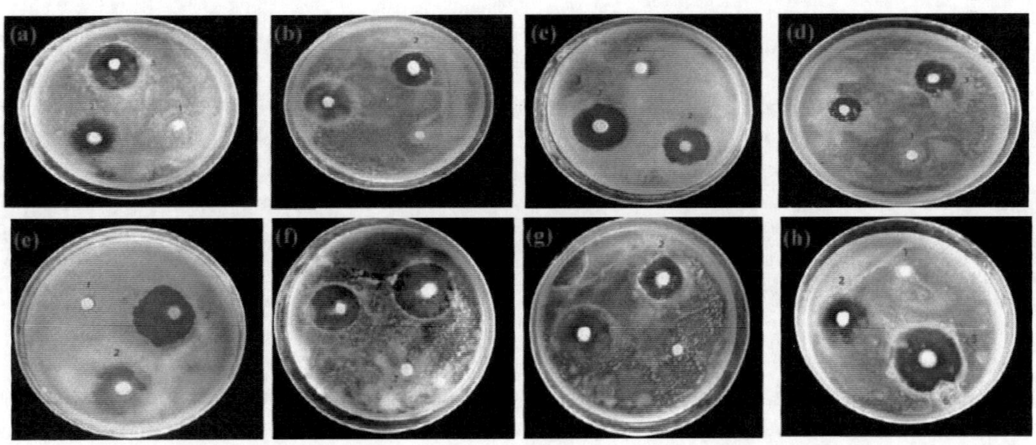

Fig. 3. Antibacterial activity of pure and Ce - doped ZrO_2Nps against (G+) bacterial pathogens such as *(i)S.aureus(ii)B.subtilis*(a,b) Pure and (e,f) Ce-doped $ZrO_2$and (G-) bacterial pathogens such as *(i) K.pneumoniae and (ii) P.aeruginosa* (c,d) Pure and (g,h) Ce-doped ZrO_2 nanomaterials

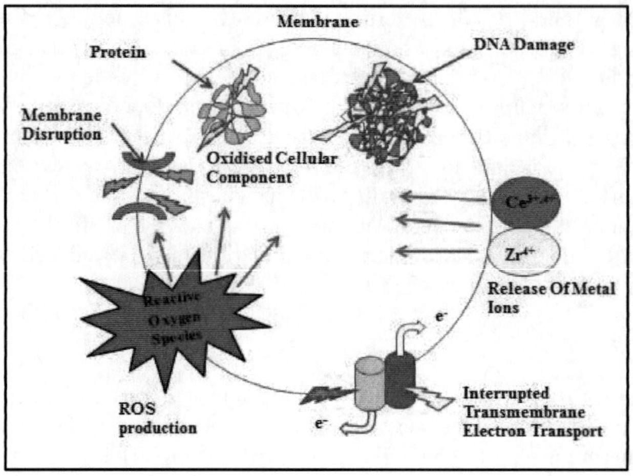

Fig.4. Antibacterial mechanism of Nps and their ions, adapted from with permission.

Samples	Zone of Inhibition (mm)			
	Bacterial Pathogens			
	Gram Positive		Gram Negative	
	Bacillus Subtilis	Staphylococcus Aureus	Klebsiella Pneumonia	Pseudomonas Aeruginosa
Pure	14	13	12	8
Ce-doped	15	16	15	11

Table 2 Zone of inhibition (mm) values against test organisms.

4. Conclusion

In summary, the pure and Ce-doped ZrO_2 Nps via a simple chemical co-precipitation technique. The XRD results have shown the formation of monoclinic and tetragonal structure of ZrO_2 for all samples. The SEM images have revealed the formation of spherical like morphology for pure and doped sample. The Ce-doped ZrO_2 Nps have been observed to be more effective than the pure ZrO_2 Nps both against (G+) i.e., S. aureus and (G-) such as K. pneumonia bacteria. Interestingly, it has been found that S. aureus bacterium can be completely eradicated by application of Ce-doped ZrO_2 Nps.

Acknowledgement

The author is grateful to the Department of Science and Technology (DST) for extending financial assistance to carry out this work.

References

[1] Gu F X, Karnik R, Wang A Z, Alexis F, Nissenbaum E L, Hong S, Langer R S and Farokhzad O C, 2007 *Nano today* **2** 14.

[2] Patil K C, Hedge M S, Rattan T, Aruna S T, 2008 *World Scientific*, ISBN-13 978-981-279-314.

[3] Li Z, Lee W E, and Zhang S, 2007 *J. Am. Ceram. Soc.,* **2** 364.
[4] Mucillo R, Buissa Netto R C, and Mucillo E N S, 2001 *Matter. Lett.* **49** 197.
[5] Costa G C C and Muccillo R, 2008 *Solid State Ionics*, **179** 1219.
[6] Martin Stefanic, Kristoffer Krnel, Tomaz Kosman, Engineering Ceramic Department, JJ. S. Institute, Ljublijana, Slovenia.
[7] Aytimur A, Uslu I, Kocyigit S, Ozcan F, 2003 *Ceramics International*, **38** 51.
[8] Feng Gu, Shu Fen Wang, Meng Kai Lu, Guang Jun Zhou, Su Wen Liu, Dong Xu, Duo Rong Yuan, 2003 *Chemical Physics Letters* **380** 185.
[9] Masonori Hirano, Hayato Dozono, 2013 *Journal of Solid State Chemistry* **204** 355.
[10] Ignacio Azocar, Esteban Vargas, Nicole Duran, Abel Arrieta, Evelyn Gonzalez, Jorge Pavez, Marcelo J. Kogan, Jose H. Zagal, Maritza A. Paez, 2012 *Materials Chemistry and Physics* **137** 396.
[11] Tarnuzzer R W, Colon J, Patil S, Seal S, 2012 *Nano Lett.* **5** 2573.
[12] Chen J, Patil S, Seal S, McGinnis J F, 2012 *Nat. Nanotechnology.***1** 142.
[13] Cai X, Sezate S A, Seal S, McGinnis J F, 2012 *Biomaterials* **33** 8771.
[14] Das M, Patil S, Bhargava N, Kang J F, Riedel L M, Seal S, 2007 *Biomaterials* **28** 1918.
[15] Karakoti A S, Kuchibhatla S V N T, Babu K S, Seal S, 2007 *J. Phys. Chem.* C **111** 17232.
[16] Korsvik C, Patil S, Seal S, Self W T, 2007 *Chem. Commun.* **10** 1056.
[17] Das S, Singh S, Dowding J M, 2012 *Biomaterials,* **33** 7746.
[18] Gambhire A B, Lande M K, Kalokhe S B, Shirsat M D, Patil K R, Gholap R S, Arbad B R, 2008 *Materials Chemistry and Physics* **112** 719.
[19] Geethalakshmi K, Prabhakaran T and Hemalatha J, 2012 *World Academy of Science, Engineering and Technology*, **6** 04.
[20] Tariq Jan, Javed Iqbal, Qaiser Mansoor, M Ismail, M Sajjad Haider Naqvi, Asmagul, S. Faizan-Ul-Hassan Naqvi and Fazal Abbas, 2014 *Journal of Physics D*: Applied Physics **47** 355301.
[21] Joanna Kolmas, Ewa Groszyk, and Dagmara Kwiatkowska-Rozycka, 2014 *Hindawi Publishing Corporation Biomed Research International*, Article ID 178123, **15** pages, http://dx-doi.org/10.1155/2014/17812.

IOP Conf. Series: Materials Science and Engineering **360** (2018) 012015 doi:10.1088/1757-899X/360/1/012015

Molybdenum Disulphide Nanoparticles Synthesis Using a Low Temperature Hydrothermal Method and Characterization

Manuja M, Sarath Krishnan V, Gijo Jose*

Post Graduate and Research Department of physics, St. Berchmans College, Changanacherry Kottayam-686101, Kerala, India

**Corresponding author Email:* gijojose@sbcollege.ac.in

Abstract. Molybdenum disulphide nanoparticles having crystallite size of 5.55nm, were synthesized by low temperature hydrothermal method. X-ray Photoelectron Spectroscopy (XPS) was employed to probe into the energy levels of the synthesized Molybdenum disulphide nanoparticle. Findings of XPS suggest the presence of Mo^{6+}, Mo^{4+} and Mo^{2+} oxidation states of Molybdenum 3d states. The Binding energy (BE) of sulphur was found to be 168.80 and 161.72 eV for S $2P_{1/2}$ and S $2P_{3/2}$ respectively, and the atomic percentage of Mo and S are in the ratio 1: 2.33, which patch up with the theoretical values.

Keywords: Molybdenum disulphide, Hydrothermal method, Oxidation state, XPS.

1. Introduction

The hydrothermal technique is one of the most important tools for advanced materials processing of mono-dispersed nanoparticles of various metal oxides and metal sulphides. Transition metal chalcogenides are now receiving researcher's attention owing to its potential advantages in catalysis [1], lubricant applications [2], super capacitor, lithium ion [3] battery storage, photo detector [4], and other optoelectronic devices [5]. Transition metal oxides like MoO_3, WO_3 etc are chromogenic and can be implemented into solar cells in order to improve their efficiency [6]. Due to the easy tunability of band gap, with size and number of layers nanostructured MoS_2 finds applications in various optoelectronic devices.

Alike graphene, MoS_2 and WS_2 have enormously deferent properties in reduced layer number limits than bulk materials [7]. MoS_2 having a controllable band gap with magnitude between 1.5–2.0 eV makes it superior over graphene of zero band gap. It follows a layered structure, where individual S-Mo-S layer (monolayer MoS_2) consists of one plane of Mo atoms sandwiched between two planes of S atoms [8]. Since Mo atoms reside between two S atoms, the localized direct bandgap states from Mo d orbitals are much less sensitive to out-of-plane interactions than the delocalized indirect bandgap states which include S p orbitals.

2. Experimental Details

The low cost, low temperature hydrothermal method enables synthesis of good quality crystals with desired composition owing to the precise control that can be established over composition and different reaction parameters during hydrothermal synthesis. Economic feasibility, relative simplicity in techniques involved and easily attainable temperature conditions are the key features of hydrothermal process which motivated us to prefer this method.

All reagents were of analytical grade and used without further purification. 1.21 gm of sodium molybdate ($Na_2MoO_4.2H_2O$), 0.14 gm of PEG 20,0000 and 1.56 gm of thiourea

IOP Conf. Series: Materials Science and Engineering **360** (2018) 012015 doi:10.1088/1757-899X/360/1/012015

(NH_2CSNH_2) were dissolved simultaneously in 30 ml of distilled water and the resulting solution was transferred into 100 ml Teflon jar and sealed in an autoclave kept in furnace at 200 °C for 11hours and then cooled to room temperature naturally. The black precipitate obtained is centrifuged and washed several times in distilled water and ethanol then dried in air. Black powder was obtained after drying the sample 80 °C for about 8 hours.

Crystal structure and phase of sample were investigated through X-ray diffraction technique using Cu Kα_1 line of wavelength 1.5406A° using Bruker X-ray diffractometer with a step size of 0.02 and a scan rate of 3^0 per minute. Spectrum is recorded between the 2θ values from 10 to 90^0. Chemical purity and the composition of the materials were examined with the Energy Dispersive X-Ray Spectroscopy (EDS) techniques using Bruker electron microscope analyser. X-Ray Photoelectron Spectroscopy (XPS) were recorded using a multiprobe system having a Mg/Al X-Ray source and a hemispherical analyser operating in a Constant Analyser Energy (CAE) mode at Omicron nanotechnology.

3. Results and discussion

The obtained XRD and corresponding JCPDS spectra are shown in Figure 1 and the data tabulated in Table1. The peaks are compared with JCPDS card no 89-5112 and the rhombohedral phase of MoS_2 was confirmed. Different peaks at 2θ values 13.69^0, 32.85^0, 35.61^0, 42.30^0, 57.54^0 correspond to hkl planes (003), (101), (012), (015), (110) respectively.

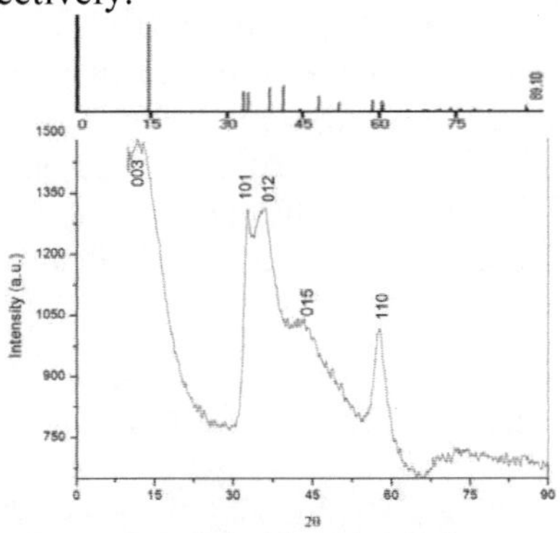

Fig 1. XRD spectra of Rhombohedral MoS_2 along with its corresponding JCPDS peaks.

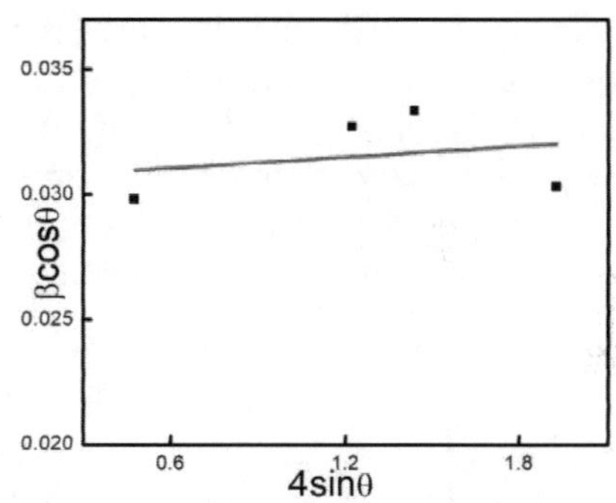

Figure 2. Williamson -Hall plot of MoS_2.

The broadened peak of the XRD spectra signifies the smaller crystallite size of the sample and is calculated by exploiting Scherrer equation which gives a value 5.55nm. The modified Scherrer formula [9]

$$\beta_{total}\cos\theta = \left(\frac{0.9\times\lambda}{L}\right) + 4d\sin\theta$$

Table 1 Calculation of crystallite size

2θ(degrees)	Intensity	d value (A^0)	hkl	FWHM	L(nm)
13.69	999	6.12	003	1.72	04.65
32.85	226	2.71	101	0.82	10.14
35.61	207	2.62	012	1.97	04.23
42.30	286	2.19	015	2.05	14.17
57.55	126	1.58	110	1.98	14.57
Average crystallite size 5.55nm					

is applied to the system to obtain a crystallite size which caters the presence of lattice strain experienced because of lattice defect. From W-H plot shown in Figure 2 it was found to be 4.53 nm with a compressive strain of 0.000749.

The energy Dispersive X-ray Spectroscopy (EDS) further confirmed the composition of the sample as molybdenum and sulphur. The graphical representation is shown in Figure 3.

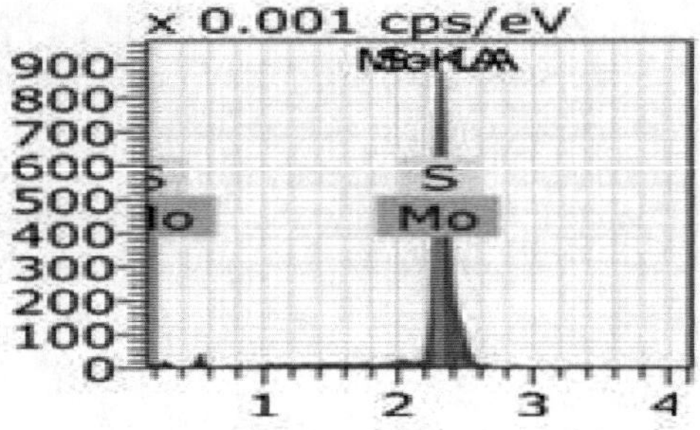

Figure 3. EDS Spectrum of MoS_2.

In order to further understand the elemental identification and chemical state of the elements, X-ray photoelectron spectroscopy (XPS) measurements were carried out and Mo 3d, S 2s, and S 2p regions were analyzed [10].The XPS recorded as such is shown in Figure 4(a), which indicates the presence of Mo, S and C atoms with the corresponding oxidation states. Elemental identification is done using the separation between the two peaks (for p, d and f states) and the peak area ratio of the XPS recordings. C 1s binding energy of carbon contamination occurs at 284 eV [11].Multiplet splitting is prominent in atoms having unpaired d electrons. The broad and strong peak at Mo $3d_{5/2}$ indicates the existence of more than one oxidation state for Mo.

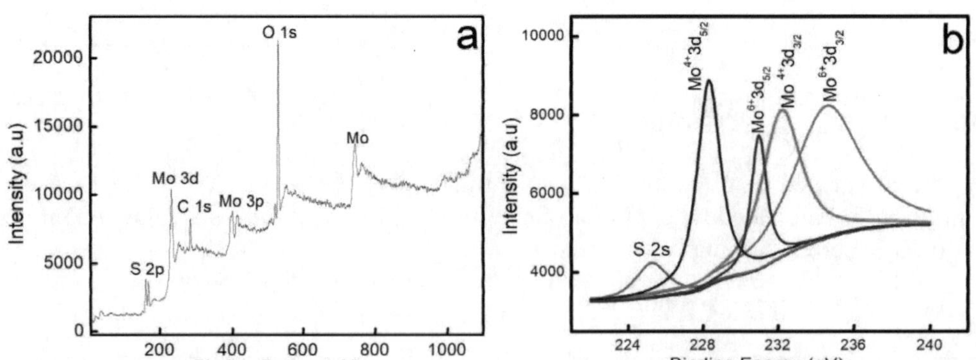

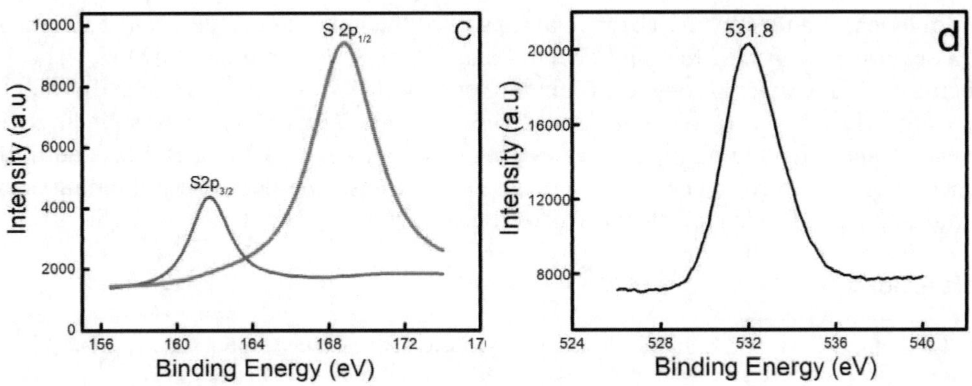

Figure 4. (a) XPS survey spectra of MoS_2 samples (b) Deconvoluted XPS spectra of MoS_2 Mo 3d states (c) Deconvoluted XPS spectra of MoS_2 S 2p states (d) XPS spectra of O 1S state.

The primary objective of studying the Mo 3d spectrum is to determine the relative percentages of different oxidation state present at the surface [12]. The deconvoluted spectra of Mo 3d are shown in Figure 4(b). Peaks assigned to Mo^{4+} $3d_{5/2}$ (228.31 eV) and Mo^{4+} $3d_{3/2}$ (232.19 eV), the expected values for Mo^{4+} in MoS_2. It shows a small peak at around 234.57 eV, and 231 eV which corresponds to Mo^{6+} $3d_{3/2}$, Mo^{6+} $3d_{5/2}$ respectively from Mo oxidation [13]. The corresponding S 2p peak consists of a doublet of S $2P_{1/2}$ (168.80 eV) and S $2P_{3/2}$ (161.72 eV), consistent with the S^{2-} type present in MoS_2. S $2P_{3/2}$ and S $2P_{1/2}$ spin orbit doublet separated by 7.08 eV with intensity ratio 1:1.1. Mo $3d_{5/2}$ and Mo $3d_{3/2}$ are separated by 3.88 eV for Mo^{4+} oxidation state and 3.57 eV for Mo^{6+} oxidation states with intensity ratio 5:6.

Table 2 BE and % Area of peaks corresponding to different oxidation states

	BE in eV	FWHM	Area in %
Mo^{2+}	225.25	2.05	02.47
Mo^{4+} $3d_{3/2}$	232.19	2.51	12.61
Mo^{4+} $3d_{5/2}$	228.31	1.48	10.38
Mo^{6+} $3d_{3/2}$	234.57	4.15	18.81
Mo^{6+} $3d_{5/2}$	231.00	1.00	04.58
S^{2-} $2p_{1/2}$	168.80	3.97	41.28
S^{2-} $2p_{3/2}$	161.72	2.50	09.86
C 1s	284.00		
O 1s	531.80		

The complete XPS analysis reports including different oxidation states, multiplet splitting and percentage area corresponding to BE is tabulated in table 2. This record confirms the fact that the addition of valence electron decreases binding energy. Trace amounts of MoO_3/MoO_2 present in the sample may reduce Mo (VI) to Mo (V) species under X-ray bombardment. The atomic percentage calculated shows that Mo and S are in the ratio 1: 2.33 and it is in good agreement with the theoretical results. Also the percentage of Mo^{6+}/ Mo^{4+} is found to be 1.02.

4. Conclusion

Molybdenum disulphide nanoparticles were prepared by hydrothermal method. The obtained sample is confirmed as 3R (Rhombohedral) structures with three repeating layers of molybdenum and sulphur. The crystallite size of the material calculated as 5.55nm. The EDS spectrum confirmed the molybdenum and sulphur compositions for MoS_2. The XPS spectrum for the material suggest a binding energy of 232.19 eV and 228.31 eV for Mo $3d_{3/2}$ and Mo $3d_{5/2}$, of Mo^{4+} state and 234.57 eV and 231 eV for corresponding Mo^{6+} states. S^{2-} states having binding energies 168.8 eV and 161.72 eV for S $2P_{1/2}$ and S $2P_{3/2}$ respectively, which reconcile with the theoretical values. The decrease in binding energy with the change of oxidation states is also studied.

References

[1]Chen, J., Li, S. L., Xu, Q., & Tanaka, K. (2002). Synthesis of open-ended MoS 2 nanotubes and the application as the catalyst of methanation. *Chemical Communications*, (16), 1722-1723.

[2]Rapoport, L., Bilik, Y., Feldman, Y., Homyonfer, M., Cohen, S. R., & Tenne, R. (1997). Hollow nanoparticles of WS2 as potential solid-state lubricants. *Nature*, *387*(6635), 791-793.

[3]Li, X. L., & Li, Y. D. (2004). MoS2 nanostructures: synthesis and electrochemical Mg2+ intercalation. *The Journal of Physical Chemistry B*, *108*(37), 13893-13900.

[4]Lopez-Sanchez, O., Lembke, D., Kayci, M., Radenovic, A., & Kis, A. (2013). Ultrasensitive photodetectors based on monolayer MoS2. *Nature nanotechnology*, *8*(7), 497-501.

[5]Wang, Q. H., Kalantar-Zadeh, K., Kis, A., Coleman, J. N., & Strano, M. S. (2012). Electronics and optoelectronics of two-dimensional transition metal dichalcogenides. *Nature nanotechnology*, *7*(11), 699-712.

[6]Simchi, H., McCandless, B. E., Meng, T., Boyle, J. H., & Shafarman, W. N. (2013). Characterization of reactively sputtered molybdenum oxide films for solar cell application. *Journal of Applied Physics*, *114*(1), 013503.

[7]Quinn, M. D., Ho, N. H., & Notley, S. M. (2013). Aqueous dispersions of exfoliated molybdenum disulfide for use in visible-light photocatalysis. *ACS applied materials & interfaces*, *5*(23), 12751-12756.

[8]Windom, B. C., Sawyer, W. G., & Hahn, D. W. (2011). A Raman spectroscopic study of MoS2 and MoO3: applications to tribological systems. *Tribology Letters*, *42*(3), 301-310.

[9]Yang, H., Shen, C., Song, N., Wang, Y., Yang, T., Gao, H., & Cheng, Z. (2010). Facile synthesis of hollow nano-spheres and hemispheres of cobalt by polyol reduction. *Nanotechnology*, *21*(37), 375602.

[10]Gopalakrishnan, D., Damien, D., Li, B., Gullappalli, H., Pillai, V. K., Ajayan, P. M., & Shaijumon, M. M. (2015). Electrochemical synthesis of luminescent MoS 2 quantum dots. *Chemical Communications*, *51*(29), 6293-6296.

[11]Peng, Y., Meng, Z., Zhong, C., Lu, J., Yu, W., Jia, Y., & Qian, Y. (2001). Hydrothermal synthesis and characterization of single-molecular-layer MoS2 and MoSe2. *Chemistry Letters*, *30*(8), 772-773.

[12]John, R. E., Chandran, A., Thomas, M., Jose, J., & George, K. C. (2016). Surface-defect induced modifications in the optical properties of α-MnO 2 nanorods. *Applied Surface Science*, *367*, 43-51.

[13]Bang, G. S., Nam, K. W., Kim, J. Y., Shin, J., Choi, J. W., & Choi, S. Y. (2014). Effective liquid-phase exfoliation and sodium ion battery application of MoS2 nanosheets. *ACS applied materials & interfaces*, *6*(10), 7084-7089.

Synthesis of nano Lead Tungstate ($PbWO_4$) by Single Step Modified Combustion Process and Characterization for their application as LTCC and Optical Material

Bincy Joseph, Bincy Abraham and Jijimon K Thomas

Electronics Material Research Laboratory, Department of Physics, Mar Ivanios College(Autonomous), Thiruvanathapuram, Kerala, 695015, India.

Corresponding author:jkthomasemrl@yahoo.com

Abstract

NanostructredScheelite$PbWO_4$were synthesized through a single step modified auto-ignition combustion technique. The X-ray diffraction, Fourier transform Raman and infrared spectroscopy studies revealed that the as-prepared samples were phase pure and the average crystallite size was ~49 nm. $PbWO_4$ crystallizes in tetragonal structure. The particle size and morphology of the sample was studied using transmission electron micrograph. The optical properties of the sample were studied using Ultraviolet-visible and photoluminescence spectra. The samples showed a maximum transmission in the visible regions but poor transmittance in the ultraviolet region. The optical band gap of the samples calculated using the Kubelka-Munk method was 3.87 eV, suggesting their semiconducting behavior. The photoluminescence spectra recorded on the samples showed that nano $PbWO_4$ has visible luminescence. The thermal stability of the sample at elevated temperatures has been studied using thermo gravimetric analysis (TGA) and differential thermal analysis (DTA). No substantive weight loss were observed for the samples in the temperature range 50-900°C. The $PbWO_4$ samples were sintered to 98% relative density at very low temperature of 787°C without using any sintering aid. The microstructure of the sintered pellets observed by Scanning Electron Micrograph (SEM) showed that the pellets were well sintered with minimum porosity. The dielectric constant (ε_r) of the sintered $PbWO_4$ sample was recorded using an LCR meter in the range 50 Hz to 5 MHz and showed a value of 86 at 1 MHz with a loss factor of 10^{-2}. The calculated temperature coefficient of dielectric constant is negative. The low sintering temperature, dielectric constant, low loss factor and negative temperature coefficient make the sample ideal for low temperature co-fired ceramic, electronic substrate materials as well as suitable for various optical applications such as UV filters, transparent films for window layers on solar cells, anti-reflection coatings, scintillators, detectors.

Key words: $PbWO_4$, LTCC, dielectric, sintering, combustion synthesis

Introduction

The emerging nano science and nanotechnology stimulateextensive research interest in nano structured metal oxidesfor potential applications. Particularly, the field of transition metal oxides represents an exciting and rapidly expandingresearch area that spans the border between the physical and engineering sciences. The main advantage of nano materials is that its properties can be fine-tuned with technical requirements. The better flexibility of nano materials to mould into different shapes and morphology compared with the bulk materials is an added advantage.Low temperature co-fired ceramic (LTCC) technology has attracted remarkable attention in fabrication of sensors and micro-fluidic devices in the last decade [1] and micro-channels are desired within the sensor module. Several works for the fabrication and on the study of LTCC material has been done[2-7].

$PbWO_4$ is an n-type semiconductor oxide material which is widely studied for their vast potential application in the various fields of science and technology. These oxides have drawn a great deal of attention of researchers due to its excellent electrochromicbehaviour [8, 9], photochromic behaviour [10,11], gasochromic behavior [12, 13] and stability towards UV and visible region which could found wide range of applications in the field ofelectrochromic devices (ECD) [14, 15], memory devices [16,17], smart windows [18] catalytic material [19] and display applications [20]. They are also good gas sensors. $PbWO_4$ is most

IOP Conf. Series: Materials Science and Engineering **360** (2018) 012016 doi:10.1088/1757-899X/360/1/012016

attractive for its high-energy physics application because of its high density (8.4gm/cm^3),short decay time(less than 10 ns for a large light output), high irradiation damage resistance(10^7 rad for undoped), small moliereradius, fast decay time, non hygroscopicity and low production cost [21-22]. $PbWO_4$ shows interesting excitonic luminescence, thermoluminescence and simulated Raman scattering behavior [15-16]. Lead tungstate crystals are fast and dense scintillators found an application for high energy electromagnetic calorimeter of Compact Muon Solenoid (CMS) detector at Large Hadron Collider (LHC) at Center of Europe for Research Nuclear(CERN).

Synthesis and characterization of snowflake-like and bamboo- leaf like $PbWO_4$nanocrystals was successfully demonstrated [23]. A study on the properties of lead tungstate crystals, the longitudinal optical transmittance and light attenuation length, light yield and response uniformity, emission spectra and decay time was investigated [24]. The radiation resistance of large crystals and possible curing with optical bleaching are discussed. The results of an in depth materials study, including trace impurities analysis, are also presented. The general conclusion from this investigation is that further research and development is needed to develop fast, radiation hard $PbWO_4$ crystals for the CMS experiment at the CERN LHC.

Experimental

Synthesis

In the present work, nanopowder was prepared through a modified combustion process. As usual procedure, in combustion technique, alcohol was used as the complexing agent and fuel respectively. In such case high temperature annealing of the as-prepared powder was required to get a phase pure powder [25-28]. But in the present modified combustion method, citric acid was used as the complexing agent instead of poly vinyl alcohol. By this change of complexing agent it is possible to obtain single phase nanoparticles in a single step combustion process without the usual calcinations for prolonged duration at high temperature. The main advantage of this method is that the as prepared powder itself shows phase formation. Single step process, i.e. no pre or post calcinations is required.Same fuel + oxidizer + complexing agent combination can be used to prepare simple oxides, perovskites, mixed oxides ,ceramics, aluminates, ferrites , hafnates, scheelitesetc. As prepared powder itself is phase pure.Narrow particle size distribution and uniform morphology.Particle size, composition and homogeneity of the products can be controlled. No particular heating rate is needed, no special crucibles or containers or arrangements are required. Easy to handle and economical both in cost and time.

In the present work we have synthesized single phase Nano Lead Tungstate using a modified combustion technique. Lead Nitrate and Ammonium para tungstate dissolved in double distilled water were used as starting reagent for the preparation of nanocrystalline $PbWO_4$. Citric acid was then added to the solutions taken in separate beakers as complexing agent. Amount of citric acid was calculated based on total valence of the oxidizing and the reducing agents for maximum release of energy during combustion. Oxidant to fuel ratio of the system was adjusted to unity (~1) by adding concentrated nitric acid which serves as oxidizer and ammonium hydroxide solution as fuel. The precursor solution of pH ~ 7.0 was stirred well for uniform mixing without any precipitation or sedimentation. The solution was then heated using a hot plate kept at ~250 ^{0}C in a ventilated fume hood. The solution boils on continuous heating and undergoes dehydration accompanied by foam. On persistent heating the foam gets auto-ignited giving a voluminous fluffy powder of nano $PbWO_4$. The structure of the as-prepared powder was examined by the powder x-ray diffraction (XRD) technique using aPhilip Pro X-ray diffractometer with CuKα radiation as the incident beam.. The infrared (IR) spectra of the samples were recorded on a Perkin

Elmer spectrum 2 instrument recording a wavelength ranging from 375-2500 cm^{-1} and it consist of an optimized KBr beam splitter. The Fourier-transform Raman spectrum of the nanocrystalline$PbWO_4$ was measured at room temperature in the wavenumber range from 50 cm^{-1} to 1200 cm^{-1} using a Bruker RFS/100S spectrometer at power level of 150mW and resolution of 4 cm^{-1} Particulate properties of the combustion product were observed using transmission electron microscopy (TEM,model JEOL-JEM 2100 with an accelerating voltage of 200 kV) .The samples were excited using a Nd:YAG laser lasing at 1064 nm, and the scattered radiation was detected using a Ge detector. The photoluminescence spectra of the samples were measured using aHoriba Scientific Fluoromax-2 Spectrofluorometer and was studied between the wavenumber 310 cm^{-1} to 600 cm^{-1}. The photons from the source were filtered by an excitation spectrometer. The monochromatic radiation was then allowed to fall on the disc samples, and the resulting radiation was filtered by an emission spectrometer and then fed to a photomultiplier detector. The variation of intensity was recorded as a function of wavelength. Optical measurements of the nanopowder were performed at room temperature using aPERKIN ELMER λ–35,Singapore, is used for recording a wavelength ranging from 150-700 nm. The same solventwas used as blank for analysis to calibrate the spectrophotometer. To study the sinterability of the nanoparticles obtained by the present combustion method, as-prepared $PbWO_4$nanoparticles were mixed with 5% polyvinyl alcohol and pressed in the form ofcylindrical pellets of 12 mm diameter and 2 mm thickness at pressure of about 150 MPa using a hydraulic press. The pellet was then sintered at 787°C for 3 h (Standard four probe technique using Keithley source meter 2440). The surface morphology of the sintered sample was analyzed using scanning electron microscopy (SEM model JEOL-JSM 5610 LV, Tokyo, Japan). For low-frequency dielectric studies, pellets were made into the form of a circular disc capacitor with thespecimen as the dielectric medium. Both of the flatsurfaces of the sintered pellet were polished and electroded by applying silver paste. The capacitanceof the sample was determined using an LCR meter (3532-50 LCR Hi-Tester; Hioki) in the frequency range from 100 Hz to 5 MHz at different temperatures from 30°C to 500°C.

Results And Discussion

Structural Ananlysis

XRD patterns of the synthesized nanoparticles using combustion method with ammonia as fuels. All the peaks were indexed for their (h, k, l) planes in the spectrum. No secondary phase was seen in the XRD pattern which confirms that powder itself is phase pure. But the XRD of the as prepared sample showed the presence of some organic and other impurity peaks and hence the sample was annealed at 700^{0}C.

The XRD pattern of $PbWO_4$nanopowder isshown in Figure1.All the diffraction peaks was assigned to Scheelite phase with tetragonal structure with space group ($I4_{1/a}$ [88]), JCPDS card No.19-0708. The lattice parameters calculated from XRD pattern are a= 5.432Å, b= 5.432Å, c=12.064Å for $PbWO_4$ which are well agreeing with the standard value reported. The crystallite size is calculated from full width at half maximum (FWHM) of the peak using Scherrer formula. The average crystallite size of the as prepared sample is 49 nm. The description about the structure, crystallie size, lattice parameter of the sample is enclosed in Table 1. The highest relative intensity is obtained for [1 1 2] crystallographic plane of tetragonal crystal structure of Scheelite phase.

IOP Conf. Series: Materials Science and Engineering **360** (2018) 012016 doi:10.1088/1757-899X/360/1/012016

Table 1:

Compound	**Crystal Structure**	**Lattice parameter**						**Unit cell volume (Å)^3**	**Crystallite size (nm)**
		Standard (Å)			**Calculated (Å)**				
		A	B	C	a	b	C		
Lead Tungstate	Tetragonal	5.461	5.461	12.049	5.432	5.432	12.064	355.96	49

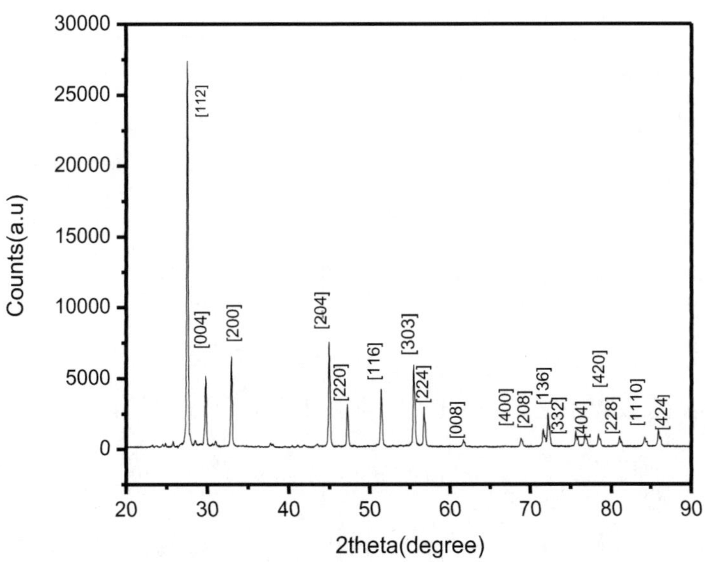

Figure 1: XRD pattern of $PbWO_4$nanopowder.

The extent of line broadening in the diffraction pattern due to the strain, which is caused by the non-uniform displacement of the atoms with respect to their reference lattice position, could be identified by Williamson–Hall (W–H) method [29].Figure 2 shows the Wiliamson–Hall plot for the $PbWO_4$

nanoparticles. The reciprocal of 'y' intercept, a measure of particle size, obtained from this plot was 45nm. The lattice strain constant 'g', proportional to the slope of the line, estimated as 5.76×10^{-5}. The positive slope of the line indicates that strain is tensile and is also a component contributing to the broadening of peaks in the XRD pattern.

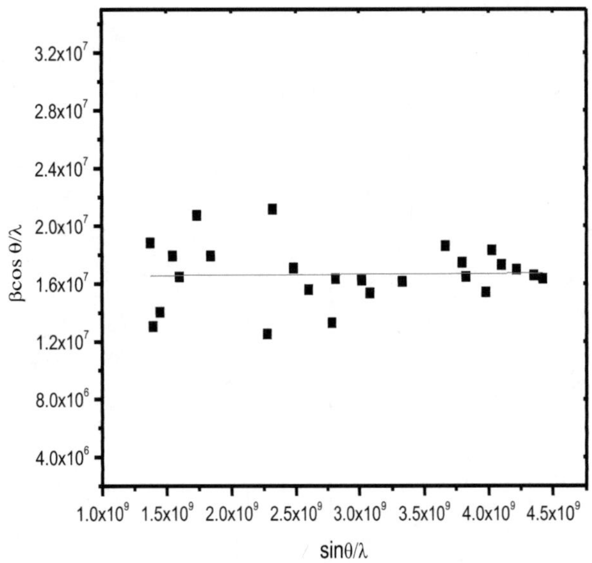

Figure:2 Williamsons Hall plot of nano $PbWO_4$

FT-IR spectra of the sample are recorded and are given in Figure 3.In the scheelite structure, W ions are in tetrahedral O^- ion cages and isolated from each other. Each Pb^{2+}cation is surrounded by eight oxygen ions. The connection between cationPb^{2+} and anion WO_4^{-2} is ionic and the connection between W^{6+} and O^{2-} in the oxyanion complex is covalent.The W atom in the scheelite structure is at the tetrahedral O-atom cages and isolated from one another and the number of O ion surrounding each Pb ion is eight. Figure 4 shows the FTIR spectra of the $PbWO_4$ lead tungstatenanoparticles. The observed absorption peaks are in good agreementwith the earlier reports [30].The FTIR spectra indicateintense band around 729 cm^{-1}. This band is due to the W-O-W stretchingvibration modes, characteristic of tetrahedral tungstate [31, 32]. The broadnessof this band is attributed to the characteristic nano nature of the lead tungstatecrystals.

$$\Gamma_{Td}=A_1(\nu_1)+E(\nu_2)+F_2(\nu_3)+F_2(\nu_4)$$

in which all four vibrational modes are Raman active ,Figure (3.13), but only the

$F_2(\nu_3, \nu_4)$ modes are IR active[33,34]. The band at 729 $cm^{-1}$are assigned to

$F_2(\nu_3)$ antisymmetric stretches vibrations for $PbWO_4$.

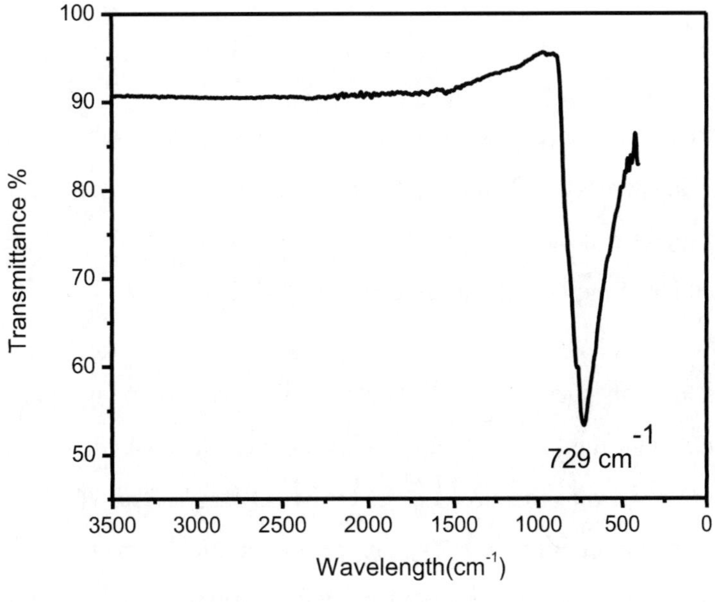

Figure 3. FT-IR Spectra of nano $PbWO_4$

$PbWO_4$ crystallizes at ambient conditions in the centro symmetricscheelite structure that has the space group ($I4_{1/a}$)[88], with four formula units per body-centered unit cell. The Pb and W atoms occupy S_4sites whereas the sixteen oxygen atoms are on general C_1sites. Group theoretical considerations[35-37]lead us to expect 13 Raman active modes at the τ point [38].

$$\tau = \nu_1(A_g) + \nu_2(A_g) + \nu_2(B_g) + \nu_3(B_g) + \nu_3(E_g) + \nu_4(B_g)$$

$$+ \nu_4(E_g) + R(A_g) + R(E_g) + 2T(B_g) + 2T(E_g). \text{---------------} (3)$$

The translational modes (T) and the rotational modes (R) are considered to be the external modes of the WO_4tetrahedraand are the lowest in frequency. The rest [(ν_1) to (ν_4)] are considered to be the internal modes of the WO_4tetrahedraand higher in frequency. The $A_g$and $B_g$modes are single, while $E_g$modes are doubly degenerated [39] .

To the best of our knowledge twelve of the thirteen modes of the scheelite-type phase of $PbWO_4$are known [40,41] the (ν_4) E_g internal mode being the only unknown one. This mode has been observed in $CaWO_4$, $SrWO_4$and $BaWO_4$ as a high-frequency shoulder of the (ν_4) B_g internal mode[37,40,41].

Figure.4 shows the Raman spectrum of Scheelitenano $PbWO_4$Raman spectrum ofscheelite is dominated by the (ν_1)A_g mode near 903 cm^{-1}. The peaks in the range 600–1000 cm^{-1} were assigned to the stretches of the W–O bands. The W–O bonds of intermediate length (about unit valency) are characteristic of bridging W–O bonds and are assigned to Raman mode stretching wavenumbers in the range 700–1000 cm^{-1}.The bands at 764, 715 $cm^{-1}$and 806, 845cm^{-1} are assigned to $F_2(\nu_3)$ antisymmetric stretches vibrationsfor $PbWO_4$[42-50].

A higher Raman stretching mode wavenumber indicates a more distorted structure, whereas a lower Raman stretching mode wavenumber indicates a more regular structure [51]. $PbWO_4$ present a strong peak at 800 cm^{-1}, the W–O bond of WO_4 tetrahedral in $PbWO_4$ catalysts [52-54].Table 2 encloses the Raman band assignments of nano $PbWO_4$.

Table.2:

$PbWO_4$assignments (cm^{-1})	Ramanband assignments
63	$T(E_g)$
75	T(Bg)
88	$T(E_g)$
176	$R(A_g)$
325	$\nu_2(A_g)$
355	$\nu_4(B_g)$
750	$\nu_3(E_g)$
903	$\nu_1(A_g)$

IOP Conf. Series: Materials Science and Engineering **360** (2018) 012016 doi:10.1088/1757-899X/360/1/012016

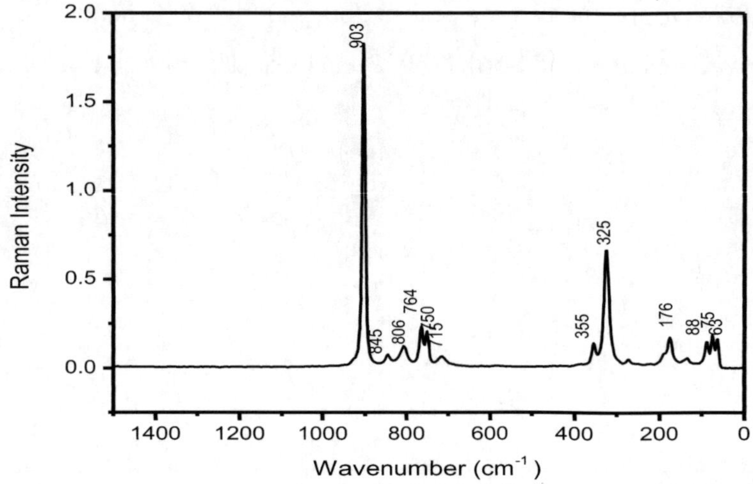

Figure.4 Raman spectra of nano $PbWO_4$

Powder Morphology

TEM images of the $PbWO_4$ samples prepared via solution combustion processes annealed at 700°C are given in Figure 5[a] and 5[b]. From the distinctly visible grains, the average grain size iscalculated as 24 nm. Corresponding to (1 1 2) plane of tetragonal crystal structure of $PbWO_4$ the interplanar spacing obtained from X-Ray diffraction plane is found to be in good agreement with the interplanar spacing, 0.322 nm for the same plane obtained from TEM image. The d-spacing for different crystallographic planes estimated from Figure 5[a] matches well with the d-spacing obtained from X-ray diffraction analysis.

IOP Conf. Series: Materials Science and Engineering **360** (2018) 012016 doi:10.1088/1757-899X/360/1/012016

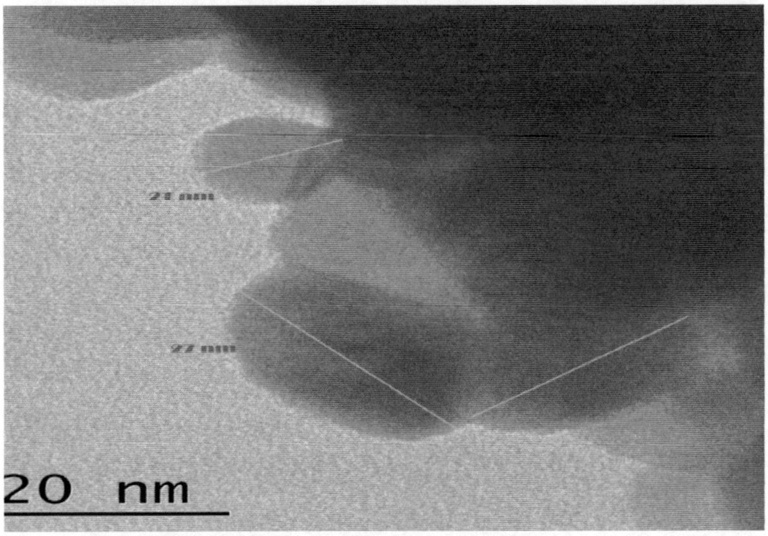

Figure.5(a) HRTEM image of a single grain

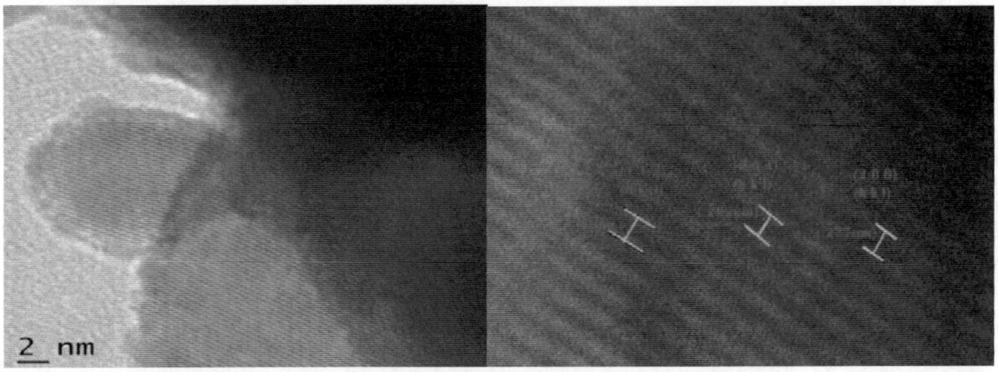

Figure.5(b) HRTEM image of different crystallographic planes.

Figure 6 shows the TGA/DTA curve of nano $PbWO_4$, in the temperature range from 30°C to 900 °C at a heating rate of 10 °C in nitrogen atmosphere. It shows no significant weight loss throughout from ambient to 900 °C. The TGA curve shows that the particles crystallize at room temperature itself and the thermal stability of the nanostructures are really good. No abrupt changes are observed in the TGA curve in the higher temperature range which indicate that no phase transitions are occurring and the $PbWO_4$ ceramics synthesized by the modified combustion synthesis is stable at elevated temperature range [23,55].

The DTA graph shows a small endothermic dip around 100 ^{0} C and 790^{0} C attributes to the evaporation of the absorbed moisture and the melting of $PbWO_4$ powder respectively, which is noted during the sintering of $PbWO_4$ powder

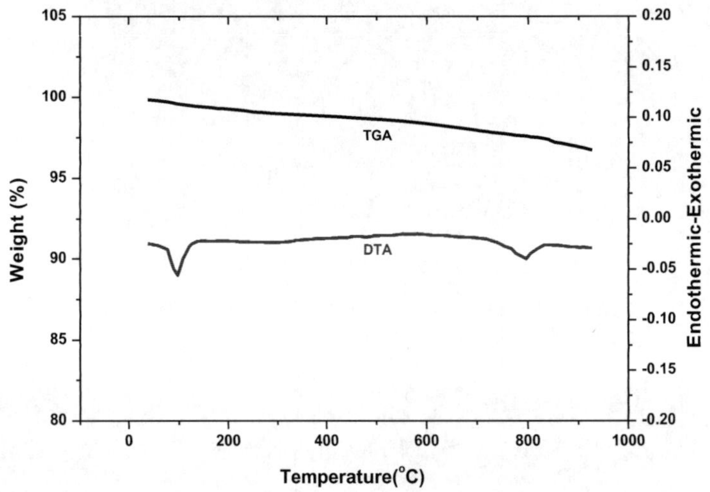

Figure: 6 TGA/DTA Curve of nano $PbWO_4$

Optical Study

The UV-Visible absorption, reflectance and transmission spectra of nano $PbWO_4$are shown in the Figure 7[a], 7[b] and 7[c] respectively. Sample shows maximum absorption in the UV region with a steep absorption edge centered at 364 nm. This absorption is attributed to a charge transfer transition in which an oxygen 2p electron goes into one of the empty tungsten 5d orbital [55]. The excitation from O_2p to Wt_2g in the WO_4^{2-} group absorbs ultraviolet irradiation in $PbWO_4$. In the excited state of the WO_4^{2-} groups, the hole (on the oxygen)and the electron (on the tungsten) remain together as an exciton because of their strong interactions [56-58].The maximum absorbance occurred within the UV region from where the absorbance decreased with the wavelength towards the visible region. $PbWO_4$ showed intense and narrow band. The transmittance increased exponentially from the UV region towards the visible region. The transmission spectrum showed that the sample exhibits very high transmittance in the visible region.The properties of poor transmittance in the UV but high transmittance in the visible regions makes this sample excellent materials in screening off UV portion of electromagnetic spectrum which is dangerous to human health.A film of this material can be used for coating eye glasses for protection from sunburn caused by UV radiations.The reflectance spectra of the samples had a maximum hike at the UV region. Maximum absorption in the UV region along with uniform transparency in the visible region makes the sample ideal for UV sensors, filters and screening applications. Also, the property of high transmittance and low reflectance in the visible regionmakes the material a good candidate for transparent windows in solar cells.

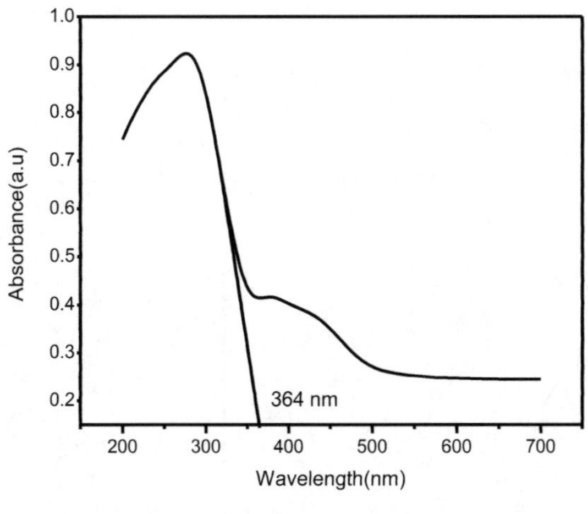

Figure 7[a]. UV-Absorbance graph of nano $PbWO_4$

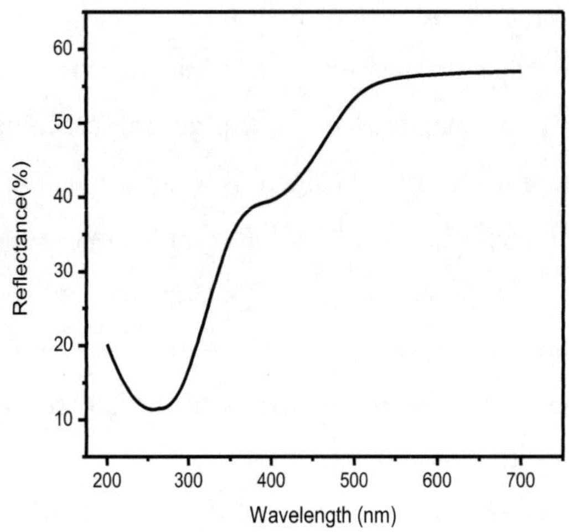

Figure: 7[b].Reflectance graph of nano $PbWO_4$

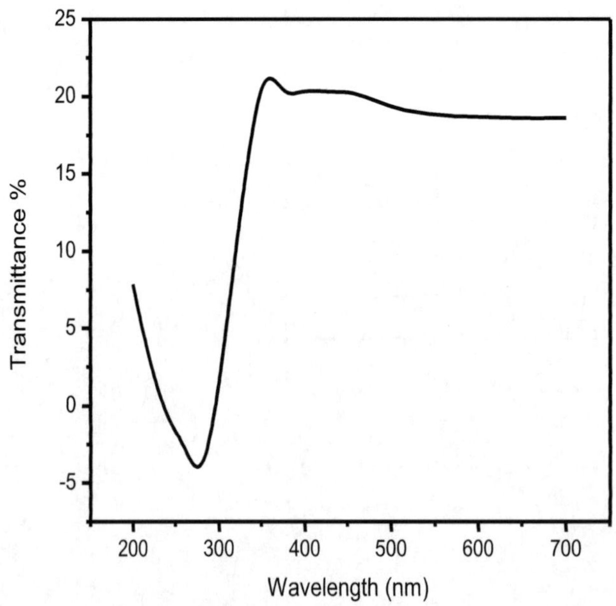

Figure: 7[c]. Transmittance spectra of nano $PbWO_4$

IOP Conf. Series: Materials Science and Engineering **360** (2018) 012016 doi:10.1088/1757-899X/360/1/012016

A semiconductor is characterized by its electronic band structure. In the high energy absorption region, the dependence on the photon energy is expressed by Tauc's equation. The Wood and Tauc [59,60] equation was used to estimate the optical band gap of nano $PbWO_4$ powder. According to this equation, the optical band gap energy is related with absorbance and photon energy by the following equation:

$\alpha h\nu = (\beta h\nu - E_g)^m$

where "β"is an energy independent constant, $\alpha$is the opticalabsorption coefficient, "h"is the Planck's constant, ν is the frequency of incident photon, "E_g"is the optical band gap and "m"is a constant which characterizes the nature of band transition. m = 1/2 and 3/2 corresponds to direct allowed and direct forbidden transitions and m = 2 and 3 corresponds to indirect allowed and indirect forbidden transitions, respectively.The optical band gap can be obtained from the extrapolation of the straight-line portion of the $(\alpha h\nu)^{1/m}$versus hν plot to$\alpha h\nu$= 0. The existence of sharp absorption band is an indication of excellent crystalline nature of the nanosample. The calculated optical band gap is illustrated in Figure 8. The band gap of the sample is found to be 3.87 eV, which is matching with the value reported [23] for $PbWO_4$. The present obtained value is lower than that reported by [61] which is 4.1eV. The lowering of band gap can be due to several factors such as changes in local atomic structure, lowering of symmetry of lattice, electro negativity of transition metal ions, connectivity of the polyhedrons, deviation in the O–X–O bonds, oxygen vacancies, distortion of the $[XO_4]^{2-}$ tetrahedrons and intrinsic surface states [23,61]. As the band gap is the energy difference between valance and conduction band, the presence of these intermediate levels resulted in the reduction of optical band gap energy. As $PbWO_4$prepared by the present method possess wide band gap along with good transmittance in the visible region is suitable for transparent conducting oxide films for window layers on solar cells.

IOP Conf. Series: Materials Science and Engineering **360** (2018) 012016 doi:10.1088/1757-899X/360/1/012016

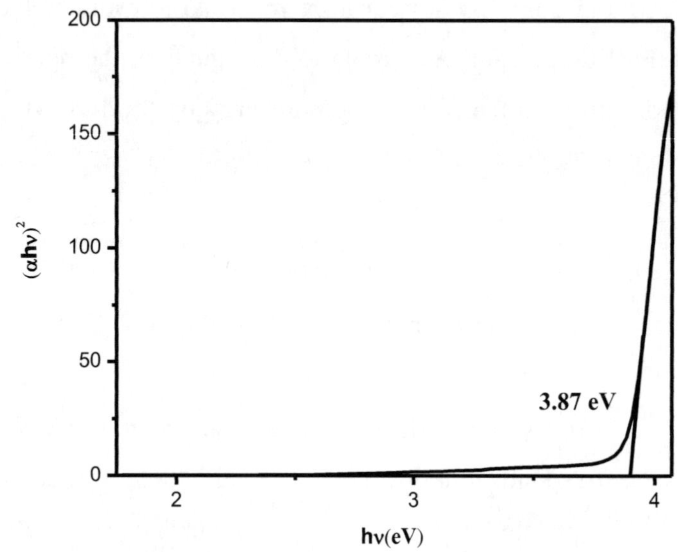

Figure:8 Tauc's plot of nano $PbWO_4$

Figure.9 shows the photoluminescence spectra of nano $PbWO_4$.Scheelite phase of $PbWO_4$ give rise to the blue luminescence at 450 nm, when excitation energy in an exciton band region around (250-330nm). This luminescence has been regarded as being an intrinsic feature of Scheelite phase of $PbWO_4$. In our case a broad emission band at370–600 nm with maximum emission at 479 nm is observed. The luminescence is ascribed to the radiative decay of self-trapped excitons in regular WO_4^{2-} group. The blue emission is due to transitions in the regular tungsten $[WO_4^{2-}]$ group in which Pb^{2+} play a role [23].There are two defects: 1) A regular lattice (WO_4^{-2}) defect and 2) Defect based on WO_3 center, which are responsible for the complex character of $PbWO_4$ emission. The blue emission is an intrinsic feature of $PbWO_4$ and is generally ascribed to the radiative decay of self-trapped exciton that locates on the regular WO_4 group [23]. The green luminescence is of

IOP Conf. Series: Materials Science and Engineering **360** (2018) 012016 doi:10.1088/1757-899X/360/1/012016

extrinsic origin and it was ascribed to a defect based WO_3 center possibly with F center. The concentration of these defects are strongly influenced by the raw material and the method used for the crystal growth. According to the band theory given [62],the bottom of conduction band is formed by both 5d state of W^{6+} ion and 6p state of Pb^{2+} ion in $PbWO_4$ and the coping of valence band is formed by both 6s state of the Pb^{2+} ion and the 2p state of O^{-2} ion .The W_{5d} states are degenerated and split in a crystal field of tetragonal symmetry in e and t_2 vibrational modes .The Pb 6s state and Pb 6p state have appreciable contribution throughout the valence band and conduction band respectively. When $PbWO_4$ is excited within fundamental absorption band charge transfer by two types of electronic transition can be observed

a. From O^{-2} to W^{6+} within WO_4^{2-} oxyanion complex
b. From O^{-2} to Pb^{+2} outside WO_4^{-2} oxyanion complex.

IOP Conf. Series: Materials Science and Engineering **360** (2018) 012016 doi:10.1088/1757-899X/360/1/012016

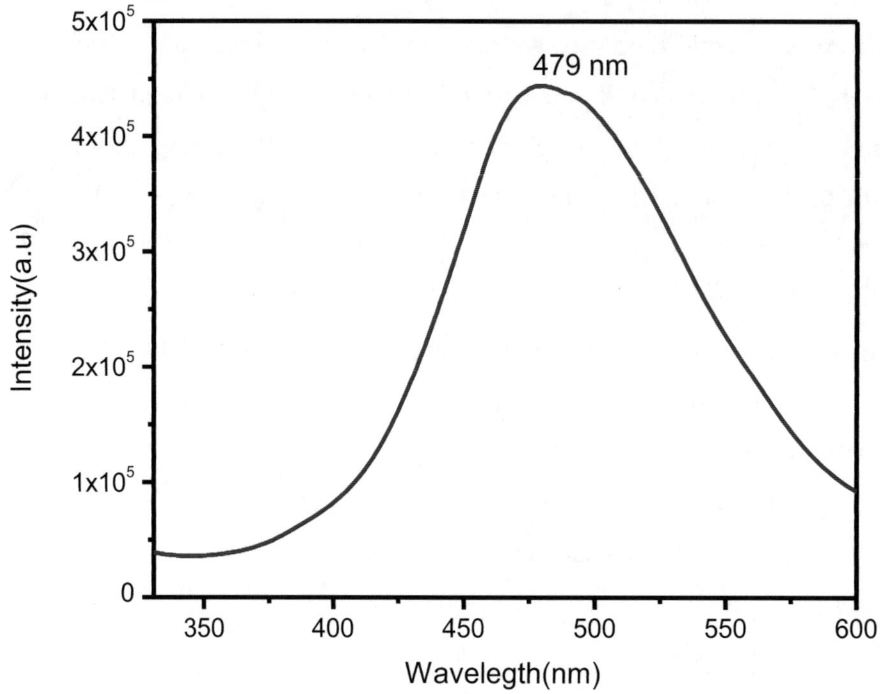

sFigure.9 Photoluminescence spectra of nano $PbWO_4$

Sintering And Dielectric Behavior

After several trials, the sintering temperature of compacted nanopowders of $PbWO_4$ were optimized.The relative green density of the specimen used for the sintering study was 72%. The sample was sintered at 787^{0}C for 3 hours at a rate of 10^{0}C/minute without using any sintering aid. The density of the sintered sample calculated by Archimedes method is 98%.Figure.10, depicts that the attempts to increase the density by increasing the temperature or soaking time led to the melting of the pellets. So the optimized sintering temperature and time is 787^{0}C for 3 h. The high sintered density for $PbWO_4$ pellets derived from nano powder was obtained through the agglomeration, ultra fine nature of the powder and grain boundary diffusion. Figure.11[a] and Figure.11[b] shows the photograph of green and sintered pellets respectively. From the photographs the change in the colour of the pellet from pale yellow to crimson yellow can be noted which is the characteristic feature of the sample nano $PbWO_4$.Lead and oxygen vacancies are generally created due to the Pb and O deficiency in the$PbWO_4$ crystal. These vacancies are responsible for colourcentres in this crystal [63]. The colour ofthe crystal is also influenced by the pre-annealing time before melting of initial powder [64].

To the best of our knowledge the sintering of the sample nano $PbWO_4$ at a lower temperature of 787°C has not been accomplished before.

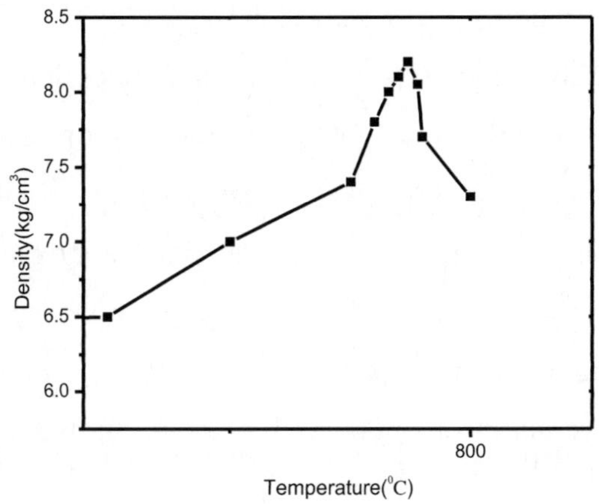

Figure.10: Variation in density with sintered temperature

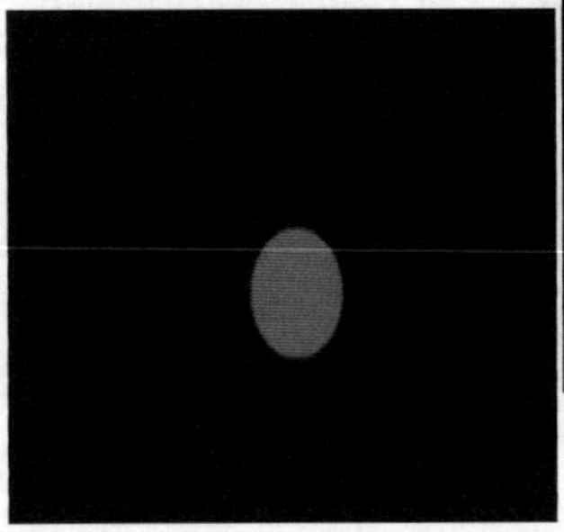

Figure.11[a]: Photograph of Green Pellet   Figure.11[b]:Photograph of Sintered Pellet

The highly dense sintered pellet was hand polished and then thermally etched at 680^{0}C for an hour in air. The surface morphology of the sintered pellet was studied using Scanning Electron Microscopy. Figure.12 is the SEM micrograph of the sintered pellet, which indicates that the pellet is well sintered with negligibly minimum porosity. It isalso noticeable that the sample has attained appreciable density and also the grain boundaries are clear cut which indicates that the sample is not in melt phaseIt is observed from the micrograph
that the size of the grains has increased from 500 nm to about 5μm due to sintering.

IOP Conf. Series: Materials Science and Engineering **360** (2018) 012016 doi:10.1088/1757-899X/360/1/012016

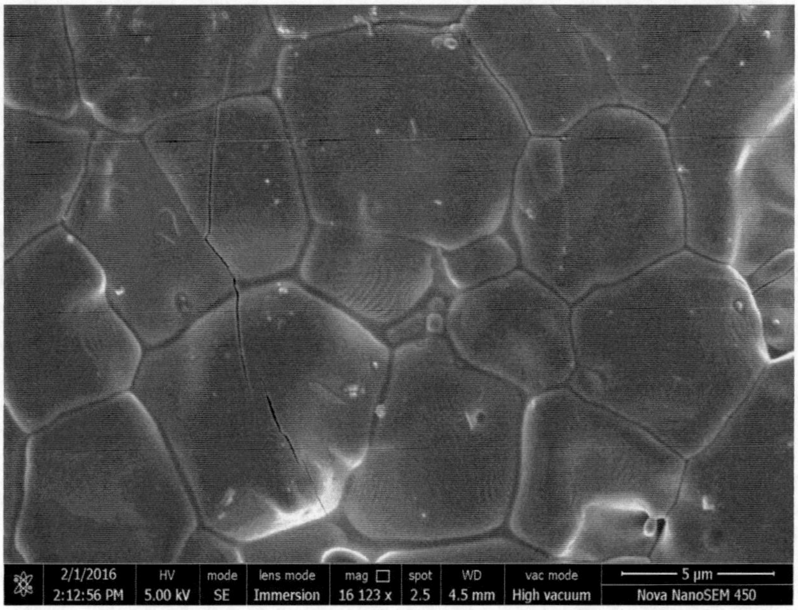

Figure.12: Scanning Electron Micrograph of nano $PbWO_4$

The dielectric constant $\varepsilon_r$and loss factor tanδ values of the pellet sintered from the nanopowder, synthesized through the above combustion process were studied in the frequency range 100 Hz to 5 MHz at room temperature with silver electrodes on both sides of the circular disc. Figure.13[a] and Figure13[b]shows variation of dielectric constant andtanδwith the frequency respectively . The dielectric constant $\varepsilon_r$and loss factor tanδvalues of the $PbWO_4$ pellets at 1MHz are 86 and $1x10^{-2}$ respectively. Figure.14 shows the variation of dielectric constant at different temperature at constant frequency of 1 MHz. The values of dielectric constant and very low loss factor indicate the suitability of the sample as a candidate for electronic and dielectric application. It is clear from the graph that the temperature dependence of dielectric constant is very minimal in the measured range. The loss factor decreases with increase of temperature and is of the order $1x10^{-2}$at temperature above 100^oC. The temperature coefficient of dielectric constant (T_ck)

IOP Conf. Series: Materials Science and Engineering **360** (2018) 012016 doi:10.1088/1757-899X/360/1/012016

is determined using the equation given below between temperature 500°C and 30°C at 5MHz.

$T_C K = [\{K_{500} - K_{30})/520/K_{30}\} \times 10^6]$ ppm/°C

whereK_{30} and K_{250} are dielectric constants at 30°C and 500°C respectively, and 220 is the temperature difference. The obtained T_c K is
-13 ppm/°Cwhich is negative. The value of dielectric constant at room temperature was found to be 86 and was observed that the value of dielectric constant at higher temperature from 30°C to 500°C remained constant without much variation. Thus nano $PbWO_4$ possess relatively low temperature coefficient of dielectric constant which makes it suitable for temperature dependent dielectric applications.

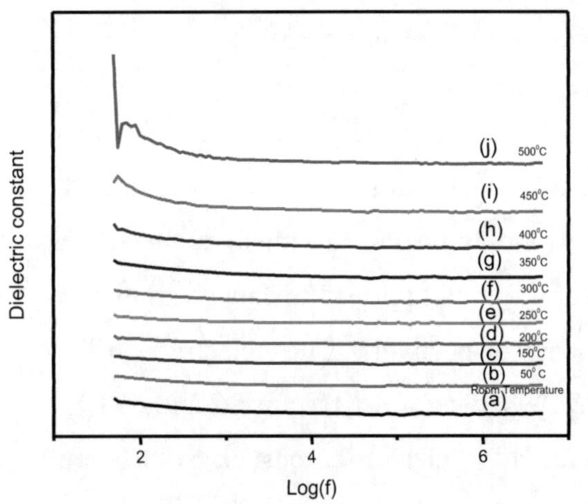

Figure.13[a]: Variation of dielectric constant with frequency of sintered $PbWO_4$ pellet at temperatures (a) Room temperature (b) 50^0C (c) 150^0C (d) 200^0C (e) 250^0 (f) 300^0C (g) 350^0C (h) 400^0C (i) 450^0C (j) 500^0C

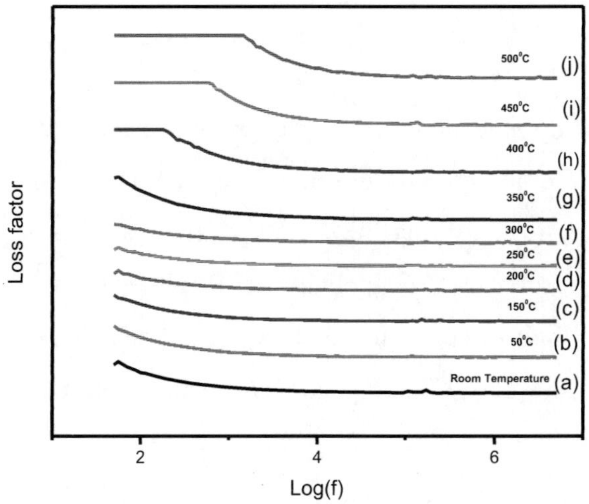

Figure.13[b]: Variation of loss factor with frequency of sintered $PbWO_4$ pellet at temperatures (a) Room temperature (b) 50^0C (c) 150^0C (d) 200^0C (e) 250^0 (f) 300^0C (g) 350^0C (h) 400^0C (i) 450^0C (j) 500^0C

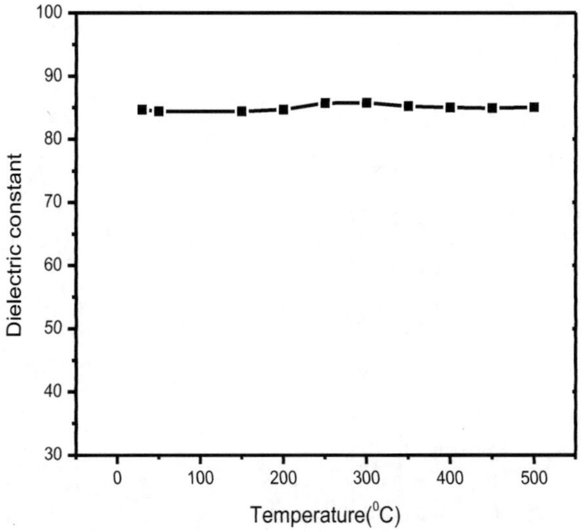

Figure.14:Variation of Dielectric constant with temperature at 1kHz.

IOP Conf. Series: Materials Science and Engineering **360** (2018) 012016 doi:10.1088/1757-899X/360/1/012016

Conclusion

Nanostructured Scheelite $PbWO_4$ compound prepared by the present technique come under the category of Low Temperature CoFired Ceramics (LTCC) materials as the sintering temperature is as low as 787^0 C. Usually glass and polymer composite materials are used for LTCC application. But their high dielectric loss compose restrictions on their effective use as LTCC materials. The low value of sintering temperature makes $PbWO_4$ samples a promising material for LTCC application, substrate application and electronic packaging materials. It may be noted in the present study we have succeeded in developing $PbWO_4$ samples, which is sintered nearly 400^0C lower than the sintering temperature reported for their conventional powders, without use of any sintering aid. Also the sample possess excellent dielectric and optical behavior which makes this material a promising candidate for LTCC and optical applications.

Reference

1. E. Khoo, P.S. Lee, J. Ma, J. Euro. Ceram. Soc. 30 ,2010.
2. S. Vidya, Sam Solomon, J. K. Thomas.,Synthesis, Characterization, and Low Temperature Sintering of Nanostructured $BaWO_4$ for Optical and LTCC Applications, 2013.
3. S. Vidya, Sam Solomon, J.K. Thomas., Synthesis of Nanocrystalline CaWO4 as Low-Temperature Co-fired Ceramic Material,2013.
4. S. Vidya, Sam Solomon, J. K. Thomas., Processing, Structural and

IOP Conf. Series: Materials Science and Engineering **360** (2018) 012016 doi:10.1088/1757-899X/360/1/012016

Physical Properties Nanocrystallinescheelite SrWO4: a low temperature Co- Fired ceramic optical material-synthesis and properties,2013.

5. S. Vidya, Sam Solomon, J.K. Thomas., Structral, optical and dielectric characterization of nanocrystallineAMo0.5W0.5O4 (where A=Ba,Sr and Ca) prepared by single step modified combustion technique,2015.
6. S. Vidya , Sam Solomon , J. K. Thomas., Synthesis and characterization of MoO_3 and WO_3nanorods for low temperature co-fired ceramic an optical applications, Springer Science, Business Media New York, 2015.
7. S Vidya, K C Mathai, Annamma John, Sam Solomon, K Joy, J. K Thomas., Optical and dielectric properties of nano BaNbO3 prepared by a combustion Technique, 2013.
8. N. Oka, M. Watanabe, K. Sugie, Y. Iwabuchi, H. Kotsubo, Y.Shigesato., Thin Solid Films, 2013.
9. Yi Shen, Rong Huang, Yang Li, Shuzhen Yao., Appl. Surf. Sci.,2011.
10. A.I. Gavrilyuk ., Appl. Surf. Sci. ,2013.
11. A. Georg, W. Graf, R. Neumann, V. Wittwer ., Sol. Energy Mater. Sol. Cells ,2000.
12. N. TahmasebiGaravand, S.M. Mahdavi, A. Irajizad, M. Ranjbar., Sens.Actuator B Chem. ,2012.
13. S. Papaefthimiou, G. Leftheriotis, P. Yianoulis., Thin Solid Films ,1999.
14. K. Kuribayashi, H. Harigae, Y. Suzuki, Y. Uchida., Energy Procedia ,2012.
15. K.S. Yook, S.O. Jeon, C.W. Joo, J.Y. Lee, S.H. Kim, J. Jang., Org. Electron.10, 2009.
16. A. Kuzmin, A. Anspoks, A. Kalinko, J. Timoshenko, R. Kalendarev., J. Non-Cryst. Solids 401, 2014.
17. K.A. Gesheva, T.M. Ivanova, G. Bodurov., Prog. Org. Coat. 74, 2012.
18. H. Liu, Z. Ma, Y. Chu, W. Sun., Colloid Surf. A 287, 2006.

19. L. Wang, X. Wei, Y. Luo, J. Yuan, Y. Ding., Displays 32, 2011.

20. P.Lecoq,IDafinei ,E Auffray, M Schneegans ,V Korzhik , O V Missevitch , V . B Pavelenko ,AAFedorov , A N Annenkov , V L Kostylev.,Nucl. Instrum.Meth. A

21. M Kobayashi ,M Ishii, Y Usuki., Nucl.Instrum Methods Phys .Res.,1998.

22. K Hara, M Ishii, M Usuki., Y NuclInstrum.methods Phys.Res,1998.

23. T. George, S. Joseph, A. T. Sunny, S. Mathew., Morphology Tuning of Tungstate Nanostructures bySoft Chemistry., Journal Of Nanoparticles

24. R.Y. Zhu a, D.A. Maa, H.B. Newman,C.L. Woodyb, J.A. Kiersteadb, S.P. Stollb, P.W. Levy ., A study on the properties of lead tungstate crystals., A California Institute of Technology, 1996

25. K. C. Patil., Advanced ceramics: combustion synthesis and properties .,Bulletin of Materials Science, vol. 16, no. 6.

26. K.C Patel ,M.SHegde ,Tanu Rattan ,S T Aruna., Chemistry of Nanocrystalline oxide materials Combustion synthesis properties and application., World Scientific Publishing, 2008.

27. Kashinath C. Patila , S.T. Aruna , TanuMimani., Combustion synthesis: An update., Current Opinion in Solid State and Materials Science 6, 2002.

28 .K.C.Mathai, S. Vidya, S. Solomon, J.K Thomas., Variation in Optical, dielectric and sintering behavior of nanocrystalline$NdBa_2NbO_6$, Adv. Mater. Res., An Intl Journal.No. 2.

29. G.K. Williamson, W.H. Hall, X-Ray line broadening from filed aluminium and wolfram, Acta Metall. 1 ,1953.

30. P.H. Bottlebergs, H. Everts, G.H.J. Broers., Mater. Res. Bull., 11 ,1976.

31. A.M. Golub, V.I. Maksin, A.A. Kapshuk, S.A. Kirillov.,Russ. J. Inorg. Chem., 21 ,1976 .

32 . D. L. Rousseau, R. P. Baumann, and S. P. S. Porto., J. Raman Spectrosc.,1981.

33. S. P. S. Porto and J. F. Scott., Phys. Rev. 157,1967.

34.G. Herzberg., Molecular Spectra and Molecular Structure: II Infra-Red and Raman Spectra _D. Van Nostrand Co. Inc., NewYork, 1945.

35. F. J. Manjón, D. Errandonea, N. Garro, J. Pellicer-Porres, P.Rodríguez-Hernández, S. Radescu, J. López-Solano, A. Mujica, A. Muñoz., Phys. Rev. B 74 ,2006.

36. F. J. Manjon,D. Errandonea, N. Garro, J. Pellicer-Porres, J. López-Solano, P. Rodríguez-Hernández,S. Radescu, A. Mujica, A., Muñoz, Lattice dynamics study of scheelitetungstates under high pressure II. $PbWO_4$.,Physical Review B 74 ,2006.

37. Burcham, L. J, Wachs, I. E., Vibrational analysis of the twonon-equivalent, tetrahedral tungstate (WO_4) units in Ce-2(WO_4)(3) and La-2(WO_4)(3). Spect. Acta Part A, 1998.

38. Nakamoto, K., Infrared, Raman Spectra of Inorganic, Coordination Compounds, 4th edn.Wiley, New York, 1996.

39. S. P. S. Porto and J. F. Scott., Phys. Rev. 157,1967.

40.G. Herzberg., Molecular Spectra and Molecular Structure: II Infra-Red and Raman Spectra _D. Van Nostrand Co. Inc., NewYork, 1945.

41. F. J. Manjón, D. Errandonea, N. Garro, J. Pellicer-Porres, P.Rodríguez-Hernández, S. Radescu, J. López-Solano, A. Mujica, A. Muñoz., Phys. Rev.

B 74,2006.

42. F. J. Manjon,D. Errandonea, N. Garro, J. Pellicer-Porres, J. López-Solano, P. Rodríguez-Hernández,S. Radescu, A. Mujica, A., Muñoz, Lattice dynamics study of scheelitetungstates under high pressure II. $PbWO_4$.,Physical Review B 74,2006.
43. J. M. Stencel, J. Springer, and E. Silberman., J. Mol. Struct., 1977.
44. S. Bastians, G. Crump, W. P. Griffith, R. Withnall., J. Raman Spectrosc,2004.
45.D. Christofilos, K. Papagelis, S. Ves, G. A. Kourouklis, C.Raptis., J. Phys.: Condens. Matter ,2002.
46.D. Christofilos, S. Ves, and G. A. Kourouklis., Phys. Status Solid B, 1996.
47 .A. N. Akimov, M. V. Nikanovich, V. G. Popov, and D. S. Umreiko., J. Appl. Spectrosc,1987.
48.N. Weinstock, H. Schulze, and A. Müller., J. Chem. Phys,1973.

49. J. M. Stencel, J. Springer, and E. Silberman., J. Mol. Struct., 1977.
50. S. Bastians, G. Crump, W. P. Griffith, R. Withnall., J. Raman Spectrosc,2004.
51.D. Christofilos, K. Papagelis, S. Ves, G. A. Kourouklis, C.Raptis., J. Phys.: Condens. Matter ,2002.
52.D. Christofilos, S. Ves, and G. A. Kourouklis., Phys. Status Solid B, 1996.
53 .A. N. Akimov, M. V. Nikanovich, V. G. Popov, and D. S. Umreiko., J. Appl. Spectrosc,1987.
54.N. Weinstock, H. Schulze, and A. Müller., J. Chem. Phys,1973.
55. H. Fu, J. Lin, L. Zhang, Y. Zhu., Appl. Catal. A, 2006.
56. M.J.J. Lammers, G. Blasé, D.S. Roberson., Phys. Status Solidi A, 1981.

57. F. Lei, B. Yan, H-H. Chen, Q. Zhang, J-T.Zhao.,Cryst. Growth & Des., 2009.
58. V.B. Mikhailik, H. J. Kraus., Appl. Phys., 2005.
59. D.L. Wood and J. Tauc., Phys. Rev. B 5, 3144 ,1972.
60. D. L. Wood and J. Tauc., Weak absorption tails in amorphous semiconductors, Physical Review B, vol. 5, no. 8, pp,1972.
61. Y. Zhang, N.A.W. Holzwarth, R.T. Williams., Phys. Rev. B, 57,1998.
62. F. Lei, B. Yan, H-H. Chen, Q. Zhang, J-T.Zhao.,Cryst. Growth & Des., 2009.
63. N. Nikl, K. Nitsch, S. Baccaro, B. Borgia, M. Martini, M.Kobayashi, M. Ishii, Y. Usuki, O. Jarolimek, P. Reiche.,J. Appl. Phys. ,1997.
64. D.H. Kim, H.S. Yang, K.D. Chang, H.Y. Park, M.B. Lee, J.H. Lee, Y.H. Song, Y.R. Cho, S.Y. Lee, J. Crystal Growth,2001.

IOP Conf. Series: Materials Science and Engineering **360** (2018) 012017 doi:10.1088/1757-899X/360/1/012017

EFFECT OF TEMPERATURE ON THE OPTICAL AND ANTIOXIDANT PROPERTIES OF BIO AND ORGANIC SYNTHESISED HEMATITE NANOPARTICLES

Malu Somaraj[1*], Nancy John[1], Nisha J Tharayil[2] and S. Usha[3]

[1]*Department of Physics, S.N College, Kollam, Kerala, India*

[2]*Department of physics, S.N College for Women, Kollam, Kerala, India*

[3]*Department of Zoology, S.N College for Women, Kollam, Kerala, India*

**Corresponding author*

*Email: **malusandram@gmail.com***

Phone Number: +919446832451

Abstract: The hematite nanoparticles with tunable band gap were synthesised through a simple and novel co-precipitation technique using bio (DNA) and organic (EDTA) templates. The obtained precursor was heated on the basis of TGA to obtain the metal oxide nanoparticles. The crystalline and morphological changes during heat treatment of synthesised sample were discussed using XRD, SEM and TEM analysis. The optical characterisations were done using UV-Vis spectroscopy and FTIR studies. The invitro antioxidant study of samples treated at two different temperatures using DPPH assay provide the possibility of improving radical scavenging activity by varying the preparation conditions.

1. Introduction

Nanosized iron oxides are of tremendous interest when compared with their bulk counterpart due to their high surface area, high rate of reactivity and biocompatibility. They have superior characters in nontoxicity, Chemical stability, durability, and low costs [1]. Among different iron oxide nanoparticles, Hematite (α-Fe_2O_3) is the most stable polymorph in nature under ambient condition and it can be easily found as mineral hematite. Nanostructured α-Fe_2O_3 powders have been prepared by different preparation techniques like sol-gel [2], hydrothermal [3], chemical vapour phase deposition [4], calcinations of hydroxides [5], radio frequency sputtering [6], gas condensation technique [7] and high energy ball-milling process [8, 9]. In this work, we describe a simple co-precipitation method for synthesising nanosized α-Fe_2O_3 particles.

High surface area to volume ratio of the nanoparticles generates surfaces with very high free energy content. For reducing this energy, surface cooperate with interactomes present in cell, and become mostly stable one. But, some interactions have the energy to overcome band gap barrier of the nanoparticles, resulting into generation of electron-hole pair. The free electron, in biological system, results in production of free reactive oxygen species [10]. Moreover, intact iron oxide nanoparticles upon contact with air lose its magnetism and monodispersibility [11]. To avoid such problems, various research groups have taken the help of different chemical and biological agents to modify the surfaces, and stop the lose. In our work, DNA, a bio- template and EDTA, an organic- template were chosen to coat hematite nanoparticles surface. Reduction of the particle size to nanometer range is accompanied by altered electrical, magnetic, electro-optical and chemical properties. Usually, these special properties are caused by the changes of the band structure with decrease of the particle size, which is usually called the quantum-confinement effect. However, as a larger fraction of the atoms in the particles comprises surface atoms, the surface layer has a significant influence on the properties of these nanoparticles. Hematite (α-Fe_2O_3) nanoparticles have a small exciton Bohr radius with favourable bandgap of 2.1-2.2 eV. The surface effect became the predominant in determining the optical properties of the nanosized α-Fe_2O_3 particles.

The high surface to volume ratio of nanostructures makes it viable candidates to act as free radical scavengers than their bulk counter parts [12].The antioxidant assessment of nanomaterials has become a crucial study in pharmaceutical science as well as nanotechnology [13]. There are previous reports on the free radical scavenging by ZnO [12] and NiO [14] nanoparticles. Also, Das et al. [13] has reported on the antioxidant activity of CuO nanoparticles. All these encouraged our present work where we studied the antioxidant activity of bio and organic synthesised Hematite nanoparticles to get an insight about the behaviour and interaction of these materials with biomolecules inside a living system.

2. Materials and methods

The chemicals used in the present investigation were of Analytical Reagent (AR) grade and obtained from Merck.

2.1 Synthesis of α-Fe_2O_3 Nanoparticles

To synthesis α-Fe_2O_3 0.1M solution of ferric nitrate and 0.4 M solution of sodium hydroxide were added drop wise into 0.01wt. % of DNA solution under constant stirring of 5 hours using a magnetic stirrer. Here DNA powder was used as capping agent to control the particle size. A reddish coloured precipitate is formed at the end of the reaction. The obtained precipitate was washed several times using distilled water to remove the impurities. The precipitate was filtered, dried in a hot air and then ground into fine powder. The same procedure was repeated with EDTA instead of DNA to get the other sample. Thus for this work we were used two different capping agents namely DNA and EDTA for the preparation of α-Fe_2O_3 nanoparticles.

2.2 Antioxidant activity test

The radical scavenging activity of hematite nanoparticles are determined using DPPH assay. The DPPH radical scavenging activity was measured as per the process described by Blois et al [15]. 0.1mM solution of DPPH solution in methanol was prepared and 1ml of this solution was mixed with 3ml of sample solutions in methanol at different concentrations. The violet colour of DPPH radical solutions becomes colourless in presence of nanoparticles. After 30 minutes, the absorbance was measured at 517nm. Buthylated hydroxyanisole (BHA) was used as standard. Decreasing of the DPPH solution absorbance indicates an increase of the DPPH radical-scavenging activity. DPPH radical-scavenging activity was calculated using the equation

$$\% \text{ Inhibition} = ((A_0 - A_1) / A_0 \times 100) \quad (1)$$

IOP Conf. Series: Materials Science and Engineering **360** (2018) 012017 doi:10.1088/1757-899X/360/1/012017

Where A_0 was the absorbance of the control (1ml of DPPH + 3ml of methanol) and A_1 was the absorbance of the sample solution.

3. Characterisation

The annealing temperature is determined by the thermo gravimetric analysis (TGA). The TGA is done in a Perkin-Elmer model Diamond TG/DT instrument in the temperature range 28-800° C at a heating rate of 15° C/minute under nitrogen atmosphere. There is no further weight loss after reaching the temperature 450° C. Hence to obtain the oxide nanoparticles, the precursors were heated in a muffle furnace at 500° C for 3 hrs. X-ray diffraction (XRD) spectra of samples were recorded with a PAN Analytical Model X'pert pro X-ray diffractometer employing Cuk$_\alpha$ radiation at 40 KV and 100 mA at a scanning rate of 8° min^{-1} in the range 20 - 80°. The surface morphology of the samples were recorded with a Hitachi-model S-3000H scanning electron microscopy and Transmission electron microscopy (TEM). The absorption spectra of the materials were taken at room temperature with the help of a Jasco V 550UV/Vis-NIR spectrophotometer. The FTIR studies of all the samples were carried out in a Perkin-Elmer FTIR Spectrophotometer between 300 $cm^{-1}$and 4000 cm^{-1}. The radical scavenging activities of samples were elucidated spectrophotometrically using UV-1601 schimadzu spectrometer.

4. Results and Discussion

4.1 Structural Characterisation

To study the thermal effects the bio (FD) and organic (FE) synthesised powder was heated 500℃ and 700° C for 3 hours. The two samples are hereafter referred as FE1 and FD1 (heated at 500℃) and FE2 and FD2 (heated at 700° C).The crystal structure of samples prepared were characterised through X-ray diffraction and the obtained pattern shows the formation of rhombohedral phase of α-Fe_2O_3 (Hematite) with no characteristic peaks of other impurities. The X-ray diffraction spectrum of hematite nanoparticles annealed at 500℃ and 700° C with prominent reflecting planes is shown in the figure 1.

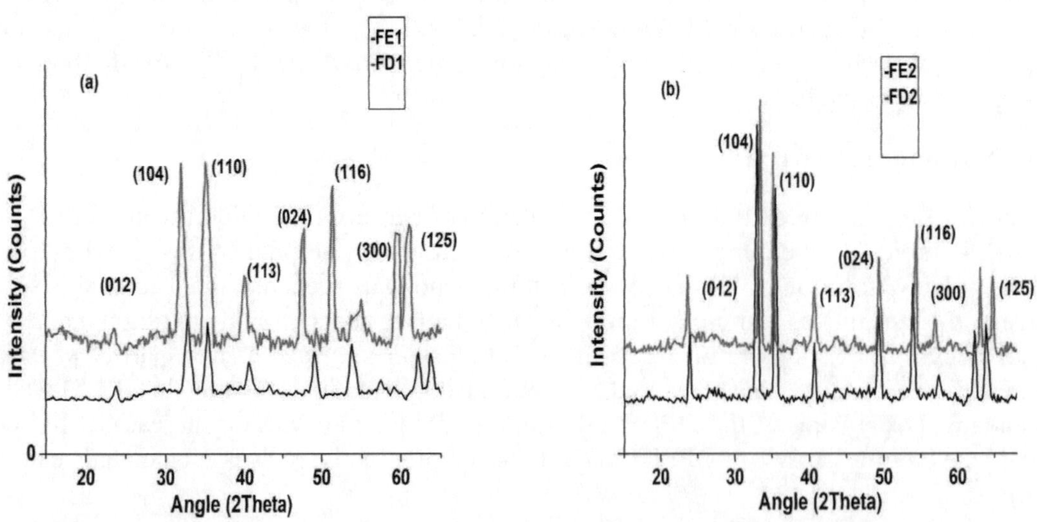

Fig. 1. XRD spectrum of samples FE and FD a) at 500℃ and b) at 700℃

Table 1. Parameters obtained from XRD of samples with particle size

Sample	h k l	$d_{observed}$	d_{jcpds}	Particle size (avg)
FE1	104 116 110	2.72745 2.53993 1.70377	2.7237 2.5560 1.7110	16.12 nm
FE2	104 116 110	2.71683 2.53242 1.69970	2.7237 2.5560 1.7110	35.02 nm
FD1	104 116 110	2.72598 2.54001 1.70332	2.7237 2.5560 1.7110	17.86 nm
FD2	104 116 110	2.71613 2.53116 1.69986	2.7237 2.5560 1.7110	28.6 nm

This broadening of the peaks could also arise due to microstrain of the crystal structures arising from defects like dislocations and twinning. These are believed to be associated with the nanocrystals synthesised chemically as they grow spontaneously and chemical ligands get less time to diffuse to an energetically favourable site. The XRD spectrum of samples (Table 1) were studied in detail, and compared with reported JCPDS values. The relative crystalline sizes were determined from the XRD line broadening using Scherrer equation [16]. The crystalline size of the samples FE1 and FD1 are 17.86 nm and 16.12 nm and it increases as a result of sintering to 35.02nm and 28.6 nm. This indicates that the size of the crystallites can be adjusted by controlling the temperature of the reaction [17]. The elastic strain of the material is calculated using the equation (1)

$$e = [(\lambda / D \cos \theta) - \beta] (1 / \tan \theta) \quad (2)$$

Here D is the crystallite size and β full width at half maximum of the peak with diffracting angle θ. The elastic strain is obtained as 3.877×10^{-4} for sample FE1 and 3.84×10^{-4} for sample FD1. This strain contributes to the broadening of XRD pattern.

4.2 Morphological studies

The general morphologies of Hematite nanoparticles were examined using TEM and SEM. Figure 2 shows the TEM images of the sample FE and FD which confirmed that the synthesised product are nanoparticles. The particles are nearly spherical, agglomerated and dumbbell in shape. The particle size calculated from XRD and that determined from a TEM are often different. This is because of the fact that Peak broadening may also be caused by other reasons, such as inhomogeneous strains, twinned structure, lattice bending or other point defects that may be present in the sample; hence Scherrer's equation may produce results that are different from the actual size [18]. The mean size of the particles FE1 and FD1 at 500°C are 17.10nm and 13.6nm and at 700℃ are 77.35nm and 23.39nm respectively. The homogeneity in shape and size of the green synthesised nanoparticles appears to be better than that of nanoparticles obtained using EDTA. It was observed that DNA can reduce the size of hematite nanoparticles as varying the temperature. The morphology of the hematite nanoparticles was not uniform, formation of large cluster and particle agglomeration resulting increased particle size was observed [19, 20].

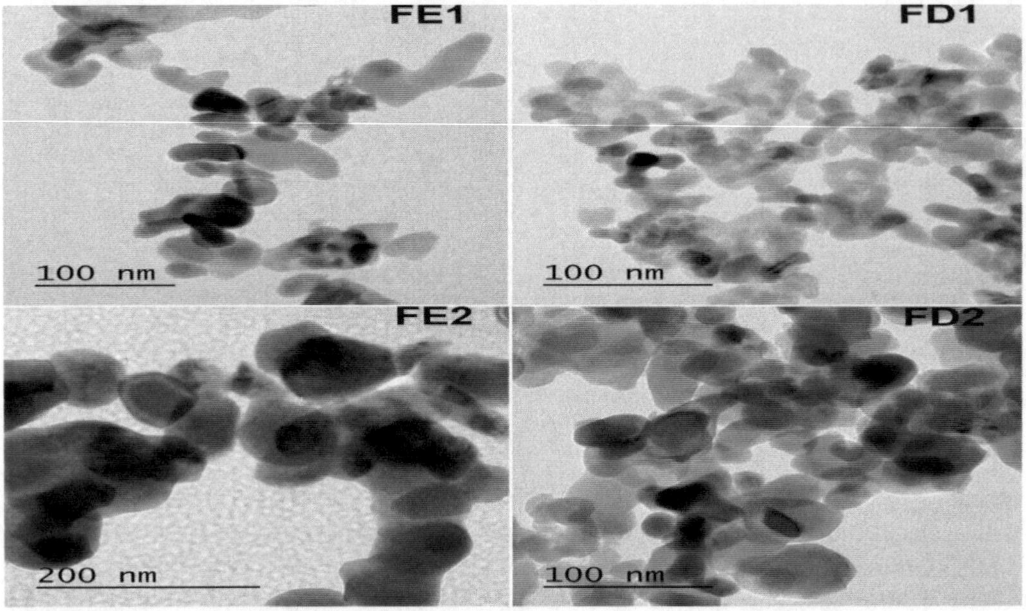

Fig. 2.TEM images of sample FE and FD

Fig. 3.SEM images of sample FE and FD

IOP Conf. Series: Materials Science and Engineering **360** (2018) 012017 doi:10.1088/1757-899X/360/1/012017

The hematite nanoparticles were found be to be aggregated into larger irregular structures with no well-defined morphologies. The agglomeration of nanoparticles is generally explained as common way to minimize their surface free energy; but some studies reported that the agglomeration is assigned to the existence of organic radicals that act as binders [21].

4.3 Absorption studies

UV-Visible spectroscopy

Figure 4 presents the UV-visible absorption spectra of the samples FE and FD annealed at 500° C and 700° C. The absorbance spectra of the two samples were composed of several absorption bands, whose positions were in good agreement with the expected absorption bands of octahedral Fe^{3+}. In fact, according to previous studies, the UV-visible absorption spectrum of hematite was composed of four intense bands located near 230, 290, 345 and 395 nm, with the major absorption being located between 210 and 230 nm [22]. All these absorption bands are due to Fe^{3+} -O^{2-} charge transfer.

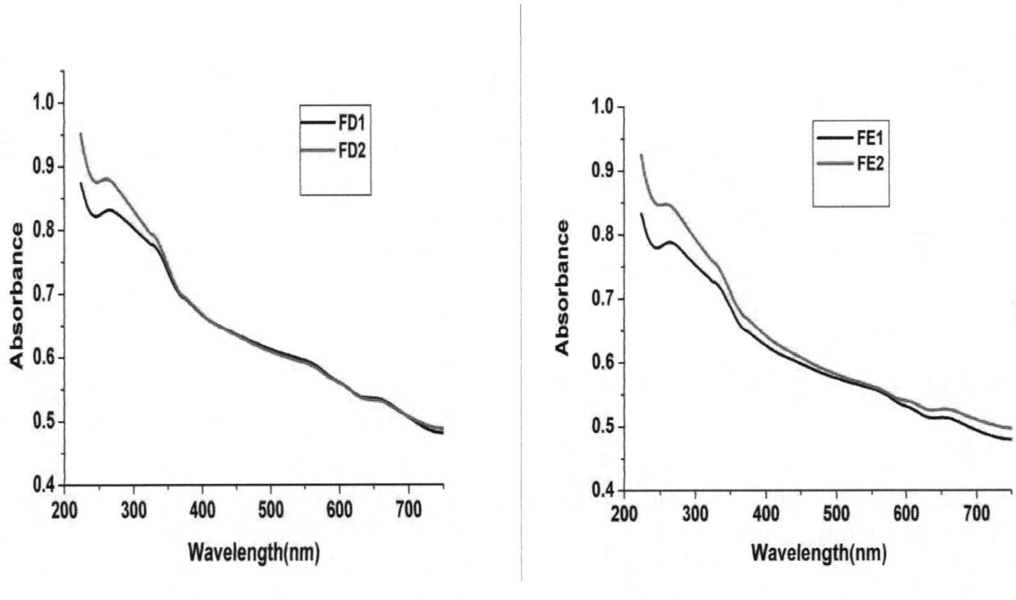

Fig. 4. UV-visible absorption spectra of the samples FD and FE at 500° C and 700° C

The presence of a blue shift of the absorption bands as the annealing temperature decreases and the consequent decrease of crystalline size was observed in this work. This blue shift can be quantified by calculating the experimental band-gap value from the absorption spectra. Moreover Hematite is an n-type semiconductor [23]. The band gap value, Eg, was determined from UV-visible absorption spectra using the relation between the absorption coefficient, α, and the energy of the incident light, hν

$$\alpha h\nu = A(h\nu - Eg)^n \quad (3)$$

e A is a constant, hν the photon energy, Eg the energy gap and n is an index which assumes different values depending on the nature of electronic transitions possible. As it was indirect transition n was taken as 2 and the graph between $(\alpha h\nu)^2$ and hν is shown in Figure 5. The intersection of the extrapolated linear portion with the abscissa axis gives the band-gap value. The band gap values were 3.85eV and 3.39eV for sample FE and 3.86 eV and 3.72eV for sample FD annealed at 500° C and 700° C. The observed values are higher and blue shifted from the reported value of bulk hematite because of the quantum size confinement in the nanoparticles.

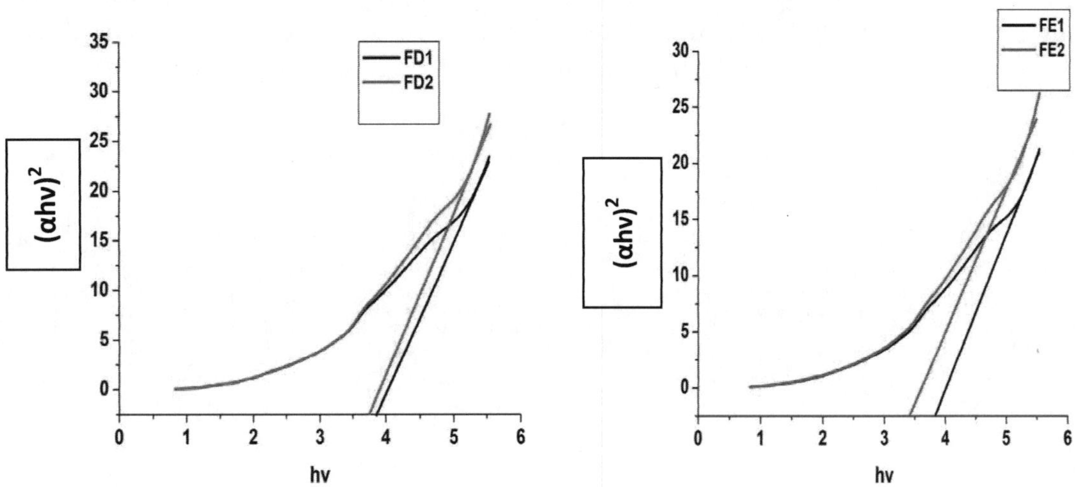

Fig. 5. Tauc plot to evaluate bandgap of FD and FE

FTIR analysis

The FTIR spectrum is a fingerprint of the structure and shows the formation of metal oxides. Two main broad metal-oxygen bands were seen in the FTIR spectrum (Fig 6) of hematite nanoparticles. The highest one $\nu_1$observed around 534 cm^{-1} corresponds to intrinsic stretching vibration of the, Fe↔O, whereas the lowest band, usually observed around 449-454 cm^{-1} is assigned to bending vibration of the O↔Fe↔O [24]. Due to the nano size of the grains the IR active modes slightly shifts due to the difference in coordination number and bond length but still the bands around 400 cm^{-1} and 500 cm^{-1} are typical metal oxides. The peak centered at 3383- 3428 cm^{-1} corresponds to the stretching vibration of intermolecular hydrogen bond (O-H) existing between the adsorbed water molecules and indicates the lower amount of hydroxyl group. The FTIR peaks at 1619-1630 cm^{-1} corresponds to the surface absorbed water. The IR spectrum of FD sample shows an extra peak around 900-1000 cm^{-1} which was the signature of deoxyribose region [25]. This indicates the interaction of DNA with Hematite particles. All the peaks in the samples FD and FE are gradually diminished as the annealing temperature increased. The FTIR results obtained in this study were coinciding well with the XRD analysis where, both analyses showed the synthesised samples are nanosized hematite.

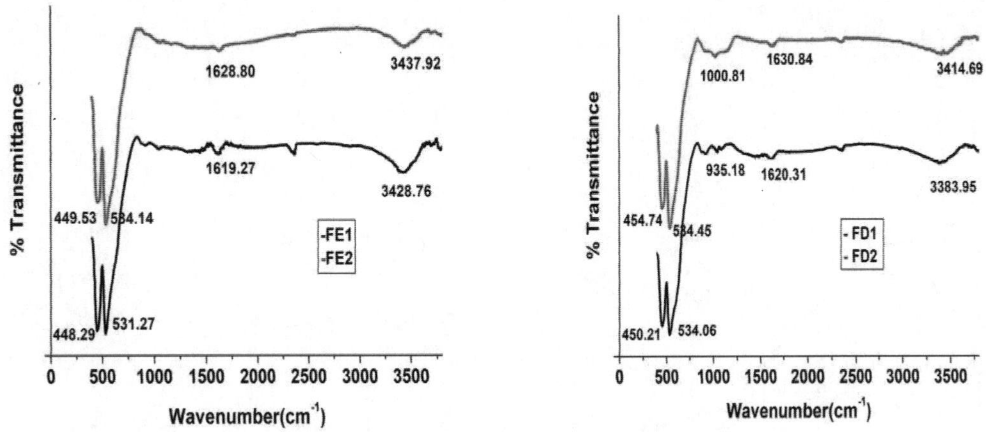

Fig. 6. FTIR spectrum of samples FE and FD

4.4 Antioxidant Activity

The antioxidant activities of the samples (FE and FD) were validated using DPPH assay. Antiradical activity assay was based on the reduction of 1, 1-diphenyl-2-picrylhydrazyl (DPPH). Due to the presence of an odd electron it gives a strong absorption maximum at 517nm.The scavenging effect of samples at 500℃ and 700℃ on DPPH radical is investigated and the variation of optical density with concentration graph showing the result on comparison with the standard is presented in fig 7. The antioxidant efficiency was found to increase with the increase of sample dosage for both samples [26-27] shown in table 2.

Table. 2. Antioxidant activity of samples FE and FD

Concentration (µg)	Antioxidant activity (%)			
	FE1	FD1	FE2	FD2
10	7.5	9	14	8.7
30	11.8	13	14.8	10.2
50	16.7	20	14.8	10.6
70	22.7	30	14.7	11.5
90	23.6	31	14.8	19.9
100	30.6	31	14.8	25

It can be noticed that the annealed samples (at 700℃) FD2 and FE2 exhibits a lower scavenging activity when compared to the samples FD1 and FE1 annealed at 500℃. The decrease in activity with increase in annealing temperature can be attributed to the decrease in surface area to volume ratio due to increased crystalline size. The scavenging activity exhibited by nano sized hematite can also be attributed to the ability of Hematite nanoparticles to transfer its electron density towards the free radical located at nitrogen atom in DPPH.

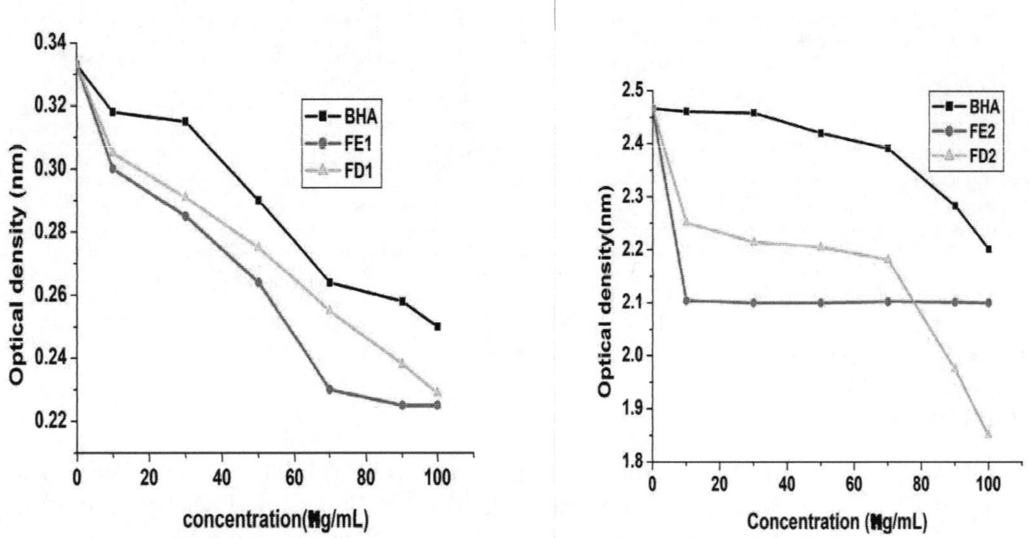

Fig. 7. Optical density-concentration graph for DPHH scavenging activity of samples FE and FD

5. Conclusion

Nanosized hematite with different crystalline sizes was successfully elaborated via co-precipitation method using bio and organic template. The XRD measurements revealed that the sample was nano sized hematite. The morphology, shape and size of the crystals improved on annealing. Compared to EDTA, DNA can be used as an effective capping agent to reduce the size of hematite nanoparticles even on annealing. This fact was confirmed from XRD and TEM measurements. The influence of crystalline size on optical properties relating to absorption was strongly confirmed. Radical scavenging activity of hematite nanoparticles showed good result at lower annealing temperature. The antioxidant activity of the prepared metal oxide nanoparticles proves it to be a potential member of bioactive materials. It is concluded that further exploration on this field needed to develop eco-friendly bio-nanomaterials for biomedicines.

Acknowledgement

The authors are pleased to acknowledge STIC, Cochin for XRD, SEM and TEM imaging. The help rendered by CEPC, Kollam for antioxidant study is thankfully acknowledged.

References

1. M. A. Legodi and D. De Waal, *Dyes and Pigments*.1, 74 **(2007)**
2. R. Pascual, M. Sayer, C. V. R. V. Kumar, and L. Zou, *J.Appl. Phys.* 4, 70 **(1991).**

IOP Conf. Series: Materials Science and Engineering **360** (2018) 012017 doi:10.1088/1757-899X/360/1/012017

3. X. Wang, X. Chen, X. Ma, H. Zheng, M. Ji, and Z. Ahang, *Chem. Phys. Lett.* 4-6,384 **(2004).**

4. E.T. Kim and S.G. Yoon, *Thin solid Films*. 1, 227 **(1993)**.

5. X. Bokhimi, A. Morales, M. Portilla, and A. Garcia-Ruiz, *Nanostruct. Mater.* 1, 12 **(1999)**.

6. W. G. Luo, A. L. Ding, X. T. Chen, and H. Li, *Integrated Ferroelect.* 9 **(1995)**.

7. R. Birringer, H. Gleiter, H. P. Klein and P. Marquardt, *Phys. Lett.A.8,102* **(1984)**.

8. D. Michel, E. Gaffet, and P. Berthet, *Nanostruct. Mater.* 5-8, **(1995)**.

9. M. Chakrabarti, D. Bhowmick, A. Sarkar, *J. Mater. Sci.*19, 40 **(2005)**.

10. M. Arakha, M. Saleem, B. C. Mallick & S. Jha, Sci rep 5, doi: 10.1038/srep09578 **(2015)**.

11. M. Mahdavi, *Molecules*.18 **(2013).**

12. D. Das, B.C. Nath, P. Phukon, A. Kalita, S. K. Dolui, *Colloids Surf. B.* 111 **(2013).**

13. . Das, B.C. Nath, P. Phukon, A. Kalita, S. K. Dolui, *Colloids Surf. B.* 101 **(2013).**

14. J. P. Saikaia, S. Paul, B. K. Konwar, S. K. Samdarshi, *Colloids Surf. B.* 78 **(2010)**.

15. M. S Blois M S, *Nature.* 181**(1958).**

16. Harol P Klug & Leroy E Alexander, X-ray powder diffraction procedure, John Wiley & Sons, New York **(1954).**

17. L. P. Wang, Hong G. Y, S. C Qu & Z.G Wang, *Nanotech.* 16 **(2005).**

18. Nisha J Tharayil, S. Sagar, R. Raveendran and A.V. Vaidyan, Physica B. 399 **(2007)**.

19. K.J. Sreeram, M. Nidhin, B.U Nair, *Colloids and Surfaces B: Biointerfaces.*71 **(2009)**.

20. K.J. Sreeram, M. Nidhin, B.U Nair, *Appl. Surf.* Sci. 258, **(2012)**.

21. N. Dhananjaya, H.C Nagabhushana, R.P.S Chakradhar, R.P.S, *J. Alloys and Compounds.* 509 **(2011)**.

22. E. A. Cloutis, K. A. McCormack, J. F. Bell, III, A. R. Hendrix, D. T. Bailey, M. A. Craig, S. A. Mertzman, M. S. Robinson, and M. A. Riner, *Icarus.* 197 **(2008).**

23. S. Zeng, K. Tang, and T. Li, *J. Colloid Interface Sci.* 312 **(2007).**

24. M. Salavati-Nissari, F. Davar, T. Mahmoudi, *Polyhedron.* 28 **(2009).**

25. Debasish Sarkar, Kalyan Mandal and Madhuri Mandal, *Nanosci. and Nanotech. Lett.* 3 **(2011).**

26. R. Bungheza, E. Barbinta patrascuc, Badead M Donceaa, Popescuc M Iona, *J. optoelect. and advanced mater.* 14 **(2012).**

27. Amit Kumar Mittal, Abhishek Kaler, Uttam Chand Banerjee, Nano Biomed.and Eng, 4 **(2012).**

IOP Conf. Series: Materials Science and Engineering **360** (2018) 012018 doi:10.1088/1757-899X/360/1/012018

Optical absorption and FTIR studies of Cu^{2+} ion doped in $25Li_2O$-$15BaO$-$(60\text{-}x)\ B_2O_3$ glasses

Ashok Bhogi,[a]* R. Vijaya Kumar[b] and P. Kistaiah[c]

[a]VNR Vignana Jyothi Institute of Engineering and Technology, Hyderabad, Telangana, India
[b]School of Physics, University of Hyderabad, Hyderabad, Telangana, India
[c]Department of Physics, Osmania University, Hyderabad,Telangana, India

E-mail: ashokbhogi82@gmail.com

Abstract. Glasses with composition $25Li_2O$-$15BaO$-$(60\text{-}x)\ B_2O_3$: x CuO (where x=0, 0.2, 0.4, 0.6, 0.8 and 1 mol %) were prepared using the conventional melt quench technique. The optical absorption spectra of studied glass samples show only one absorption band which is attributed to the $2B_{1g} \rightarrow 2B_{2g}$ characteristic transition of Cu^{2+} ions in the distorted octahedral sites. The cut-off wavelength (λ_c), optical band gap (E_{opt}) and Urbach energy (ΔE) were determined. The FTIR studies show different stretching and bending vibrations of alkaline earth lithium borate glasses. The relative variations in the intensities of BO_3 and BO_4 units in IR spectra with the concentration of CuO indicate that the copper ions mostly occupy modifying positions in the glass network when the concentration of CuO ≤ 0.8mol%.

1. Introduction

Oxide glasses incorporated with transition metal ions have attracted a great deal of attention, because of their potential applications as lasers, photo-conducting devices, magnetic materials, especially for tunable solid state lasers, efficient phosphors etc. [1]. Among the oxide glasses, borate glasses have attractive considerable scientific interest because of their potential applications such as electro-optic switches, electro-optic modulators, solid state laser material and non-linear optical parametric converters. In addition, they are often used as dielectric and insulating materials and it is known that borate glass constitutes a good shield against IR radiation [2].Incorporation of Alkaline earth cations in the structure of these glasses are responsible for decrease of expansion coefficient, the glass transition temperature increases, strengthen the glass structure and may raise the chemical resistance of the glass [3]. The alkaline earths have been added to borate glasses were found to be good stabilizers of borate glasses. Copper oxide is used as a significant colorant in glasses for optical applications. Copper ions exist in various valence state in glass matrix such as monovalent Cu^{1+} and divalent $Cu^{2+}$and that it can also represent as CuO_4 and CuO_6 in the glass network. The co-existence of copper ions in more than one valance state i.e., Cu^{2+} and Cu^{1+} is a general condition for semiconducting and super paramagnetic behavior of borate glasses [4].

The fact that the addition of copper ions in glass matrix helps to bring significant changes in its structural, magnetic, electric, thermal and optical behavior, a careful optical and structural survey of glasses doped with CuO becomes quite. The present work is taken up with an objective to characterize the glass system with higher B_2O_3 content and constant BaO concentration by means of Optical absorption and Infrared studies and to understand the structural details of the Li_2O-BaO-B_2O_3 glasses containing copper ions.

2. Experimental

A series of glass samples with chemical formula $25Li_2O$-15BaO -(60-x)B_2O_3:xCuO (where x = 0, 0.2, 0.4, 0.6, 0.8 and 1 mol%) have been prepared using conventional melt quenching technique. The requisite mixture of pure reagent grade chemicals, i.e. $BaCO_3$, Li_2CO_3, H_3BO_3 and CuO is grinded to obtain a fine power and is melted in a porcelain crucible at a temperature of 1050°C for 2h and stirred frequently to ensure homogeneity. Then the melt is quenched by pouring on to a steel mould plate maintained at 200°C and pressing immediately with another steel plate in air. The prepared glasses were immediately transferred to another furnace maintained at 300°C for 3 hours, for annealing to obtain strain free glasses.

3. Results and discussion

Optical absorption

The optical absorption spectrum of prepared glass system in the wavelength region 300-1000 nm is shown in Fig. 1. LBBC0 pure glass sample has not shown any peak in the observed region and introduction of Cu^{2+} ions into glass matrix a single broad absorption band has been observed at around 764 nm which is due to the characteristic transition of $2B_{1g} \rightarrow 2B_{2g}$ in a distorted octahedral site. Further increase of CuO concentration the broadness and intensity of the characteristic band of Cu^{2+} ions has been increased up to 0.8 mol% and beyond this concentration the intensity of the characteristic band is found to decrease.

The non-sharp edge in the absorption spectra of these glass samples gives a clear indication of the amorphous nature of the prepared samples. The main feature of absorption edge of the amorphous materials at the lower values of absorption co-efficient is an exponential behavior of the absorption coefficient $\alpha(\nu)$ with the photon energy $h\nu$ as shown in figure 3 and it usually follows the Urbach rule [5].

$$\alpha(\nu)=C\exp(h\nu/\Delta E) \quad (1)$$

where C is a constant and ΔE is measure of the band tailing and is also known as Urbach energy.

The absorption co-efficient $\alpha(\nu)$ near the fundamental absorption edge can be determined at different wavelengths by using the formula

$$\alpha(\nu)=(1/d)\ln(I_0/I) \quad (2)$$

where A is absorbance at frequency ν and d is the thickness of the sample.

For the amorphous materials, the variation of absorption coefficient $\alpha(\nu)$ with photon energy $h\nu$ was given by the equation [3]

$$\alpha(\nu)=[B(h\nu-E_{opt})^n]/[h\nu] \quad (3)$$

where B is an energy independent constant and n is an index.

According to Tauc's relation, for most of the amorphous materials indirect transitions (n=2) are valid [6]. The values of indirect optical band gap energy E_{opt} can be measured from the linear regions of the curves by extrapolating them to meet the hv axis at $(\alpha h\nu)^{1/2} \rightarrow 0$ as shown in figure 2. From the plots $(\alpha h\nu)^{1/2}$ vs $h\nu$, the slopes are calculated. The optical band gaps values obtained in the present work are given in Table 1. From Table 1, it has been observed that the absorption edge observed at 378 nm for LBBC0 (pure glass) is observed to shift slightly towards higher wavelength with increase in the concentration of CuO up to 0.8 mol% and beyond this concentration the edge is shifted towards lower wavelength. The value of E_{opt} is found to decrease with the increase in concentration of CuO up to 0.8mol%. Beyond this concentration the value is however observed to increase.

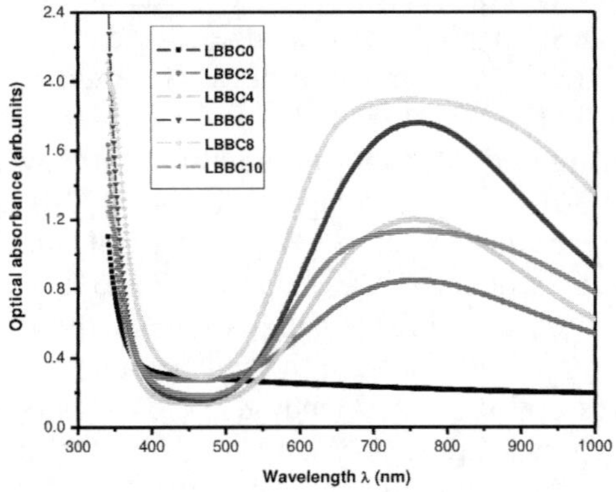

Figure 1. Optical absorption spectra of LBBC glass system.

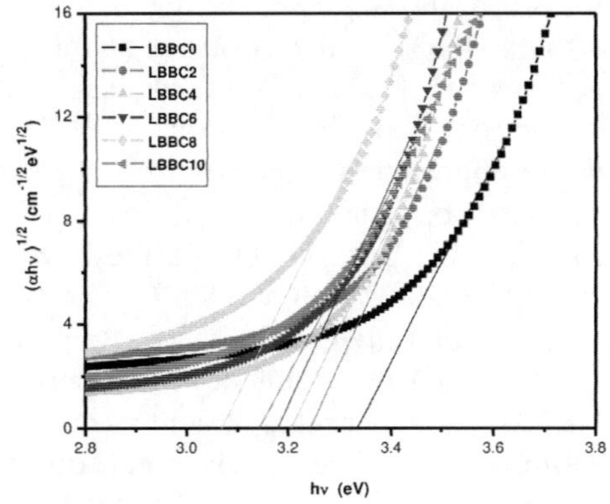

Figure 2.$(\alpha h\nu)^{1/2}$ vs. hν of LBBC glass system.

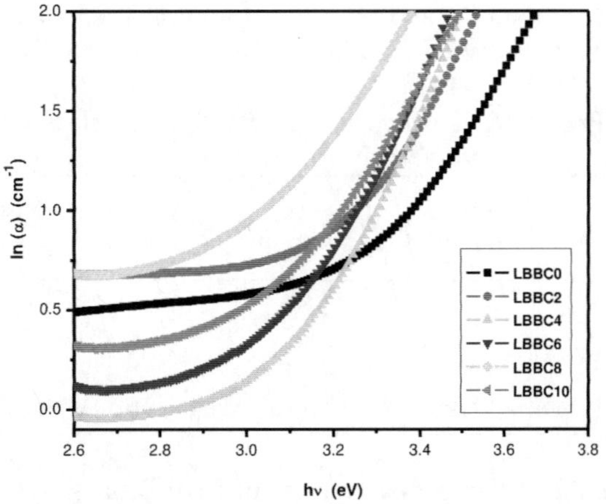

Figure 3.ln α vs. hν of LBBC glass system.

When the concentration of CuO is increased beyond 0.8 mol%, a gradual decrease in the intensity of optical absorption band has been observed. This observation indicates a gradual decrease in the concentration of Cu^{2+} ions that take modifying positions in the glass network. The lower concentration of modifying Cu^{2+} ions, the lower is the concentration of NBO's in the glass matrix. This leads to decrease in the degree of localization of electrons thereby decreasing the donor centers in the glass matrix. The presence of smaller concentration of these donor centers increases the optical band gap and shifts the absorption edge towards the lower wavelength side as observed for the glass LBBC10 [7].

FTIR

The FTIR is used to study the local structural modes in inorganic glassy materials. FTIR spectra in the glasses provide significant information of various bonds present between different ions in the glass network and these are identified by comparing the experimental data of glasses with those

IOP Conf. Series: Materials Science and Engineering **360** (2018) 012018 doi:10.1088/1757-899X/360/1/012018

of crystalline compounds. Fig. 4 shows the FTIR spectra of Cu^{2+} ion doped in $25Li_2O$-15BaO-(60-x) B_2O_3 glasses in the wavenumber region 400-1600 cm^{-1} at room temperature. The addition of alkaline earth oxides (BaO) to the borate glass network is accompanied by a change in the co-ordination number from 3 to 4. The vibrational peak around ~460 cm^{-1} is assigned to the angles modification of B-O-B linkages / specific vibrations of Ba-O bonds [8]. The position of this peak is observed to shift towards the higher frequencies up to 0.8 mol% of CuO and after that is found to shift towards the lower frequencies. Another absorption band around ~704 cm^{-1} is assigned to the B-O-B bending vibrations. This band is found to shift towards the lower frequency up to 0.8 mol% of CuO and beyond this concentration it is found to shift towards the higher frequency. The absorption band around ~775 cm^{-1} is due to the O_3B-O-BO_4 bending vibrations. The position of this band is remained in the same position. The intensity of this peak is observed to increase with increase in the concentration of CuO up to 0.8 mol% and later it is found to decrease. An absorption band around ~873 cm^{-1} is present in all the prepared glass samples and is assigned to the stretching vibrations of B-O bonds in BO_4 units from diborate groups. The position of this peak is moved towards the higher frequencies till 0.8 mol% of CuO and later it is moved towards the lower frequencies. The absorption band around ~1010 cm^{-1} which is attributed to the B-O symmetric stretching vibrations of BO_4 units. The area of this band is found to decrease up to 0.8 mol% of CuO and later it is found to increase. This band is observed to shift towards the higher frequencies up to x=0.8 mol% and later it is observed to shift towards the lower frequencies. The absorption band around ~1364 cm^{-1} is present in all the prepared glass samples and is attributed to the asymmetric stretching vibrations of the B-O of triangle $(BO_3)^{3-}$ units in meta-, pyro- and ortho-borate units. The intensity of this peak is observed to increase with increase in the concentration of CuO up to 0.8 mol% and later it is found to decrease. The another absorption band around ~1390 cm^{-1} is attributed to the B-O^- stretch in BO_2O^- units from varied types of borate groups.

Figure 4.FTIR spectra of LBBC series glass system.

Glass Composition	X=0 LBBC0	X=0.2 LBBC2	X=0 LBBC4	X=0 LBBC6	X=0 LBBC8	X=0 LBBC10
Cut-off wavelength(nm)	378	385	391	397	407	404
Indirect Optical bandgap(eV)	3.324	3.225	3.203	3.141	3.101	3.163
Urbach energy(ΔE)	0.560	0.539	0.496	0.406	0.384	0.421

Table 1. Optical parameters of LBBC glass system.

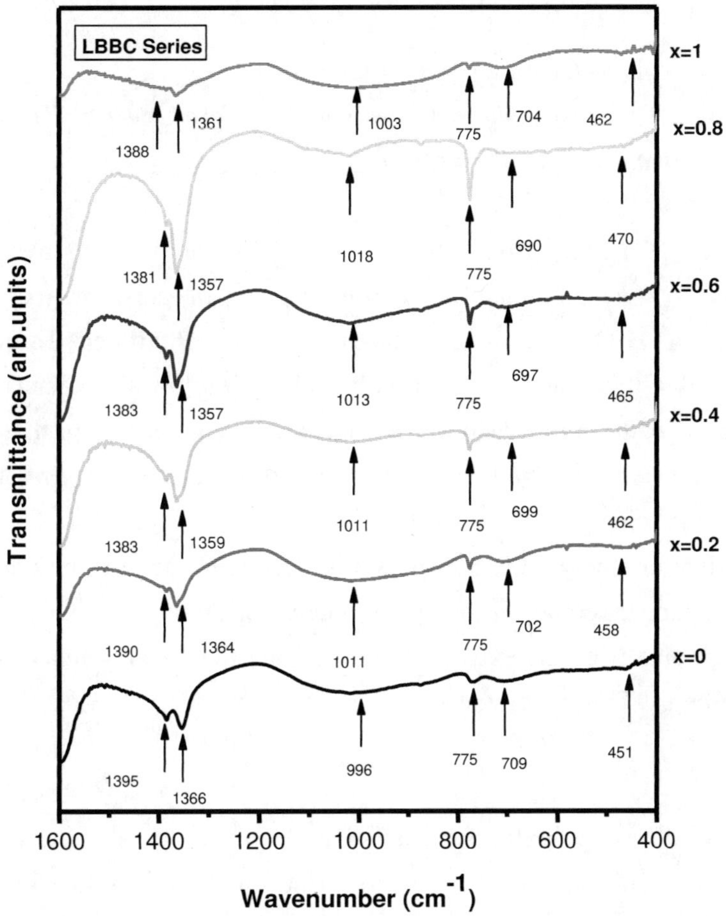

4. Conclusion

The optical absorption studies of these glasses indicate the presence of copper ions mostly in tetragonally distorted octahedral positions. The FTIR study shows that the glass network consists of BO_3 and BO_4 units.

Acknowledgments

One of the authors Ashok Bhogi thanks the management of VNR Vignana Jyothi Institute of Engineering and Technology, Hyderabad for their constant encouragement and support.

References

1. N. J. Kreidl, J. Non-Cryst. Solids 123 (1990) 377.
2. R. P. S. Chakradhar, A. Murali, J.L. Rao, J. Alloys and Compd. 265 (1998) 121.
3. H.F. Shermer, J. Res. Nat. Bur. Stand. (US) 56 (2) (1956) 73.
4. Ashok Bhogi, R. Vijaya Kumar, P. Kistaiah, J. Non- Cryst. Solids 426 (2015) 47-54.
5. F.Urbach, Phys.Rev. 92(1953) 1324.
6. J.Tauc, Amorphos and Liquid Semiconductors, Plenum Press, New York 1974.
7. G. Naga Raju, M. Srinivasa Reddy, N. Veeraiah, Opt. Mater. 29 (2007) 1467-1474.
8. G.E. Walrafen, S.R. Samanta, P.N. Krishnan, J. Phys. Chem. **72** (1980) 113.

Thermal Phases in Gel-Grown Cadmium Oxalate Single Crystals

K M Chauhan* and Punit Suthar

Physics Department, C U Shah Science college, Ashram Road, Ahmedabad 380 014, Gujarat, India

Corresponding author Email: **kmchauhan07@yahoo.com*

Thermogravimetric (TG) and Differential thermal (DT) analysis carried out for studying thermal kinetics and mechanism of phase changes of gel-grown crystals of cadmium oxalate ($CdC_2O_4 \cdot 3H_2O$). The material turns out to be trihydrated. The obtained experimental data show that the material is thermally stable up to 64 $^{\circ}C$, beyond which it exhibits marked tendency to structural decomposition. The process occurs in three phases until, ultimately, cadmium oxide is obtained. The energies of the reactions involved and the mechanism of phase change at each stage have been examined. The kinetic parameters such as activation energy, order of reaction, frequency factor and entropy are evaluated from non-mechanistic equation for thermal dehydration.

Keywords: Thermogravimetric analysis, Differential thermal (DT) analysis, Phase transition, Activation Energy, Frequency factor

1. Introduction

Thermal methods of investigations, generally referred to as thermoanalytic techniques,[1] have found wide application in material characterization, since these methods essentially depend on measurement of the changes in physicochemical property as a function of increasing temperature with time.[2,3] An organometallic compound $CdC_2O_4.3H_2O$ being sparingly soluble in water is crystallized successfully by gel method. The study of thermal analysis[4] is significant for knowing different phases of stability, analyzing and characterizing wide variety of materials for chemical reactions, decomposition, reaction kinetics and mechanisms etc. Hence the grown crystals of cadmium oxalate have been subjected to thermal treatment in air and nitrogen atmosphere, using gravimetric and differential thermal techniques.

2. Experimental Details

Single crystals with average dimension 3x4x1 mm^3 were grown using the controlled three dimensional diffusion of aqueous $CdCl_2$ through agar gel impregnated with oxalate ions.[5]

Thermogravimetric (TG) and differential thermal (DT) data were obtained from ambient temperature to 600°C in steps of 10 °C/min. with 'Perkin Elmer-TGA 7' and 'Perkin Elmer- DTA 7' analyzer, respectively. The thermograms have been used for evaluation of some important kinetic parameters in respect of the decomposition phases of cadmium oxalate single crystals.

3. Results and Discussion

3.1 Evaluation of kinetic parameters using TG and DTA Thermogram

The recorded thermogram (Fig. 1a) is first analysed here to obtain information about the percentage weight loss at different temperatures and hence about the thermal stability and kinetics of dissociation of the crystals. Methods for determination of the activation energy from thermogravimetric curve are available in literature.[6,7,8] A basic method for the interpretation of thermogravimetric traces[9] is described quantitatively by Eq. 1

$$-\frac{df}{F^n} = \frac{k_0}{a}\exp\left(-\frac{E}{RT}\right)dT \qquad (1)$$

Where F denotes the fraction of reactant remaining at the end of pyrolysis, n is the reaction order, k_0 is the pre-exponential part of the rate constant, a is the rate of temperature rise, R is the gas constant and T is the absolute temperature.

When a crystal is heated at a uniform rate of 10 $^{o}Cmin^{-1}$, its weight is found to be lost as a function of the temperature applied. Figure 1a shows the TG thermogram for Cadmium oxalate in air atmosphere. Evidently the phase change begins at the 64 °C and the course of dehydration went on upto 360 °C. During the temperature range weight loss of 49.6% had occurred and the crystal reduced ultimately to the basic oxide of cadmium as the final stable product. The first stage of transformation is, in fact, the dehydration, which starts at 64 °C and continues upto, or terminates at 150 °C. The net weight loss (21.4%) as observed between 64 and 150 °C corresponds to loss of three water molecules, suggesting that the grown crystal were initially trihydrated and hence the chemical formula $CdC_2O_4.3H_2O$. The coordinated three molecules of water lost during this phase change leads to the formation of anhydrous cadmium oxalate as per the reaction (i) given in Table 1. The second phase transition sets in at around 220 °C and is completed at around 310 °C, resulting into the formation of an intermediate product, viz, cadmium carbonate with the loss of carbon monoxide, represented by reaction (ii). This stage of reaction corresponds numerically to 11.08% of the total weight loss. Then, in the third phase of decomposition which seemingly occurs in the temperature interval of 310 °C to 360 °C, the cadmium carbonate is turned into basic oxide of cadmium as the final product, following the reaction (iii), as confirmed by 17.12% of the total mass observed to be lost. The final product CdO remains fairly, reasonably stable.

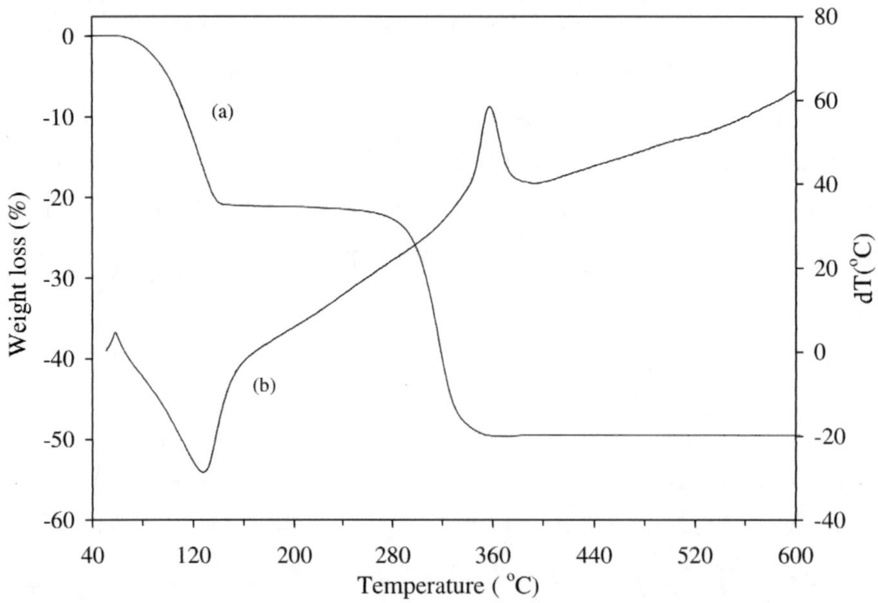

Figure 1. Simultaneous (a) TGA-(b) DTA curve recorded for $CdC_2O_4.3H_2O$ crystal in air atmosphere.

In the DTA thermogram of Fig. 1b the first well-defined, distinct endothermic peak at 127 °C within the temperature range 60-160°C may be attributed to the removal of the co-ordinated water molecules. The second endothermic peak at 360°C is due to the decomposition of anhydrous cadmium oxalate, accompanying the removal of carbon monoxide and then after the formation of cadmium oxide as the final product compound, accompanied by issuing out of carbon dioxide (third step).

3.1 Thermal Kinetics

The following models are used to evaluate the kinetic parameters of solid-state phase changes with the help of TGA thermogram.

3.1.1 Piloyan - Novikova (PN) relation[7]

Using van Krevelen's[9] interpretation given in Eq. 1, Piloyan - Novikova simplified by assuming that E is determined at $\alpha \leq 0.5$, to

$$\log\left(\frac{\alpha}{T^2}\right) = \log\left(\frac{ZR}{E\beta}\right) - \frac{E}{2.30RT} \quad (2)$$

where β is the heating rate, Z is the Frequency factor (min^{-1}).

A graphical plot of log (α/T^2) against 1/T (Fig. 2) yields a straight line. From the slope and intercept, activation energy and frequency factor respectively are calculated (Table 2). The main advantage of this method is that reaction order is not needed be determined first.

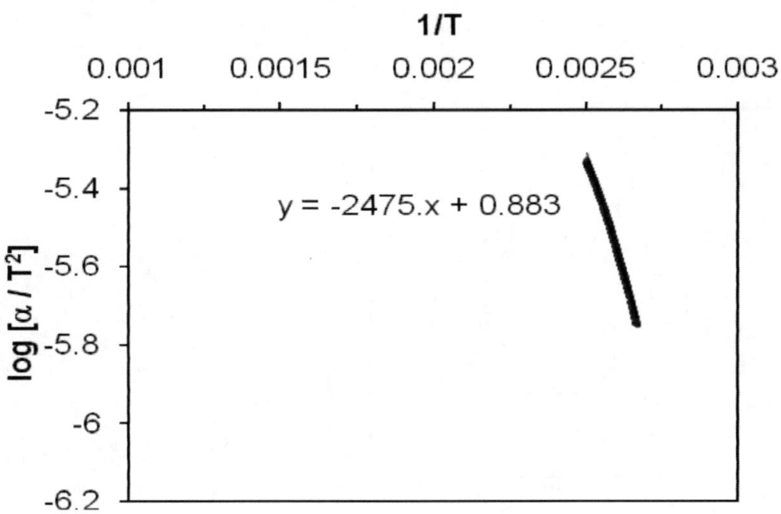

Figure 2. Graphical plot of Piloyan-Novikova (PN) relation.

IOP Conf. Series: Materials Science and Engineering **360** (2018) 012019 doi:10.1088/1757-899X/360/1/012019

3.1.2 Coats – Redfern (CR) relation[8]

With the reaction aA(s)→a B(S) +cC(g), the rate of disappearance of A may be expressed as

$$\log\left(\frac{g(\alpha)}{T^2}\right) = \log\left(\frac{ZR}{E\beta}\right) - \frac{E}{2.30RT} \tag{3}$$

where $g(\alpha) = [-\log(1-\alpha)^{1-n}]$

The activation energy, E is calculated from the slope of the straight line obtained on plotting L.H.S. of the Eq. (2) against 1/T (Fig. 3) and Z from the intercept (Table 2). Hence the values of entropy S are obtained, using the equation

$$Z = \frac{kT_m}{h} \cdot \exp\frac{S}{R}$$

where k is Boltzmann's constant, h is Planck's constant, Z=Pre-exponential factor.

The following model is used to evaluate the thermal kinetics in respect of the grown crystals with the help of DTA thermogram.

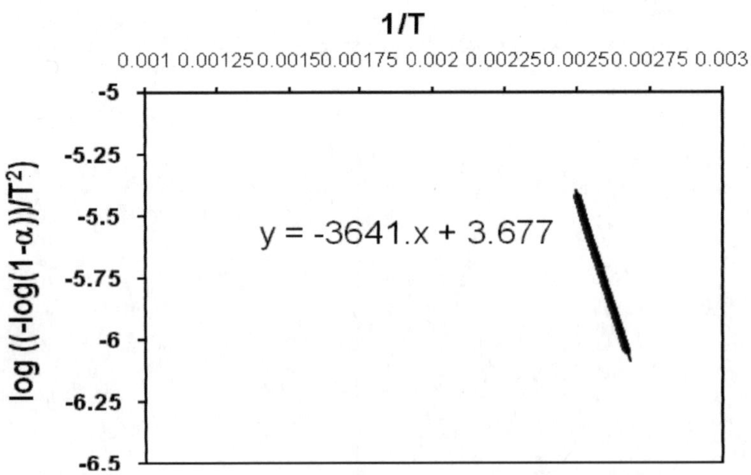

Figure 3. Graphical plot of Coats-Redfern (CR) relation.

IOP Conf. Series: Materials Science and Engineering **360** (2018) 012019 doi:10.1088/1757-899X/360/1/012019

3.1.3 Borchardt and Daniels (BD) relation[2]

For the case of a first order dissociation reaction, we have

$$k = \frac{C_p \left({d\Delta T}/{dt} \right) + K\Delta T}{K(A-a) - C_p \Delta T} \tag{4}$$

The quantities $C_p \, {d\Delta T}/{dt}$ and $C_p \Delta T$ are usually an order of magnitude smaller than the quantities to which they are added and subtracted. Neglecting these smaller terms, Eq. 4 leads to

$$k = \frac{\Delta T}{(A-a)} \tag{5}$$

$$\ln k = \ln Z - \frac{E}{RT} \tag{6}$$

where k is the Specific reaction rate constant (min^{-1}), A is the Total area of peak (min·°C), ΔT is the Peak height at any temperature T(°C), a is the Area of the peak at temperature T(min·°C).

The plot of ln k against 1/T (Fig. 4) is used to calculate the activation energy from the fitted slope and the frequency factor from the intercept (Table 2).

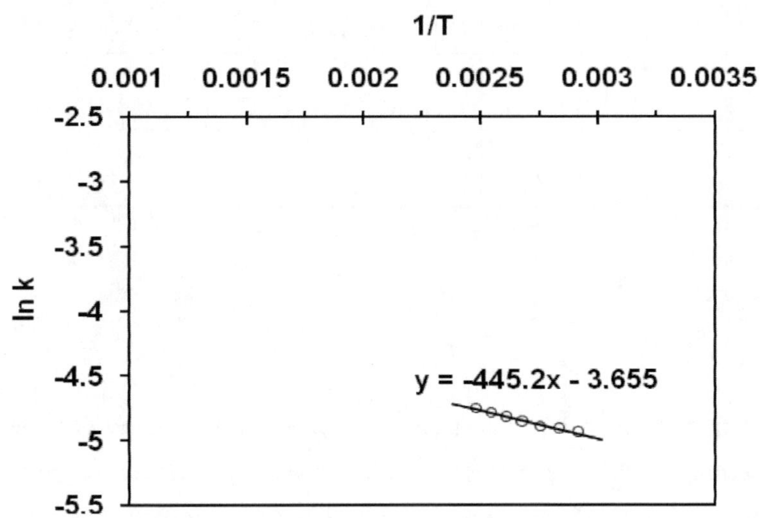

Figure 4. Graphical plot of Borchardt-Daniels (BD) relation.

IOP Conf. Series: Materials Science and Engineering **360** (2018) 012019 doi:10.1088/1757-899X/360/1/012019

Table 1. TGA results of phases of Cadmium oxalate in air.

Step	Temp. Range (°C)	DTA Peak Temp. (°C)	Mass loss %		Reaction involved
			Obs.	Cal.	
I	64-150	127.37	21.4	21.338	$CdC_2O_4.3H_2O \rightarrow CdC_2O_4+3H_2O$(i)
II	220-310	-	11.08	11.007	$CdC_2O_4 \rightarrow CdCO_3 + CO$ (ii)
III	310-360	357	17.12	17.294	$CdCO_3 \rightarrow CdO + CO_2$(iii)

Table 2. Kinetic parameters evaluated from non-mechanistic equations for thermal phases $CdC_2O_4.3H_2O$ for the first step of transformation.

Step	Temp. Range (°C)	DTA Peak Temp(°C)	Mass loss %		Reaction involved
			Obs.	Cal.	
I	64-150	127.37	21.4	21.338	$CdC_2O_4.3H_2O \rightarrow CdC_2O_4+3H_2O$(i)
II	220-310	-	11.08	11.007	$CdC_2O_4 \rightarrow CdCO_3 + CO$ (ii)
III	310-360	357	17.12	17.294	$CdCO_3 \rightarrow CdO + CO_2$(iii)

4. Conclusion

In cadmium oxalate crystal there are three watermolecules present which render the crystal unstable at higher temperature. The thermal phase changes in the air atmosphere suggest the material to be unstable even at moderate temperatures. The activation energy (also the frequency factor) of dehydration step is the lowest in all the other stages of decomposition, implying low energy involved for loosening of the bonded water molecules. Activation energy of the third stage of decomposition is the highest, implying that oxalate to oxide conversion is indeed a harder reaction.

References and Notes

1. J. P. Redfern, *Thermal Analysis Review, Stanton Instruments, London.* 4 **(1967)**.
2. H. J. Borchardt and F. J. Daniels, *J. Am. Chem. Soc.* 79, 41**(1957)**.
3. A. Broido, *J. Polym. Sci.* 7A(2), 1761 **(1961)**.
4. E. B. Michel. *Hand Book of Thermal Analysis: Principles & practice, Elsevier publication, New York.* 509 **(1998).**
5. K. M. Chauhan, *Studies on Crystal Growth and Characterization of Cadmium Oxalate Trihydrate Single Crystals, Ph.D. Thesis, S. P. University, India* **(2009).**
6. W. W.Wendlandt, *Thermal Methods of Analysis, Interscience, NewYork* **(1964).**
7. G. O. Piloyan and O. S. Novikova, *Russ. J. Inorg. Chem.*12**,** 313**(1967).**
8. A.W. Coats and J. P. Redfern, *Nature* 201**,** 68**(1964)**.
9. D.W. van Krevelen, C van Heerden, and F. J. Huntjens, *Fuel* 30, 253**(1951)**

IOP Conf. Series: Materials Science and Engineering **360** (2018) 012020 doi:10.1088/1757-899X/360/1/012020

Structural and Collective Dynamics of $Bi_{50}Zn_{50}$ of binary metallic glass

P. H. Suthar[1] and K. M. Chauhan[2]

[1,2]Department of Physics, C. U. Shah Science College, Ashram Road, Ahmedabad 380014, Gujarat, India.

Email: sutharpunit@rediffmail.com

Abstract: In the present paper, structural properties, collective dynamics and elastic properties have been studied of $Bi_{50}Zn_{50}$ liquid metallic glass at a wide range of temperatures. Ashcroft and Langreth binary structure factor have been used to study for static structure factor and pair correlation function for present alloy. The collective excitation has been studied for longitudinal and transverses phonon modes based on our well-established model potential along with Hubbard and Beeby (HB) approach. These frequencies are expressed in terms of many-body correlation functions of atoms as well as of interatomic potential. The important ingredients in the present study are the pair-potential and pair-correlation functions. The local field correlation functions due to Hartree (H) and Farid et al (F) are employed to investigate the influence of the screening effects on the vibrational dynamics of $Bi_{50}Zn_{50}$ liquid metallic glass. The results for bulk modulus (B_T), modulus of rigidity (G), Poisson's ratio (ξ), Young's modulus (Y), Debye temperature (Θ_D), the propagation velocity of elastic waves and dispersion curves are reported. Good agreement with the experiment is observed for the structure factors and elastic properties results are consistent and confirm the applicability of model potential and self-consistent phonon theory for such studies.

1. Introduction

Binary metallic glasses show the wide range of application in the filed of engineering and material science during the several decades. They made up of components of simple metals provide us with physically interesting systems for theoretical investigations[1-4]. The simple structure of binary alloys lets us to use the pseudopotential theory to calculate the interatomic potentials among simple metal components to a high degree of precision[2-4]. Based on the knowledge of interatomic interactions we can know the mechanical, optical thermodynamic and electronic transport properties of Bi-Zn materials. The research on binary metallic glasses put on more momentum in the early ninety eighties when the continuous casting process for commercial production of metallic glasses sheets, lines and ribbons was developed [5]. The present Bi- based liquid alloy is consisting of polyvalent metal and it's still inspiring to academician and researchers on temperature variation of physical properties such as viscosity, structural behavior, thermodynamics, mechanical, sound velocity and so many [1-6]. Bi-Zn binary alloy has occupied an important position due to their technological important and industrial application.

H Ueno et al [6] was successfully reported the structure factors and pair correlation functions of binary metallic glass $Bi_{50}Zn_{50}$ using x-ray diffraction, neutron and Moteno Carlo method at four different temperature [6].This has motivated us to undertake a first-time theoretical investigation of the structural properties, phonon modes and its elastic properties of a liquid $Bi_{50}Zn_{50}$ binary alloy at various temperatures [6]. To incorporate the electron ion interaction we have used well recognized model potential [7, 8]. In momentum space (q-space) the unscreened bare-ion form factor is given by [7, 8]. The model potential used to describe electron-ion interaction is form in q-space [7,8]

$$V_b(q)=\frac{-4\pi Ze^2}{\Omega_0 q^2}\left[\cos(qr_c)-\frac{\exp(-1)q\cdot r_c}{1+q^2r_c^2}\frac{q^2r_c^2}{\left(1+q^2r_c^2\right)}\right] \tag{1}$$

Here Z, Ω, **q**, e and r_c are the valency, atomic volume, wave vector, charge of an electron and the parameter of the potential respectively. In the present paper, the parameter of the potential for binary alloys has been determined following the procedure $q=q_o$ method. Two local field correction functions viz. Hartree (H) [9] and Farid et al (F) [10] are used to study the screening influence on longitudinal and transverse sound velocity, isothermal bulk modulus, modulus of rigidity, Poisson's ratio, Young's modulus and Debye temperature at wide range of temperature (793K, 873K, 1023K 1123K) [6] .

2. Theory

One of the earliest attempts to compute binary structure factor S(q) of binary metals are the system of neutral Hard Sphere as a reference system[11]. Among these the straight forward and most extensively used model is obtained from the exact solution of the Percus-Yevick (PY) equation for Hard-Sphere diameter σ, which is obtained, to get best fit of S(q) with experimental data also known as Ashcroft-Langreth model [11].The total structure factor $S(q)$ of a binary mixture in terms of partial structure factors $S_{11}(q)$, $S_{22}(q)$ and $S_{12}(q)$ is written by Ashcroft et al [11] and the final expression for the structure factor is written as under.

$$S(q)=xS_{22}(y)+2\{x(1-x)\}^{\frac{1}{2}}S_{12}(y)+(1-x)S_{11}(y) \tag{2}$$

These structure factors have been used for the computation of the pair distribution function g(r). The expression for the pair distribution function g(r) is written as,

$$g(r)=1+\left(\frac{1}{2\pi^2\rho r}\right)\int_0^\infty q[S(q)-1]\sin(qr)dq. \tag{3}$$

The fundamental ingredient, which goes into the calculation of the phonon dynamics of binary metallic glass, is the effective pair potential, which consists of two contributions. One is the direct interaction between ion given by (Ze^2/r). Other contribution is due to the indirect interaction between ions through the electron cloud and this interaction is computed using the normalized energy wave number characteristics.

In the present study the effective core radius r_c is calculated using the expression

$$r_c^{eff}=\frac{0.51\,r_{eff}}{Z_{eff}^{\frac{1}{3}}} \tag{4}$$

and r_{eff} is given by

$$r_{eff}=(x\cdot r_A^3+(1\text{-}x)\cdot r_B^3)^{1/3} \tag{5}$$

where r_A and r_B are the Wigner-seitz radii of A and B atoms respectively.

The effective ion-ion interaction is given as,

$$V(r)=\frac{Z^2e^2}{r}+\frac{2}{\pi}\int dqF(q)\exp(iq\cdot r), \tag{6}$$

where $F(q)$ is the energy wave number characteristic. The first term gives the coulomb interaction between ions and the second term is due to the indirect interaction through the conduction electrons. This indirect interaction can be calculated using the energy wave number characteristics $F(q)$. The effective pair potential for the binary system will be obtained if the pair potentials for the single components are known. It is given by the expression

$$V_{eff}(r) = C_A^2 V_{AA}(r) + 2C_A C_B V_{AB}(r) + C_B^2 V_{AA}(r) \tag{7}$$

Where $V_{AA}(r)$ is the pair potential for the A-A component, $V_{BB}(r)$ is that for the B-B component and $V_{AB}(r)$ is that for the A-B component in the binary system. C_A and C_B are the concentration of the A and B type components respectively,

$$C_A = \frac{(1-x)\Omega_A}{(1-x)\Omega_A + x\Omega_B} \tag{8}$$

$$C_B = \frac{(1-x)\Omega_B}{(1-x)\Omega_A + x\Omega_B} \tag{9}$$

Where Ω_A and Ω_B are the molar volumes of A and B components, respectively.

Another way of finding the pair potential for the binary alloy is to consider the System $A_x B_{1-x}$ is considered as a system of the effective atoms. In the present Study we have considered the effective atoms approach for computing the dispersion curves.

Two different model potentials mentioned in earlier chapters are used for the present study also. The usual form of the effective pair potential in the liquid metals and their alloys is

$$V_{eff}(r) = \frac{(Z_{eff}e)^2}{r} + \frac{\Omega_{0eff}}{\pi^2}\int dq \cdot F_{eff}(q) \cdot q^2 \left[\frac{\sin(qr)}{qr}\right] \tag{10}$$

Where Z_{eff} is the effective valence, Ω_{0eff} is the atomic volume of the liquid metals and binary alloys, respectively.

The energy wave number characteristics in expression is given by,

$$F(q) = -\frac{\Omega_{0eff}q^2}{16\pi}\left|V_B^{eff}(q)\right|^2 \times \frac{\left[\varepsilon_H^{eff}(q)-1\right]}{\left[1+(\varepsilon_H^{eff}(q)-1)(1-f_{eff}(q))\right]} \tag{11}$$

Here $V_B^{eff}(q)$ is the effective bare ion potential, $\varepsilon_H^{eff}(q)$ the Hartree dielectric response function and $f_{eff}(q)$ the local field correction function.

In the present investigation of computing the phonon dispersion relation of alkaline earth metals, transition metals and their binary alloys, the most frequently used approach of Hubbard and Beeby [12] is taken up. With the physical argument that the product of the static pair correlation function "g(r)" and the second derivative of the interatomic potential "$V_{eff}(r)$" is peaked near the heard sphere diameter 'σ' , Hubbard and Beeby have derived the expression longitudinal phonon frequencies $\omega_L(q)$ and transverse phonon frequencies $\omega_T(q)$ as [7,8,12]

IOP Conf. Series: Materials Science and Engineering **360** (2018) 012020 doi:10.1088/1757-899X/360/1/012020

$$\omega_L^2(q) = \omega_E^2\left[1 - \frac{3\sin(q\sigma)}{q\sigma} - \frac{6\cos(q\sigma)}{(q\sigma)^2} + \frac{6\sin(q\sigma)}{(q\sigma)^3}\right] \tag{12}$$

$$\omega_T^2(q) = \omega_E^2\left[1 + \frac{3\cos(q\sigma)}{(q\sigma)^2} - \frac{6\sin(q\sigma)}{(q\sigma)^3}\right] \tag{13}$$

Here, ω_E represents the maximum phonon frequency and is given as

$$\omega_E = \frac{4\pi n_{eff}}{3M_{eff}} \int_0^\infty g(r) r^2 V''_{eff}(r) dr \tag{14}$$

With

$$V''_{eff}(r) = \frac{4Z^2}{r^3} + \frac{\Omega_0}{\pi^2}\int_0^\infty F(q) q^2 \left[\frac{2\sin(qr)}{qr^3} - \frac{2\cos(qr)}{r^2} - \frac{q\sin(qr)}{r}\right] dq \tag{15}$$

Here ρ, M, g(r), Ω_0, F(q) and S(q) be the number density, atomic mass, pair correlation function, atomic volume, energy wave number characteristic and the structure factor of the element, respectively. The expression for structure factor and pair correlation function is given in equation (2) and equation (3) respectively.

In the long-wavelength limit, the phonon dispersion curve shows an elastic behaviour. Hence, the longitudinal v_l and transverse v_t sound velocities are also calculated by

$$\omega_L \,\alpha\, q \text{ and } \omega_T \,\alpha\, q, \; \therefore \omega_L = v_l q \text{ and } \omega_T = v_t q \tag{16}$$

Various elastic properties are then determined by the longitudinal and transverse phonon frequencies.

The bulk modulus B, Poisson's ratio 'ξ', modulus of rigidity G, Young's modulus Y and the Debye temperature θ_D are calculated using the expression below[7]

$$B = \rho\left(v_l^2 - \frac{4}{3}v_t^2\right) \tag{17}$$

ρ is the isotropic density of the solid,

$$\xi = \frac{1 - 2(v_t^2/v_l^2)}{2 - 2(v_t^2/v_l^2)} \tag{18}$$

$$G = \rho v_t^2 \tag{19}$$

$$Y = 2G(\xi + 1) \tag{20}$$

and

$$\theta_D = \frac{h}{k_B}\left[\left(\frac{9\rho}{4\pi}\right)^{1/3}\left(\frac{1}{v_l^3} + \frac{2}{v_t^3}\right)^{-1/3}\right] \tag{21}$$

where 'h' is Plank constant and k_B is the Boltzmann constant.

IOP Conf. Series: Materials Science and Engineering **360** (2018) 012020 doi:10.1088/1757-899X/360/1/012020

3. Results and Discussion

Table 1: Input parameters and constants for $Bi_{50}Zn_{50}$ binary metallic glass

T[K]	$Z_{eff}\ (au)$	$\Omega_{oeff}\ (a.u)^3$	$r_c^{eff}\ (au)$	$M_{eff}\ (a.m.u)$
793		173.351	1.1613	
873	3.5	174.973	1.1655	137.185
1023		178.003	1.1721	
1123		180.017	1.1766	

The constant and parameters used in the present study of collective excitation are given in Table 1. In the fig 1 and fig 2 show the structure factors S(q) and pair distribution functions g(r) along with the experimental data [6] respectively. Good agreement with the experiment is observed for the structure factors and very good agreement is observed for pair correlation function at various temperatures. When temperature increases the position of the first peak points are shifted toward q=0. The present pair correlation function is used to compute the phonon modes and elastic properties for Bi-Zn binary alloy.

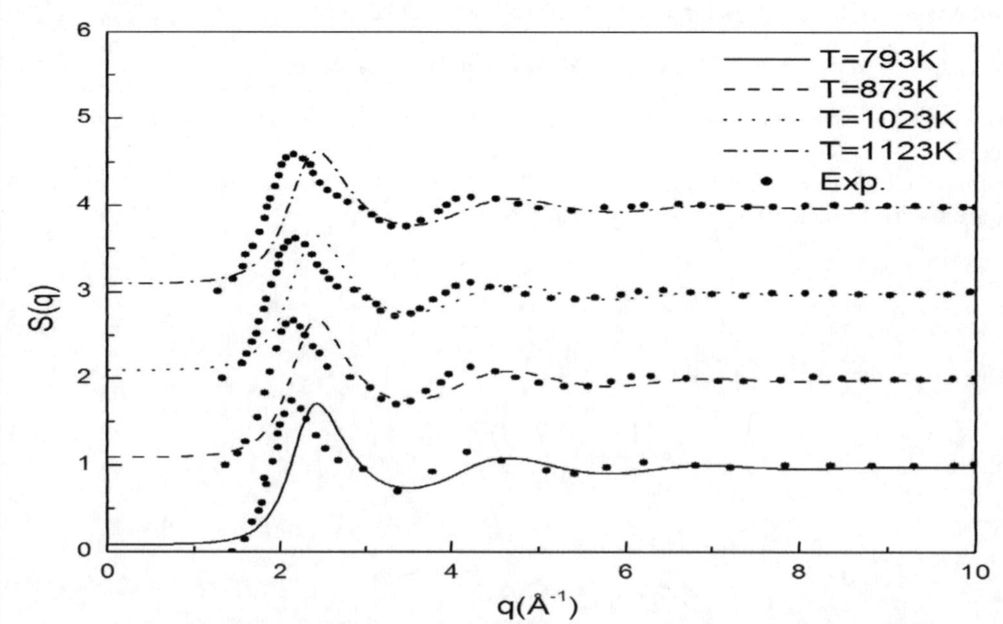

Figure 1. Structure factor, S(q) for $Bi_{50}Zn_{50}$ alloys at different temperature along with Exp. data [6]

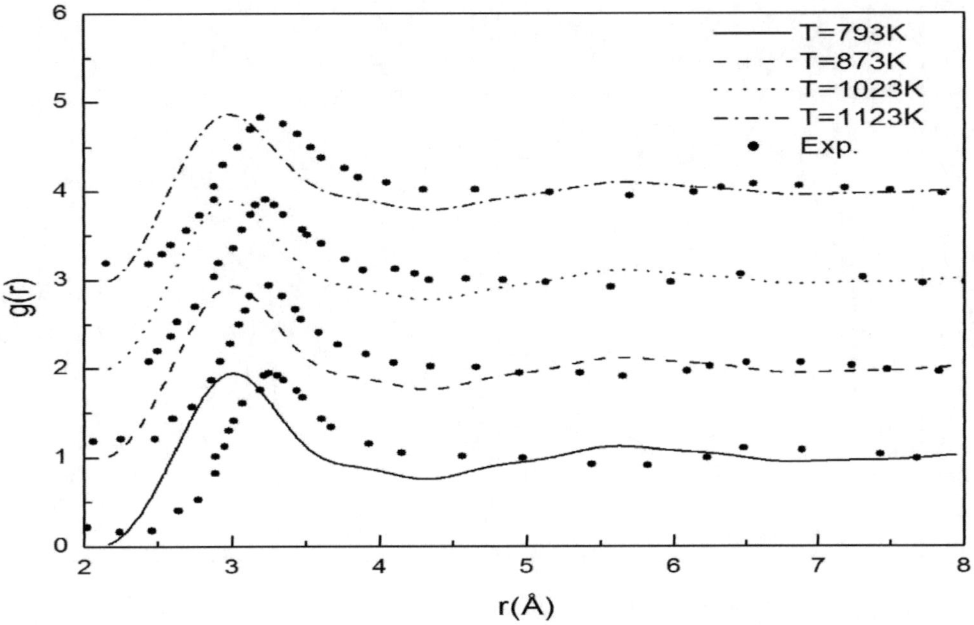

Figure 2. Pair distribution functions, g(r) for $Bi_{50}Zn_{50}$ alloys at different temperature along with Exp. data [6]

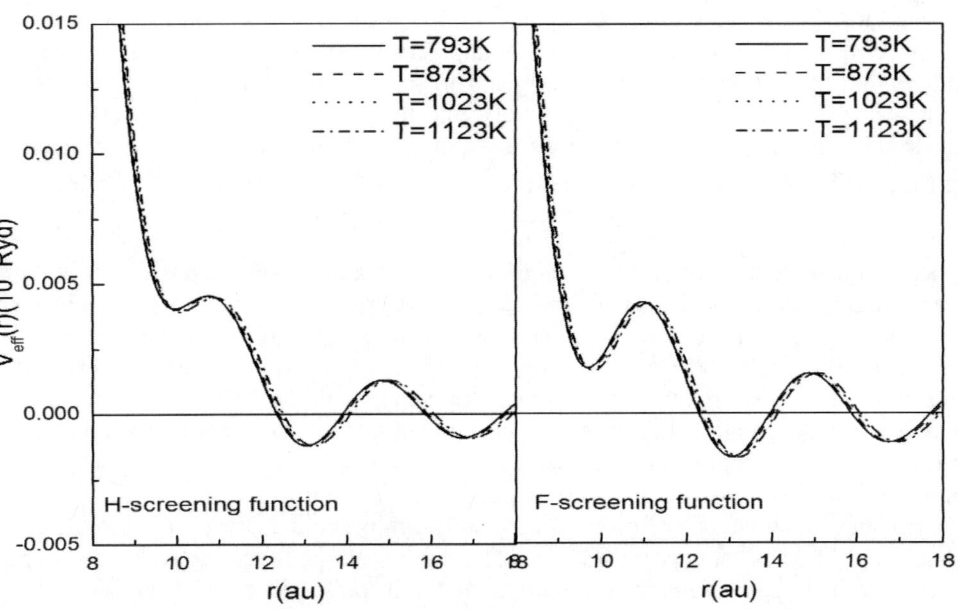

Figure 3. Effective pair potential for $Bi_{50}Zn_{50}$ binary metallic glass at various temperatures.

IOP Conf. Series: Materials Science and Engineering **360** (2018) 012020 doi:10.1088/1757-899X/360/1/012020

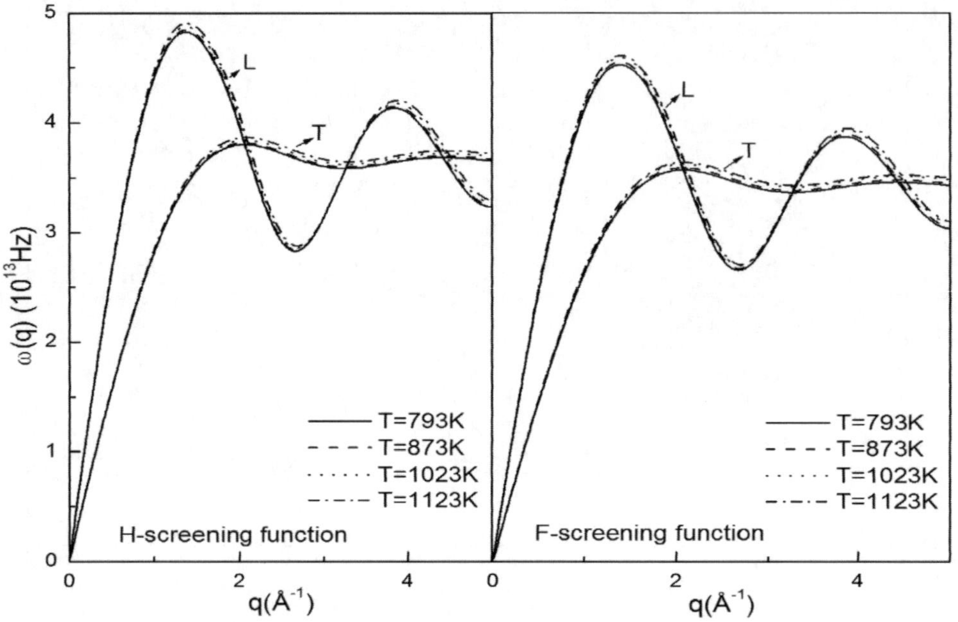

Figure 4. Longitudinal and Transverse phonon mode of $Bi_{50}Zn_{50}$ binary metallic glass at various temperatures.

Our well established model potential [7,8] is used along with two different local field correction functions Hartree [9] and Farid et al [10] for the first time to produce effective pair potentials for $Bi_{50}Zn_{50}$ at various temperatures shown in fig 3. From figure 3 it is observed that position of the first minima is affected by the different types of screening functions along with various temperatures. The first minimum in the pair potential occurs nearly at same interatomic spacing. However, the magnitude of potential energy at first neighbour is greatly affected by type of screening as well as temperatures used in the system. The maximum depth in the pair potential in the present study is obtained due to F screening function. It is also noticed that the first zero for $V(r = r_0)$ due to two local field correction functions occurs at $r_0 \approx 12.39$Å and $r_0 \approx 12.39$Å for the H and F screening functions along with different temperatures, respectively. It is also observed that maximum depth position is not affected by the different temperatures. The oscillatory nature is also seen even at large *r*-region. Fig 4 shows the collective modes of present metallic glass. H gives the higher magnitude compared to F screening function and it is observed that the magnitude is increased as temperature increases. It is also observed that the position of the first minima approximately coincides with the first peak in the static structure factor of respective different temperatures for present alloy. We have not superimposed S(q) $\rightarrow$ q curve on $\omega \rightarrow$ q to avoid complication in the figures. It is also noticed that the related magnitude of first peak of PDC increasers as a temperature increasers but the position of first peak is not affected by different screening functions. The first minimum in the longitudinal branch of the dispersion curve falls at a value $q \approx 2.68$ Å^{-1} and $q \approx 2.68$Å^{-1} for H and F screening functions at different temperatures respectively. The first peak heights of the longitudinal and transverse phonon modes of this metallic glass are nearly the same at the different temperatures. so that, the dispersion curves of present alloys is found to be similar. Phonons of transverse branch undergo large thermal modulation due to anharmonicity of lattice vibrations in this branch. In the long wave length limit, the dispersion curves are linear and confirming characteristics of elastic wave.

Table 2: Elastic properties of $Bi_{50}Zn_{50}$ binary metallic glass at various temperatures along with local field correction function

properties	T=793K		T=873K		T=1023K		T=1123K		Cal.
	H	F	H	F	H	F	H	F	
$\upsilon_L \times 10^5$ (cm/s^2)	5.579	5.217	5.589	5.256	5. 604	5.265	5.618	5.277	--
$\upsilon_T \times 10^5$ (cm/s^2)	3.221	3.012	3.227	3.034	3.235	3.039	3.244	3.047	--
$B\times10^{12}$ (dyne/cm^2)	1.435	1.255	1.428	1.263	1.414	1.247	1.406	1.240	0.505
$G\times10^{12}$ (dyne/cm^2)	0.861	0.752	0.857	0.758	0.848	0.748	0.843	0.744	0.275
$Y\times10^{12}$ (dyne/cm^2)	2.152	1.882	2.143	1.893	2.120	1.871	2.109	1.860	0.70
ξ	0.25	0.25	0.25	0.25	0.25	0.25	0.25	0.25	0.29
$\theta_D(K)$	361.1	337.7	360.5	339.0	359.4	337.6	358.9	337.1	224.5

Presently, calculated bulk modulus (B_T), modulus of rigidity (G), Poisson's ratio (ξ), Young s modulus (Y), Debye Temperature (θ_D) and sound velocities are shown in Table 2. For this composition of Bi-Zn system, no experimental data of elastic constants are available in literature. As a result, we have average the elastic constants of pure Bi and Zn and the average values are shown in Table 2 along with the presently calculated values. It is observed in the present case that bulk modulus, modulus of rigidity and Young modulus are overestimate. The presently calculated values of elastic constants using H and F functions show overestimated with the average values. Thus, it is observed that proper local field correction function is necessary in order to estimate the elastic properties of metallic glasses like Bi-Zn.

4. Conclusion

Bi-Zn binary metallic glass is containing wide range of application in the field of technology and material science. So that, we conclude that results of structure factors and pair correlation functions found satisfactorily good agreement with available experimental data. The PDC generated from HB approach reproduces satisfactorily general characteristics of dispersion curves. The magnitude of longitudinal and transverses curves are increases as increases the temperatures. Present study of elastic properties will remain very useful to the set of theoretical data. This local model potential depends on effective core radius which is used here to describe the ion-electron interaction. The present study are also confirms the applicability of local model potential in the aforesaid properties. Our well known model potential along with H and F local field correction functions generates consistent results regarding the phonon modes of the binary metallic glasses.

Acknowledgement

The author is thankful to Prof. P. N. Gajjar, Head, Department of Physics, Gujarat University, Ahmedabad, India, for valuable guidance.

References

[1] B. Y. Thakore, S. G. Khambholja, P. H. Suthar, N. K. Bhatt and A. R. Jani 2010 *Chin. Phys. Letts.* **27** 096203
[2] B. Y. Thakore, H. Dabhi, M. Joshi and A R Jani 2009 *Phys. Scr.* **79** 0250007
[3] B. Y. Thakore, P. N. Gajjar and A. R. Jani 2000 *Bull. Mat. Sci.* **23** 5
[4] A. M. Vora 2008 *Rom. J. Phys.* **53** 517
[5] W. H. wang 2006 *J. Apply.Phys.* **99** 093506

[6] H.Ueno, Y.Kawakita, K. Ohara, S. Tahara, S. Kohara, M. Itou and S. Takeda 2012 Proc. of the 5th European Conf. on neutron scattering *J.Phys.Conf. Serie* **340** 012081
[7] P. H. Suthar, P. N. Gajjar, B. Y. Thakore and A. R. Jani 2013 *Proc. of the ScieTech-2013*, *J Phys. Conf. Series* **423** 012030
[8] P. B. Thakore, P. N. Gajjar and A. R. Jani 2009 *Pramana J. Phys.* **72** 1045
[9] W. A. Harrison 1967 *Pseudopotential in the theory of metals* W A Benjamine Publisher, New York
[10] B. Farid, V. Heine, G. E. Engel, I .J. Robertson 1993 *Phys. Rev. B* **48** 11602
[11] N. W. Ashcroft and D. C. Langreth 1967 *Phys. Rev.* **156** 685
[12] J. Hubbard and L. Beeby 1969 *J. Phys. C* **2** 556

IOP Conf. Series: Materials Science and Engineering **360** (2018) 012021 doi:10.1088/1757-899X/360/1/012021

Synthesis and Characterization of ceria co-doped with Samarium(Sm) and Antimony: a potential candidate as a solid electrolyte for intermediate temperature solid oxide fuel cells

Sandhya K[a*], Chitra Priya N S[a], Deepthi N Rajendran[a]

[a] Department of Physics, Govt. College for Women, Thiruvananthapuram. Kerala, India.

* sandhyakprabeesh@gmail.com

Abstract. The huge potential of solid oxide fuel cells (SOFCs) in providing a clean and reliable energy source has ignited enormous interest in them. The electrolyte is one of the key components of SOFC. Samarium doped ceria oxide is one of the promising materials as an electrolyte for IT–SOFCs. In this paper, ceria co-doped with samarium and antimony have been prepared and characterized to be used as solid electrolyte for intermediate temperature fuel cells. The composite was prepared by Solid state reaction and the effect of sintering temperature on the microstructure for a new composite based on ceria were studied. All samples were characterized using X-ray diffraction, Scanning electron microscope, FTIR, Atomic Force Microscope and Impedance Spectroscopy. Results of analysis of the structural and electrical properties of the present composite indicate that present research is good as a solid electrolyte for IT-SOFCs.

1. Introduction

Solid Oxide Fuel Cells are currently generating enormous worldwide interest due to their widespread application in power generation systems [1]. Yttrium stabilized Zirconia has been favored by researchers traditionally due to its higher ionic conductivity at its high operating temperature around 1000°C·The high operating temperature limits its application due to its degraded solid oxide fuel cell application at its low operating temperature. However, this drawback can be overcome by using electrolyte operating at moderate temperature [2]. In this context, doped ceria emerged as potential candidate operating at intermediate temperature [3]. Minimum dopant level is the main factor, which stabilizes the high temperature cubic face centered (FCC) phase which in turn results in higher conductivity. Ceria Oxide has a cubic fluorite structure with their ability to get reduced from Ce^{4+} to Ce^{3+} at elevated temperature. This leads to formation more oxygen vacancies needed for higher ionic conductivity at moderate temperature [4]. Various dopant such as Sm^{3+}, Dy^{3+}, Y^{3+}, $Pr^{3+}$and Gd^{3+}showed good solubility in ceria resulting in better ionic conductivity. There was a remarkable enhancement in the ionic conductivity of doped ceria electrolyte obtained by adopting different synthesis routes [5]. Various composites such as $Ce_{0.85}Gd_{0.15}O_2$ [6], $Ce_{0.8}Sm_{0.2}O_{1.9}$ [7], $Ce_{0.75}Gd_{0.25}O_{1.875}$ [8], $Ce_{0.8}Sm_{0.2}O_{1.9}$, $Ce_{0.83}Sm_{0.17}O_{1.915}$ [9], $Ce_{0.8}Y_{0.2}O_{1.9}$ [10], $Ce_{0.8}Y_{0.2}O_{1.9}$ [11], $Zr_{.7}Ce_{0.08}Nd_{0.1}O_{1.9}$ [12], $Ce_{0.8}Sm_{0.1}Nd_{0.1}O_{1.9}$ [13], and $Ce_{0.85}La_{0.125}Sr_{0.025}O_{1.9125}$ [14] are reported having better conductivity than pure ceria.

Various synthesis routes are reported for the synthesis of samarium-doped ceria to enhance the ionic conductivity at low operating temperatures. The new composition of the electrolyte is first time prepared by the solid state method to attain good ionic conductivity at reduced sintering temperature. In the present investigation, detailed studies on structural and electrical properties of the new composition of the electrolyte are presented.

2. Experimental

2.1. Synthesis procedure

Solid state reaction method was used to synthesize ceria co-doped with Samarium(Sm) and Antimony sample. Stoichiometry calculated amount of CeO_2, Sm_2O_3 and Sb_2O_3 were weighed and mixed with acetone manually using an agate mortar to obtain $Ce_{0.8}Sm_{0.1}Sb_{0.1}O_2$ compositions. The mixtures were grinded using agate-mortar for 48 h to obtain uniform and finely ground powder. The prepared mixture was calcined at 600°C for 3 h to remove any carbon residues remaining in the oxide powder. The foamy, pale yellow in color, the residue was crushed and then pelletized using a hydraulic press by applying 2.5MPa pressure. The pellets were sintered at 800°C for 10 h.

Structural characterizations were carried out using X-ray diffraction analysis to determine phase purity, lattice parameter and crystalline size of the sample. Qualitative analysis regarding different functional group present in the compound are identified using FTIR images. The micron morphologies of the sintered sample were examined using Scanning electron microscope (SEM). Atomic force microscopes were employed to characterize topographical information about the topography of the sample. Electrical impedance as a function of temperature in the range 250°C-750°C was measured using impedance spectroscopy technique.

3. Results and Discussion

3.1. Characterization by XRD

The XRD pattern of the prepared electrolyte sample is shown in the figure 1. Diffraction peaks recorded at different angles corresponding to the crystallographic phases of cubic fluorite type cerium oxide phase in the space group of Fm3m with reference to the JCPDS file (34-0394). From the Scherrer's equation [15], the average crystallite size was found to be 15 nm with a lattice parameter of 5.47334 Å. The peak of the diffraction pattern appeared broad pointing out the smaller crystallite size of the sample leading to high ionic conductivity. The prepared electrolyte sample is pure as there are no secondary phases of samarium oxide and antimony oxide in the diffraction pattern. A shift in the XRD pattern is observed in the co-doped sample compared to the samarium doped ceria oxide due to the incorporation of the samarium and antimony ions in the ceria lattice. A qualitative analysis of XRD results agrees with reported literature.

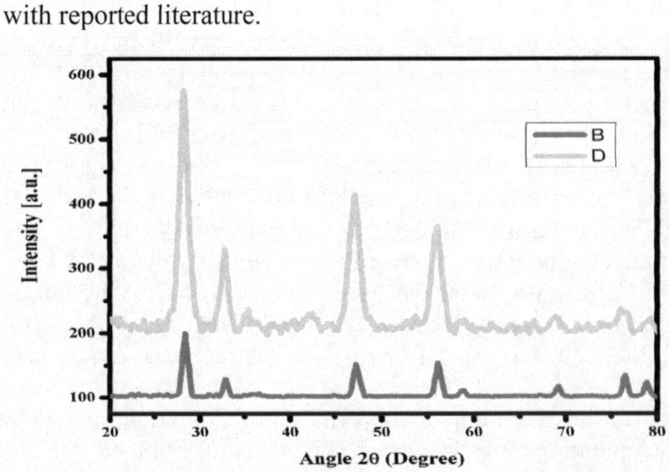

Figure 1. XRD Pattern of (B) $Ce_{0.8}Sm_{0.1}O_{2-\delta}$ and (D) $Ce_{0.8}Sm_{0.1}Sb_{0.1}O_{2-\delta}$

3.2. Characterization by FTIR

Fourier transform infrared ((FT-IR) spectroscopy provide a better insight into the phase and structural transformation of the prepared sampled calcined at 600°C for 3 h. From the FTIR spectrum, the strong band near in the range 3000-3750 cm^{-1} corresponds to O-H stretching of physically adsorbed H_2O or surface -OH group. The solid peak centered at 1631 cm^{-1} corresponds to the stretching vibrations of carbonyl containing functional groups. Sharp and fine peak spotted at 1496 cm^{-1} corresponds to Sb-O bond. The band designated in the range 433 to 1120 cm^{-1} resembles the stretching mode vibrations of Ce–O bond. [16].

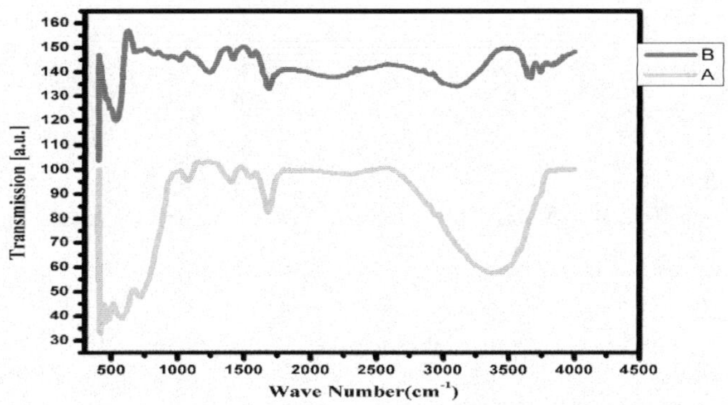

Figure 2. FTIR Spectra (A) $Ce_{0.8}Sm_{0.1}O_{2-\delta}$ and (B) $Ce_{0.8}Sm_{0.1}Sb_{0.1}O_{2-\delta}$

3.3. Characterization by SEM

SEM analysis was performed to determine the average grain size in the sintered pellet of the $Ce_{0.8}Sm_{0.1}Sb_{0.1}O_{2-\delta}$ electrolyte. The grain is uniform in size with more agglomeration. It is evident from the reported work that for an efficient charge carrier separation of the porous structure, the particle distribution should be narrower along with a tightly packed surface. Few pores are found in the SEM images of the $Ce_{0.8}Sm_{0.1}Sb_{0.1}O_{2-\delta}$ electrolyte.

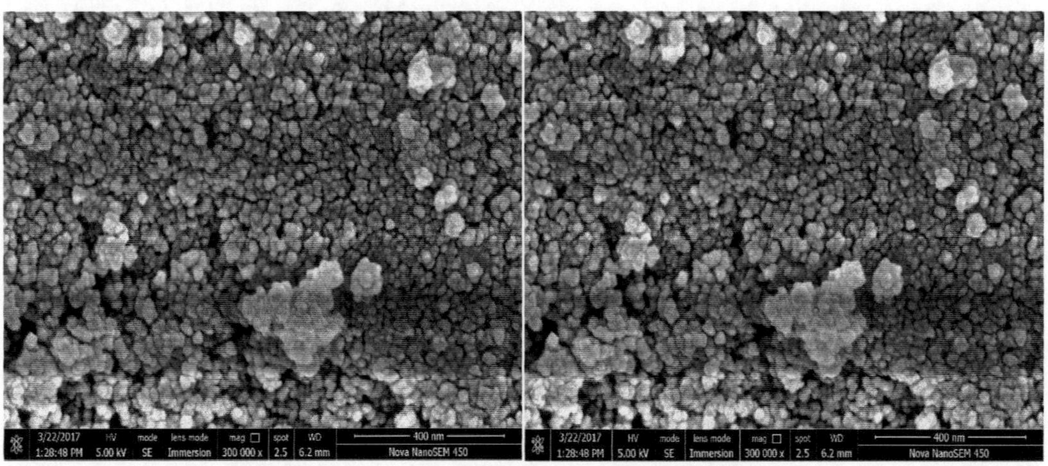

Figure 3. SEM images of $Ce_{0.8}Sm_{0.1}O_{2-\delta}$

3.4. EDAX analysis

The concentration of the dopant affects the electrical properties of the electrolyte. EDAX analysis report shows the presence of the element Ce, Sm and Sb. Absence of any other extra element confirms the formation of pure phase of $Ce_{0.8}Sm_{0.1}Sb_{0.1}O_{2-\delta}$ electrolyte.

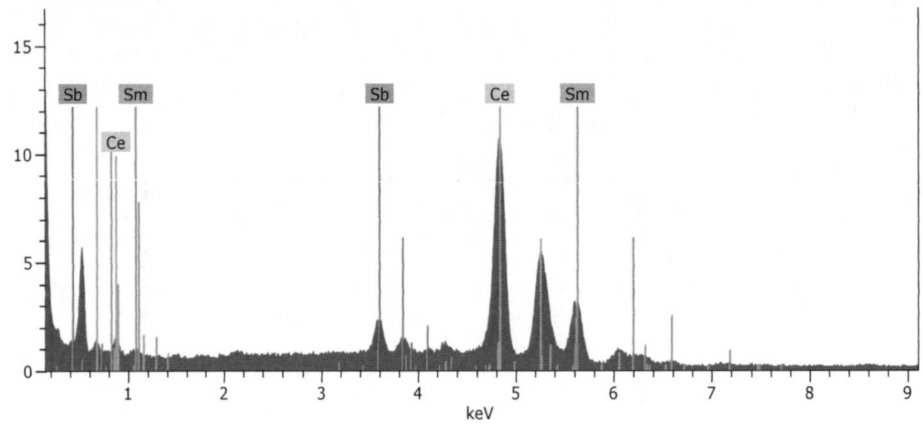

Figure 4. Energy dispersive spectrum of $Ce_{0.8}Sm_{0.1}Sb_{0.1}O_{2-\delta}$.

3.5. Atomic Force Microscope

The surface roughness of the electrolyte effects performance of fuel cell performance. The average AFM surface roughness of the prepared sample was measured using atomic force microscope operated in non –contact mode [17]. The sample was measured at multiple spot to determine the average grain size. Ionic hopping mechanisms are hindered with large number of grain boundaries for electrolyte material. Improvement in fuel cell performance can be obtained with a large grain with less number of grain boundaries which leads to better ionic conductivity. The topography and grain formation of the prepared sample are shown in AFM images. The agglomeration of the grains is more in the AFM images.

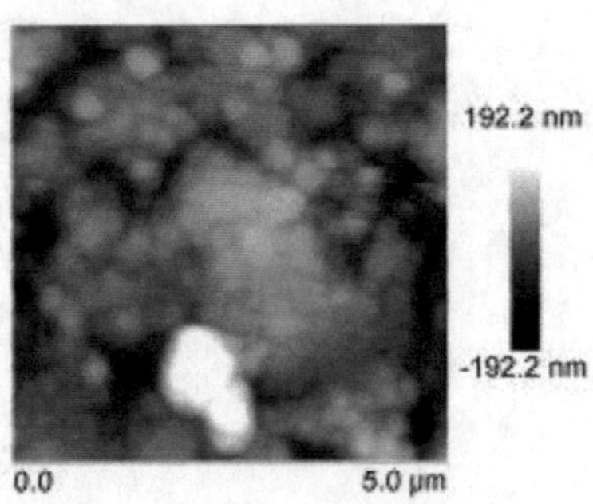

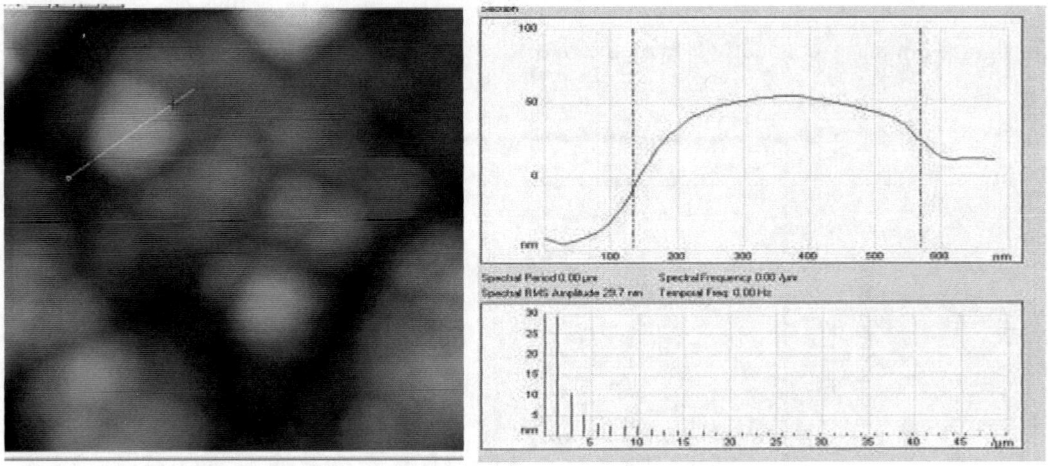

Figure 5. AFM image of $Ce_{0.8}Sm_{0.1}Sb_{0.1}O_{2-\delta}$.

3.6. Impedance analysis

Impedance spectroscopy is an effective method to determine the distribution of the conductivity of the prepared powders. Figure 6 shows the plot of z′ versus z′′ at different temperature which produces semicircles with a diameter R. As the temperature increases, it is spotted that the diameter of the arc reduces and as a result shrinking from right to left of the axis. This indicates non-Debye type of relaxation with distribution of relaxation time. The reduction in the diameter of the arc is due to the decrease of impedance at high temperature. An increase in conductivity is yielded at high temperature due to the reduction in the resistance of the sample.

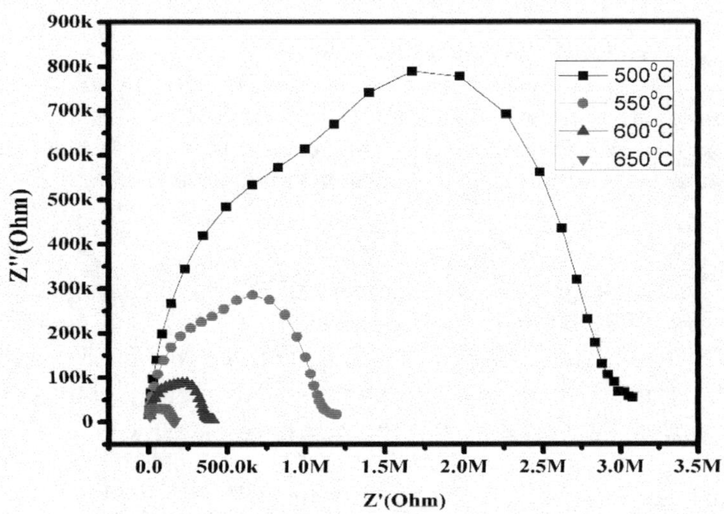

Figure 6. Impedance plot of $Ce_{0.8}Sm_{0.1}Sb_{0.1}O_{2-\delta}$.

Figure 7 depicts the variation of imaginary part of impedance z′′ with frequencies at different temperature. From the plot, it can be seen there are relaxation peaks corresponding to the bulk conduction. There is a shift in the peak towards higher frequency side and broadening of peak is noted with increase in the temperature. These are the clear evidence of temperature dependent relaxation phenomena in the material. It also observed that as temperature increases, the reactance value is low even at low frequencies. This means that capacitance value is greater in the microfarad range which indicates grain-boundary conductivity with the rise of temperature at low frequencies.

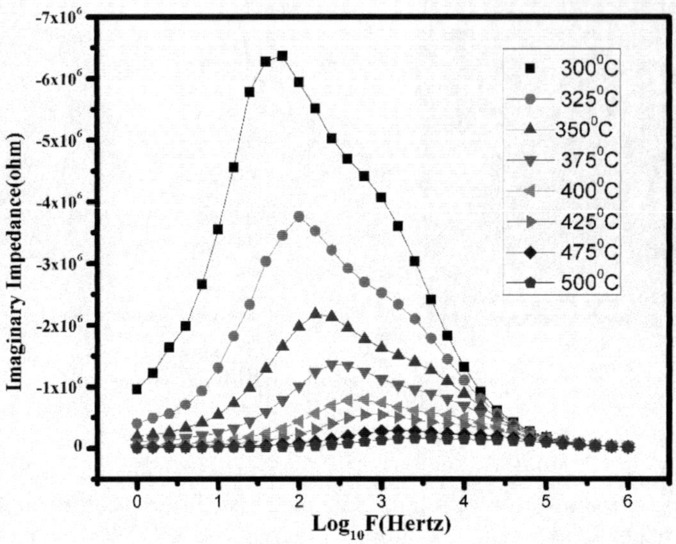

Figure 7. Variation of capacitive reactance with frequency for $Ce_{0.8}Sm_{0.1}Sb_{0.1}O_{2-\delta}$.

3.7. Electrical Studies

Bulk conductivity was computed from the impedance spectroscopy by

$$\sigma = L / RA \qquad (1)$$

Where L is the thickness of the pellet, A is the electrode area and R is the resistance as calculated with the impedance spectroscopy [18]. The conductivity is elucidated from the Arrhenius plot of log (бT) vs 1000/T shown in figure 8. The conductivity σ is obtained using the equation

$$\sigma = (\sigma_0 / T)\exp^{\left(-\frac{E_A}{kT}\right)} \qquad (2)$$

Here σ_0 is the high temperature limit of conductivity, E_A is activation energy and kT is associated with temperature variation in the measurement where k is Boltzmann constant and T is temperature [19]. The activation energy is figured out from the linear portion of the slope is .934 eV. From the graph 8, it is evident that the conductivity lightly deflects from the Arrhenius behavior. There is a deviation in the Arrhenius behavior due to the formation of inhomogeneity in the grain. This is caused by dopant-oxygen vacancy interaction [20]. The ac conductivity is estimated to be $3.58x10^{-05}$ Scm^{-1} for the pellet sintered at 800°C for 10 h. It is noteworthy that the better ionic conductivity is reported with the addition of trivalent ions at lower sintering temperature. Better cell performance is confirmed by high ionic conductivity at lower temperature.

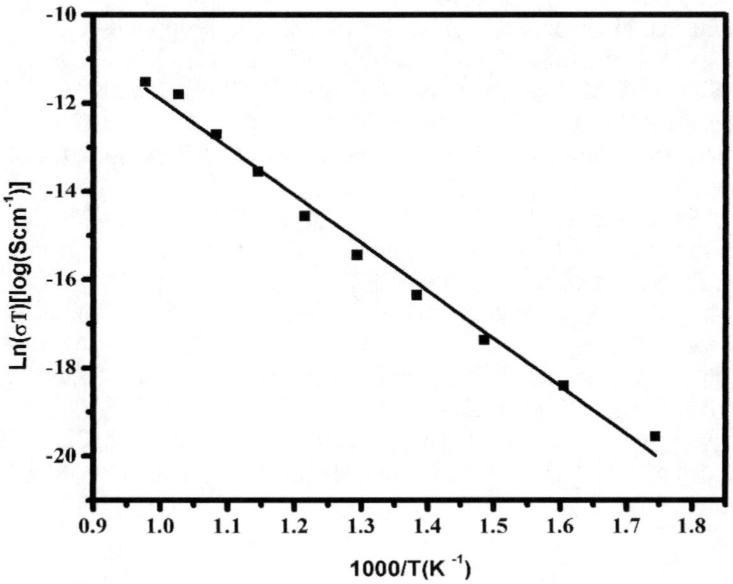

Figure 8. Arrhenius plot of $Ce_{0.8}Sm_{0.1}Sb_{0.1}O_{2-\delta}$.

4. Conclusion

Nanocrystalline $Ce_{0.8}Sm_{0.1}Sb_{0.1}O_{2-\delta}$ electrolyte was prepared by simple solid state reaction method capable for operating at intermediate temperature. The crystallite size of the sample is few of nanometers. In the synthesized sample, the grain appears to be round and with a uniform size in nanometer range. It is evident that the lower sintering temperature resulted change in structural and electrical property suitable for the electrolyte material in the field of SOFC applications.

5. Acknowledgement

The authors would like to express heartfelt gratitude towards the University of Kerala for providing research facilities and for financial support as a University Junior Research Fellowship. We are also thankful to STIC Cochin and NIIST Trivandrum for providing characterization facilities.

References

[1] N. Shehata, K. Meehan, M. Hudait, N. Jain, S. Gaballah, 2014 *J. Nanometre* ,**401798,** pp. 1–7.
[2] R. Waser, M. Aono, 2007 *Nat. Mater.* **6**, pp.833–840.
[3] M. Biswas, P. Kumar Ojha, E. Moses Jayasingh, C. Durga Prasad, 2011 Nanomater. Nanotechnol. **1**, pp.55–58.
[4] A. Venkatasubramanian, P. Gopalan, T.R.S. Prasanna, 2010 *Int. J. Hydrogen Energy* ,**35**, 4597–4605.
[5] S. Ji, I. Chang, Y.H. Lee, J. Park, J.Y. Paek, M.H. Lee, S.W. Cha, 2013 *Nanoscale Res. Lett.* **8**, pp. 1–7.
[6] Zhang TS, Ma J, Cheng H, Chan SH., 2006 *Mater Res Bullet*, **41**, pp.563–8.
[7] Huang W, Shuk P, 1997 *Solid State Ionics* **100**, pp. 23–7.
[8] Seo DJ, Ryu KO, Park SB, Kim KY, Song RH., 2006 *Mater Res Bullet*,41, pp.359–66.
[9] Huang W, Shuk P, Greenblatt M..,1997 *Chem Mater*;**9**, pp.2240–5.

[10] Zhu B. 1999 *Solid State Ionics*,**119**, pp.305–10.
[11] Kimpton J, Randle TH, Drennan, 2002 *J. Solid State Ionics*,**149**, pp.89–98.
[12] Liu Y, Li B, Wei X, Pan W., 2008 *J Am Ceram Soc*, **91**, pp.3926–30.
[13] W. Huang, a P. Shuk, a,d, M. Greenblatt,a,z M. Croft,b F. Chen,c, and M. Liuc, 2000 *Journal of The Electrochemical Society*, **147**, pp.4196-4202.
[14] Nandini Jaiswal & Devendra Kumar & Shail Upadhyay & Om Parkash.,2014 *Ionics* ,20, pp.45–54.
[15] A.G.M. Silva, T.S. Rodrigues, A. Dias, H.V. Fajardo, R.F. Gonc¸alves, M. Godinho, P.A. Robles-Dutenhefner, 2014 *Catal. Sci. Technol.*, **4**, 814–821.
[16] E.S.I. El-Shafey, *2014 J. Med. Bioeng.*, **3**, pp.292–296.
[17] S. Kundu, N. Sutradhar, R. Thangamuthu, B. Subramanian, A.B. Panda, M. Jayachandran, 2012, *J. Nanopart. Res.* **14**, 1–16.
[18] R.V. Mangalaraja, S. Ananthakumar, K. Uma, R.M. Jimenez, S. Uthayakumar, M. Lopez, C.P. Camurri, 2012 *J. Ceram. Process. Res.* **13**, pp.15–22.
[19] N. Jaiswal, S. Upadhyay, D. Kumar, O. Parkash, 2013 *J. Power Sources* 222, 230–236
[20] T. Chalermkiti, M. Panapoy, N. Chaiyut, B. Ksapabutr, 2013 *Energy Procedia* **34**, pp.471–478.

IOP Conf. Series: Materials Science and Engineering **360** (2018) 012022 doi:10.1088/1757-899X/360/1/012022

Physical, Optical and FT-IR studies of Bismuth-Boro-tellurite Glasses containing BaO as modifier

B Srinivas, Abdul Hameed, M N Chary and Md Shareefuddin
Department of physics, Osmania University, Hyderabad, TS, India

E-mail: srinivas.physicsscholar@gmail.com

Abstract. The addition of glass modifier changes the properties of host glass network. The role of Ba^{2+} as glass modifier has been investigated using physical, UV-Visible and FT-IR studies on xBaO-(30-x)TeO_2-35Bi_2O_3 -35B_2O_3 (where x =5,10,15,20 and 25 mole%) glasses. X-ray diffractograms of crushed glass samples were recorded at room temperature. The absence of sharp peaks in X-ray diffractograms confirms the amorphous nature of the glass samples. Density (ρ) of glass samples were estimated using Archimedes principle. Physical properties such as molar volume (V_M), oxygen packing density (OPD), optical basicity etc. were determined. The optical absorption spectra were recorded in the wave length range of 200–1000 nm at room temperature. From the optical absorption spectra indirect allowed optical band gap, Urbach energy, refractive index (n), dielectric constant (ε), reflection loss (R), molar refractivity (R_m) and metallization criteria values were evaluated. The refractive index values were decreasing while optical band gap values were increasing with increasing concentration of BaO. IR spectra were recorded in the spectral range 400 - 4000 cm^{-1}.

1. Introduction

Borate glasses containing heavy metal oxides (HMO) such as Bismuth oxide, tellurium oxide and Barium oxide have been attracting researchers to synthesis and to study structural and physical properties due to their low melting temperature, widespread glass formation range, high density, high physical and chemical stability, and nonlinear optical properties. Generally heavy metal oxides do not form glasses on their own compare to the conventional glass forming oxides due to their low field strength. In HMO containing glasses, bismuth oxide glasses have shown wide range of applications in the area of glass ceramics, reflecting windows, γ-ray absorbers, thermal and mechanical sensors, optoelectronic devices and scintillation detectors [1, 2]. High polarizable Bi^{3+} ions and asymmetry of their oxygen coordination polyhedra can assist the non-crystallization of melts [3]. On the other hand covalently bonded B_2O_3 is a very good glass former with interesting physical and chemical properties. Basically B_2O_3 glasses are insulating in nature and the B^{3+} ion in borate crystals and glasses usually coordinates with either three or four oxygen atoms forming [BO_3] or [BO_4] structural units. Borate glass is well-known and best choice as optical material for its high transparency, low melting point and high thermal stability [4]. A.M. Abdelghany et.al [5] have prepared and characterized the borate glasses containing SrO substituting both CaO and NaO for their bioactivity or bone bonding ability. They have confirmed the formation of nodular texture of calcium phosphate (hydroxyapatite) and transform to micro-grains with increase of SrO. Tellurite glasses have paid extensive attention as promising candidates for optical fibers, ultra-fast optical switching devices and high non-linear optical devices because of their low-phonon energy, high refractive index, high polarisability of the non-bonding electron lone pair of Te^{4+} ions and low crystallization ability [6-8]. However, it must be noted that pure TeO_2 does not possesses its own glassy network, under normal melting conditions. Hence addition of network modifiers or glass forming oxides is needed to form tellurite-based glasses [9]. Structural and physical properties of glasses can be altered by the addition of a network modifier such as alkali or alkaline earth oxides and sometimes both alkali and alkaline earth oxide combinations. These network modifier oxides generate non-bridging oxygens (NBO) in the glass network by breaking the bridging oxygen bonding.

2. Experimental

Bismuth-Boro-tellurate glasses containing BaO as network modifier were prepared with the composition xBaO-(30-x)TeO_2-35Bi_2O_3-35B_2O_3 (with the glass codes as, x=5; BTBiB1, x=10;BTBiB2, x=15; BTBiB3; x=20; BTBiB4 and x=25;BTBiB5) by melt quenching technique. Analytic reagent grade chemicals of highest purity BaO, TeO2, Bi_2O_3 and B_2O_3 were used as starting materials which are thoroughly mixed in an agate mortar for 20 min. The batch materials were taken in porcelain crucibles and placed in an electrically heated muffle furnace kept around 1200 to1280 K for about one hour, until a bubble free, clear and homogeneous melt was obtained. The homogeneous melt was then quickly poured onto a stainless steel plate and pressed with another stainless steel plate, both were maintained at 500 K. The obtained glass samples were kept at 500K for 5h to relieve the thermal strains in the glass. Density measurements were carried out at room temperature using the Archimedes method with xylene as the immersion liquid. The X-ray diffractograms were recorded at room temperature using the Philips X-ray diffractometer with copper K_α tube target and nickel filter operated at 40 kV, 30 mA. The optical absorption spectra of all the glass samples were recorded at room temperature on a UV–VIS spectrophotometer (model JASCO V 570) in the wavelength range 20 0–1000 nm. The FTIR spectra were recorded at room temperature by using a Perkin Elmer Frontier FTIR in the wave number range (400–4000 cm^{-1}).

3. Results and discussion

3.1 X-ray diffraction

XRD patterns of xBaO-(30-x)TeO_2-35Bi_2O_3-35B_2O_3 glasses are shown in figure 1The absence of characteristic peaks indicated that all the glass samples are amorphous in nature.

3.2. Physical properties

The Densities of all of the glass samples were estimated using the formula

$$\rho = \left(\frac{w_{air}}{w_{air} - w_{xylene}} \times 0.865 \right) \tag{1}$$

where W_{air} and W_{xylene} are the weights of the glass sample in air and xylene, respectively, and 0.865 g/cm^3 is the density of the xylene at room temperature. The molar volume of the glass composition was calculated using the formula

$$V_m = \frac{\sum x_i M_i}{\rho} \tag{2}$$

where x_i is the molar fraction, M_i is the molecular weight of the ith component of the glass and ρ is the density of the glass. The Molar volume relates directly to the spatial distribution of the oxygen in the glass network. The Density and molar volume values are given in table 1. From the table 1it was found that the density increases from 5.73 to 6.12 g/cm3 with the increase of BaO content. This could be due to the higher density of BaO (ρ = 5.72 g/cm^3) than that of TeO_2 (ρ = 5.67 g/cm^3). Figure 2 shows the variation of Density (ρ) and Molar volume (V_m) as a function of BaO content. From the figure 2 it was found that the density and molar volume shows opposite behavior [10, 11].

Oxygen packing densities (OPD) of glass samples were calculated using the expression given below and the values obtained were tabulated in table 1.

$$\text{OPD}= \frac{\rho}{M} \times O_n \tag{3}$$

Here O_n is total number of oxygen atoms per chemical formula. From the table 1 it was observed that the OPD values were found to decrease from 59.28 to56.86 mol/l. This decrease in OPD values may be attributed to the increase in molecular weight of the glass.

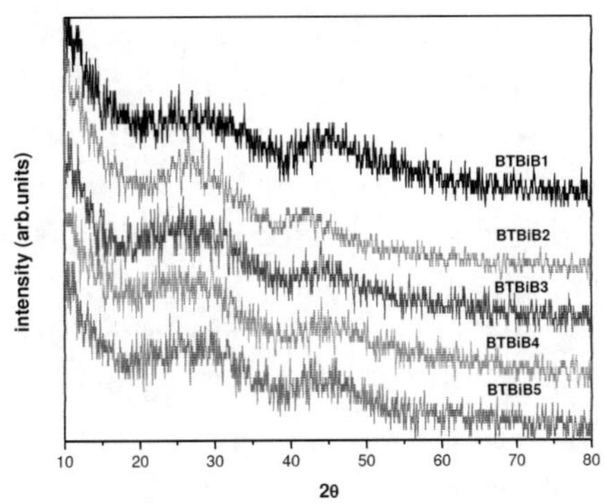

Figure 1. XRD patterns of $xBaO-(30-x)TeO_2-35B_2O_3-35Bi_2O_3$ glasses

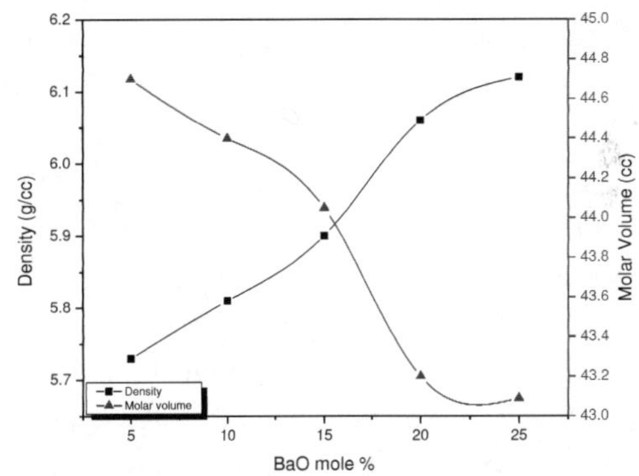

Figure 2. Variation of density and molar volume with BaO content

3.3. Optical Band Gap and Urbach Energy

The optical absorption spectra of the glass samples are shown in figure 3. The power law region is clearly observed in absorption spectra and fundamental absorption edges are not sharp. This is the characteristic feature of amorphous glass samples. From table 1, it was observed that that the fundamental absorption edge shifts to higher wavelength side with an increase of BaO content. Absorption coefficient near the edge of each curve was determined by using the relation

$$\alpha(\upsilon)=(1/t)\log(I_0/I) \tag{4}$$

where log (I_o/I) is the absorbance, I_0 and I are the intensities of incident and transmitted light respectively. According to Davis and Mott[12], the absorption co-efficient as a function of photon energy is given by the relation

$$\alpha(\upsilon)= B(h\upsilon - E_{opt})^n / h\upsilon \tag{5}$$

where E_{opt} is the optical band gap energy, B is the constant related to the extension of the Urbach tail and 'n' take the values of 2, 1/2, 3, and 3/2 depending on the nature of electronic transitions responsible for absorption. For amorphous materials, indirect transitions are valid according to Tauc's relation [13]. The values of E_{opt} were obtained by extrapolating to $(\alpha h\nu)^{1/2}=0$ for indirect transitions. [$(\alpha h\nu)^{1/2}$ Vs hν are plotted in (figure 4)], to find optical band gap energies. E_{opt} values were found to increase with an increase of BaO content. In glass and amorphous materials there exist band tailing in

the forbidden energy band gap. The extent of this band tailing is a measure of the disorder in the material and can be estimated by using the Urbach equation [14, 15]
given by,

$$\alpha(\upsilon) = C\exp(h\upsilon / \Delta E) \tag{6}$$

where, C is a constant and ΔE is the width of band tail also known as Urbach energy. The lack of crystalline long range order or the presence of defects in the amorphous materials is associated with a tailing of density of states [16]. Urbach energies (ΔE) are calculated by taking the reciprocals of slopes of linear portion in lower photon energy regions of these curves and are shown in figure 5 Urbach energy (ΔE) provides a measure of disorder in the amorphous and crystalline solids. Smaller is the value of Urbach energy, greater is the structural stability of the glass system. In the present glass system Urbach energies are in the range of 0.19eV to 0.27eV. The smaller values of Urbach energies show high stability of the prepared glasses. Refractive index of the $xBaO-(30-x)TeO_2-35Bi_2O_3$ - $35B_2O_3$ glasses is decreasing with increase of BaO content. The polarizability of barium ion is larger than the other cations [17]. High refractive index is attributed to the higher polarization of BaO. According to the Herzfeld theory [18] of metallization the valence electrons, which before have been quasi-elastically bound to their atoms, are now set free via mutual polarization in the condensed state and the solid material becomes a metallic conductor. In this regard the following criterion appears to be a necessary condition to predict the metallic or non-metallic nature of solids i.e $R_m / V_m > 1$ (for metals) and $R_m/V_m < 1$ (for nonmetals). The present reported values of R_m / V_m are in the range of 0.3624 to 0.3745. As BaO concentration is increasing in glass network an increased tendency to metallization was noticed. The smaller value of M (metallization) of these glasses indicated that the width of both conduction band and valence band increases resulting in narrow band gap.

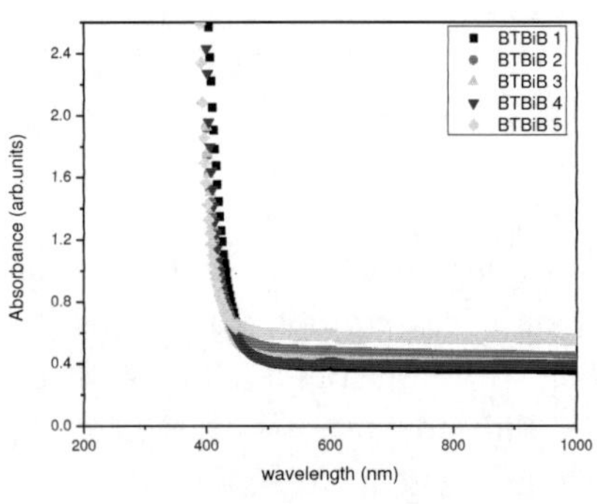

Figure 3. Optical absorbance spectra of $xBaO-(30-x)TeO_2-35Bi_2O_3-35B_2O_3$

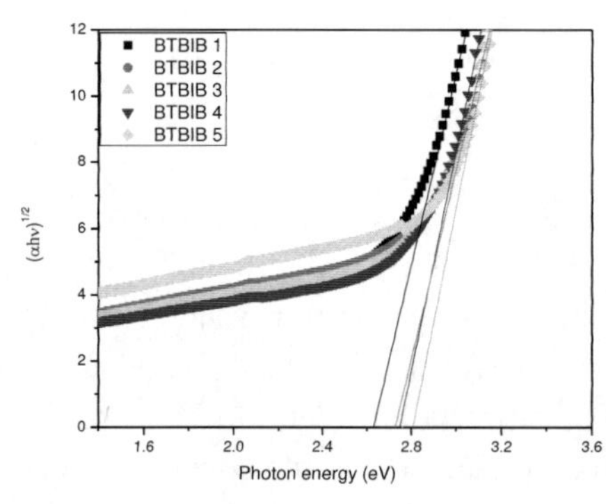

Figure 4. Plot of $(\alpha h\upsilon)^{1/2}$ as a function hυ of for the present glasses

3.4. Optical Basicity

In the development of the Lewis acid-base principles Duffy and Ingram have made great contribution by introducing the "optical basicity" concept. This concept is very useful for quantitative evaluation of the acid-basic properties of glasses, alloys, slags, molten salts, etc. The degree of polarization plays an important role to transfer electron density from oxygen to surrounding cations. High polarizable cations such as P^{5+}, Si^{4+}, B^{3+} ions decreases the donor nature of oxygen by influencing the electron cloud of oxygen. According to the Duffy [19], the theoretical optical basicity (Λ_{th}), for the present glasses is calculated using the following formula

$$\Lambda_{th} = X_{BaO}\Lambda_{BaO} + X_{TeO}{}^{2}\Lambda_{Teo}{}^{2} + X_{Bi2O3}\Lambda_{Bi2O3} + X_{B2O3}\Lambda_{B2O3} \quad (7)$$

Where X_{BaO}, X_{TeO2}, X_{Bi2O3} and X_{B2O3} are the contents of individual oxides in mole%. Λ_{BaO} , Λ_{Teo2}, Λ_{Bi2O3} and Λ_{B2O3} are the theoretical optical basicity values corsponding to the each oxides have been taken from the literature . The theoretical optical basicity Λ_{th} values calculated using above equation for the brium bismuth boro-tellurate glasses are presented in table 1. It is found that the theoretical optical basicity decreases slightly with increasing BaO content.

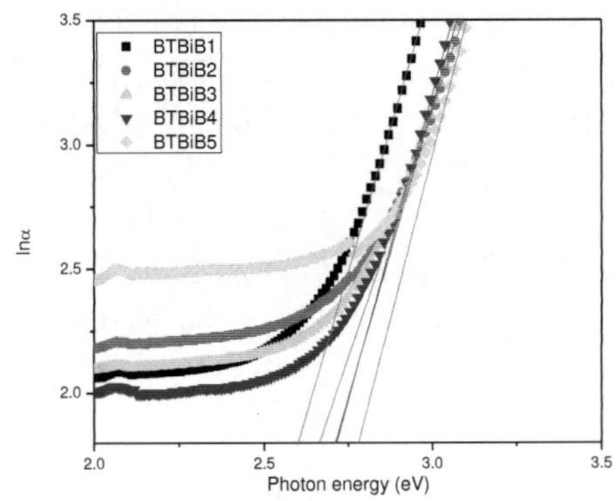

Figure 5. Urbach energy plot of xBaO-(30-x)TeO_2-35Bi_2O_3-35B_2O_3glasses

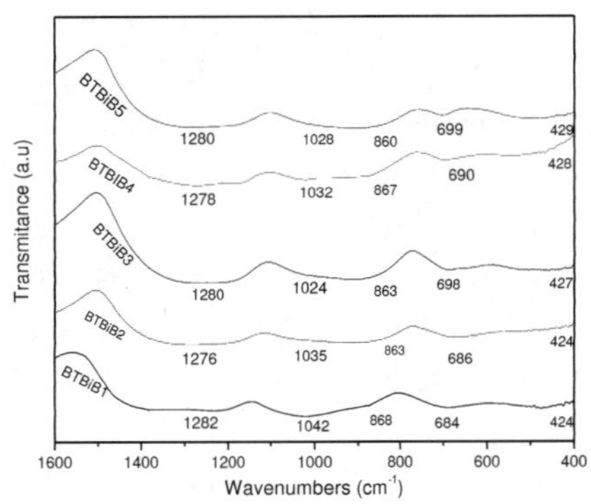

Figure 6. FTIR spectra of xBaO-(30-x)TeO_2-35Bi_2O_3-35B_2O_3glasses

Table 1. Physical and optical parameters of xBaO-(30-x)TeO_2-35Bi_2O_3 -35B_2O_3 glass system.

Physical & optical Parameters	Glass codes				
	BTBiB-1	**BTBiB-2**	**BTBiB-3**	**BTBiB-4**	**BTBiB-5**
Density (ρ) (g/cc)	5.73	5.81	5.90	6.06	6.12
Molar volume (V_m) (cc/mol)	44.70	44.40	44.05	43.20	43.09
Oxygen packing density (mol/l)	59.28	58.56	57.89	57.87	56.86
Cut-off wavelength (nm)	441	442	445	451	458
Indirect optical band gap (eV)	2.63	2.72	2.75	2.75	2.81
Urbach energy E (eV)	0.20	0.27	0.20	0.19	0.20
Refractive index (n)	2.51	2.48	2.47	2.47	2.45
Dielectric constant (ε)	6.28	6.13	6.08	6.09	6.01
Reflection loss (R)	0.1844	0.1803	0.1790	0.1792	0.1769
Molar refractivity (R_m) ($cm^{-3)}$	28.50	28.01	27.70	27.19	26.95
Metallization (M)	0.3624	0.3690	0.3711	0.3707	0.3745
Theoretical optical basicity (Λ_{th})	0.1602	0.1587	0.1571	0.1556	0.1540

3.5 IR spectra

The infrared transmittance spectra of the glasses with composition xBaO-(30-x)TeO_2-35Bi_2O_3 -35B_2O_3 are shown in figure 6 From the figure it was observed that the absorption bands were found around 428 cm^{-1}, 690 cm^{-1}, 863 cm^{-1}, 1030 cm^{-1},and 1280 cm^{-1} for all the glass samples. The small and feeble absorption band found around 428 cm^{-1} is assigned to the Bi-O bond in BiO_6 polyhedra [10]. The prominent band found around 690 cm^{-1} is due to TeO_3 (tp) trigonal pyramidal units [20]. Another prominent band observed around 863 cm^{-1} is attributed to the vibration of tri, penta and diborate groups of BO_4 tetrahedra [21]. A shallow broad absorption band found around 1030 cm^{-1} is attributed to B-O stretching vibrations of tetra hedral BO_4 units [22]. Another shallow broad absorption band observed in the region 1280 cm^{-1} is attributed to the asymmetric stretching vibrations of B-O in trigonal BO_3 units [22].

4. Conclusions

The density values increases from 5.73 to 6.12 g/cm3 with the increase of BaO content. The density and molar volume shows opposite behavior. Refractive index of the present glasses is decreasing with increase of BaO content. High refractive index values are due to the higher polarization of BaO. From the infrared transmittance spectra of the glasses, it is observed that the absorption bands around 428 cm^{-1} is assigned to the Bi-O bond in BiO_6 polyhedra, 690 cm^{-1} is [1] is due to TeO_3 (tp), 863 $cm^{-1}$is attributed to the vibration of tri, penta and diborate groups of BO_4 tetrahedra, 1030 $cm^{-1}$is attributed to B-O stretching vibrations of tetra hedral BO_4 units and 1280 cm^{-1} band is due to asymmetric stretching vibrations of B-O in trigonal BO_3 units.

Acknowledgements

The authors would like to thank The Head, Department of Physics, Osmania University for providing laboratory facilities. One of the authors B.Srinivas would like to thank the UGC, BSR-RFSMS, New Delhi, for providing financial assistance.

References:

1. E. Culea, L. Pop, P. Pascuta, M. Bosca 2009 *J. Mol. Struct.* **924** 192
2. A. Agarwal, V.P. Seth, P.S. Gahlot, S. Khasa, P. Chand 2003 *J. Phys. Chem. Solids.* **64** 2281
3. L. Baia, R. Stefan, J. Popp, S. Simon, W. Kiefer 2003 *J. Non-Cryst. Solids*, **109** 324
4. S. Mohan, K.S. Thind, G. Sharma, Braz 2007 *J. Phys.* **37 (4)** 1306.
5. A.M. Abdelghany, M.A. Ouis, M.A. Azooz, H.A. ElBatal, G.T. El-Bassyouni 2016 *Spectrochim. Acta A Mol. Biomol. Spectrosc.* **152** 126
6. M.A. Villegas, J.M. Fernández Navarro 2007 J. Eur. Ceram. Soc. **27** 2715
7. X. Shen, Q.H. Nie, T.F. Xu, Y. Gao 2005 *Spectrochim. Acta A Mol. Biomol. Spectrosc.* **61** 2827
8. Abdeslam Chagraoui, Abdelmjid Tairi, Kaltoum Ajebli, Hanane Bensaid 2010 *J. Alloys Compd.* **495** 67
9. El-Mallawany RAH. 2002 Tellurite glasses handbook. Boca Raton/London/NewYork/Washington, DC: CRC Press
10. M. Subhadra, P. Kistaiah 2012 *Vibrational Spectroscopy* **62** 23
11. P.P. Pawar, S.R. Munishwar, R.S. Gedam 2016 *J. Alloys Compd.* **660** 347
12. E A Davis and N F Mott 1970 *Phila. Mag.,* **22** 903
13. N Elkhoshkhany, R Abbas, R El-Mallawany and A J Fraih 2014 *Ceramics International* **40** 14477.
14. F Urbach 1953 *Phys. Rev. B.* **92** 1324.
15. S Sanghi, S Sindhu, A Agarwal and V P Seth 2014 *Radiations Effects & Defects in solids :Incorporating Plasma Science and Plasma Technology* **159** 369.
16. N Elkhoshkhany 2014 *J. Chem. Chem. Eng.* **8** 11.
17. V Dimitrov and S Sakka, 1996 *J. Appl. Phys.,* **79** 1736.
18. Herzfeld K 1970 *Phys. Rev. B*, **2** 2045
19. J.A. Duffy 1989 *Phys. Chem. Glasses* **30** 1
20. S. Suresh , P. Gayathri Pavani , V. Chandra Moulim 2012 *Materials Research Bulletin* **47** 724
21. I. Kashif, A.A. Soliman, H. Farouk, M. El-Shorpagy, A.M. Sanad 2008 *Physica B* **403** (2008) 3903
22. Chandkiram Gautam, Avadhesh Kumar Yadav, and Arbind Kumar Singh 2012 ISRN Ceramics **2012** 17

IOP Conf. Series: Materials Science and Engineering **360** (2018) 012023 doi:10.1088/1757-899X/360/1/012023

Microwave assisted sintering of Nanostructured Infrared-Transparent $Nd_{0.1}Y_{1.9}O_3$ ceramics Synthesized by a Modified Combustion Technique

C.T Mathew, Ancy Das, Jacob Koshy, Jijimon K.Thomas*

Department of Physics, Mar Ivanios College, Thiruvananthapuram 695015, India

**Corresponding author e-mail: jkthomasemrl@yahoo.com ,Ph: +91 471 2530887, Fax: +91 471 2532445*

Abstract

Synthesis of nano particles of $Nd_{0.1}Y_{1.9}O_3$ ceramics (~15nm) by a single step auto-igniting combustion technique, followed by microwave sintering to optimum density, and their remarkable infrared transmission characteristics are presented in this paper. The XRD pattern of the as prepared sample clearly indicates that the Nd^{3+} ions are effectively replacing Y^{3+} ions in the yttria cubic structure and it matches very well with the JCPDS file no-89-5591. The size of the crystallites was calculated using Scherrer formula and they found to be in the size range of 10-19 nm and the average size of the crystallites was 15 nm. The HRTEM analysis confirms the nano nature and the well-defined lattice structure of the sample. The UV-Visible and FTIR spectroscopy were effectively used to characterise the powder and to obtain the transmittance properties of the sintered samples. Microwave heating was successfully employed to sinter the sample pellets. A soaking duration of 20 minutes at 1500°C yielded well sintered (98.8%) translucent pellet with reduced average grain size of 500nm and it showed a transmittance of 77% in the UV-Visible region and 78% in the mid infrared region. Microwave assisted sintering of the combustion synthesized powder is therefore a promising fabrication technique, especially for infrared transparent ceramics, which considerably reduces the sintering temperature, soaking duration, the grain size and enhances the transmittance properties. These results clearly indicate that by prudently tuning the concentration of the additives, the IR transmittance properties can be improved and they can be used very effectively for the fabrication of infrared transparent windows and domes.

Keywords : Microwave sintering, combustion synthesis, infrared transparent

1. Introduction

Infrared transparent ceramics found to have applications as windows in homing missiles. In addition to the strategic defense and space applications they also have large spectra of civilian applications [1-6]. In most of the cases cubic phase of transparent ceramics are selected by the researchers due to the absence of birefringence. Cubic Y_2O_3 is a potential infrared transparent candidate due to its large infrared cut-off, high corrosion resistance, absence of birefringence, high mechanical strength, high thermal stability and low emissivity at elevated temperatures [1, 7-11]. A large number of factors adversely affect the efficacy of yttria as an infrared transparent window. The sintering temperature of yttria is above 2000°C[12,13]. But there are reports in which nano phase yttria is sintered at a comparatively low temperature of 1700°C by different methods where the fabrication methods are cumbersome[14-16]. Nanostructured high quality starting powder may enhance the mechanical properties without affecting the infrared transmittance in the mid IR range. In addition to the

IOP Conf. Series: Materials Science and Engineering **360** (2018) 012023 doi:10.1088/1757-899X/360/1/012023

quality of the starting powder, doping and fast sintering mechanisms effectively improve the performance of infrared windows.

A number of fabrication methods [17-22] have been developed successfully so far to synthesize ultrafine nanostructured polycrystalline Y_2O_3. Among them combustion synthesis is evolved as the economic and efficient method in terms of the cost, time and the complexities in the synthesis of nanomaterial. In the present work, a modified combustion technique is used to synthesize nano structured $Nd_{0.1}Y_{1.9}O_3$. The modified combustion technique is a relatively simple method viz. with the omission of high temperature calcinations for prolonged duration; we could obtain phase pure nano particles of extremely small size (5-25 nm) and further compacted to an optimum density at a much lower temperature with high thermal stability and good transparency to infrared radiation using microwave heating technique.

In this paper we report the synthesis of nanostructrured poly-crystalline $Nd_{0.1}Y_{1.9}O_3$ ceramics by an auto-ignited combustion technique [23]. The structure, vibrational spectra and surface morphology of the combustion synthesized powder are also studied and presented. The sintering behaviour of compacted pellets using resistive heating and the transmittance of infrared radiations through the sintered pellets in the UV-Visible and mid infrared ranges are also reported.

2. Experimental Procedure

A single step auto-igniting combustion synthesis is used to prepare nano structured $Nd_{0.1}Y_{1.9}O_3$ [23-26]. Stoichiometric amount of high purity $Y(NO_3)_3.6H_2O$ and Nd_2O_3(99.99%, Alfa Aesar, USA) were dissolved in double distilled water to make a clear solution. Citric acid was then added to it as a complexing agent. Amount of citric acid is calculated based on total valence of the oxidizing and the reducing agents for maximum release of energy during combustion[26]. Nitric acid was used as oxidising agent and ammonia as fuel and the pH of the solution was monitored till it became 7. The solution containing the precursor mixture was heated using a hot plate at 250 °C in a ventilated fume hood. The solution boils on heating and undergoes dehydration accompanied by foam. The foam then ignites by itself on persistent heating giving voluminous and fluffy product of combustion.

The as-prepared nanopowders obtained from the combustion process are characterized by different powder characterization techniques. As-prepared samples are characterized by X-ray diffractometer (Xpert-pro) with Cu Kα radiation in the range of 20–80° in steps of 0.0840 for the determination of crystalline structure and phase of the nanomaterials. The average crystallite size is estimated for all the samples from Scherrer's equation. The Wiliamson-Hall plot is used for studying lattice strain in the sample. The absorption spectrum of as-prepared sample is recorded using a Shimadzu spectrophotometer (UV-1700). The uv–vis spectra are recorded to resolve the correlation between the grain size and absorption edge of the nanoparticles. Additional information regarding the phase purity and the presence of any inorganic impurity are obtained using FTIR spectroscopy using Fourier Transform Infrared spectrometer (FTIR, Perkin elmer spectrum 2) in the range 400-4000 cm^{-1} using ATR mode. Particulate properties of the combustion product are examined using high resolution transmission electron microscopy (TEM, Model-Hitachi H600 Japan) operating at 200 kV. The samples for Transmission Electron Microscope (TEM) are prepared by ultrasonically dispersing the powder in methanol and allowing a drop of this to dry on a carbon-coated copper grid.

The pure white powder is uniaxially compacted in to pellets in a 14mm diameter steel die at 200 MPa. The pellets are heat treated at 600 °C for half an hour to remove the binder and absorbed gases. The pellets are sintered at 1500 °C for a soaking time of 20 minutes in a in a microwave furnace with silicon carbide susceptors (VBCC/MF/86,VB Ceramics Consultants, India). The morphological investigations of the sintered pellets are performed in Scanning Electron Microscope (SEM, Jeol 6390LV). The transmittance of radiation in the uv-visible and in the IR range is measured using UV-Vis and FTIR spectrometers.

3. Results & Discussion

Figure 1 shows the powder XRD patterns of as-synthesised $Nd_{0.1}Y_{1.9}O_3$. All the peaks are indexed for a body centred cubic structure of Y_2O_3 which agrees very well with the XRD data reported in JCPDS file no-89-5591. The XRD pattern of the as prepared sample matches very well with that of pure yttria which clearly indicates that Nd^{3+} ions are effectively replacing Y^{3+} ions in the yttria cubic structure. Thus the XRD results confirm the formation of pure cubic $Nd_{0.1}Y_{1.9}O_3$ nanoparticles without any post annealing or calcination process.

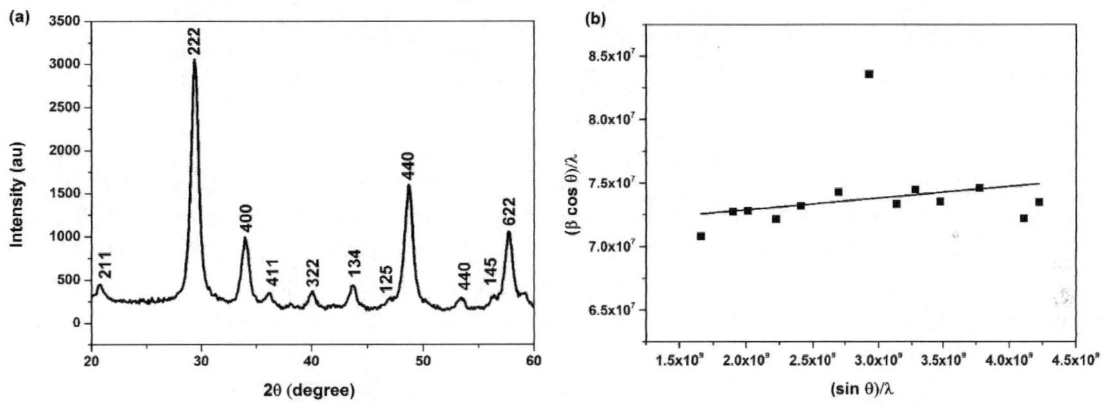

Figure 1: (a) XRD pattern of as prepared nano $Nd_{0.1}Y_{1.9}O_3$ **(b)** Hall-Williamson plot of as prepared nano $Nd_{0.1}Y_{1.9}O_3$

The crystallite size calculated using Scherrer formula [27] and for the major (*222*) peak, it is found to be 12.4 nm and the interfacial spacing corresponding to the plane is 0.3040 nm which matches very well with the JCPDS value of 0.3615 nm for the same plane. The particles were found to be in the size range of 10-19 nm and the average size of the crystallites is 15 nm. Hence the modified auto-igniting combustion method offers an excellent and economic way for the preparation of phase pure $Nd_{0.1}Y_{1.9}O_3$ nano powder.

The extent of line broadening in the diffraction pattern due to the strain, which is caused by the non-uniform displacement of the atoms with respect to their reference lattice position, could be identified by Hall–Williamson (W-H) method [28]. In this method, the reciprocal peak width $\beta cos\theta/\lambda$ is plotted against the reciprocal lattice distance $sin\theta/\lambda$, and the resulting plot was linearly fit. Figure 1(b) shows the Hall–Williamson plot for the $Nd_{0.1}Y_{1.9}O_3$ nanoparticles. The reciprocal of y intercept which was a measure of particle size obtained from this plot was 14.1 nm. The lattice strain constant η is proportional to the slope of the line, and from this plot the estimated value of η is 9.24x10^{-4}. The positive slope of the line indicates that strain is tensile strain and even though it was very small also a component contributing to the broadening of peaks in the XRD pattern.

Figures 2(a) and (b) are the HRTEM images of the as-prepared powder sample. From the distinctly visible grain in figure 2(a) the grain size was calculated as 12.0 nm corresponding to the *(222)* plane of cubic crystal structure of $Nd_{0.1}Y_{1.9}O_3$ and was found to be in good agreement with the interfacial spacing for the same plane obtained from X-ray diffraction analysis.

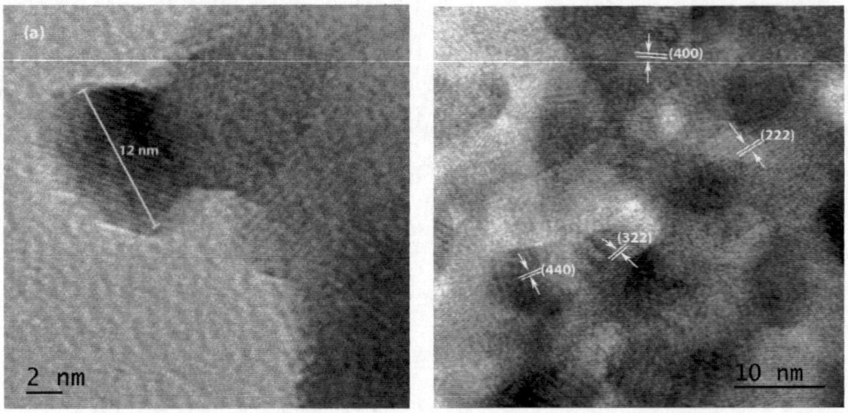

Figure 2 :(a) HRTEM image of a single grain and (b) HRTEM image of different crystallographic planes

Structural analysis in detail was done using FT IR shown in figure 3. The graph confirms that the $Nd_{0.1}Y_{1.9}O_3$ powder transmits IR in the range 2.5µm-20µm. The peak at 564 cm^{-1} corresponds to metal-oxide stretching vibration [18,19,29,30]. Broad peak around 3400 cm^{-1} correspond to O-H stretching modes of water adsorbed by the powder. No other absorption peaks are observed in this range which shows that no residual nitrate or organic matter is present in the precursor powder.

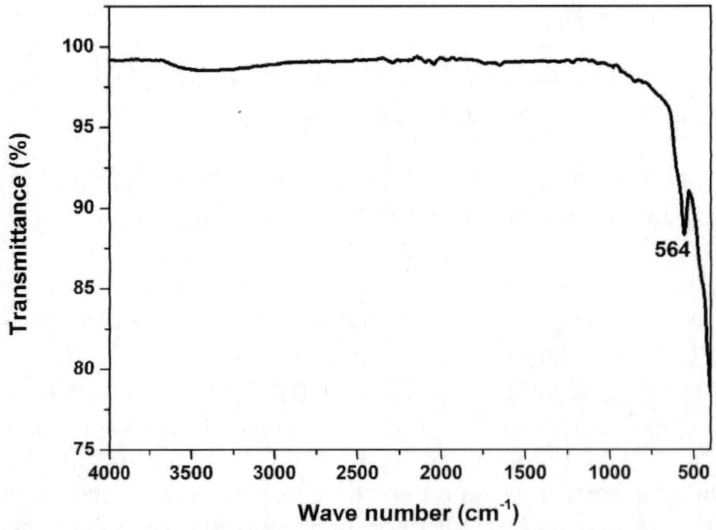

Figure 3 : FTIR transmittance spectrum of as synthesised $Nd_{0.1}Y_{1.9}O_3$

IOP Conf. Series: Materials Science and Engineering **360** (2018) 012023 doi:10.1088/1757-899X/360/1/012023

The UV-visible spectrum was recorded in the range 200-800nm and is shown in figure 4. It is observed that the sample shows maximum absorption in the UV region and decreases towards the visible region. The maximum absorption is at 242 nm, such materials were potential candidates for UV filters, which filter higher energies of the ultraviolet spectrum from 10nm to 120nm which are ionising and also find application as microwave filters. From the absorption spectrum it is clear that the sample exhibit transmittance of above 90% in the visible region. The properties of poor transmittance in the UV region but high transmittance in the Vis–NIR regions makes nano $Nd_{0.1}Y_{1.9}O_3$ an excellent material for screening off the hazardous UV portion of the electromagnetic spectrum . The optical band gap energy could be determined from the excitonic peak at 205 nm wavelength using the equation $E=hc/\lambda$ where h is the Plank's constant, c is the velocity of electromagnetic radiation and λ is the wavelength corresponding to the absorption. The corresponding band gap for the wavelength 205 nm is 6.05 eV. The refractive index of the sample can be calculated from the band gap [31,32].

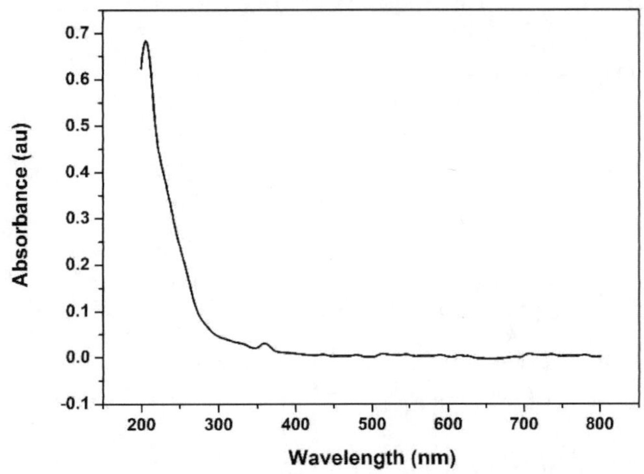

Figure 4 : UV-Visible absorption spectrum of as synthesised $Nd_{0.1}Y_{1.9}O_3$

Sintering behaviour was studied on the disc shaped yttria pellets using a silicon carbide susceptor aided microwave furnace. The relative densities of sintered pellets at different sintering temperatures for a soaking time of 20 minutes are calculated. In the microwave furnace the pellets were heated at a constant rate of 40 $^oCmin^{-1}$ and were sintered to 98.8 % of theoretical density by holding for two hours at 1500°C, which is considerably low sintering temperature relative to the recent reports[33].

The sintered pellet which achieved 98.8 % of the theoretical density was hand polished and thermally etched at 1400°C for 1 hour in air. The surface morphology of the sintered sample was studied using Scanning Electron Microscopy. The SEM micrograph of the $Nd_{0.1}Y_{1.9}O_3$ ceramics after the initial polishing is shown in Figure 5. The grain boundaries were sharp with minimum porosity without showing any abnormal growth or melting at the grain boundaries and the full-grown grains are in the size range of 100-700 nm. The grain sizes are accurately measured using the Image J software and the average grain size is found to be 600 nm.

Figure 5: SEM micrograph of conventionally sintered $Nd_{0.1}Y_{1.9}O_3$ pellet sintered to 98.8 % of theoretical density

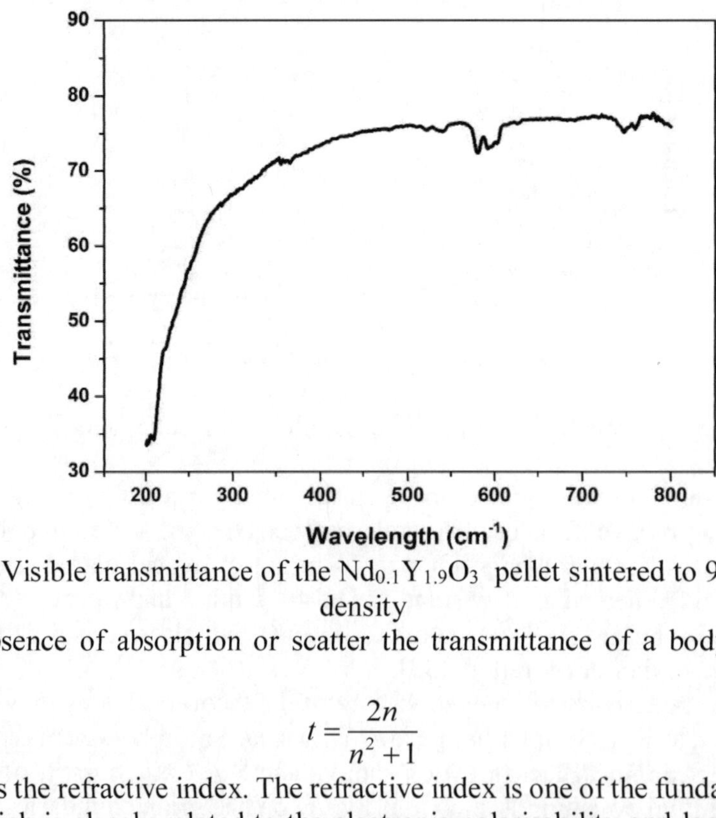

Figure 6: UV-Visible transmittance of the $Nd_{0.1}Y_{1.9}O_3$ pellet sintered to 98.8% theoretical density

In the absence of absorption or scatter the transmittance of a body is given by the relation,

$$t = \frac{2n}{n^2 + 1}$$

where n is the refractive index. The refractive index is one of the fundamental properties of a material which is closely related to the electronic polarisability and local field inside the material. Reddy et al.[31,32] formulated an empirical relationship between the refractive index and band gap energy,

$$E_g e^n = 36.3$$

Using the above relation the refractive index of the present sample with band gap energy of 6.05 eV is calculated as 1.79. The theoretical limit of transmittance through the sintered pellet according to the relation is calculated as 85.11%. The pellets after the initial hand polishing followed by polishing using diamond pastes of fine grades in a self-designed lapping and polishing machine and the final mirror polished pellets are used for the transmittance studies. The UV-Visible spectrum of the sintered pellet of density 98.8% of the theoretical density and thickness 0.5mm is shown in the figure 6. The pellet shows a maximum transmittance of 77.0% at 800 nm and is found to be translucent.

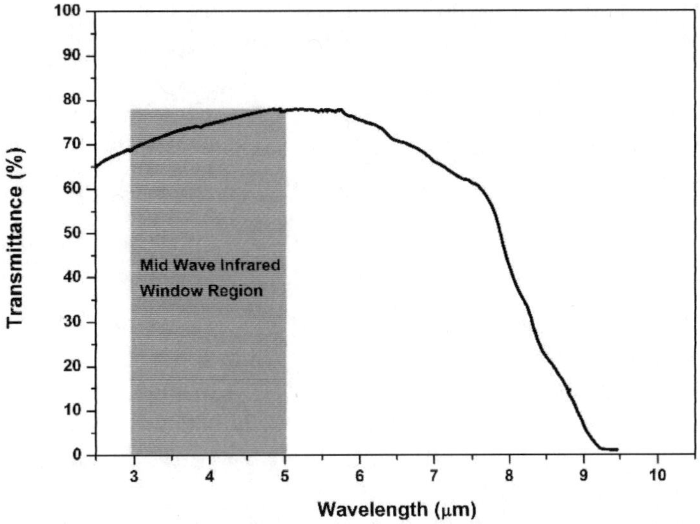

Figure 7: IR spectrum of $Nd_{0.1}Y_{1.9}O_3$ pellet sintered to 98.8% of theoretical density

The IR spectrum of the conventionally sintered sample in figure 7 shows that it has a maximum transmittance of 78.0 % at 5 μm . The good transmittance in the Mid IR range makes it ideal for infrared transparent windows and domes.

4. Conclusions

A single step auto-igniting combustion synthesis is used to prepare nano structured $Nd_{0.1}Y_{1.9}O_3$ X-Ray Diffraction (XRD) analysis of the as prepared powder confirms the formation of phase pure cubic $Nd_{0.1}Y_{1.9}O_3$ nanoparticles . All the peaks in the XRD were indexed to a body centred cubic structure with a calculated lattice parameter a=10.53Å agreeing very well with the XRD data reported in Joint Committee on Powder Diffraction Standards (JCPDS) file no. 89-5591. The XRD analysis and High Resolution Transmission Electron Microscopic(HR-TEM) studies reveal that the average crystallite size is ~15 nm and the lattice planes are well defined. Fourier transform Infrared spectrum of the as prepared sample is recorded in the range 400-4000 cm^{-1} .The vibrational spectroscopic studies confirms the structure of the sample. The absorption spectra of the as prepared $Nd_{0.1}Y_{1.9}O_3$ is measured in the range 200-800nm and it shows that the material absorbs heavily in the shorter wavelength region of the ultraviolet spectrum and transmit long wavelength uv and visible light and hence found applications in the filters and sensors of UV radiations. The pure white $Nd_{0.1}Y_{1.9}O_3$ powder is uniaxially compacted in to pellets in a 14 mm diameter steel die and sintered to 98.8% theoretical density without any sintering additives or applying pressure at 1500 °C for 20 minutes in a microwave furnace. The SEM micrograph gives the surface morphology of the well sintered pellet and from the SEM it is found that the average grain size is ~600nm. The transmission spectrum of the sintered pellet in the UV-Vis region reveals that the material is showing 77.0% of transmittance in the visible region and 78% in the Mid IR region. The better transmittance of the pellets attribute to the quality of the nano powder synthesised by the modified combustion technique, high sintered density with minimum porosity and the smaller grain size of the sintered pellet. The results clearly indicate that the ultra fine $Nd_{0.1}Y_{1.9}O_3$ nanopowder synthesized by the modified auto-igniting combustion technique synthesised using single step combustion method followed by resistive heating can be used very effectively for the fabrication of improved infrared transparent windows and domes.

Acknowledgments

This work is supported by the Department of Science and Technology- Science and Engineering Research Board, Government of India under Grant No. SB/S2/CMP-0021/2013.

References

1. D.C.Harris, *Materials for Infrared Windows and Domes, Properties and Performance*. SPIE Press, Bellingham, WA, 1999.
2. Y.H. Lan , J.Z. Jun, M.X. Jian, and W.S. Wei, *J. Inorg. Mater.*, 25 [8] 795-800(2010).
3. M. Grujicic, W.C. Bell, and B. Pandurangan, *Mater. Design.,* 34 808–819 (2012).
4. X.Qin,H.Yang,D. Shen, H. Chen,G. Zhou, S. Wang,D. Luo, D. Tang,J. Zhang,and J. Ma, *Int. J. Appl. Ceram. Tec.,*10 [1] 123-128 (2013).
5. T. Tetsui, T. Kobayashi, T. Mori,T. Kishimoto,and H. Harada, *Mater. Trans.,* 51[9] 1656 - 1662 (2010).
6. D.D. Silva, and A.R. Boccaccini, *Rec. Pat.Mater. Sci.* 1 56-73 (2008).
7. D.C. Harris, *Infrared Phys.Techn.* 39 185-201 (1998).
8. J. Chen, Z.P. Gao, J.M. Wang, and D.H. Zhang, *J.Ele. Tech. Chi.* 5 [4] 320-324(2007).
9. A.C Pastor, R.C Pastor, *Mater.Res.Bull.* 2 555-559 (1967).
10. Y. Tsukuda, *Mater. Res. Bull.* 16 453-459 (1981).
11. R.M. Sova, M.J. Linevsky, M. E. Thomas, and F. F. Mark, *Infrared Phys. Techn.* 39 251-261 (1998).
12. Y.TSUKUDA, A. MUTA, J. Ceram. Soc. Jpn. 84 (1976) 141-147.
13. Y.TSUKUDA, A. MUTA, Sintering of Y_2O_3 at High Temperatures, J. Ceram.Soc.Jpn.84 (1976) 585-589.
14. L.Wen, X. Sun , Q. Lu , G. Xu , X. Hu , Opt. Mater. 29 (2006) 239-245.
15. L. An, A.Ito, T. Goto, J. Eur. Ceram. Soc. 32 (2012) 1035-1040.
16. R. Chaim, A.Shlayer, C. Estournes, J.Eur. Ceram. Soc. 29 (2009) 91–98.
17. Z. Huang, X. Sun, Z. Xiu, S. Chen, and C.T. Tsai, *Mater. Lett.* 58 2137-2142 (2004).
18. R.Srinivasan, R.Yogamalar, and A. C. Bose, *Mater. Res.Bull.* 45 1165-1170 (2010).
19. Y. Huang, D. Jiang , J. Zhang , and Q. Lin, *Opt. Mater.* 311448-1453 (2009).
20. H.Gong, D.Y. Tang, H. Huang, T.S. Zhang, and J.Ma, *Mat. Chem. Phys.* 112 423-426 (2008).
21. R.V. Mangalaraja, J. Mouzon, P. Hedstrom,K.V.S. Ramam, . P. Camurri, M.Oden, and I. Kero, *J. Mater. Process. Tech.* 208 415-422 (2008).
22. N. Dasgupta , R. Krishnamoorthy , and K. T. Jacob, *Inter. J.Inorg. Mater.* 3 143-149 (2001).
23. Mathew.C.T, Sam Solomon , Jacob Koshy and Jijimon K Thomas, Ceram. Int. 41[8] (2015) 10070-10078.
24. J.K. Thomas, H. Padma Kumar, R. Pazhani, S. Solomon, R. Jose and J. Koshy, *Mater. Lett.* 61 1592–1595 (2007) .

25. Mathew.C.T, Vidya S, Jacob Koshy, Sam Solomon and Jijimon K Thomas, Infrared Phys. Techn. 72 153-159 (2015) .
26. J.K. Thomas, H. Padma Kumar, S. Solomon, C.N.George, K.Joy and J.Koshy, *phys. status. solidi. (a)* 204 [9] 3102–3107 (2007).
27. J.I.Langford, and A.J.C.Wilson, *J.Appl.Cryst,* 11 102-113(1978).
28. G.K.Williamson, and W.H.Hall, *Act.Meter.*1 22-31(1953).
29. T.K.Vishnuvardhan,V.R.Kulkarni,C.Basavaraja, and S.C.Raghavendra, *Bull.Mater.Sci.*29 [1] 77-83 (2006).
30. Y.Repelin, C.Proust,E.Husson, and J.M.Beny, *J.Solid state Chem.* 118 163-169(1995).
31. R.R.Reddy, and S.Anjaneyulu, *Phys.Status Solidi.(b).* 174 1091-1093 (1992).
32. R.R.Reddy, K.R.Gopal, K.Narasimhulu, L.S.S.Reddy, K.R.Kumar, C.V.K.Reddy, and S.N.Ahmed, *Opt.Mater.,* 31 209-212 (2008).
33. R Mann, K Laishram, S A Ahmed and N Malhan, Ceram.Inter. 38 4131-4135 (2012).

IOP Conf. Series: Materials Science and Engineering **360** (2018) 012024 doi:10.1088/1757-899X/360/1/012024

Raman Spectroscopy and Low Temperature Dielectric Properties of $(Bi_{0.5}Na_{0.5})TiO_3$ Ceramics

Srinivas Pattipaka[1] and D. Pamu [1]

[1]Department of Physics, Indian Institute of Technology Guwahati, Guwahati-781039, India

Email: pamu@iitg.ernet.in

Abstract. In the present study, lead-free $Bi_{0.5}Na_{0.5}TiO_3$ (BNT) ceramics were prepared by solid state reaction route and the structural, microstructural and dielectric properties have been reported. The Rietveld refinement method and Raman spectroscopy were used to analyze the structural properties of BNT ceramics. The Rietveld refinement analysis of all the sintered samples revealed the rhombohedral crystal symmetry with R3c space group. The sintered (1100 °C) BNT ceramics exhibited homogeneous grain distribution with an average grain size of 1.40 μm. The dielectric properties of BNT ceramics were investigated over the frequency range of 1 MHz-100 MHz in the temperature range of 133K – 653K. The obtained values of dielectric constant (ε_r), dielectric loss (tanδ) and Curie temperature (T_C) of the ceramics are 525, 0.067 and 603K respectively, as measured at 1 MHz. The temperature dependent AC-conductivity (σ_{ac}) of the BNT ceramics was analyzed using Arrhenius law and the activation energies (E_A) were estimated in different regimes. The obtained E_A values are in the range of 12.37 – 23.99 meV in the temperature regime 1 (R1: 133K – 278K), whereas in regime 2 (R2: 320K – 603K) in the range 51.5 – 126.2 meV. The frequency (1 MHz – 100 MHz) dependent σ_{ac} measured at different temperatures was analyzed using Jonscher's power law and super linear power law (SLPL). The obtained results indicate the BNT ceramics have potential applications in high frequency tunable devices.

1. Introduction

Lead zirconate titanate ($Pb(Zr,Ti)O_3$, PZT) is a well-known piezoelectric material and has been widely used in many applications, such as transducers, filters, actuators and sensors owing to their excellent piezoelectric properties [1]. Nevertheless, lead is toxic and harmful for human health and environment [2]. Therefore, it is necessary to replace the PZT with lead-free piezoelectric materials. The lead-free piezoelectric materials with perovskite structure, such as $(Bi_{0.5}Na_{0.5})TiO_3$ (BNT) [3], $BaTiO_3$ (BT) [4] and $(K_{0.5}Na_{0.5})NbO_3$ (KNN) [5] have been widely studied because of their excellent piezoelectric properties. Amongst all lead-free piezoelectrics, BNT based ceramics have been considered as a good candidate because of its large remnant polarization (P_r =38 μC/cm^2), coercive field (E_C = 73 kV/cm) at room temperature and high Curie temperature (T_C = 320 °C) [3]. A high dielectric constant and low dielectric loss are the key parameters to develop the high performance piezoelectric devices. The BNT ceramics were synthesized by various methods and the studies were mainly focused on ferroelectric, piezoelectric and dielectric properties [6, 7]. However, there are no reports available on the dielectric and AC-conductivity analysis of BNT ceramics measured in the frequency range of 1MHz – 100 MHz and in the low temperature range of 133K – 653K. The significance of studying the AC-conductivity of ferroelectric ceramics in the low temperature regime lies in unveiling the dielectric relaxations and hopping nature of charge carriers. These studies are useful for considering the applicability of BNT ceramics in high frequency tunable devices. This has motivated us to pursue the systematic analysis of structural, microstructural and dielectric properties of BNT ceramics in the broad frequency and temperature range.

2. Experimental Details

Lead-free $Bi_{0.5}Na_{0.5}TiO_3$ ceramics were prepared by using conventional solid state reaction method. Highly pure Bi_2O_3 (Alfa-Aesar, 99.975%), Na_2CO_3 (Sigma Aldrich, 99.99%), TiO_2 (Sigma-Aldrich, 99.99%) were chosen as starting materials and weighed according to the desired stoichiometry. These powders were mixed for 5 hr using a planetary ball mill (M/s Fritsch GmbH, Germany, Pulverisette 6) using deionised water as a grinding media. All the mixtures were dried at 120 °C and calcined at 800 °C for 3 hr. These calcined powders were again ground for 10 hr to reduce the initial particle size. Organic binder poly vinyl alcohol was added to the finely ground powder and finally pressed into pellets of 10 mm diameter and 1 – 2 mm thickness. The prepared samples were sintered in the range of 1050 – 1150 °C for a constant duration of 3 hr. The density of sintered pellets was measured using Archimedes's principle. The samples sintered at 1100 °C displayed the maximum density. Hence, these samples are chosen for further studies. For dielectric measurements, both sides of the samples were coated with silver paste to be used as electrodes and the pellets were heated at 150 °C for 10 minutes.

The structural properties of the BNT ceramics have been characterized by using x-ray diffractometer (M/s Rigaku, TTRAX III 18 KW) with Cu-K_α radiation (λ=1.5406 Å). Raman spectra of the BNT ceramics was collected using (M/s Horiba Jobin Yvon, LABRAM HR) Raman spectrometer employing a He-Ne source (514 nm as the excitation wavelength). Microstructure and surface morphology of the ceramics were analyzed with Field emission scanning electron microscope (FESEM) (M/s Zeiss, Sigma). The temperature and frequency dependence of dielectric properties were measured over a frequency (1 MHz to 100 MHz) and temperature range (133K to 653K) by using RF impedance material Analyzer (M/s Agilent Technologies, E4991A and M/s Novocontrol, BDS 2200).

3. Results and Discussion

The Rietveld refined XRD pattern of the sintered BNT ceramics is shown in Figure 1. The XRD analysis revealed the formation of a single phase with rhombohedral crystal symmetry at room temperature.

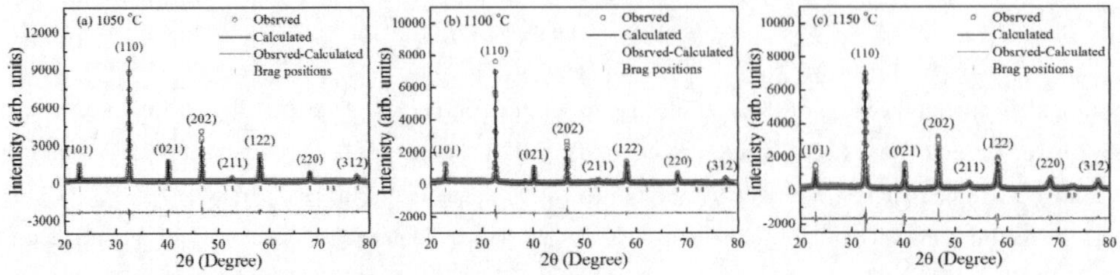

Figure 1. Reitveld refined XRD pattern of the BNT samples sintered at (a) 1050 °C, (b) 1100 °C and (c) 1150 °C for 3 hr.

The lattice parameters, atomic positions and occupancy of the Bi, Na, Ti, and O atoms were refined by using Fullprof software, considering R3c space group. The obtained lattice parameters and unit cell volumes are listed in Table 1. The lattice volume and density increases with sintering temperature up to 1100 °C and decreases beyond that. The density of the prepared samples was found to be in the range of 90-95 % of the theoretical density. The density and lattice volume decreases as the sintering temperature increases to 1150 °C. The grain boundary movements increase with sintering temperature. As a result, some closed pores remain inside the grains. Hence, the density of the ceramics sintered at 1150 °C is lower than 1100 °C.

Table 1. Rietveld refined parameters of BNT ceramics at different sintering temperatures.

Sintering Temperature (°C)	χ^2	a = b (Å)	c (Å)	Volume (Å^3)
1050	3.57	5.4980	13.4370	351.7562
1100	2.33	5.5010	13.4486	352.4488
1150	5.17	5.4988	13.4099	351.1523

Raman spectroscopy is the most useful technique to study the structural deformation of perovskite systems. Figure 2 shows the Raman spectra of BNT ceramics (sintered at (a) 1050, (b) 1100 and (c) 1150 °C for 3 hr) measured over the range of 50 – 1000 cm^{-1}. The Raman spectrum is fitted with Origin software by using Gaussian function. There are nine vibrational modes observed at 83, 131, 227, 291, 371, 530, 599, 772 and 862 cm^{-1}, which are shown in Figure 2. The obtained results are in good agreement with the earlier reports [8, 9].

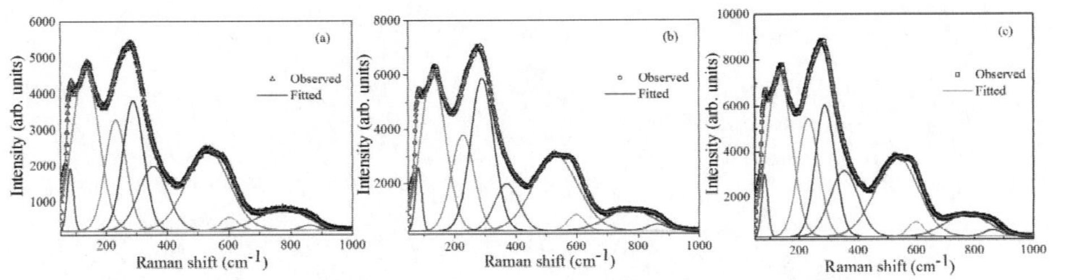

Figure 2. Raman spectra of BNT samples sintered at (a) 1050 °C, (b) 1100 °C and (c) 1150 °C.

The BNT ceramics possess rhombohedral crystal symmetry with R3c space group. According to the group theory, R3c space group consist of 13 Raman active optical modes ($\Gamma_{Raman} = 4A_1 + 9E$), where A_1 and E modes are Raman and IR active. These Raman active optical modes are due to disorder in the A-site of BNT system associated to BiO_6 and NaO_6 clusters [10]. The scattering of light with respect to incoming polarization is responsible for splitting of optical modes into transverse (TO) and longitudinal (LO) components. The Raman active A_1 (TO_1) modes at 83 cm^{-1} and 131 cm^{-1} are related to the BiO_6 and NaO_6 clusters, respectively. The Raman active E (TO_2) modes at 227, 291 and 371 cm^{-1} are associated to the octahedral TiO_6 clusters and were observed in many perovskites [11]. The Raman-active LO_2 modes are responsible for low intensity peaks and are related to short range electrostatic forces associated with the lattice iconicity [9, 12]. The TO_3 mode located around 530 cm^{-1} and 599 cm^{-1} corresponds to the (O-Ti-O) stretching symmetric vibrations of the TiO_6 clusters. The LO_3 modes observed at 772 and 862 cm^{-1} are due to the presence of sites within the rhombohedral lattice containing distorted octahedral TiO_6 clusters [9]. These results indicate that intensity of all Raman modes increased with sintering temperature due to enhancement in disorder of A-site in BNT ceramics. However, no significant change was observed in the Raman shift as well as full width at half maximum (FWHM) in the spectra for sintered samples.

Figure 3. FESEM micrographs of the BNT samples sintered at (a) 1050 °C, (b) 1100 °C and (c) 1150 °C.

Figure 3 shows the FESEM micrograph of the BNT ceramics, sintered at 1050, 1100 and 1150 °C for 3 hr. The average grain size has been calculated using linear intercept method and is found to increase from 1.09 to 1.98 μm with rise in sintering temperature. The sample sintered at 1100 °C, exhibits rectangular shape grains with average grain size of 1.40 μm. This sample showed well defined grain boundaries and less porosity, which indicates that the sintered (1100 °C) BNT ceramics having a high dense microstructure.

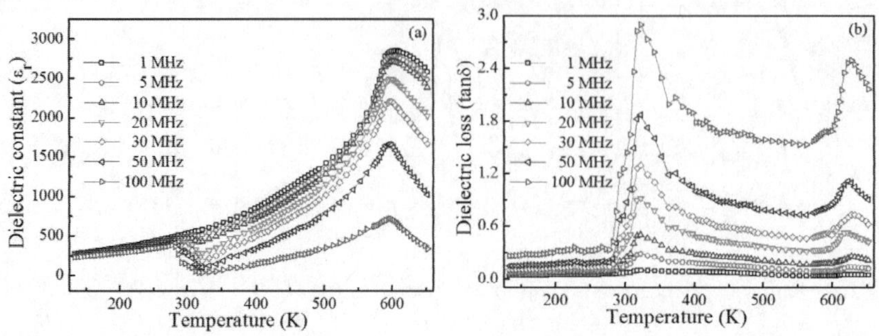

Figure 4. The temperature variation of (a) ε_r and (b) tanδ of BNT ceramics measured at different frequencies.

Figure 4 shows the temperature variation of dielectric constant (ε_r) and dielectric loss (tanδ) of BNT ceramics, sintered at 1100 ºC for 3 hr, measured at different frequencies from 1 MHz to 100 MHz. The ε_r increases with temperature from 133K to 300K, which is almost independent of frequency but has a frequency dependency at high temperature. The BNT ceramics undergo structural transitions from ferroelectric rhombohedral to antiferroelectric tetragonal structure around 473K, which is also known as depolarization temperature (T_d), and there is another transition from antiferroelectric tetragonal to paraelectric cubic structure around 603 K, which is the Curie temperature (T_C). The T_C of the BNT ceramics was estimated based on the temperature at which the maxima of dielectric constant is reached (~603K), which is in good agreement with our earlier report [9]. However, a weak anomaly in ε_r (T_d ~ 473k) is obtained at 1 MHz and further disappears with increasing frequency due to the stabilization of non-polar phase in the BNT samples [13]. The obtained ε_r and tanδ are 277, 525 and 0.040, 0.067 at 133K and 298K (1MHz), respectively. Above T_C, ε_r is found to decrease due to the increasing randomness of dipoles with temperature and it tends to weaken the dielectric properties in cubic phase. The decrease in ε_r with increasing frequency measured over a temperature range (133K – 653K) is due to the reduction in the polarization. The dielectric loss also exhibited similar behaviour at low temperature and rises at the high temperatures with increase in the frequency and is attributed to the enhancement in ionic conductivity in the BNT system. The tanδ also exhibited clear anomalies at phase transition temperatures.

The frequency dependence of AC-conductivity of BNT ceramics can be explained on the basis of Jonscher's power law at different temperatures,

$$\sigma_{ac}(\omega) = \sigma_{dc} + A\omega^{s} \qquad (1)$$

Where, $\omega = 2\pi f$ is the angular frequency, σ_{dc} is the dc conductivity, 'A' is the temperature dependent pre-exponential factor and 's' is the frequency exponent [14]. However, the term σ_{dc} disappeared at high frequencies, and it follows the super linear power law (SLPL) or nearly constant loss (NCL) ($\sigma \propto \omega^s$), ($s \geq 1$) [15]. If $s<1$, it corresponds to the translational short-range hopping motion, which is a characteristic of low or intermediate frequencies and is associated with the grain boundary conductivities. On the other end, $s \geq 1$ correspond to the re-orientational or localized hopping mechanism, and it represents the grain conductivity at higher frequencies [16]. The AC-conductivity response is more complex and there are no clear theoretical details so far. Lunkenheimer et al. [15] analyzed the AC-conductivity of different disordered materials at high frequencies (> 1MHz) using SLPL and NCL. Shanming et al. [17] observed the NCL and SLPL behaviour either at low temperatures/low frequencies or high temperature/high frequencies.

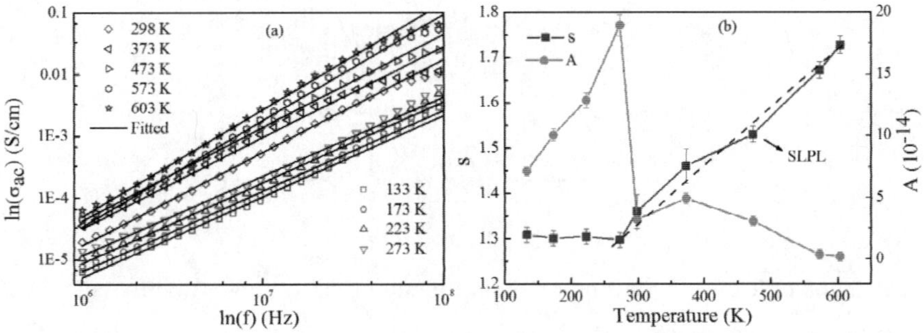

Figure 5. (a) The frequency dependence of electrical conductivity log(σ_{ac}) fitted by using Jonscher's power law and (b) Temperature dependence of the exponent 's' and 'A' of BNT ceramics at different temperatures.

Figure 5 (a) shows the frequency dependence of electrical conductivity σ_{ac} (ω) of BNT ceramics measured at different temperatures (133K - 603K) and analyzed using Jonscher's power law. The symbols indicate the experimental data, and the solid line represents the fitting of the experimental data by using equation (1). The fitting parameters such as 'A' and 's' were extracted for all the measured temperatures and are shown in Figure 5 (b). From the plots, two regimes were clearly identified, which gives us two different slopes. The exponent 's' is almost constant (~1.30) below 273K, which shows a weak dependence on temperature. Above the 273K, 's' varies linearly/super-linearly with temperature, which is a characteristic of the SLPL [17]. The obtained results (s>1) indicate that hopping motion is localized or re-orientational via small polarons. The small polaron hopping mechanism in BNT ceramics is thermally activated beyond the low temperature (>298K). In the low frequency and low temperature region, the conductivity is mainly due to the successful hoping between different sites but this hoping reduces with a rise in the frequency. Also, as the frequency increases, the ratio of successful to unsuccessful hoping changes with frequency. Therefore, the dispersion is observed at high frequencies. On the other hand, the coefficient 'A' exhibited the maximum value at 273K where the 's' shows the minimum and it further decreases with increasing temperature. This might be due to the large extent of interaction among the charge carriers and polarization.

IOP Conf. Series: Materials Science and Engineering **360** (2018) 012024 doi:10.1088/1757-899X/360/1/012024

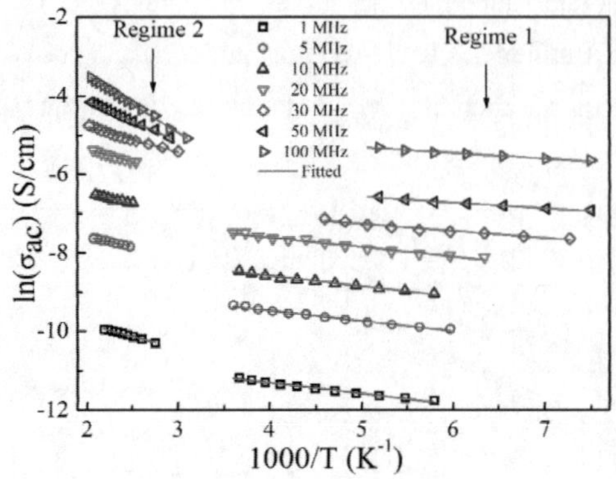

Figure 6. The variation of $\ln(\sigma_{ac})$ with reciprocal temperature of BNT ceramics in different regimes (R1, R2).

The activation energy (E_A) of the BNT ceramics have been calculated by using Arrhenius relation [18],

$$\sigma_{ac} = \sigma_0 \exp(-E_A / k_B T) \tag{2}$$

Where, σ_0 is the pre-exponential factor, k_B is the Boltzmann constant, and T is the absolute temperature. The E_A values were estimated from the slope of $\ln(\sigma_{ac})$ versus 1000/T plot, where the linear variation observed in regime R1 and R2, and are shown in Figure 6. The E_A values were found to decrease from 23.99 meV to 12.37 meV with increasing the frequency from 1 MHz to 100 MHz in R1, whereas the E_A values were found to be enhanced from 51.5 meV to 126.20 meV in R2. The different activation energies at distinctive temperature regions are due to a difference in the conductivity and domain mechanism [19]. The lower E_A values in R1 are attributed to the carrier transport mechanism associated with the hopping between localized states. The higher values of E_A in the R2 are due to the formation of small polarons and oxygen vacancies [19]. The small polarons are created by the interaction between the charge carriers and lattice, and this deformation will increase with the temperature. The oxygen as well as A-site vacancy would be created in the sample at higher sintering temperatures, which could be the reason for increment in the conductivity. The higher conductivity observed near the phase transition temperature is due to the change in domain structure, which can cause the release of charge carriers and enhance the conductivity within the sample.

4. Conclusion

$Bi_{0.5}Na_{0.5}TiO_3$ (BNT) ceramics have been prepared by conventional solid state reaction method. The structural properties of BNT ceramics were explained using Rietveld refinement method and Raman spectroscopy. BNT ceramics sintered at 1100 °C, exhibited the uniform grain size with less porosity. The dielectric properties of BNT ceramics were examined over the temperature range 133K – 653K and broadband frequency range of 1 MHz – 100 MHz. The temperature dependent dielectric properties of the BNT sample revealed the T_C around 330 ºC and $\varepsilon_r = 525$ and $\tan\delta = 0.067$ at 1 MHz. The frequency

IOP Conf. Series: Materials Science and Engineering **360** (2018) 012024 doi:10.1088/1757-899X/360/1/012024

dependence of AC-conductivity measured at distinctive temperatures follows the SLPL and hopping motion is localized or reoriented via small polarons. The temperature dependent AC-conductivity of the BNT ceramics was analyzed using Arrhenius law and activation energies (E_A) were estimated in different regimes. The obtained high dielectric constant, low dielectric loss and nearly constant loss over a wide range of frequency makes this material suitable for high performance piezoelectric and microwave tunable device applications.

Acknowledgments
The authors acknowledge the financial support from DAE BRNS [37(1)/14/33/2015/BRNS] India.

References

[1] G. H. Haertling 1999 J. *Am. Ceram. Soc.* **82** 797
[2] L. E. Cross 2004 *Nature* **432** 24
[3] G. A. Smolenskii, V. A. Isupov, A. I. Agranovskaya and N. N. Krainik 1961 *Sov. Phys. Solid State* **2** 2651
[4] Q. Xu, S. T. Chen, W. Chen, S. J. Wu, J. Lee, J. Zhou, H. J. Sun and Y. M. Li 2004 *J. Alloys Compd.* **381** 221
[5] L. Egerton and D. M. Dillon 1959 *J. Am. Ceram. Soc.* **42** 438
[6] Y. Hiruma, H. Nagata and T. Takenaka 2009 *J. Appl.Phys.* **105** 084112
[7] M. Cernea, E. Andronescu, R. Radu, F. Fochi and C. Galassi 2010 *J. Alloys Compd.* **490** 690
[8] H. Lidjici, B. Lagoun, M. Berrahal, M. Rguitti, M. A. Hentatti and H. Khemakhem 2015 *J. Alloys Compd.* **618** 643
[9] Srinivas Pattipaka, Mahesh Peddigari and Pamu Dobbidi 2017 *Ceram. Int.* **43** S151–S157
[10] J. Petzelt, S. Kamba, J. Fabry, D. Noujni, V. Porokhonskyy, A. Ashkin, I. Franke, K. Roleder, J. Suchanicz, R. Klein and G. E. Kugel 2004 *J. Phys: Condensed Matter* **16** 2719
[11] M. Zannen, A. Lahmar, M. Dietze, H. Khemakhem, A. Kabadou and M. Es-Soun 2012 *Mater. Chem. Phys.* **134** 829
[12] A. Chaves, R.S. Katiyar and S.P.S. Porto 1974 *Physical Review B* **10** 3522
[13] A. Hussain, A. Zaman, Y. Iqbal and M.H. Kim 2013 *J. Alloys Comp.* **574** 320
[14] A. K. Jonscher 1983 *Dielectric Relaxation in Solids* Chelsea Dielectric Press London
[15] P. Lunkenheimer and A. Loidl 2003 *Phys. Rev. Lett.* **91** 207601
[16] K. Funke and Prog 1993 *Solid State Chem.* **22** 111
[17] K. Shanming, H. Huang, Y. Shuhui and Z. Limin 2010 *J. Appl. Phys.* **107** 084112
[18] W. S. Kang, S. K. Lee, J. H. Koh 2015 *Ceram. Int.* **41** 6925–32
[19] O. Raymond, R. Font, N. S. Almodovar, J. Portelles and J. M. Siqueiros 2005 *J. Appl. Phys.* **97** 084107

IOP Conf. Series: Materials Science and Engineering **360** (2018) 012025 doi:10.1088/1757-899X/360/1/012025

Microalgae as Sustainable Energy and Its Cultivation

S. S. Sawant[a], B. D. Gajbhiye[a], C. S. Mathpati[a*], Reena Pandit[b], A. M. Lali[a,b]
[a] *Department of Chemical Engineering, Institute of Chemical Technology, Matunga, Mumbai 400019, India*

[b]*DBT-ICT Centre for Energy Biosciences, Institute of Chemical Technology, Matunga, Mumbai 400019, India*

*Corresponding author Email: cs.mathpati@ictmumbai.edu.in

Ph. No. +91-22-33612017 Fax: +91-22-33611020

Abstract:

Growth of every country is depended on the ability to supply energy as per demand. Weather we notice it or not, the fact is energy is important in our daily life. The predicted expansion of world's population is at least 9 billion by 2050, correlates increase of energy use globally to 812 quadrillion (10^{15}) kJ by 2035 from 533 quadrillion kJ in 2008. A key challenge is to meet the growing demand for energy in a safe and environmentally responsible manner. Sustainable energy serves the needs of the present without compromising the ability of future generations to meet their needs. Technologies that promote sustainable energy include renewable energy sources, and bioenergy is one of the types of renewable energy. Bioenergy is any organic material which can store sun light in the form of chemical energy. Algal biofuel has a potential to reduce the demand of fossil fuels for energy requirements. Algae production do not have a stringent requirement of environmental conditions and can be produced in aqueous suspensions where sufficient light and carbon dioxide is available. Biofuels from algae have a high lipid density (15-300 times higher than traditional crops) as compared with other terrestrial crops. Algae have a harvesting cycle of 1–10 days. The amount of biomass is doubled typically within 24 hours under optimal growth conditions (enough nutrients, sunlight etc.). The two major bioreactor classes of algae growth systems are open ponds and closed reactors. The most common growth systems used for commercial purposes are open pond. Open ponds are relatively inexpensive compared to closed photo bioreactor. Algae cells are sensitive to hydrodynamic stress. The concept of cell damage correlates interaction between free cells and isotropic turbulence eddies based on Kolmogorov's theory. When the micro-eddies size is close to the cell dimension, cell damage is higher; therefore, one should adjust the fluid flow in the open pond to maintain length of micro-eddies larger than the cell size. In the present investigation CFD code was used to study shear stress in open pond with side entry axial flow impeller. Three different channel lengths to width ratio were investigated with respect to size of micro eddies and shear stress.

Keywords: microalgae, raceway pond, sustainable energy, bioenergy, biofuel, computational fluid dynamics.

1: Introduction:

1.1: Energy: For Today and Future

Energy is inevitable for human life. It's safe and accessible supply is not only important but crucial for the modern societies. Problems such as depletion of fossil fuel reserves, geopolitical & military conflicts indicate an unsustainable situation. Renewable energy is the key solution to the growing energy challenges. Renewable energy resources such as biomass, solar, wind is abundant, environmentally friendly and inexhaustible. Global warming and energy crisis are two major problems human being faces today. Population growth, fast industrialization, and increased use of fossil fuels contributes in the consumption of the available energy. The need for energy is rising with economic wealth(M. Asif, 2007) (J. Rupprecht, 2009). Worldwide supply of liquid fuels is completely dependent on petroleum. Due to present growth of prosperity and economics, petroleum will become more expensive as well as will cause political and military conflicts (C. Posten, 2009). Weather we notice it or not, the fact is energy is very important in our daily life as growth of every country depends on the ability to supply energy as per demand. The predicted expansion of world's population is at least 9 billion by 2050, correlates increase of energy use globally to 812 quadrillion (10^{15}) kJ by 2035 from 533 quadrillion kJ in 2008(J. N. Rogers, 2013).

1.2: Microalgae

Microalgae are the largest autotrophic microorganisms of plant life. Microalgae are known to synthesize and can rapidly accumulate higher amounts of lipids than terrestrial plants due to their high growth rates. Microalgae, recognized as one of the oldest living organisms. While the mechanism of photosynthesis in these microorganisms is similar to that of higher plants, they are generally more efficient converters of solar energy because of their simple cellular structure. In addition, Microalgae have more efficient access to water, CO_2, and other nutrients as they grow in aqueous suspension. Microalgae can be either autotrophic or heterotrophic. Microalgae can fix CO_2 efficiently from different sources, including the atmosphere, industrial exhaust gases, and soluble carbonate salts. Fixation of CO_2 from atmosphere is probably the most basic method to sink carbon, and relies on the mass transfer from the air to the microalgae in their aquatic growth environments during photosynthesis (F. A. K. Alparslan,). Microalgae use light and carbon dioxide (CO_2) and grow in aquatic environments to create biomass. They are basically a large and diverse group of simple, typically autotrophic organisms, ranging from unicellular to multi-cellular forms. They do not compete with food or other crops as they can be cultivated on arid and unhospitable land (J. Singh, 2016)' (L. A. Ribeiro, 2015). Algae production on large commercial scale do not have a stringent requirement of environmental conditions and can be produced in aqueous suspensions where sufficient light and carbon dioxide is available.

1.3: Microalgae as Biomass

Algal biomass contains three main components: carbohydrates, proteins, and lipids/natural oils. Microalgae grow very quickly compared to terrestrial crops. Algae have a harvesting cycle of 1–10 days. They commonly double in size every 24 hours. They are capable to produce greater amounts of biomass per hectare than any kind of terrestrial crop. During the peak growth phase, some microalgae can double every 3.5 hours under optimal growth conditions (enough nutrients, sunlight etc.). Biofuels from algae have a high lipid density (15-300 times higher than traditional crops) as compared with other terrestrial crops (V.Andersson,

2012), (A. F. Clarens, 2010), (Y. Chisti, 2008). In addition to producing biofuel, algae can also be considered for a variety of other uses, such as fertilizer, pollution control, and human nutrition. Considering increase in oil prices and climate changes, biofuels have become more important as potential alternative energy sources. Energy can be generated by burning biomass to generate heat and electricity. Some algae can even produce hydrogen gas under specialized growth conditions (W. Z. Johnson, 2009).

1.4: Micro algal Biofuel

Biofuels are the most important types of renewable energy. They can be categorized into first, second and third generation biofuels. Micro algal biofuels are thought as third generation biofuels. They are one of the most promising biofuel feed- stock to solve global energy crisis and climate changes. The advantages of microalgae as a biofuel feed stock include potentially high photosynthesis efficiency and biomass yields, lack of competition for arable land and fresh water resource, and carbon mitigation (R. E. Reviews, 2015). An integrated system for a steam power plant and cogeneration processes with an algae cultivation system for bio fixation of CO_2 was proposed by Lira-Barragan et al with the purpose of reducing the overall CO_2 emissions and producing biodiesel that replaces fossil fuels used as liquid transportation fuels (L. F. Lira-barraga, 2015). Biodiesel and bioethanol are two most promising candidates for biofuels. All algae have the capacity to produce energy-rich oils. Oil content of some microalgae can exceed 80% of the dry weight. The most important aspect of microalgae is their ability to produce several different types of biofuels such as biodiesel, bioethanol, biohydrogen, and biogas. Interestingly, biofuels derived from algae provide a significant reduction in vehicular exhaust emissions versus standard fossil fuels (T. Taparia). Biodiesel from microalgae is a clean-burning alternative fuel, biodegradable, user-friendly, non-toxic, and free from Sulphur and odor. It has a higher flashpoint and greater lubricity as compared with traditional diesel fuel. It can be easily used by blending with petroleum fuel to create biofuel blend, rather than fully dependent on petroleum fuel (C. Hong, 2015), (C. N. Dasgupta , 2015).

1.5: Bioreactors for cultivation of Microalgae: closed photo-bioreactor and open raceway pond

Design of photo-bioreactor is deciding factor to control capital and operating cost for cultivation of microalgae on large scale. The two major bioreactor classes of microalgae growth systems are open ponds and closed reactors. The process parameters for growth of micro algal biomass can be controlled more efficiently in closed photo-bioreactors than the open ponds. But the cost of building (Capital) and operating such systems is too high for production of algal biomass especially for biofuels and bioenergy. Open ponds are more cost effective but they suffer from lesser control over growth parameter (Y. Chisti, 2008). The most common growth systems used for commercial purposes are open pond. Open ponds are frequently designed like raceway ponds (RWP), which feature paddle wheels and baffles to promote mixing. RWP are very popular as they are relatively inexpensive as compared to closed photo-bioreactor(J. G. G. Jonker, 2013), (A. Ozkan, 2012). RWP are further classified as convectional raceway pond and modified raceway pond. In convectional raceway pond mixing is carried out by using one or more paddlewheel (A. Darzins, 2010). Where modified raceway pond is operated using side entry axial flow impeller like Hydrofoil (HF) (Pub. No: WO/2013/186626, International Application No: PCT/IB2013/001224, 2013). Modified raceway pond is more energy efficient than the convectional raceway pond as power number of HF is considerably less than the paddle wheel (J. B. Joshi, 2011). HF impeller generates strong convective motion and lower turbulence compared to paddle wheels and pitched blade turbines(B. N. Murthy, 2008). Algae cells are sensitive to hydrodynamic stress. The concept of cell damage correlates interaction between free cells and isotropic turbulence eddies based on Kolmogorov's theory. When the micro-eddies size is close to the cell dimension, cell damage is higher; therefore, one should adjust the fluid flow in the open pond to maintain length of micro-eddies larger than the cell size (H. Hadiyanto, 2013).

Generally, RWP consists of rectangular tank which is divided into two compartments by central baffles (Figure 1). In the present work, an attempt has been made to study modified design of RWP with respect ratio of Length (L) to Width (W) of channel of RWP. Computational Fluid Dynamics (CFD) analysis of RWP is carried out for optimization of L/W ratio and power consumption. CFD provides detailed flow pattern in the reactor which helps to understand the mixing patterns as well as optimization of configuration. Attempt has been made to study effect of various L/W ratio on important parameters like circulation velocity (V_c), Dead Zone, Power Consumption and Shear Stress which decide performance of RWP.

2: Pond Description:

The design and optimization of adequate bioreactors to cultivate microorganisms is a major step in the strategy that focus on converting scientific findings into a feasible product.

Experiments were carried out in a pilot scale RWP. Volume of pilot scale RWP is 4500 liters. It consists of rectangular tank (Figure 1) with dimensions (325 cm (length) x 100 cm (depth) x 150 cm (width)). Rectangular tank is divided into two compartments of 65 cm width by a central baffle having dimensions of (195 cm (length) x 100 cm (depth) x 20 cm (width). Tank consists of HF impeller having three blades of 45 cm diameter.

Figure 1: Modified RWP

3: Simulation:

The simulations were carried out in ANSYS CFX 14. The impeller baffle interaction was modeled using multiple reference frame approach. The inner zone was rotated at 125 rpm. Space discretization was carried out using second order upwind method. The simulations were carried out using steady state approach up to residuals of 10^{-4}. SIMPLE algorithm was used for pressure velocity coupling. CFD simulations of raceway pond were carried out using standard

k-ε model. The set of equations for single phase flow can be found in Murthy and Joshi (2008)(B. N. Murthy, 2008).

The k - ε turbulence model includes two transport equations to represent turbulence properties of the fluid flow. The first equation is for turbulent kinetic energy (k), and the second is for turbulent kinetic energy dissipation rate equation (ε) (H. Hadiyanto, 2013). These equations can be written as follows:

The turbulent kinetic energy (k) is calculated from its governing equations:

$$\frac{\partial}{\partial t}(\rho\, k) + \nabla(\rho\, u\, k) - \nabla\left(\left(\mu + \rho\, C_{\mu}\, \frac{k^2}{\varepsilon}\right)\nabla k\right) = \frac{1}{2}\,\rho\, C_{\mu}\frac{k^2}{\varepsilon}(\nabla u + (\nabla u)^T)^2 - \rho\,\varepsilon \quad (1)$$

The rate of dissipation of turbulent kinetic energy is calculated from its governing equation:

$$\frac{\partial}{\partial t}(\rho\, \varepsilon) + \nabla(\rho\, u\, \varepsilon) - \nabla\left(\left(\mu + \rho\, C_{\mu}\, \frac{k^2}{\varepsilon}\right)\nabla \varepsilon\right) = \frac{1}{2}\,\rho\, C_{\varepsilon 1}\, k(\nabla u + (\nabla u)^T)^2 - \rho\, C_{\varepsilon 2}\,\frac{\varepsilon^2}{k} \quad (2)$$

The standard model parameters have been considered in present simulations such as C_{μ} = 0.09, $C_{\varepsilon 1}$=1.44, $C_{\varepsilon 2}$=1.92.

The grid independence study has been carried out with three grid resolutions. CFD prediction for three different grid sizes of 0.7, 0.8 and 0.9 million were compared with results obtained by ultrasonic velocity profiler (U V P). In that comparison, it was observed that with an increase in number of grids from 0.8 to 0.9 million, improvement in CFD predictions is minimal. Thus, grid with 0.8 million cells was selected for further CFD simulations. Detail study of grid independency and validation of CFD prediction of U and W velocities with UVP results can be find out in same authors work as Computational studies of pilot scale high depth algal photo-bioreactor which is under review in Engineering in Life Science Journal (manuscript id: elsc.201600067).

3.1: Domain for Computational Study:

The CFD simulations were carried out for different L/W ratios using finite volume approach. Hydrodynamic investigation was carried out for three different L/W ratios such as 5, 10 and 15. Geometrical details and volumes of three designs are described in Table 1.

Table 1: Reactor Dimensions and Volumes of three different L/W Ratios.

	Unit	L/W = 5	L/W = 10	L/W = 15
Channel Length (L)	(cm)	325	650	975
Channel Width (W)	(cm)	65	65	65
Depth	(cm)	100	100	100
Central Baffle Length	(cm)	195	520	845
Central Baffle Width	(cm)	20	20	20
Volume	(liter)	4485	8710	12935

Channel Width, Depth and width of the central baffle were kept constant for all L/W ratios. It was found that volume of the tank increases with the length of the tank. For three different L/W ratio simulation were carried out to find velocity distribution, dead zone, power consumption and shear stress in RWP.

3.2: Velocity Distribution:

Velocity between 15 cm/s to 30 cm/s is commonly required velocity for cultivation of algae in commercial RWP. Type of flow such as Laminar or turbulent is depend upon velocity, density and viscosity of fluid. For higher velocities, flow will be turbulent which will enhance operational conditions of RWP. Uneven distribution of velocity can reduce efficiency of RWP and it was significantly found in lower L/W ratio. Velocity distribution study was carried out in different geometries using CFD simulations.

3.3: Dead Zone:

Dead zone is the volume of stagnant flow. It reduces the RWP's volumetric capacity as well as the residence time of fluid. It causes uneven flow in the RWP and sedimentation of the algal cell (Hadiyanto et al., 2013). Lowest fluid velocity required to avoid settling of algae cell is 10 cm/s. To quantify dead zone Hadiyanto et.al used following formulae.

$$\text{Dead Zone} = \frac{\text{Volume of tank with less than 10 cm/s velocity}}{\text{Total volume of tank}} * 100 \quad (3)$$

Presence of dead zone affects the productivity of microalgae in the RWP. Dead zone is generally form at the sharp turn (ends of central baffle) where possibilities of formation of boundary layer separation is maximum. Dead zones must be considered as an important parameter while designing RWP and it should be minimized.

3.4: Power Consumption:

Power consumption determines the operating cost of RWP. Power represents as energy consumed by the impeller for mixing and to keep biomass in suspension. It can be calculated by multiplying Torque (T) on the blades with angular velocity. Torque can be obtained by solving k - ε turbulence model for different geometries of RWP. Feasibility of production of biomass and biofuels from the cultivation of algae depends on power consumption. To make system feasible, designs should be modified in such a way that power consume as Watt / dry weight of biomass should be minimum.

3.4: Shear Stress:

The pond should be designed in order to accommodate a stress range that gives the optimal cell response. It is necessary to compute the area of RWP where value of shear stress is more than optimal value. The concept of cell damage correlates interaction between free cells and isotropic turbulence eddies based on Kolmogorov's theory. When the micro-eddies size is close to the cell dimension, cell damage is higher; therefore the fluid flow should be controlled in the photo-bioreactor to maintain length of micro-eddies larger than the cell size (F. Garcı, 2000).

4: Results of C F D Simulation and Discussions:

In order to evaluate optimum design of RWP with respect to Length (L) to width (W) ratio, flow characteristics of three different L/W ratios such as 5,10 and 15 were investigated.

4.1 Velocity Distribution:

Figure 2 shows the results of the simulation under constant agitation speed (125 rotation per minute) for three different L/W ratios. Contour plot of velocity shows liquid flows along the channel and take 180° turns at the bends. As per the contour plot, Figure 2 it was clear that uneven distribution of velocity was more significant in lower L/W ratio and by increasing L/W ratio more uniformity of velocity can be maintained. Minimum required velocity in commercial cultivation plant of algae is 15 cm/s.

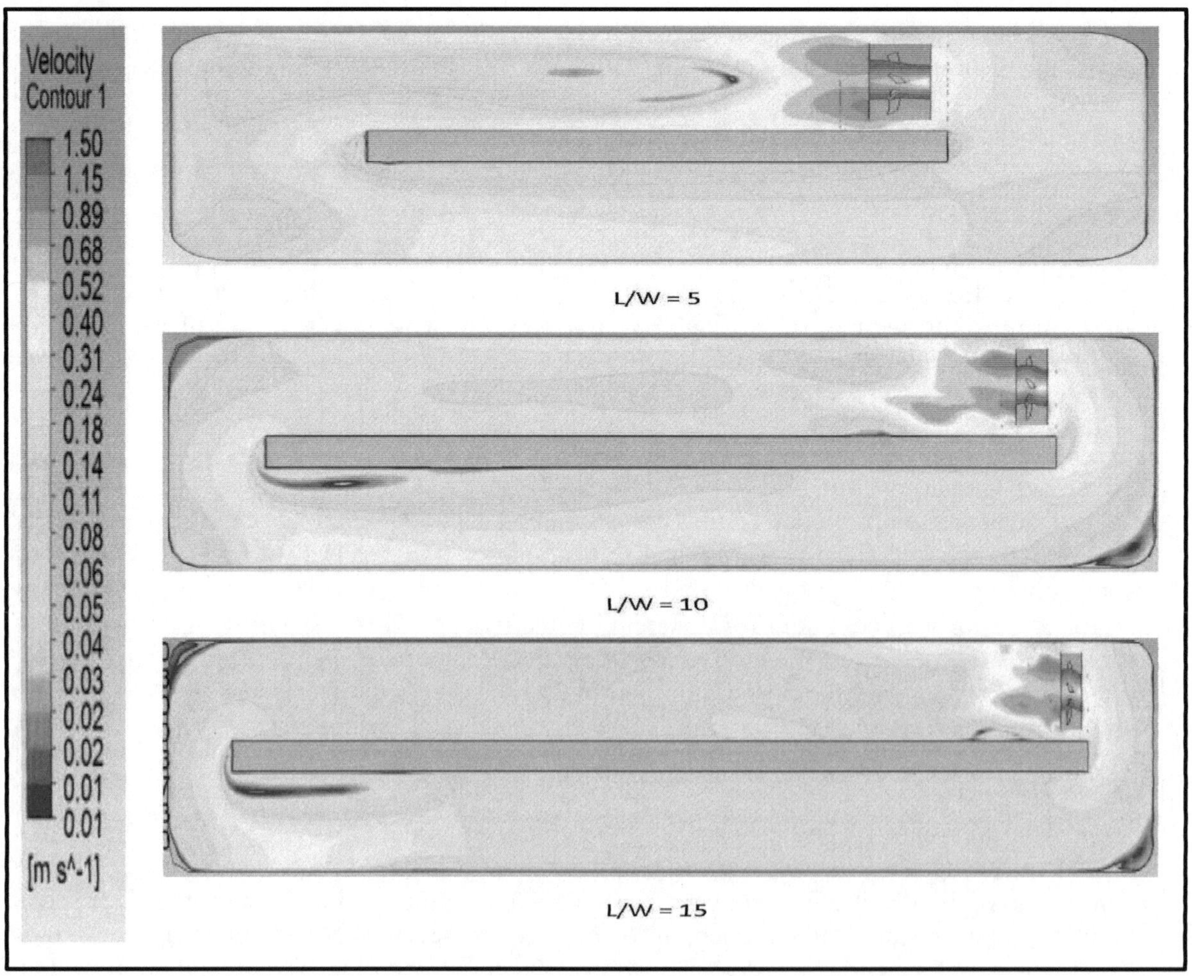

Figure 2: Velocity Contour

4.2 Dead Zone:

Dead zone for all three L/W of RWP were shown in figure 3. Figure 3 indicates decrease in dead zone with the increase in L/W ratio. L/W, 15 shows minimum dead zone and L/W, 5

IOP Conf. Series: Materials Science and Engineering **360** (2018) 012025 doi:10.1088/1757-899X/360/1/012025

shows maximum dead zone. Dead zones were calculated by equation no 3 where volume of region having velocity less than 10 cm/s is evaluated by CFD code.

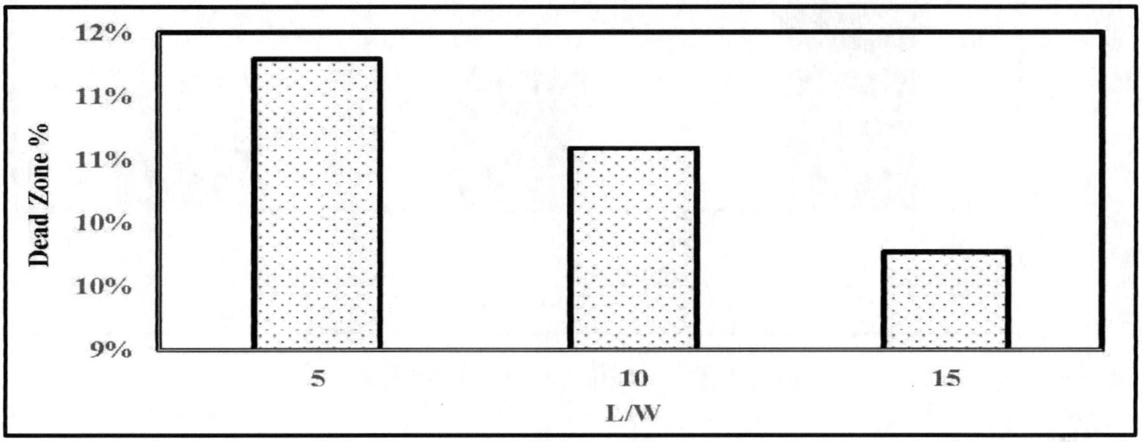

Figure 3: Dead Zone

4.3 Power Consumption:

Power consumption for different L/W ratios was shown in the Table 2. Ratio of power to volume decreases with increasing L/W ratio.

Table 2: Power Consumption

Sr No	L/W	Volume m^3	Torque N m	Power Watt	Watt/ m^3
01	5	4.4	2.88	47	10.7
02	10	8.7	4.53	71	8.2
03	15	12.9	5.71	75	5.8

As per the Table 2, Power required for constant agitation speed (125 rotation per minute) in the case of L/W, 15 was minimum where for L/W, 5 it was maximum. Therefore higher L/W design is more energy efficient. Calculated power consumption for various L/W was shown in Figure 4.

IOP Conf. Series: Materials Science and Engineering **360** (2018) 012025 doi:10.1088/1757-899X/360/1/012025

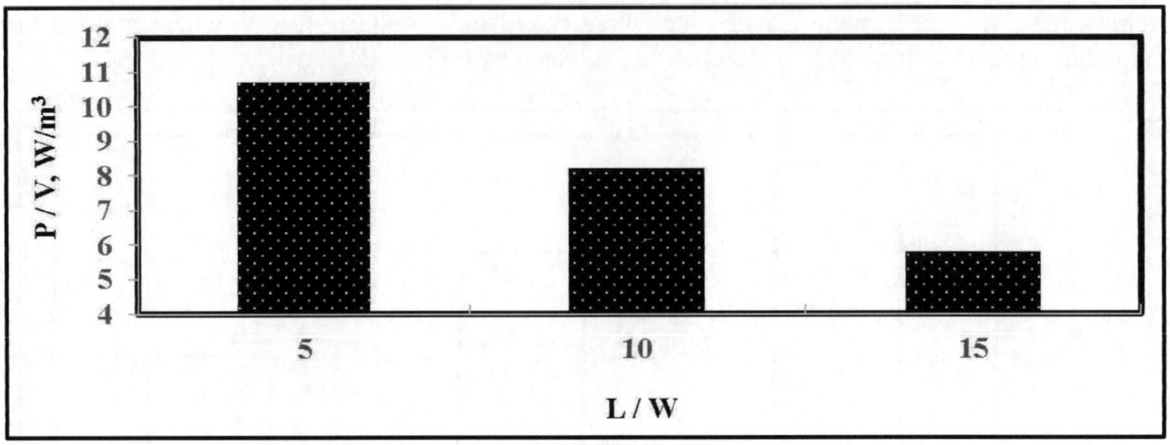

Figure 4: Power for various L/W Ratio

4.4: Shear Stress:

Size of micro-eddies can be find out by using formula given by (H. Hadiyanto, 2013) Hadiyanto et al.

$$\boldsymbol{\lambda} = \left(\frac{\eta}{\rho}\right)^{0.75} \varepsilon^{-0.25} \tag{4}$$

Where $\boldsymbol{\lambda}$= Size of micro-eddy, m

η = Dynamic viscosity, Pa.s

ρ = Density kg m^{-3}

ε = Rate of energy dissipation, W kg^{-1}

Figure 5 indicates size of micro-eddies for all designs in micro meter. Size of algal cell is approximately 100 μm, so to minimize damage to algal cell, RWP should be designed to maintain size of micro-eddies more than 100 μm. In all modification size of micro-eddy was found to be more than 100 μm. In, L/W = 10, length of micro-eddies is found to be 114 μm which is slightly greater than 100 μm.

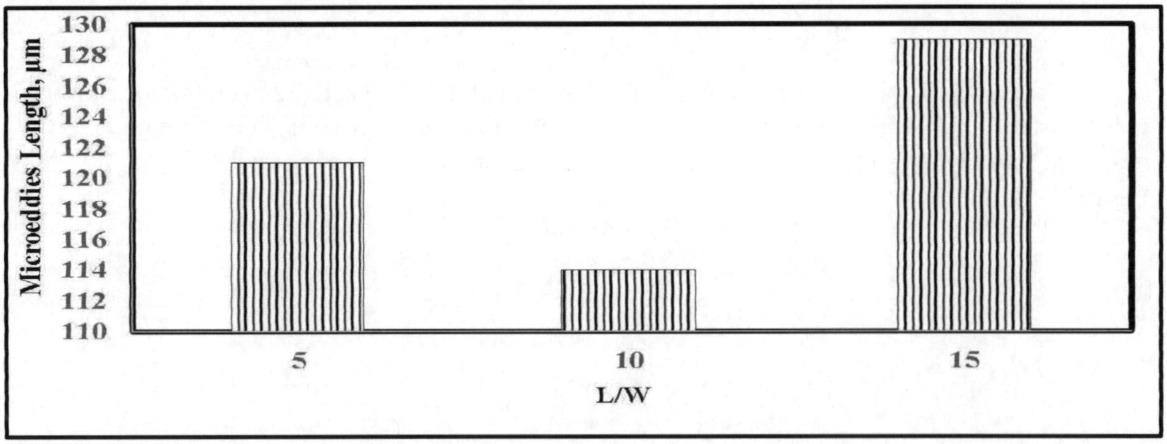

Figure 5: Length of micro-eddies μm

Figure 6 shows magnitudes of shear stress in all three designs. Shear stress can be formulated using k - ε turbulence model. From the figure 6, it was clear that, shear stress decreased with increase in L/W ratio. Maximum shear stress is observed in L/W, 5 where it decreases with increasing L/W ratio.

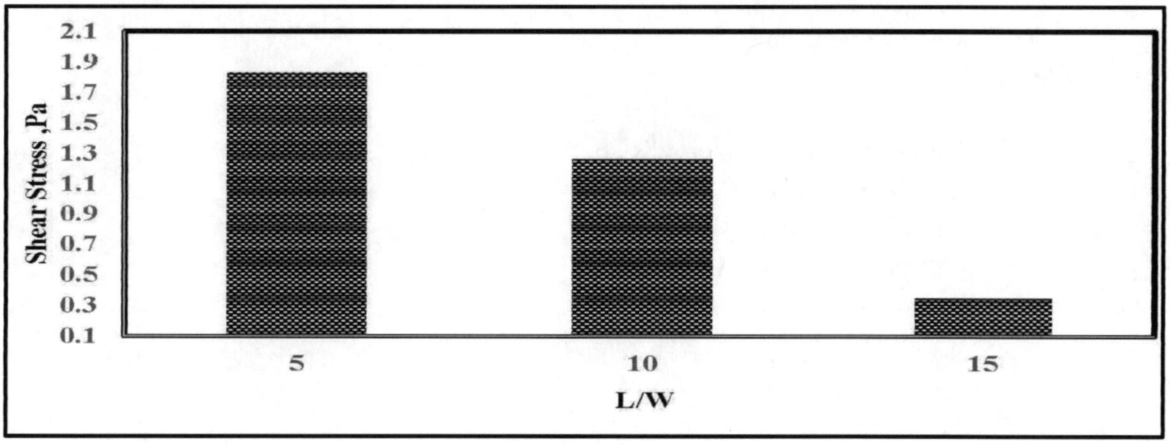

Figure 6: Shear Stress, Pa

5. Conclusion:

The CFD model was developed to study hydrodynamics of modified RWP with different L/W ratios. An increase in L/W ratios showed increase in uniformity of velocity distribution and thus dead zone in the RWP photo-bioreactor was also decreased with increase in L/W ratio. The power consumption per unit volume was lower when L/W ratio was increased from 5 to 15. Higher ratios of L/W, RWP photo-bioreactor design were more energy efficient. The modified RWP design enabled algae production for liquid depth of 1 m. Further studies involving optimization of bend geometry, sparger design, sparger location to improve the performance of modified RWP design are ongoing.

Acknowledgements

This work was supported by the Department of Biotechnology, Ministry of Science and Technology, Government of India (No. BT/PR13796/PBD/26/139/2010). Authors, Mr. Shekhar S Sawant and Mr. Bhavesh Ghajbhiye would like to acknowledge the fellowship support from UGC and DAE-ICT Centre, respectively.

Nomenclature

C_μ	Constant in k–ε turbulence model
$C_{\varepsilon 1}$	Model parameter in dissipation rate of turbulent kinetic energy in Equation 2
$C_{\varepsilon 2}$	Model parameter in dissipation rate of turbulent kinetic energy in Equation 2
k	Turbulent kinetic energy (m^2/s^2)
L	channel length (m)
W	channel width (m)
m	Mass (kg)
N_P	Power Number (-)

p	Pressure (N/m^2)
P	Power (W)
t	Time (s)
T	Torque (N.m)
u	Velocity vector (m/s)
v	Mean velocity (m/s)

Greek Symbols

ε	Turbulent kinetic energy dissipation rate (m^2/s^3)
μ_{eff}	Effective viscosity (Pa.s)
μ_T	Turbulent viscosity (Pa.s)
μ	Molecular viscosity (Pa.s)
ν	Kinematic Viscosity (m^2/s)
ρ	Density (kg/m^3)

Abbreviations

CFD	Computational Fluid Dynamics
HF	Hydrofoil
RWP	Raceway Pond
UVP	Ultrasonic Velocity Profiler

References

F. A. K. Alparslan. "Microalgae for renewable energy: biodiesel production and other practies", The 4 th International Symposium of Sustainable Development.

Viktor Andersson, Sarah Broberg, Roman Hackl, 2011. "Integrated algae cultivation for municipal wastewater treatment and biofuels production in industrial clusters", Energy Forum 2011, 1–8. Retrieved from https://ases.conference-services.net/resources/252/2859/pdf/SOLAR2012_0347_full paper.pdf

M. Asif & T. Muneer, 2007. "Energy supply, its demand and security issues for developed and emerging economies", Renewable and Sustainable Energy Reviews, 11(7), 1388–1413. https://doi.org/10.1016/j.rser.2005.12.004

Yusuf Chisti, 2007. "Biodiesel from microalgae beats bioethanol. Trends in Biotechnology", 126–131. https://doi.org/10.1016/j.tibtech.2007.12.002

A. F. Clarens, E. P. , Resurreccion, M. A. White, L. M. Colosi, 2010. "Environmental Life Cycle Comparison of Algae to Other Bioenergy Feedstocks", Environmental Science & Technology, 44(5), 1813–1819. https://doi.org/10.1021/es902838n

A. Darzins, P. Pienkos, L. Edye, 2010. "Current Status and Potential for Algal Biofuels Production", National Renewable Energy Laboratory, A report to IEA BioEnergy Task 39, Report T-39-T2, 6 August,2010 .

C. N. Dasgupta, M. R. Suseela, S. K. Mandotra, P. Kumar, M. K. Pandey, K. Toppo, J. A. Lone, 2015. "Dual uses of microalgal biomass : An integrative approach for biohydrogen and biodiesel production". Applied Energy, 146, 202–208. https://doi.org/10.1016/j.apenergy.2015.01.070

F. Garcıa Camacho, A. Contreas Gomez, T. Mazzuca Sobczuk, E. Molina Grima, 2000. "Effects of mechanical and hydrodynamic stress in agitated , sparged cultures of Porphyridium cruentum", Process Biochemistry, Volume 35, Issue 9, Pages 1045–1050.

H. Hadiyanto, S. Elmore, T. Van Gerven, A. Stankiewicz, 2013. "Hydrodynamic evaluations in high rate algae pond (HRAP) design". Chemical Engineering Journal, 217, 231–239. https://doi.org/10.1016/j.cej.2012.12.015

C. H. Tang, W. Y. Cheah, T. C. Lian, S. P. Loke, J. C. Juan, J. S. Chang, 2015. "Algae Cultivation in Wastewater for Biodiesel – A Review", 45, 1393–1398. https://doi.org/10.3303/CET1545233

W. Z. Johnson, 2009. "Microalgae as a Feedstock for Biofuel Production". Virginia Cooperative Extension, 442–886. https://doi.org/10.1016/j.nbt.2009.06.630

J. G. G. Jonker, A. P. C. Faaij, 2013. "Techno-economic assessment of micro-algae as feedstock for renewable bio-energy production". Applied Energy, 102, 461–475. https://doi.org/10.1016/j.apenergy.2012.07.053

J. B. Joshi, N. K. Nere, C. V. Rane, B. N. Murthy, C. S. Mathpati, A. W. Patwardhan, V. V. Ranade, 2011. "CFD simulation of stirred tanks: Comparison of turbulence models (Part II: Axial flow impellers, multiple impellers and multiphase dispersions". Canadian Journal of Chemical Engineering, 89(August), 754–816. https://doi.org/10.1002/cjce.20465

A.M. Lali, R.A. Pandit, P. Gunjan, C.S. Mathpati, S.P. Gangal, C.P. Vira, J.A. Palkar, S.D. Patil, S. P. Gaikwad, 2013. Pub. No: WO/2013/186626, International Application No: PCT/IB2013/001224. India.

L. F. Lira-barraga, F. A. Jose, M. M. El-halwagi, 2015. "Reduction of greenhouse gas emissions from steam power plants through optimal integration with algae and cogeneration systems", 2401–2415. https://doi.org/10.1007/s10098-015-0982-1

B. N. Murthy, J. B. Joshi, 2008. "Assessment of standard k - ε, RSM and LES turbulence models in a baffled stirred vessel agitated by various impeller designs". Chemical Engineering Science, 63(22), 5468–5495. https://doi.org/10.1016/j.ces.2008.06.019

A. Ozkan, K. Kinney, L. Katz, H. Berberoglu, 2012." Reduction of water and energy requirement of algae cultivation using an algae biofilm photobioreactor". Bioresource Technology, 114, 542–548. https://doi.org/10.1016/j.biortech.2012.03.055

C. Posten, G. Schaub, 2009." Microalgae and terrestrial biomass as source for fuels-A process view". Journal of Biotechnology, 142(1), 64–69. https://doi.org/10.1016/j.jbiotec.2009.03.015

R. E. Reviews, 2015. "Biogas production from algal biomass : A review". https://doi.org/10.1016/j.rser.2014.11.052

L. A. Ribeiro, P. Pereira, T. M. Mata, 2015. "Prospects of using microalgae for biofuels production : Results of a Delphi study", 75, 799–804. https://doi.org/10.1016/j.renene.2014.10.065

J. N. Rogers, J. N. Rosenberg, B. J. Guzman,, V. H. Oh, L. E. Mimbela, A. Ghassemi, M. D. Donohue, 2013. "A critical analysis of paddlewheel-driven raceway ponds for algal biofuel production at commercial scales". Algal Research, 4, 76–88. https://doi.org/10.1016/j.algal.2013.11.007

J. Rupprecht, 2009. "From systems biology to fuel-Chlamydomonas reinhardtii as a model for a systems biology approach to improve biohydrogen production". Journal of Biotechnology, 142(1), 10–20. https://doi.org/10.1016/j.jbiotec.2009.02.008

J. Singh, S. Gu, 2016. "Commercialization potential of microalgae for biofuels production" .

Renew Sust Energ Rev,. https://doi.org/10.1016/j.rser.2010.06.014
T. Taparia, M. Mvss, R. Mehrotra, P. Shukla. "Developments and challenges in biodiesel production from microalgae : A review" , 1–12. https://doi.org/10.1002/bab.1412

Structural and Spectroscopic Studies of Nanostructured Alumina Doped $LaFeO_3$ a Photo catalyst Ceramics Synthesized Through an Auto Igniting Combustion Synthesis.

Bincy Abraham[a] and Jijimon K Thomas[a,*]

[a]*Electronic Materials Research Laboratory, Department of Physics, Mar Ivanios College, Thiruvananthapuram 695015, Kerala, India*

**Corresponding author Email: jkthomasemrl@yahoo.com (J.K. Thomas)*

$LaFeO_3$ with an orthorhombic phase of the ABO_3 type perovskite structure has become a currently attractive research topic because it is proposed for various applications in several advanced technologies such as catalysts, various kinds of chemical and gas sensors and electrode materials in solid oxide fuel cells. In the present work $La_{1-x}Al_xFeO_3$ (x = 0.01) nanopowder was prepared through an auto ignited combustion technique. In order to study the phase stability, the structural, vibrational and optical properties, the as prepared material was heat treated at 600^0C. The phase purity of the powder was examined using X-ray diffraction (XRD) and Fourier transform infrared spectroscopy (FTIR) studies. XRD studies show that the powder possesses an orthorhombic structure with lattice constants a=5.55, b=7.85 and c=5.55. The transmission electron microscopic study confirmed the ultrafine nanocrystallite nature of the powder. The thermal stability of the nanopowder studied using thermo gravimetric analysis (TGA) and differential thermal analysis (DTA) shows that the nanopowder has ~7% weight loss in the temperature range 100-600°C. The UV-Visible diffuse reflectance spectroscopy study carried out on the alumina doped $LaFeO_3$ ceramics indicate that the sample has narrow optical band gap of ~2.08 eV. These results show the strong visible-light absorption characteristics of the sample and it perfectly satisfies the band gap requirements of an ideal visible light active photo catalyst.

Keywords: Combustion synthesis, bandgap, Photocatalyst.

1. Introduction

Photocatalysis has long been studied for clean energy and environmental applications. Over the past two decades, the number of applications based on photocatalysis has increased sharply, while a wide range of material systems have been developed[1-3]. Photocatalysis has been of particular interest in the production of hydrogen from water using solar energy[4]. Further, conversion of carbondioxide to hydrocarbons (fuels) is also of significant interest, as it is a solution to reduce carbondioxide emissions across the globe[5]. Apart from the clean energy generation, photocatalysis has several promising applications in the environmental field. Some of the applications include degradation of volatile organic compounds (VOC) for water treatment[6], germicide and antimicrobial action[7], de-coloration of industrial dyes[8], nitrogen fixation in agriculture[9] and removal of NO_x/SO_x air pollutants[10]. Perovskite based photocatalysts are of significant interest in the field of photocatalysis. Titanates, tantalates, niobates, vanadates and ferrites are promising photocatalytic materials in visible light driven photoreactions[11].

The visible light activity still remains a challenge in the field of photo catalysis since most of the active photo catalysts have a wide intrinsic band gap greater than 3 eV. This allows such photo catalysts to absorb only UV light. Extending the band gap of photo catalysts in to the wide visible light region is an important branch of research. Transition metal ion doping and rare-earth metal ion doping have been extensively investigated for enhancing photo catalytic activities of photo catalyst under visible light.

As a kind of important functional material, $LaFeO_3$ with a typical ABO_3-type perovskite structure has many applications, such as catalytic oxidation, electrode materials and chemical sensors for the humidity and alcohols, and it is also an ideal potential semiconductor material for visible light photocatalyst because of its small band gap energy[12]. A number of fabrication methods such as hydrothermal synthesis[13], combustion synthesis[14,15], sol–gel[16] and precipitation[17] have been developed successfully so far to synthesize ultrafine nanostructured ceramic powder $LaFeO_3$. Among them combustion synthesis evolved as the economic and efficient method in terms of the cost, time and the complexities in the synthesis of nanomaterials.

In the present work a single step auto igniting modified combustion method is used to synthesize nanostructured ceramic powder $La_{0.99}Al_{0.01}FeO_3$. The modified combustion technique used in the present study is a relatively simple method through the omission of high temperature calcinations for prolonged duration, we could obtain phase pure nano particles of extremely small size. However, there are no reports on the synthesis of nanocrystalline $LaFeO_3$ with this method up to now. The structure, vibrational spectra, surface morphology, optical properties, thermal stability and weight loss are recorded and analysed. The effect of Aluminium doping on the optical properties of nanostructured $LaFeO_3$ for their application as photocatalytic material is studied.

2. Experimental Details

Stoichiometric amount of high purity Al $(NO_3)_3.9H_2O$, La_2O_3 and $Fe(NO_3)_3.9H_2O$ was dissolved in double distilled water to make a clear solution. Citric acid was then added to the solutions taken in separate beakers as complexing agent. Amount of citric acid was calculated based on total valence of the oxidizing and reducing agents for maximum release of energy during combustion[18]. Oxidant to fuel ratio of the system was adjusted to unity (~1) by adding concentrated nitric acid which serves as oxidizer and ammonium hydroxide solution as fuel. The precursor solution of pH ~7.0 was stirred well for uniform mixing without any precipitation or sedimentation. The solution was then heated using a hot plate kept at ~250^0C in a ventilated fume hood. The solution boils on continuous heating and undergoes dehydration accompanied by foam. On persistent heating the foam gets auto-ignited giving a voluminous fluffy powder of nano $La_{0.99}Al_{0.01}FeO_3$. The sample is then heated at 600 ^{0}C.

The prepared nanopowder obtained from the combustion process are characterized by different powder characterization techniques. The prepared samples are characterized by X-ray diffractometer (Xpert-pro) with Cu Kα radiation for the determination of crystalline structure and phase of the nanomaterials. The average crystallite size is estimated for the sample from Scherrer'sequation. The absorption spectrum of prepared sample is recorded using a UV-Visible spectrophotometer (Perkin Elmer lamda-35). The uv-vis spectra are recorded to resolve the correlation between the grain size and absorption edge of the nanoparticles. Additional information regarding the phase purity and the presence of any inorganic impurity are obtained using FTIR spectroscopy using Fourier Transform Infrared spectrometer (FTIR, Perkin elmerspectrum 2) in the range 350-4000 cm^{-1} using ATR mode.

Particulate properties of the combustion product are examined using high resolution transmission electron microscopy (TEM, Model-Hitachi H600 Japan) operating at 200 kV. The samples for Transmission Electron Microscope (TEM) are prepared by ultrasonically dispersing the powder in methanol and allowing a drop of this to dry on a carbon-coated copper grid. The thermal stability of the nanopowder $La_{0.99}Al_{0.01}FeO_3$

studied using thermal gravimetric analysis (TGA) and differential thermal analysis (DTA).

3. Results and Discussion

The powder XRD patterns of the synthesized $LaFeO_3$ and $La_{0.99}Al_{0.01}FeO_3$ with 2θ values between 10^0 and 90^0 are shown in the Figures 1 (a) and (b) respectively. XRD studies show that the powders possess an orthorhombic structure of space group Pnma (62) with lattice constants a=5.55, b=7.82 and c=5.53 for $LaFeO_3$ and a=5.55, b=7.85 and c =5.55 for $La_{0.99}Al_{0.01}FeO_3$ which agrees well with the XRD data reported in JCPDS file number-88-0641.The average crystallite size was calculated using the Scherrer formula[19] $D = 0.9\lambda/(\beta\cos\theta)$, where λ is the wavelength of CuKα radiation, β is the full width at half maximum and θ is the Bragg's diffraction angle.The average crystallite size of $LaFeO_3$ and $La_{0.99}Al_{0.01}FeO_3$ estimated by the Scherrer's equation are 35 nm and 32 nm respectively.

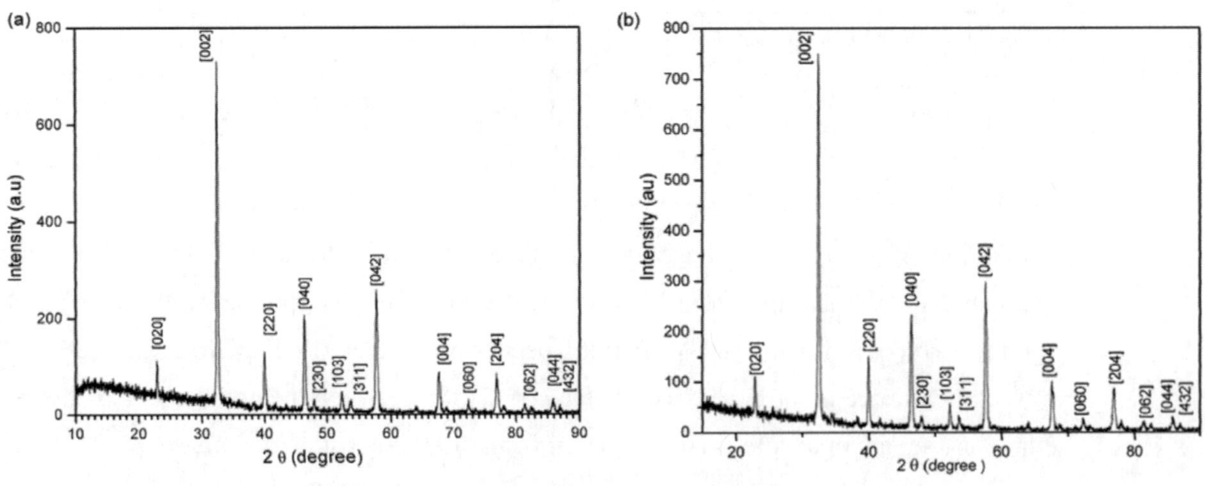

Figure .1. XRD pattern of (a) $LaFeO_3$(b) $La_{0.99}Al_{0.01}FeO_3$

The description about the structure, crystallite size and lattice parameters of the nanopowders $LaFeO_3$ and $La_{0.99}Al_{0.01}FeO_6$ are enclosed in Table 1.

Compound	Crystal structure	Lattice Parameter						Crystallite size (nm)
		Standard (A^0)			Calculated (A^0)			
		a	b	c	a	b	c	
$LaFeO_3$	Orthorhombic	5.56	7.85	5.55	5.55	7.82	5.53	35
$La_{0.99}Al_{0.01}FeO_3$	Orthorhombic	5.56	7.85	5.55	5.55	7.85	5.55	32

Table 1: Crystal structure, Lattice parameter and Crystallite size of samples

Though the XRD pattern of the sample $LaFeO_3$ when it is doped with 0.01% of Aluminium shows no detectable change in their peak positions, their particle size decreases which in turn increases their photocatalytic activity.

The extent of line broadening in the diffraction pattern due to the strain, which is caused by the non-displacement of the atoms with respect to their reference lattice position, could be identified by Williamson-Hall (W-H) method[20]. In this method, reciprocal peak width $\beta\cos\theta/\lambda$ are plotted versus the reciprocal lattice distance $\sin\theta/\lambda$ and the resulting plot is linearly fit.

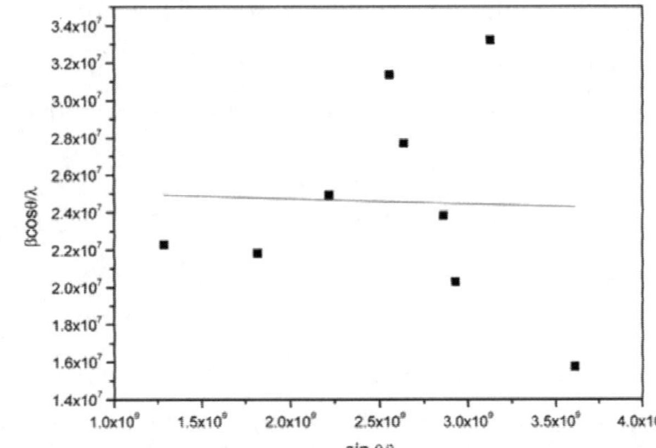

Figure.2. Williamson- Hall plot of $La_{0.99}Al_{0.01}FeO_3$

Figure.2 shows the Williamson-Hall plot for $La_{0.99}Al_{0.01}FeO_3$ nanoparticles. The reciprocal of y intercept, a measure of particle size obtained from this plot was 39 nm. The lattice strain constant η, proportional to the slope of the line estimated as 2.8039 x10^{-4}. The negative slope

IOP Conf. Series: Materials Science and Engineering **360** (2018) 012026 doi:10.1088/1757-899X/360/1/012026

of the line indicates that strain is compressive and the broadening of peaks due to the strain is very small.

Figure. 3 shows the TGA/DTA curve of $La_{0.99}Al_{0.01}FeO_3$ powder in the temperature range from 30 ^{0}C to 950 ^{0}C at a heating rate of 10 ^{0}C in nitrogen atmosphere. It was observed that the nanopowder $La_{0.99}Al_{0.01}FeO_3$ has approximately 7% weight loss in the temperature range 100-600°C after which there is no considerable weight loss in the TG curve. The fractional decrease in weight below 600 ^{0}C is due to liberation of the adsorbed moisture present in the combustion synthesised sample.

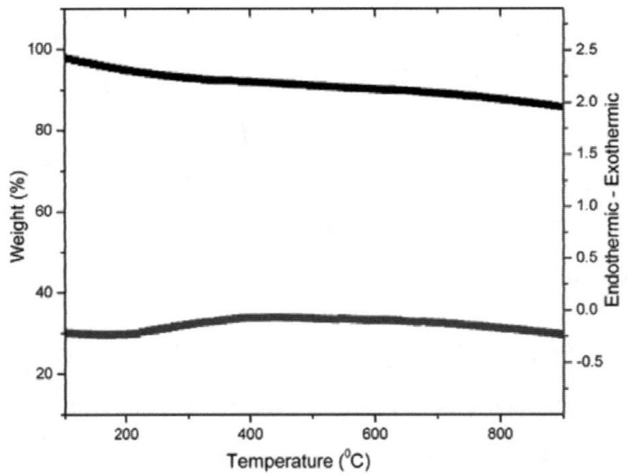

Figure. 3. TGA and DTA curve of $La_{0.99}Al_{0.01}FeO_3$

No abrupt changes are observed in the DTA curve in the higher temperature range which indicate that no phase transitions are occurring and the $La_{0.99}Al_{0.01}FeO_3$ ceramics synthesised by the modified combustion synthesis is stable at elevated temperature ranges.

Absorption spectrum of $La_{0.99}Al_{0.01}FeO_3$ was shown in the Figure. 4. It is observed that the sample $La_{0.99}Al_{0.01}FeO_3$ shows maximum absorption at 607 nm. The property of high absorption in the visible region makes the material a good candidate as photocatalytic material. It has been found that $La_{0.99}Al_{0.01}FeO_3$ is visible light photocatalytic active due to its unique optoelectronic properties and narrow bandgap[21].The strong absorption can generally be related to the electronic transition from the valence band to conduction band ($O_{2p} \rightarrow Fe_{3d}$). Consequently, this indicates that the $La_{0.99}Al_{0.01}FeO_3$ nanoparticles prepared by this method could be a kind of photocatalytic material. In otherwords, due to their optical properties, these nanoparticles may find valuable applications in photocatalysis and magneto-optical devices[22].

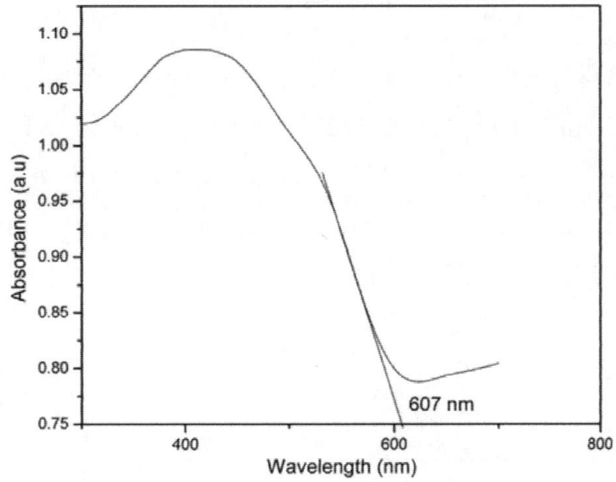

Figure.4. Absorption Spectrum of $La_{0.99}Al_{0.01}FeO_3$

A semiconductor is characterized by its electronic band structure. In the high-energy absorption region, the dependence on the photon energy is expressed by Tauc's equation. According to this relation, the absorption coefficient α is related to the band gap of a material by the relation, $\alpha h\nu = \beta(h\nu - E_g)^m$. where β is an energy independent constant, α is the optical absorption coefficient, h is the Planck's constant, ν is the frequency of incident photon, E_g is the optical band gap and m is a constant which characterize the nature of band transition. m=1/2 and 3/2 corresponds to direct allowed and direct forbidden transitions while m=2 and 3 corresponds to indirect allowed and indirect forbidden transitions, respectively. The optical band gap can be obtained from extrapolation of the straight-line portion of the $(\alpha h\nu)^{1/m}$ vs hν plot to hν =0. The obtained value of band gap is 2.08 eV for $La_{0.99}Al_{0.01}FeO_3$, which was shown in the Figure 5.

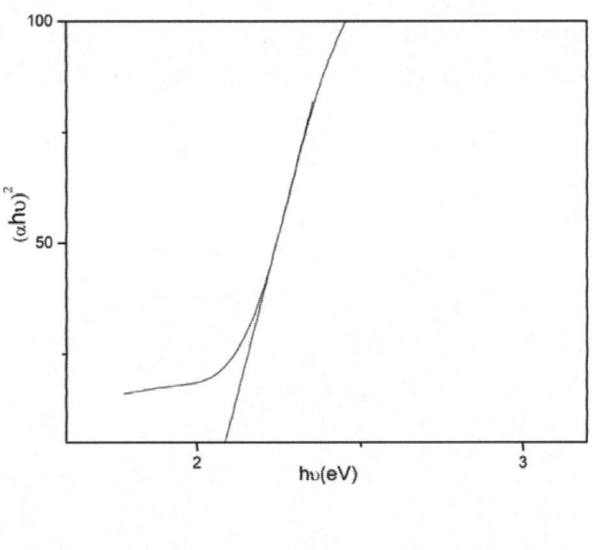

Figure.5. Tauc's plot of $La_{0.99}Al_{0.01}FeO_3$

Figure.6 shows the IR spectra of $La_{0.99}Al_{0.01}FeO_3$ recorded at room temperature, over the range 350-1000 cm^{-1}. The peak at 372 cm^{-1} corresponds to O-Fe-O bending mode. The peaks ranging from 400 cm^{-1} to 450 cm^{-1} corresponds to deformation of FeO_6 octahedra. Broad band ranging from about 500 cm^{-1} to 650 cm^{-1} corresponds to asymmetrical stretching vibration of Fe-O-Fe bonds [23-26].

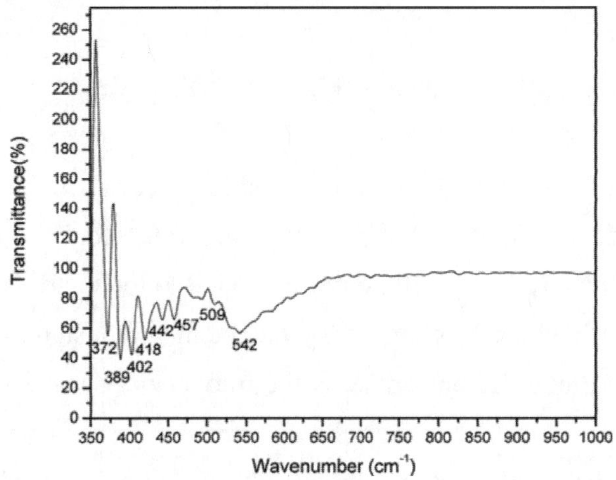

Figure. 6 IR Spectra of $La_{0.99}Al_{0.01}FeO_3$

Figure.7 shows the TEM and HRTEM images of the prepared powder sample $La_{0.99}Al_{0.01}FeO_3$. From the distinctly visible grain in Figure.7(a), the average grain size was calculated as 32 nm. From the Figure.7(b) and (c), the interfacial spacing was calculated as 0.395 nm and 0.273 nm for [0 2 0] and [0 0 2] planes respectively. From XRD, it is calculated as 0.393 nm and 0.277 for [0 2 0] and [0 0 2] planes respectively.

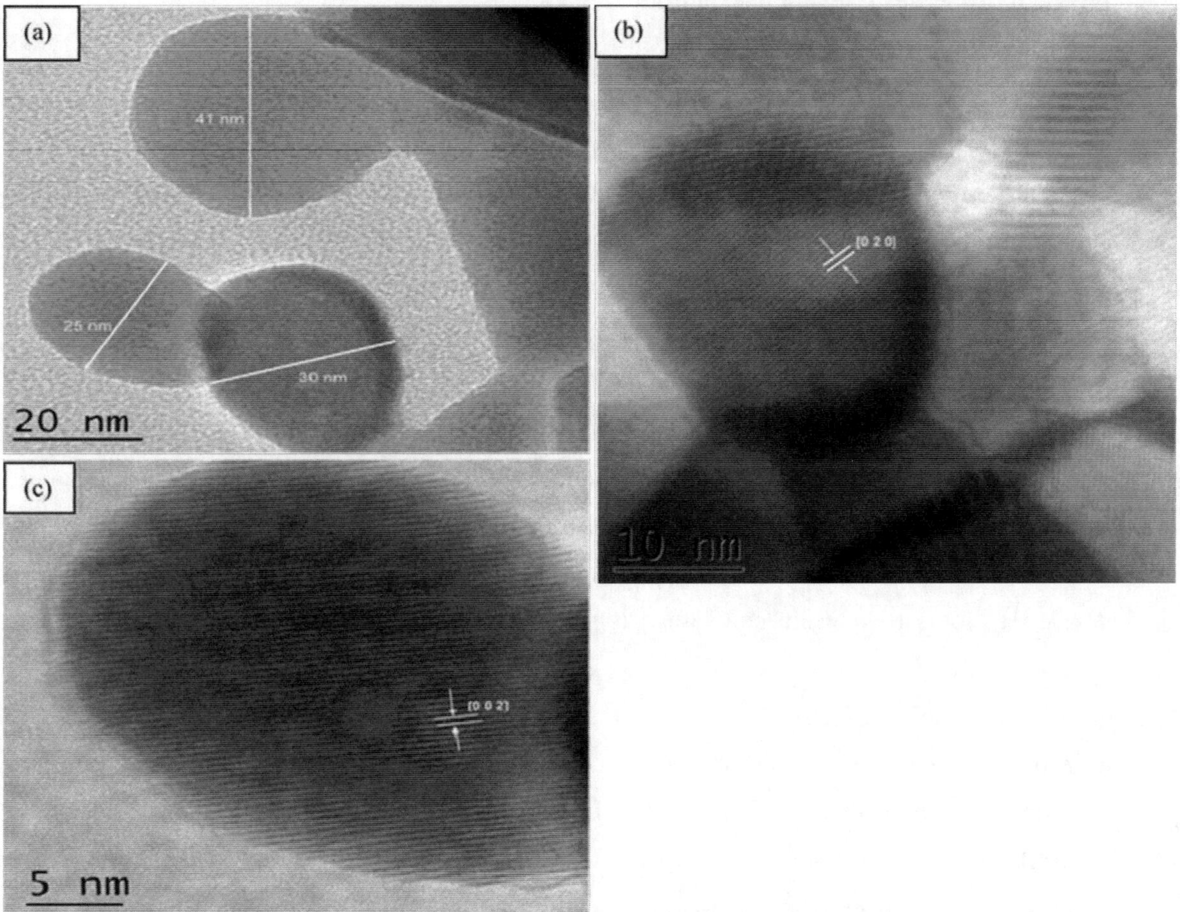

Figure.7. (a) TEM image of $La_{0.99}Al_{0.01}FeO_3$ (b) HRTEM image of $La_{0.99}Al_{0.01}FeO_3$

(c) HRTEM image of $La_{0.99}Al_{0.01}FeO_3$

Hence it is concluded that the average particle size and interfacial spacing from the HRTEM result was found to be in well agreement with result obtained from XRD analysis.

These results indicated that $La_{0.99}Al_{0.01}FeO_3$ nanoparticles are efficient photocatalyst under visible light. In general, the photocatalyst with a smaller particle size has more active surface sites and exhibits high efficiency of surface charge carrier transfer in photocatalysis. In otherwords, higher surface to volume ratio of the catalyst enhances the photocatalytic activity[23]. Consequently, it is anticipated that $La_{0.99}Al_{0.01}FeO_3$ nanoparticles has promising applications in the optoelectronic devices and photocatalysis. It is also expected that the nanoparticles produced in this system might show other interesting physical properties relevant to potential applications.

4. Conclusion

A single step modified combustion synthesis is used to prepare nano structured ceramic powder $La_{0.99}Al_{0.01}FeO_3$. This synthesis route is an economic and efficient method in terms of the cost, time and the complexities in the synthesis of nanomaterials. It does not require expensive high temperature calcination process. X-ray diffraction (XRD) analysis of the prepared powders confirms the formation of phase pure orthorhombic structured nanoparticles. All the peaks in the XRD of the sample was indexed to an orthorhombic structure of space group Pnma (62) with calculated lattice parameters agreeing well with the XRD data reported in Joint Committee on Powder Diffraction Standards (JCPDS) file no.88-0641. The XRD analysis, TEM and HRTEM studies reveal that the average crystallite size is approximately 32 nm for $La_{0.99}Al_{0.01}FeO_3$ and the lattice planes are well defined. From the W-H plot, lattice strain constant η is proportional to the slope of the line and the estimated value of η is -0.00028039. The vibrational spectroscopic studies confirm the structure of the sample. In IR spectra, most of the fundamental bands are indexed and interpreted. The absorption spectrum of $La_{0.99}Al_{0.01}FeO_3$ was recorded and the maximum absorption is at 607 nm for $La_{0.99}Al_{0.01}FeO_3$. The obtained band gap of $La_{0.99}Al_{0.01}FeO_3$ are direct band gap of E_g value 2.08 eV. The TGA and DTA curve shows that the nanopowder $La_{0.99}Al_{0.01}FeO_3$ has approximately 7% weight loss in the temperature range of 100-600°C which is due to the presence of adsorbed moisture in the samples. The results clearly indicate that the ultra-fine nanopowder $La_{0.99}Al_{0.01}FeO_3$ synthesised by the modified auto-igniting combustion technique can be used effectively as a photocatalytic material. Aluminium doping in the nanostructured $LaFeO_3$ reduces the particle size and increases the bandgap. It is hoped that this study will represent a useful material in the field of photocatalysis, particularly as a photocatalyst for industrial applications.

Acknowledgments

The authors acknowledge National Centre for Earth Science and Studies (NCESS), Trivandrum and Sophisticated Test and Instrumentation Centre (STIC), Ernakulam for the XRD and TEM analysis.

References

1. K.Maeda , *J. Photchem. Photobiol.* 12, 237–268 **(2011)**

2. Qu,Y.Duan and X. Progress, *J.Chem. Soc. Rev.* 42, 2568–2580 **(2013)**

3. W. Hou, S.B. Cronin, *J. Adv. Funct. Mat.* 23, 1612–1619 **(2013)**

4. T.R.Cook, D.K. Dogutan, S.Y.Reece, Y.Surendranath, T.S.Teets and D.G. Nocera. J. *Chem. Rev.*110, 6474–6502 **(2011)**

5. M. Tahir and N.S. Amin, *J. Renew. Sust. Energy Rev.* 25, 560–579 **(2013)**

6. Y.N.Tan, C.L. Wong and A.R. Mohamed. *J.Mat. Sci.* 18, **(2011)**

7. O.K.Dalrymple, E.Stefanakos, M.A.Trotz and D.Y. Goswami, *J. Appl. Catal. B.* 98, 27–38 **(2011)**

8. S.Ahmed, M.G.Rasul, W.Martens, R.Brown and M.A.Hashib, *Water Air Soil Pollut.* 215, 3–29 **(2011)**

9. R.I.Bickley and V.Vishwanathan , *Nature.* 280, 306–308 **(1979)**

10. J. Zhao and X.Yang, *Build. Environ.* 38, 645–654 **(2008)**

11. PushkarKanhere and Zhong Chen, *Molecules.* 19, 19995-20022 **(2014)**

12. H. J. Su, L. Q. Jing and K. Y. Shi, *J. Nanopart. Res.*12, 967-974 **(2010)**

13. W.J. Zheng, R.H. Liu, D.K. Peng and G.Y. Meng, *Mater. Lett.* 43, 19 **(2000)**

14. S.S. Manoharam and K.C. Patil, *J. Solid State Chem.*102, 267 **(1993)**

15. A.Chakraborty, P.S. Devi and H.S. Maiti, *J. Mater. Res.*10, 918 **(1995)**

16. J. Xiao, G.Y. Hong, D.C. Yu, X.T. Dong, *Acta Chim. Sin.* 52, 784 **(1994)**

17. H.N. Pandya, R.G. Kulkami and P.H. Parsania, *Mater. Res. Bull.* 25, 1073 **(1990).**

18. J.K. Thomas, Padma Kumar, S.Solomon, C.N George, K.Joy and J.Koshy, *phys.stat.sol.(a)*, 204 **(2007)**

19. J.I. Langford and A.J.C. Wilson, *J. Appl. Crystallogr.* 11,102–113 **(1978)**

20.G.K. Williamson and W.H. Hall, *Acta Metall.* 1, 22–31 **(1953)**

21. J. Yang, H.Zhong, M.Li, L.Zhang and Y.Zhang, *React.Kinet.Catal.Lett.* 97, 269–274 **(2009)**

22 S.Chakrabarti, D.Ganguli, S.Chaudhuri, *Physica E:Low-Dimens.Sys. Nanostructures* 24,333–342 **(2004).**

23. RezaAbazari ,SoheilaSanati and Lotf Ali Saghatforoush, *Materials Science in Semiconductor Processing.* 1-6 **(2014)**

24. R. Andoulsi, K. Horchani Naifer and M. Férid,*Ceramica.* 58, 126-130 **(2012)**

25. P.A. Desai and Anjali A. Athawale, *Defence Science Journal.* 63, 285-291 **(2013)**

26. Manjunath B.Bellakki, Brandon J. Kelly and V.Manivannan, *Journal of Alloys and Compounds*. 489, 64–71 **(2010)**

IOP Conf. Series: Materials Science and Engineering **360** (2018) 012027 doi:10.1088/1757-899X/360/1/012027

Improvement of efficiency in CZTSSe solar cell by using back surface field

Atul Kumar, Ajay D Thakur

Department of Physics, Indian Institute of Technology Patna,
Bihta 801118, India.

E-mail: atul.pph14@iitp.ac.in

Abstract. A kesterite CZTS based solar cell is studied with back surface field. In order to achieve this, a p^+ layer of SnS is added to the standard cell configuration. The resulting p^+-p-n configuration of SnS/CZTS/CdS is studied using Solar Cell Capacitance Simulator (SCAPS). Low V_{oc} is responsible for lower efficiencies of kesterite solar cells. The proposed p^+-p-n structure increases the V_{oc} and thus the efficiency. Simulation of CZTSSe and CZTS with back surface p^+ layer has shown considerably higher efficiency than the highest reported value for kesterite cell of CZTSSe from 12.3% to 15.7%. A comparative summary is presented with literature reported performance and our simulated performance for kesterite cell.

1. Introduction

Solar cells based on $Cu_2ZnSnS_{4-x}Se_x$ (CZTSSe) can have Shockley Quisser limit of 32% as per their bandgap, and is a potential low cost photovoltaic (PV) material. The reported highest efficiency of 12.6% for kesterite CZTSSe based solar cells [1] is much less than SQ limit and chalcopyrite $CuIn_xGa_{1-x}S_2$ (CIGS) counterpart (21.6% for) [2]. This inferiority in performance of kesterite suggests complex defect physics resulting in Lower V_{oc} and overall performance [3,4]. Secondary phase and point defect in absorber layer, non optimized interfaces [3] and possible flaws in cell structure are issues which require careful attention for kesterite solar cells. For optimization of cell structure simulation tools can be utilized where cell structures/configuration, interface and their performance can be optimized. A simulation, studying the effect of various parameters on photovoltaic performance is an easy step to predict optimized conditions without actually fabricating a solar cell. AMPS, PC1D, SILVACO, SCAPS [5,6] are some extensively used solar cell simulation software. SCAPS-1D simulation software is used in this study, which solve Poisson equation, relating the charge to the electrostatic potential (φ) and Continuity equation for electrons (n_i) and holes (p_i) for simulation of devices. In one dimension, the cell is divided into N intervals and solved for (φ, n_i, p_i) at each interval. The SCAPS can simulate $CuInSe_2$, CdTe, c-Si a-Si [7], CZTSSe [6, 8] and perovskite [9]. The present work focuses on the role of back surface field. The back surface field has been reported for enhanced PV performance in a variety of solar cells [10,11]. A high-low junction p^+-p at back cause conducive band banding, an electric field in direction of p-n junction is setup decreasing electrons current at back contact. H. Cui *et al* [12] reported the increase in efficiency by adding a back surface layer in CZTSSe. In this regard we propose an SnS back surface layer to modify the overall cell structure. Cell configuration of p^+(SnS)/p(CZTSSe)/CdS/ZnO is simulated with SCAPS-1D software to study and optimize overall performance. A comparative study of p-n junction and p^+-p-n junction is presented in following section.

IOP Conf. Series: Materials Science and Engineering **360** (2018) 012027 doi:10.1088/1757-899X/360/1/012027

2. Solar cell description with back surface field

A p-n junction based photovoltaic device has built in voltage at the junction is given by

$$V = \frac{KT}{q} \ln \frac{N_a N_d}{n_i^2}$$

When a p^+ layer (layer with higher hole density) is added at back, a p^+-p high low junction is formed giving rise to a built-in voltage

$$V = \frac{KT}{q} \ln \frac{N_{a+}}{N_a}$$

Where Na, Na^+, is the hole density in the p, p^+, N_d is electron density in n layer respectively.
For a p^+-p junction, by virtue of acceptor concentration variations, band bending occurs at junction and an electric field in direction of p-n junction is set up. The generated voltage at the p^+-p junction produces a barrier for electron current at rear contact. Such high-low junction can be employed in solar cell structure at appropriate locations to induce directional flow of charge carriers in solar cell for selective contacts. This arrangement can provide higher V_{oc} and less recombination at junction and bulk. A higher V_{oc} is observed in case of an applied p^+-p in solar cell as discussed in the next section.

3. Simulation Details

A p-type SnS is used as a back surface layer. The configuration of standard cell is Mo/CZTSe/CdS/ZnO and of that for the modified cell is Mo/SnS/CZTSe/CdS/ZnO as shown in Fig. 1. A comparative study of the two cell configurations is performed using SCAPS-1D [7]. SnS is used since it has similar material composition as CZTS, suitable p-type nature and easily synthesized by various methods [13]. The material parameters used are listed in the Table I [5,8]. The device is illuminated from window layer (ZnO) side with AM1.5 spectrum using relevant SCAPS option at temperature 300 K. For the series resistance (R_s) and shunt resistance (R_{sh}), we have used typical values of 0.72 ohm-cm^2 and 620 ohm-cm^2, respectively which are conservative estimates considering the experimentally reported values of 0.72 ohm-cm^2 and 620 ohm-cm^2, respectively [1].The Schematic of the two configuration is shown in the Fig.1.

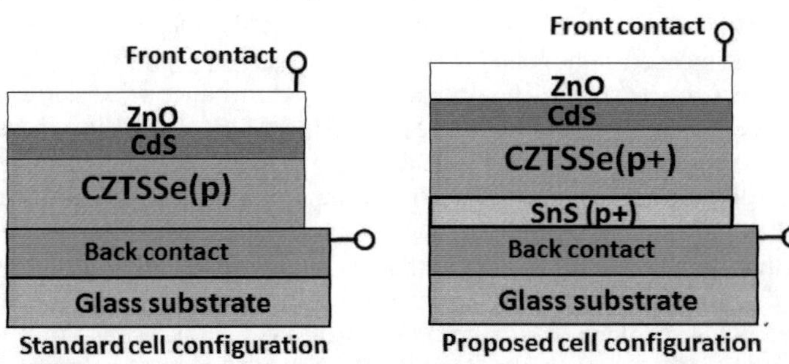

Figure 1: Schematic diagram of the conventional cell configuration and the proposed cell configuration for CZTSSe based solar cells.

Table 1. Material parameters used in simulations [6,7,14].

Parameters	SnS	*CZTSSe*	CdS	*ZnO*
Thickness(nm)	100	2000	80	200
Bandgap(eV)	1.25	1.3	2.4	3.3
Electron affinity	3.9	4.2	4.5	4.6
Dielectric permittivity	13.7	13	10	9
Electron/hole mobility(cm^2/Vs)	30/15	100/25	350/50	100/25
Donor/acceptor density(cm^{-3})	10/10^{18}	10/10^{15}	10^{17}/10	10^{18}/10

4. Results and Discussion

4.1. Absorber layer: CZTSSe

The quality of interface between the CZTSSe absorber layer and the CdS buffer layer is crucial in the separation of generated electron-hole pairs. The recombination in bulk leading to losses is primarily caused by radiative processes, non-radiative (Shockley-Read-Hall mechanism) processes, and the interface recombination which increases saturation current and thus decreases V_{oc}. In case of polycrystaline absorber, inhomogeneities in band gap variations and electrostatic fluctuations increases the saturation current. Here, CZTSSe passivated with SnS is observed to have a better performance as shown in Fig. 2. The gain in performance is seen along with higher V_{oc}, fill factor (FF) and J_{sc} in cell with SnS. Typical thicknesses for SnS and CdS were chosen as 100 nm and 80 nm, respectively. The p^{+}-p and the p-n junctions at either side of absorber layer creates an asymmetry for charge flow. The efficiency with the thickness of the absorber layer is shown in Fig.2. The V_{oc} for the SnS/CZTSSe/CdS/ZnO configuration is observed to be higher than the unpassivated cell due to the generation of the built-in voltage at both the ends of the absorber layer. Short circuit current, I_{sc} of the passivated cell is higher than the unpassivated counterpart due to the presence of a directed field for charge flow in absorber layer.

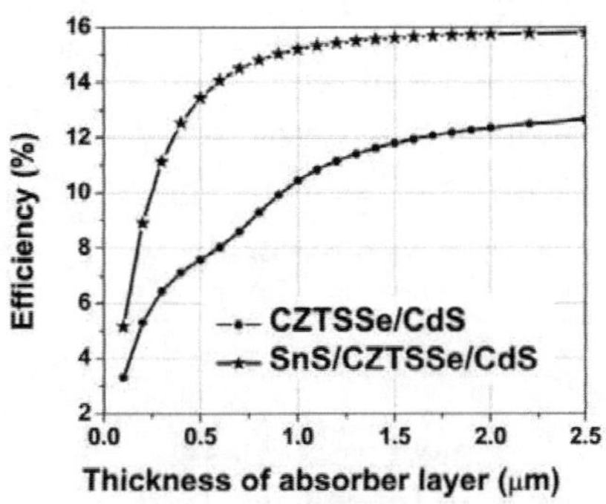

Figure 2: Variation of efficiency with absorber layer thickness in the investigated cell configurations.

4.2. Band-diagram for CZTS solar cell.

Simulated band diagram shown in Fig. 3 convey band bending at junction/interfaces and energy levels in bulk. The Fermi level splitting at p-n junction gives rise to a light induced voltage. The addition of SnS back surface layer leads to the formation of a high-low junction by virtue of difference in acceptor concentration in SnS and CZTSSe. The higher energy level of SnS than CZTSe in band diagram in Fig. 3 is a favorable condition as an electric field is set up at both ends of the absorber layer. Generated charge carrier in absorber will feel a drift field thereby increasing its probability of collection at junction. A high low junction at the rear also acts as the near ohmic contact.

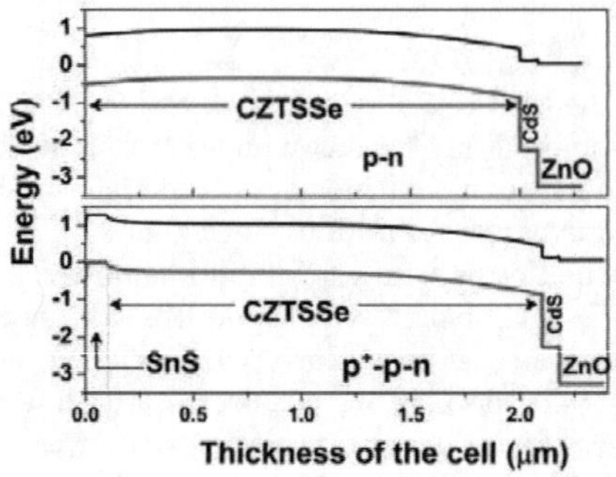

Figure 3: Simulated band diagram for the two configuration of cells.

4.3. Quantum efficiency.

Quantum efficiency shows the charge pair generated by absorption of different wavelength. Different wavelengths are absorbed at different thickness. Higher wavelengths are absorbed in deeper in bulk of absorber and low wavelength light is absorbed at surface. The band gap of CZTSSe (1.3eV) corresponds to 954nm wavelength. The QE is observed to improve at higher wavelength i.e. in depth at rear side in the modified cell configuration. This is due to SnS back surface field which efficiently created at field thus efficient collection of charge carriers.

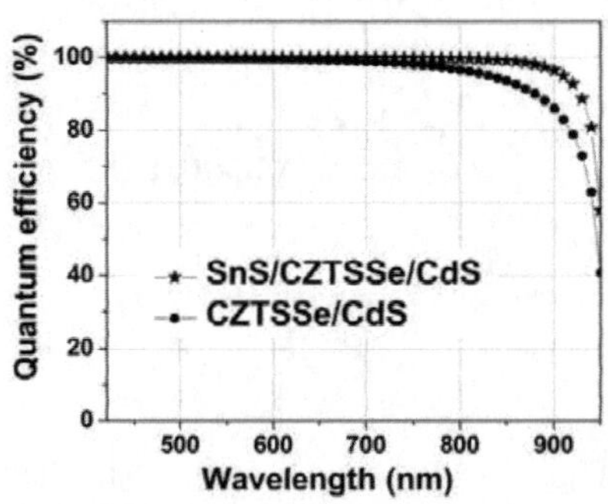

Figure 4: Quantum efficiency measurements for the solar cell configurations under investigation.

5. Conclusions

The back surface field in CZTSSe can improve V_{oc} in cell by creating a built in voltage at rear, It also improve the contact of metal semiconductor making a near ohmic metal-p^+-p. In our work we simulated back surface using an SnS layer and showed that a structure of p^+-p-n is useful for solar cell performance improvement. A higher V_{oc} is obtained in CZTSSe cell along with improved FF and QE. We have increased the efficiency of cell from 12.3 to 15.7% by incorporation of back surface field. The comparative performance of all the simulated and experimentally reported cell of CZTSSe cell is shown in Fig. 5 and tabulated in Table 2.

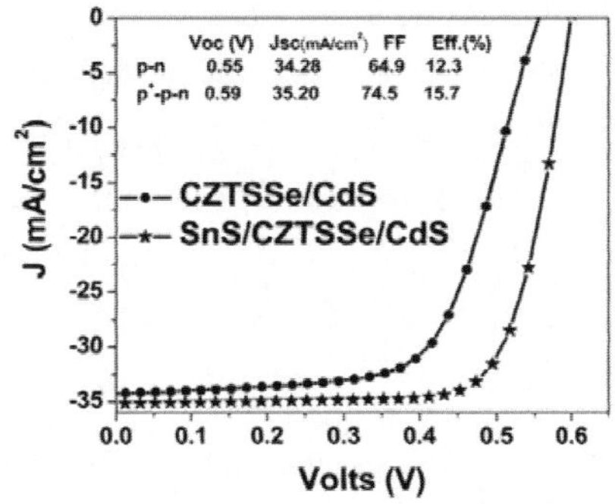

Figure 5: I-V curve of the two cell configuration and the simulated parameters.

Table 2. Comparative performance of different configuration.

Configuration	V_{oc} (V)	J_{sc} *(mA/cm²)*	FF (%)	η (%)	Reference
CZTSe/CdS/ZnO	0.513	35.2	69.8	12.6	[1]
CZTSe/CdS/ZnO	0.55	34.28	64.9	12.3	Our simulation
SnS/ CZTSe/CdS/ZnO	0.59	35.20	74.5	15.7	Our simulation

Acknowledgments
The solar cell capacitance simulator SCAPS-1D used here is generously provided by Dr. Marc Burgelman from the University of Gent.

References

[1] Wang W, Winkler M T, Gunawan O, Gokmen T, Todorov T K, Zhu, Y., Mitzi D B 2014. Device Characteristics of CZTSSe Thin-Film Solar Cells with 12.6% Efficiency, *Adv. Energy Mater* **4** 1301465.

[2] Jackson P, Hariskos D, Wuerz R, Kiowski O, Bauer A, Friedlmeier T M and Powalla M, 2015 Properties of Cu(In,Ga)Se_2 solar cells with new record efficiencies up to 21.7% *Phy. Status Solidi RRL* **9** 28–31

[3] Polizzotti A, Repins I L, Noufi R, We, Su-Huai and Mitzi DB, 2013 The state and future prospects of kesterite photovoltaics. *Energy Environ. Sci.*, **6** 3171-3182.

[4] Mark T W, Wang W, Gunawan O, Harold J. Hovel, Todorov T K and Mitzi D B, 2014 Optical designs that improve the efficiency of $Cu_2ZnSn(S,Se)_4$ solar cells, *Energy Environ. Sci.* **7** 1029.

[5] Patel M, Ray A 2012 Enhancement of output performance of Cu_2ZnSnS_4 thin film solar cells—A numerical simulation approach and comparison to experiments, *Physica B: Condensed Matter*, **407** 4391-4397.

[6] Burgelman M, Nollet P, Degrave S 2000 Modelling polycrystalline semiconductor solar cells. *Thin Solid Films* **361**, 527-532

[7] Niemegeers A, Burgelman M, Decock K, Verschraegen J, Degrave S, SCAPS Mannual, Version Date: June 27, (2015).

[8] Simya O K, Mahaboobbatcha A, Balachander K 2015 A comparative study on the performance of Kesterite based thin film solar cells using SCAPS simulation program. *Superlattices and Microstructures* **82**, 248.

[9] Minemotoa T and Murata M 2014 Device Device modeling of perovskite solar cells based on structural similarity with thin film inorganic semiconductor solar cells, *Journal of Applied Physics*, **116** 054505.

[10] Fossum J G1977 Physical operation of back-surface-field silicon solar cells *IEEE TRANSACTIONS ON ELECTRON DEVICES* **24** 4.
[11] Roos O V 1978 A simple theory of back surface field (BSF) solar cells, *Journal of Applied Physics* **49**, 3503.
[12] Cui H, Chang-Yeh Lee, Li W, Liu X, Wen X, and Hao X 2015 Improving Efficiency of Evaporated Cu_2ZnSnS_4 Thin Film Solar Cells by a Thin Ag Intermediate Layer between Absorber and Back Contact *International Journal of Photoenergy* Article ID 170507
[13] Antoine de Kergommeaux, Jérôme Faure-Vincent, Adam Pron, Rémi de Bettignies, Peter Reiss 2013 SnS thin films realized from colloidal nanocrystal inks, *Thin Solid Films*, **535** 376-379.

IOP Conf. Series: Materials Science and Engineering **360** (2018) 012028 doi:10.1088/1757-899X/360/1/012028

Molecular structure and vibrational analysis of 2,5-pyridine-dicarboxylic acid using experimental and theoretical methods

J. Laxman Naik and B. Venkatram Reddy*

Department of Physics, Kakatiya University, Warangal – 506009, Telangana, India.

*bvreddy67@yahoo.com

Abstract. The FTIR and FT-Raman spectra of 2,5-pyridine-dicarboxylic acid (PDA) has been recorded in the range 4000-450 cm^{-1} and 4000-50 cm^{-1}, respectively. The optimized geometry and vibrational frequencies were computed using density functional theory (DFT) employing B3LYP functional with 6-311++G(d,p) basis set. The rms error between observed and calculated frequencies is obtained as 10.8 cm^{-1}. A normal coordinate analysis was carried out for all the vibrations of this molecule by solving inverse vibrational problem (IVP) using a 74-parameter modified valence force field employing overlay technique. This reproduced 35 observed frequencies of this molecule with an error of 9.28 cm^{-1} in the zero-order calculations, demonstrating the transferability of the force field obtained in our earlier work on some related pyridine-dicarboxylic acids. PED and eigen vectors calculated in the process were used to make unambiguous vibrational assignments of all the fundamentals of the molecule.

1. Introduction

We have been investigating the molecular structure and vibrational analysis of pyridine-carboxylic acids by recording their infrared and Raman spectra; and subjecting them to normal coordinate analysis using density functional theory (DFT) and solving inverse vibrational problem (IVP). The force constants thus obtained in the process were utilized to demonstrate their transferability to the related pyridine-carboxylic acids. In our earlier paper, we reported the results of experimental and theoretical study of 2,3-, 2,4- and 3,4-pyridine-dicarboxylic acids; and evaluated their optimized geometries and valence force field [1]. In continuation of this work, we now investigated 2,5-pyridine dicarboxylic acid (PDA) for its structure and vibrational assignments.

Pyridine-dicarboxylic acids have immense importance in several fields of scientific research such as coordination chemistry, biochemistry, pharmaceuticals and medicine. They are used in coordination chemistry as multifunctional ligands capable of forming extended networks of different dimensionalities with transition metals, yielding materials having interesting topologies and physical properties [2, 3]. In biochemistry, they find application as enzyme inhibitors [4], plant preservatives [5] and food sanitizers [6]. NMR, UV and IR absorption spectra of 2,3-, 2,4-, 2,5-, 2,6-, 3,4- and 3,5-pyridine-dicarboxyilic acids were reported by Wasylina et al [7]. They proposed tentative assignments for very few IR bands on the basis of qualitative considerations, leaving many other observed vibrational frequencies unassigned. Monomeric and dimeric structures of 3,5-pyridine-dicarboxylic acid or dinicotinic acid were studied by Nataraj et al by recording FT-Raman, FTIR and UV spectra and making DFT and ab initio calculations [8]. Dinicotinic acid and its dianions were investigated, both experimentally (Raman and infrared spectra) and theoretically, by Kathleen McCann and Jaan

IOP Conf. Series: Materials Science and Engineering **360** (2018) 012028 doi:10.1088/1757-899X/360/1/012028

Laane [9], whereas its Raman spectra in microcrystalline form was reported by Alexandre et al [10]. Thermochemical and theoretical studies of 2,3-; 2,5- and 2,6-pyridine-dicarboxylic acids are also available along with that of dimethyl pyridine-2,6-dicarboxylate [11].

From the above, it is clear that the experimental (Raman and infrared) and theoretical investigation of 2,5-pyridine-dicarboxylic acid (PDA) is not yet appear in the literature. Hence, a systematic investigation of vibrational spectra of this molecule employing experimental and theoretical methods is undertaken. The purpose of this investigation is:

i) To record FTIR and FT-Raman spectra of PDA to get complete information on its vibrational frequencies.
ii) To make DFT (Density Functional Theory) calculations in order to make geometry optimization so as to identify the most stable rotational isomer of the molecule in the ground state and calculate theoretical vibrational frequencies using scaling.
iii) To test the transferability of force constants obtained for the related molecules by solving inverse vibrational problem using optimized geometry by Wilson's GF-matrix method employing overlay least-squares technique.
iv) To make unambiguous vibrational assignment of all the fundamental vibrations of the molecule using PED and eigen vectors obtained in the results of both DFT and IVP computations.

2. Experimental Details

The compound PDA was obtained from TCI Chemical Company, Japan and used as such for the spectral measurements. The room temperature FTIR spectra is recorded using Bruker IFS 66V spectrometer employing KBr optics in the spectral range 4000-450 cm^{-1} with a scanning speed of 30 cm^{-1} min^{-1} with spectral width 2.0 cm^{-1}. The FT-Raman spectra is recorded in the range 4000-50cm^{-1} using FRA 106 Raman module equipped with Nd:YAG laser source operating at 200 mw power with spectral resolution of 2 cm^{-1}. The wavelength of the exciting radiation used was 1064 nm.

3. Computational details

3.1. The Quantum Chemical Calculations – Molecular geometry

As the determination of the most stable conformation is the starting point for further calculations, the molecule was subjected to a rigorous conformational analysis around free rotation bonds using Gaussian 09w software package [12]. The standard density functional triply-parameter hybrid model DFT/B3LYP [13, 14] with 6-311++G(d,p) basis set was employed for the computations. The optimized geometry of the molecule corresponding to minimum on the potential energy surface was obtained by solving self-consistent field equation iteratively. These computations yielded eight stable rotational conformers for PDA as shown in Fig.1.

The conformer in Fig.1(d) has the lowest global minimum energy for PDA. Hence, this is accepted as the most stable conformer for optimizing the geometry of the molecule PDA. Subsequent calculations were performed with this optimized structure as shown in Fig. 2. The same figure contains numbering of atoms also. The optimized structure parameters in the most stable conformation are given in Table 1.

As per the computations, the molecule PDA C_s point group symmetry. The molecule consists of 17 atoms. Hence, it has 45 fundamentals, distributed as 31 in-plane vibrations of a′-species and 14 out-of-plane vibrations of a″-species in C_s symmetry. All the vibrations of C_s symmetry are active in both IR absorption and Raman scattering.

Detailed description of vibrational modes using the full set of 58 primitive internal coordinates containing 13 redundancies, normal coordinate analysis using the MOLVIB 7.0 program and scaling of the force constants was made according to scaled quantum mechanical procedure as discussed in our earlier work [1]. For plotting simulated IR and Raman spectra, a pure Lorentzian band shape was

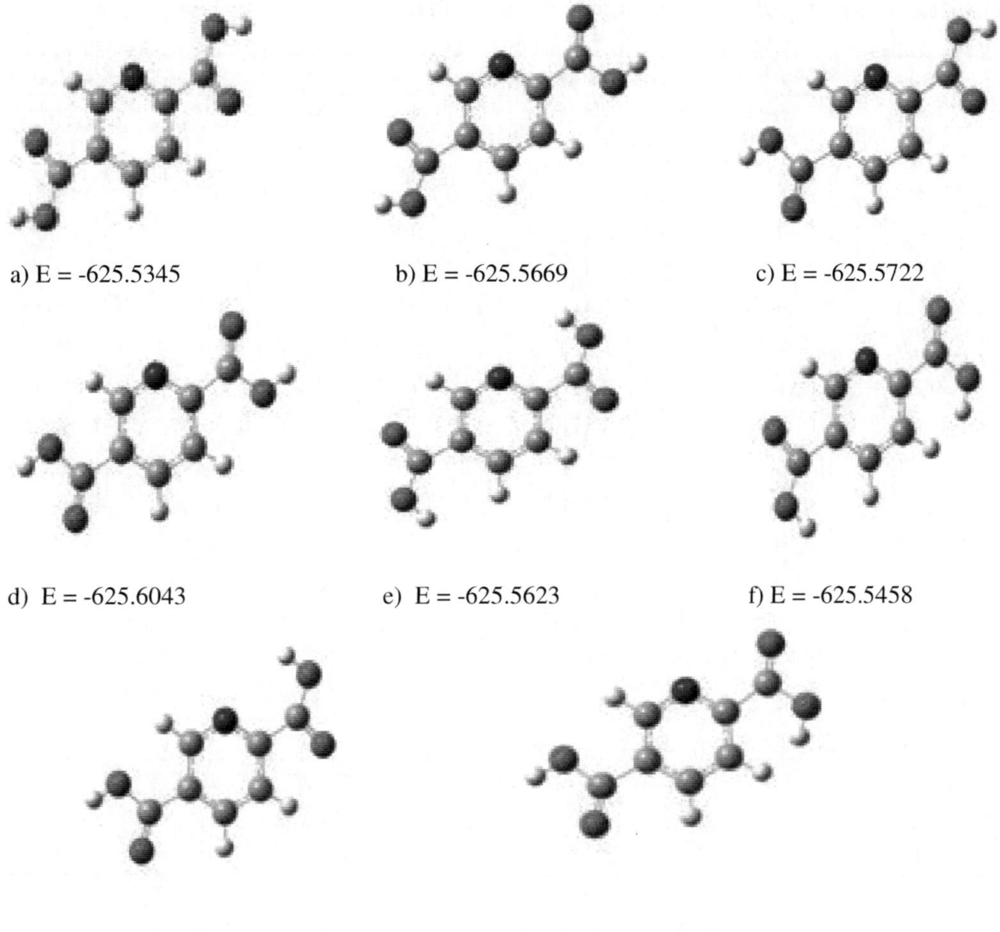

a) E = -625.5345 b) E = -625.5669 c) E = -625.5722

d) E = -625.6043 e) E = -625.5623 f) E = -625.5458

g) E = -625.5749 h) E = -625.5863

Fig.1. Conformers of PDA (Energy, E in Hartree)

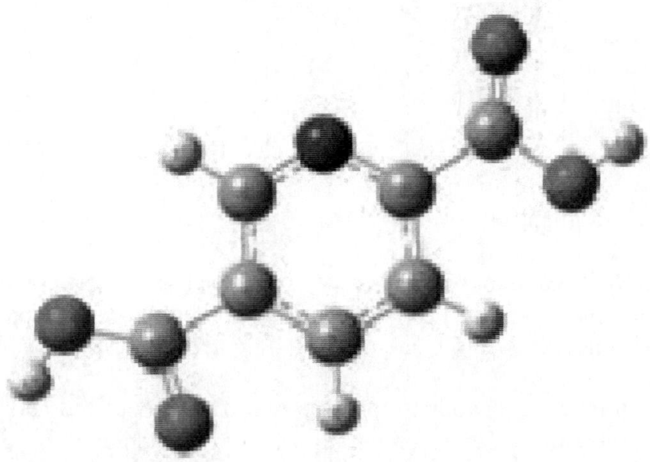

Fig.2. Optimized structure of PDA

IOP Conf. Series: Materials Science and Engineering **360** (2018) 012028 doi:10.1088/1757-899X/360/1/012028

Table 1. Optimized geometrical parameters of PDA

Bond	Bond length Å	Bond angle	Value (in °)	Torsional angle	Value (in °)
N1-C2	1.336	N1-C2-C3	123.34	N1-C2-C3-C4	0.00
C2-C3	1.398	C2-C3-C4	118.32	C2-C3-C4-C5	0.00
C3-C4	1.388	C3-C4-C5	118.81	C3-C4-C5-C6	0.00
C4-C5	1.394	C4-C5-C6	118.33	C6-N1-C2-C7	180.00
C5-C6	1.402	N1-C2-C7	115.32	N1-C2-C3-H8	180.00
C6-N1	1.334	C2-C3-H8	120.34	C2-C3-C4-H9	-180.00
C2-C7	1.507	C3-C4-H9	121.66	C3-C4-C5-C10	180.00
C3-H8	1.081	C4-C5-C10	118.89	C4-C5-C6-H11	180.00
C4-H9	1.083	C5-C6-H11	120.24	N1-C2-C7-O12	0.0
C5-C10	1.488	C2-C7-O12	125.60	N1-C2-C7-O13	180.0
C6-H11	1.084	C2-C7-O13	111.49	C2-C7-O13-H14	-180.0
C7-O12	1.200	C7-O13-H14	106.93	C4-C5-C10-O15	0.0
C7-O13	1.359	C5-C10-O15	124.48	C4-C5-C10-O16	180.0
O13-H14	0.968	C5-C10-O16	112.71	C5-C10-O16-H17	-180.00
C10-O15	1.207	C10-O16-H17	107.23	*	
C10-O16	1.352	*		*	
O16-H17	0.968	*		*	

* Not applicable

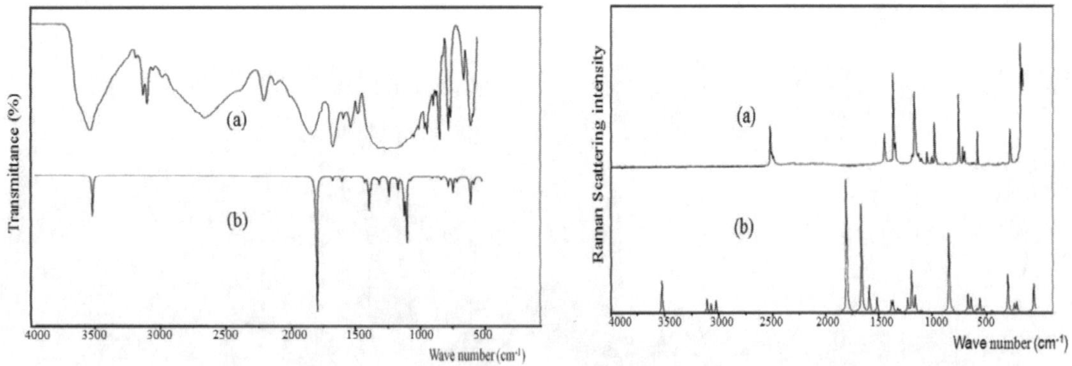

Fig.3. FTIR and FT-Raman spectrum of PDA: (a) Observed, (b) Simulated with DFT

used with a full width at half maximum (FWHM) of 10 cm^{-1}. The observed and simulated FTIR and FT-Raman spectra of this molecule are presented in Fig. 3. The observed frequencies (both IR and Raman) and calculated frequencies (both un-scaled and scaled) and vibrational modes of PDA are reported in Table 2.

It is necessary to determine the goodness of fit between the calculated and observed frequencies to lend credence to the reliability of the computations. This is achieved by calculating root mean square (rms) error between the calculated and experimental frequencies using the following relation

$$\text{rms} = \sqrt{\{(1/n-1) \sum_{i=1}^{n}[(\nu_i)^{cal} - (\nu_i)^{exp}]^2\}}$$

where, $(\nu_i)^{cal}$ is the ith calculated frequency, $(\nu_i)^{exp}$ is the ith experimental frequency, and n is the number of experimental frequencies. The rms error between un-scaled frequencies and observed

IOP Conf. Series: Materials Science and Engineering **360** (2018) 012028 doi:10.1088/1757-899X/360/1/012028

frequencies for PDA was found to be 35.8 cm^{-1}. On using the refined scaling factors, this deviation was reduced to 10.8 cm^{-1} for this molecule.

3.2. Calculations based on Classical Mechanics

A zero-order normal coordinate analysis of all the vibrations of PDA was made by solving inverse vibrational problem (IVP) using Wilson's GF-matrix method employing overlay technique as discussed in our earlier work on pyridine-dicarboxylic acids [1]. The initial set of 74 force constants was transferred there from for this molecule. This reproduced 35 observed frequencies of this molecule with an error of 9.28 cm^{-1} in the zero-order calculations. This should be considered as good, as the force constants are not refined. This demonstrates the transferability of force constants obtained in our earlier work [1]. Observed frequencies, calculated frequencies using the above set of force constants by IVP method, frequencies computed by DFT and the vibrational modes of PDA are presented in Table 2.

Table 2. Observed, calculated frequencies (in cm^{-1}) with DFT/B3LYP/6-311++G and IVP; and vibrational modes of PDA

		Obs. frequency		Cal. frequency		
S.No	**Mode[a]**	**IR**	**Raman**	**DFT**		**IVP**
				Un-scaled	**Scaled**	
a′- species						
1	ν(C-N)1	807	813	814	804	802
2	ν(C-C)8a	-	1622	1617	1614	1628
3	ν(C-N)8b	1609	1602	1608	1611	1610
4	ν(C-C)14	1270	-	1250	1261	1267
5	ν(C-C)19a	1519	-	1541	1520	1521
6	ν(C-C)19b	1464	-	1471	1480	1468
7	ν(C-H)2	3071	3082	3273	3065	3060
8	ν(C-H)7a	3108	3100	3220	3108	3110
9	ν(C-H)20b	-	3064	3120	3076	3072
10	ν(C-C$_{\alpha}$)7b	-	1020	1041	1024	1027
11	ν(C-C$_{\alpha}$)20a	-	-	1358	1312	1304
12	β(CH)3	-	1228	1241	1224	1221
13	β(CH)9b	-	1138	1152	1134	1137
14	β(CH)18a	1187	1172	1160	1184	1185
15	β(CC$_{\alpha}$)9a	-	110	139	111	112
16	β(CC$_{\alpha}$)18b	-	263	240	267	264
17	β(CCC)6a	-	360	331	352	351
18	β(CCC)6b	496	-	462	498	486
19	β(CCC)12	693	-	711	698	694
a″- species						
20	π(CH)5	863	-	871	861	
21	π(CH)11	787	792	785	780	787
22	π(CH)17a	930	-	912	932	937

23	$\pi(CC_\alpha)$ 10b	-	-	181	165	163
24	$\pi(CC_\alpha)$17b	-	128	141	124	125
25	$\tau(CCCC)$4	764	-	751	761	763
26	$\tau(CCCC)$16a	563	-	585	565	563
27	$\tau(CCCC)$16b	-	440	422	441	442
Vibrations of acid groups						
a′- species						
28	$\nu(C_\alpha=O)_1$	1790	-	1763	1786	1791
29	$\nu(C_\alpha=O)_2$	-	1730	1741	1735	1728
30	$\nu(C_\alpha\text{-}O)_1$	-	1328	1341	1321	1327
31	$\nu(C_\alpha\text{-}O)_2$	-	1328	1338	1320	1337
32	$\delta(OH)_1$	-	1372	1342	1370	1371
33	$\delta(OH)_2$	-	1372	1348	1375	1376
34	$\nu(O\text{-}H)_1$	3521	-	3652	3522	3531
35	$\nu(O\text{-}H)_2$	3521	-	3652	3524	3538
36	$\delta(C_\alpha=O)_1$	676	667	681	671	666
37	$\delta(C_\alpha=O)_2$	-	639	671	648	641
38	$\gamma(C_\alpha\text{-}O)_1$	-	-	581	562	571
39	$\gamma(C_\alpha\text{-}O)_2$	464	-	438	467	461
a″- species						
40	$\tau(CC_\alpha)_1$	-	75	91	78	73
41	$\tau(CC_\alpha)_2$	-	-	80	84	81
42	$\omega(OH)_1$	516	-	530	511	517
43	$\omega(OH)_2$	-	-	415	428	424
44	$\omega(C_\alpha=O)_1$	-	857	842	856	860
45	$\omega(C_\alpha=O)_2$	-	857	851	861	864

[a]: Mode in Wilson's notation [15]. C_α represents Carbon atom of acid group; ν, stretching; β, in-plane bending; δ, deformation; γ, in-plane rocking; π, out-of-plane bending; τ, torsion; ω, wagging

4. Results and Discussion

The bond-distance between the pyridine ring and each of the two acid groups in the present study is found as 1.507 and 1.488 Å in PDA. The calculated bond length of C-N in this molecule is 1.336 Å. This value is in good agreement with the experimental values available in literature [7, 9]. The optimized structure of the molecule and the torsional angles presented in Table 1 reveal that the two acid groups are within the plane of pyridine ring due to the presence of non-bonding electrons on the nitrogen atom and thus, possess C_s symmetry. Wasylina et al [7] studied the NMR, UV and IR spectra and made vibrational assignments for some vibrations of the molecule under investigation and the symmetry of the molecule was not reported. The other structure parameters computed in the present study are in good agreement with the literature values [7, 9].

4.1. Vibrational assignments

Vibrational assignments of the molecule are made by using density functional theory (DFT) as well as solving classical inverse vibrational problem. In DFT computations, the assignments are arrived at by referring to potential energy distribution (PED) and deriving considerable help from Gauss View program, which shows oscillations (or displacements) of atoms of the molecule in each mode of vibration. PED and eigen vectors were used to make vibrational assignments in classical inverse vibrational problem. There are no significant differences between the PED and hence the vibrational assignments, arrived at on the basis of both the methods. Hence, results of DFT computations are taken as reference in the discussion of vibrational assignments, as they are obtained from general valence force field, derived from quantum chemical principles, which is the best at the present state of our knowledge. The results presented in Tables 2 are self-explanatory. Hence, the detailed discussion is unwarranted.

5. Conclusion

Geometry optimization has yielded structure parameters that agree well with their experimental counter parts. A complete vibrational analysis of 2,5-pyridine-dicarboxylic acid is performed using the DFT method at B3LYP/6-311++(d,p) level of theory and solving inverse vibrational problem using optimized molecular geometry. All the fundamental frequencies of the molecule are assigned unambiguously based on the PED and eigen vectors obtained from normal coordinate analysis. The rms error between un-scaled frequencies and observed frequencies for PDA was found to be 35.8 cm^{-1}. On using the refined scaling factors, this deviation was reduced to 10.8 cm^{-1} for this molecule. Transferability of 74-parameter force field obtained in our earlier work for the related molecules has been demonstrated for the molecule under investigation and reproduced 35 experimental frequencies with an error of 9.28 cm^{-1} in the zero-order calculations. This should be considered as good, as the force constants are not refined.

Acknowledgements

The financial support from University Grants Commission, New Delhi, India (F. No. 41-960/2012 (SR), dt.26/7/2012) is gratefully acknowledged. The authors are thankful to the Sophisticated Analytical Instrumentation Facility (SAIF), IIT Madras, Chennai, India for the spectral measurements.

References

[1] J. L. Naik, B. Venkatram Reddy and N. Prabavathi, *J. Mol. Struct.* 1100 (2015) 43.
[2] J. F. Eubank, R. D. Walsh, P. Poddar, H. Srikanth, R. W. Larsen and M. Eddaoudi, *Cryst. Growth Des.* 6 (2006) 1453.
[3] M. L. Ton, J. Wang and S. Hu, *J. Solid State Chem.* 178 (2005) 1518.
[4] T. A. Slieman and W. L. Nicholson, *Appl. Environ. Microbiol.* 67 (2001) 1274.
[5] J. E. Rington, *Microbiol. Rev.* 57 (1993) 1.
[6] K. I. Murkami, Y. I. Tnemura and M. J. Yoshino, *Nutr. Biochem.*14 (2003) 99.
[7] L. Wasylina, E. Kucharska, Z. Weglinski and A. Puzsko, *Chem. of Heterocyclic Compounds* 35(2) (1999) 186**.**
[8] A. Nataraj, V. Balachandran, T. Karthick, M. Karabacak and A. Atac, *J. Mol. Struct.* 1027 (2012) 1.
[9] Kathleen McCann and Jaan Laane, *J. Mol. Struct.,* 890 (2008) 346.
[10] Alexandre A Kolemenskii, Sergei N Jerebstov, Tomas Opatrny, Hauss A Schuessler and Marlan O Scully, *J. Modern Optics* 50 (2003) 2369.
[11] M. Agostinha, R. Matos, Victor M F Marais, Emanuel A Sausa, Jorge P Castimeiras, Claudia P Santos and W. E. Acree Jr, *J. Chem. Eng. Data* 50 (2005) 1184.
[12] M. J. Frisch, G. W. Trucks et al, Gaussian, Inc., Wallingford CT, 2009.
[13] A. D. Becke, *J. Chem. Phys.* 98 (1993) 5648.
[14] C. Lee, W. Yang and R. G. Parr, *Phys. Rev. B.* 37 (1998) 785.
[15] E.B. Wilson, *Phy. Rev.* 45 (1934) 706

Synthesis, Growth, Optical and Dielectric Characterization of 8-Hydroxyquinolinium Hydrogen Fumarate Single Crystal

G.Ahila,[a] M.Divya Bharathi,[a] J.Mohana,[a] R.Bhuvaneswari,[a] P.Illayaraja,[a] G.Anbalagan[b*]

[a]Department of Physics, Presidency College, Chennai-600 005, Tamil Nadu, India

[b]Department of Nuclear Physics, University of Madras, Chennai-600 025, Tamil Nadu, India

*anbu24663@yahoo.co.in

Abstract. An organic nonlinear optical single crystal, 8-hydroxyquinolinium hydrogen fumarate (8HQF) was grown from mixed solvent of ethanol and water using slow evaporation solution growth technique. The single crystal XRD studies confirmed that the grown crystal belongs to the triclinic crystal system with the centrosymmettric space group P $\bar{1}$. FT-IR spectral analysis confirms the various functional groups present in the compound. The optical absorption studies shows that the crystal is transparent with a lower cut-off wavelength of 439 nm and the optical energy band gap Eg is found to be 2.66 eV. The variation of dielectric properties of the grown crystal with respect to the frequency has been investigated at different temperatures.

Keywords: Solution growth, Optical transmittance, Vibrational analysis.

1. Introduction

The explore of new advanced materials is an essential area of current research in many branch of knowledge in science and the development of new technologies. The inception of nonlinear properties is well grasped phenomenon and the advancement of optical research relies on the development of new materials compatible with various devices [1]. A non linear optical material unveiled a wide range of applications in telecommunication, optical information processing and integrated optics [2]. These materials play a vital role in the development of modern technology and industrial applications [3].In this context, organic molecules materialize as promising candidates since they have several advantages such as low dielectric constant with electronic polarizability, refractive index, etc., over inorganic nonlinear optical crystals. Organic 8-Hydroxyquinoline is a proton donor which forms variety of charge transfer complexes and their complexes are well known for their anti microbial and antifungal activities. Generally aromatic compounds containing electron releasing substituent group act as donors and those containing highly electron with drawing group act as acceptors [4-6].Franklin *et al.* has reported the crystal structure of 8-hydroxyquinolinium hydrogen fumarate (8-HQF) using the compound 8-hydroxyquinoline and fumaric acid as precursors [7]. In this present investigation we report the growth, vibrational, optical and dielectric studies of an organic 8-hydroxyquinolinium hydrogen fumarate single crystal for the first time.

2. Experimental Details

The title compound 8-hydroxyquinolinium hydrogen fumarate was synthesized by the reaction of commercially available 8-hydroxyquinoline (Merck) with that of Fumaric acid (Sigma Alderich, AR Grade) in the stochiometric ratio of 1:1 in a mixed solvent of ethanol and water. It is a proton transfer reaction where a proton is transferred from the electron donor group of fumaric acid to the electron acceptor group of 8-hydroxyquinoline. A yellow precipitate was formed on continuous stirring. The title crystal was obtained by repeated recrystallization of the combined compound. The reaction scheme of the 8HQF is shown in Fig.1. Yellow color crystals of size 5 x 5 x 3 mm^3 were harvested in 20 days of evaporation at room temperature from the saturated solution. The as grown crystal is shown in Fig.2.

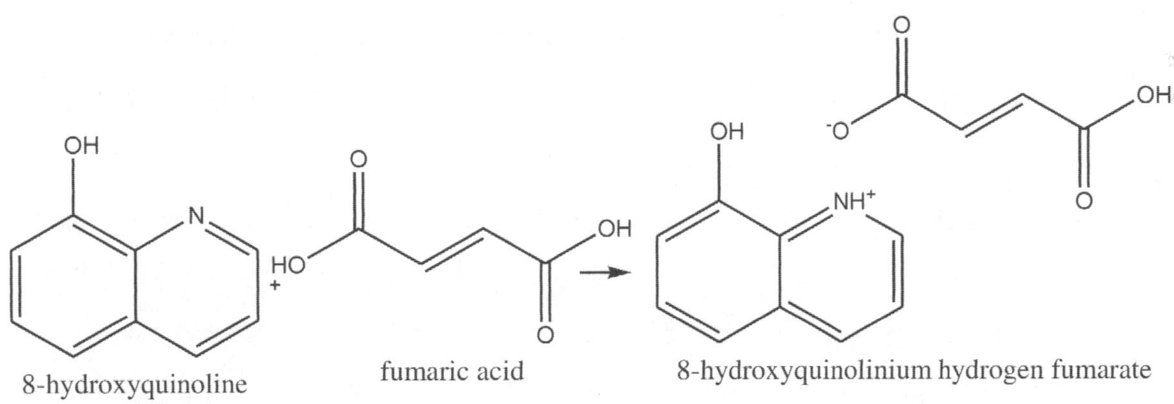

Figure 1. Reaction scheme of 8HQF crystal

Figure 2. As grown crystal of 8HQF

3. Results and Discussion

3.1 Single X-Ray Diffraction Studies

The single crystal diffraction data of 8HQF crystal were collected using Bruker Kappa Apex II CCD diffractometer with Mo–Kα radiation (λ=0.71073 Å: graphite monochromator). It is observed that the crystal belongs to the triclinic system with the centrosymmettric space group P $\bar{1}$. The obtained lattice parameters are presented in Table 1. These values are in good agreement with the literature [7].

Table 1. Crystallographic data of 8HQF

Parameter	Presented study	Reported values [7]
a	7.36Å	7.3282(3)Å
b	7.48Å	7.4363 (3)Å
c	11.64Å	11.5680(5)Å
α	79.39°	79.349 (2)°
β	74.85°	74.994 (2)°
γ	89.32°	89.337 (2)°
V	608Å^3	597.97 (4) Å^3
System	Triclinic	Triclinic
Space group	P $\bar{1}$	P $\bar{1}$

3.2 Spectral Analysis

FT-IR spectroscopy is used to identify the functional groups of the synthesized compound and hence to elucidate its molecular structure [8].The FTIR spectrum of 8-hydroxyquinolinium hydrogen fumarate crystal was recorded using Perkin Elmer FT-IR spectrometer in the range 4000-450 cm^{-1} by the KBR pellet method. The FT-IR spectrum is depicted in Fig.3. The related assignments of the functional groups are tabulated in Table 2. The 8HQF compound includes weak to strong intermolecular hydrogen bonding. The peak observed at 3419 cm^{-1} was attributed to O-H stretching vibration attached to quinoline group in FT-IR spectrum [9].The medium intensity sharp band observed at 2949 cm^{-1} is due to the N-H stretching vibration. The absorption bands occurred in the regions 2851, 2776 and 2671 cm^{-1} corresponds to the C-H stretching vibration [10].The C=O stretching vibrations of 8HQF complex is observed at 1711 and 1387 cm^{-1} [4]. The bands occurred at 1636, 1491 cm^{-1} in FT-IR corresponds to C-C stretching vibration [11].The aromatic C=C stretching vibrations are exhibited at 1598 and 1553 cm^{-1}. The C-H in plane bending vibration is observed vat 1096 and 1136 cm^{-1}. The C-O symmetrical stretching vibrational peak is observed at 1302, 1310 and 1065 cm^{-1}. The band at 1265 cm^{-1} is due to

IOP Conf. Series: Materials Science and Engineering **360** (2018) 012029 doi:10.1088/1757-899X/360/1/012029

the phenolic C-O stretching. The C-N deformation is observed at 927 cm^{-1} [12]and C-H out of plane bending vibration is observed at 794 cm^{-1} and 753 cm^{-1}. The C-H rocking vibration appears as a weak band in IR at 884 cm^{-1}. The C=O group deformation stretching vibration appears at 824 cm^{-1}. The bands at 641 cm^{-1} and 578 cm^{-1} is due to the O-H and C-O deformation vibration respectively.

Table 2. Vibrational frequency assignments of 8-HQF crystal

FT-IR (cm^{-1})	Assignments
3419	O-H stretching vibration
2949	C-H stretching vibration
2851, 2776, 2671	N-H stretching vibration
1711,1387	C=O stretching vibration
1636,1491	C-C stretching vibration
1598, 1553	C=C stretching vibration
1302, 1310, 1065, 1265	C-O symmetric stretching vibration
1096, 1136	aromatic C-H in plane bending
927	C-N deformation
884	C-H rocking vibration
824	C=O deformation
794, 753	C-H out of plane bending
578	C-O deformation

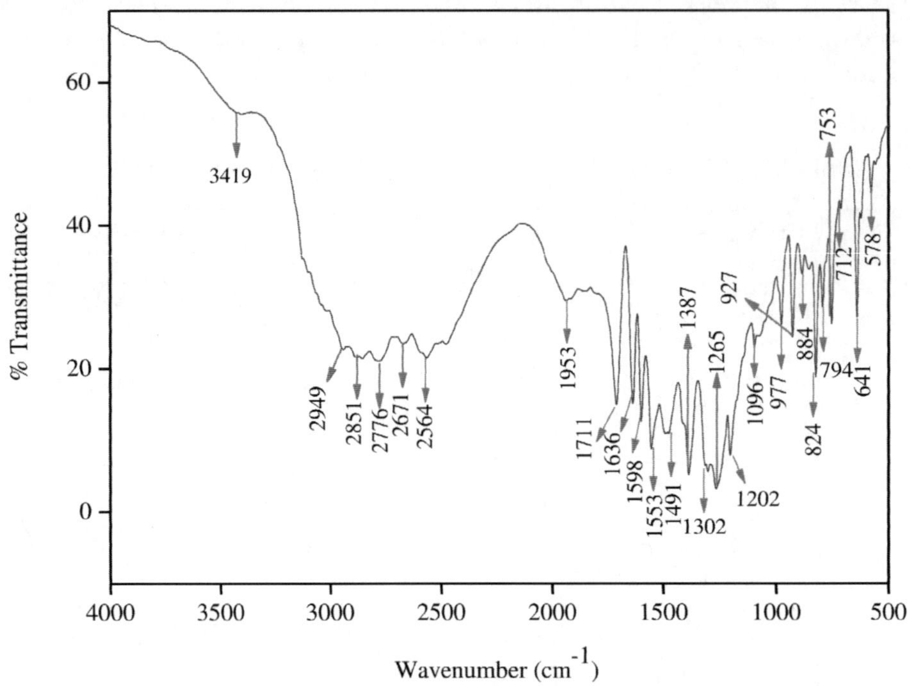

Figure 3. FT-IR spectrum for 8HQF compound

3.3 Uv-Vis Studies

The optical transmission spectrum of 8HQF single crystal was recorded in the wavelength region from 190nm to 900nm as shown in Fig.4 (a). The Uv absorption edge of the 8HQF crystal was observed to be around 442nm. The co-relation of optical absorption coefficient with the photon energy helps to study the band structure and the electron transitions [13]. Using the transmittance 'T'and the thickness of the crystal't', the optical absorption coefficient (α) is calculated using the following relation:

$$\alpha = \frac{1}{t}\log\left[\frac{1}{T}\right]$$

For high photon energies the crystal under study obeys the following relation:

$$\alpha = \frac{A(h\nu - Eg)^{1/2}}{h\nu}$$

where Eg is the optical band gap of the crystal and A is constant. The Eg is evaluated by extrapolation of the linear portion of the curve to x-axis [14]in the plot of variation of $(\alpha h\upsilon)^2$ vs. hυ as depicted in Fig.4 (b) and the bandgap is found to be 2.66eV. As a consequence of wide bandgap, the 8HQF has large transmittance in visible region and therefore it proves to be a suitable candidate for optoelectronic applications[15].

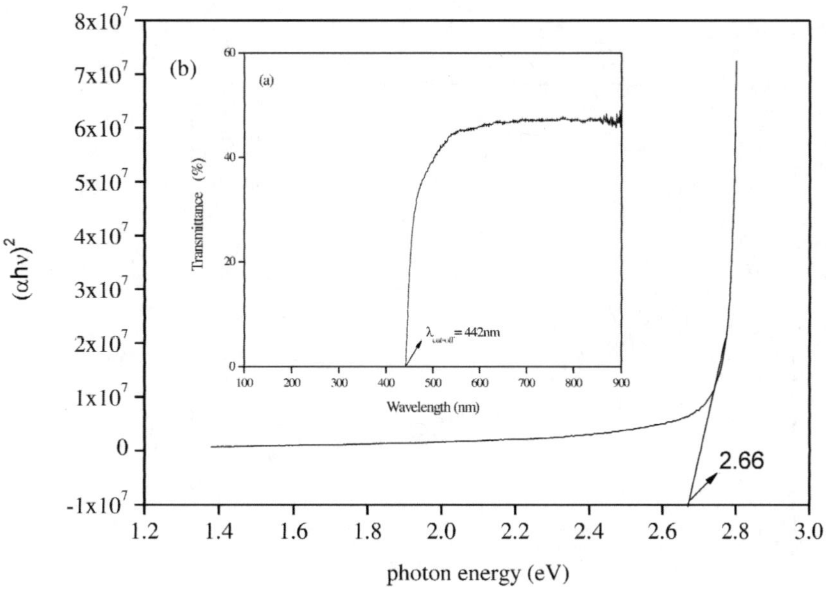

Figure 4. **(a)** Uv-Vis spectrum of 8HQF crystal and **(b)** Tauc's plot for 8HQF crystal

3.4 Dielectric studies

The dielectric analysis is measured to avail the details about the electrical properties of the material at different frequencies. The dielectric studies were carried out using LCR meter in a frequency range 50Hz-5MHz. Silver paste was applied over the cut and polished 8HQF sample and placed in the cell which is connected with a thermocouple. A thermocouple attached to the electrode is used to measure the temperature. Local displacement of electrons in the direction of the applied field takes place due to the electronic exchange of number of ions in the crystals. Fig.5 shows the plot between dielectric constant and log frequency. The dielectric permittivity occurs because of the four polarization mechanisms of materials such as electronic polarization, ionic polarization, orientation polarization, and space polarization. At very low frequencies the dielectric characteristics are affected by interfacial polarization or at the microscopic level by space charge polarization [16].Thus, the high value of permittivity at low frequency is mainly due to space charge polarization,[17] hence this type of dielectric constant values are good candidates for heating devices [18]. The value of dielectric constant can be calculated using the formula:

$$\varepsilon_r = \frac{Ct}{\varepsilon_0 A}$$

where 'C' is the capacitance,'t' is the thickness of the crystal, 'ε_0'is the permittivity of free space and 'A' is the area of the crystal. Fig.6 shows the plot between the dielectric loss and log frequency. It is observed that the dielectric loss decreases with the increase in frequency at different temperatures. The low value of dielectric loss at high frequency of 8HQF crystal sample shows that the grown crystal possesses enhanced optical quality with lesser defects [19]. Thus the good quality crystals are used for extending the applications towards, photonic, electro-optic and NLO devices [20].

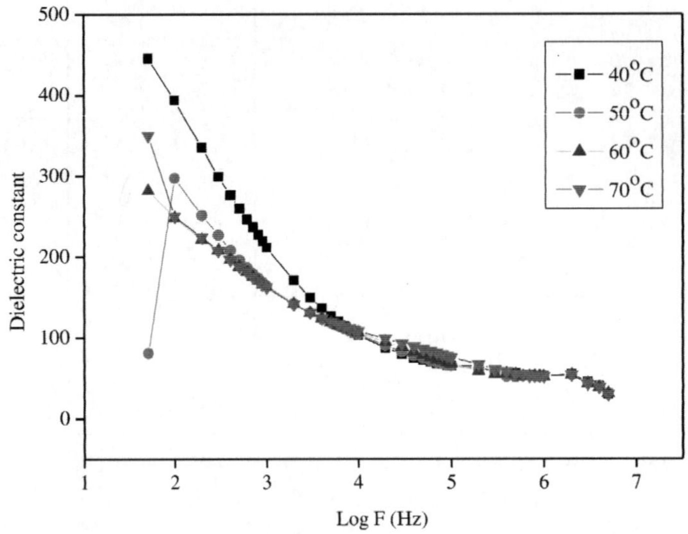

Figure 5. Plot between dielectric constant and log frequency of 8HQF

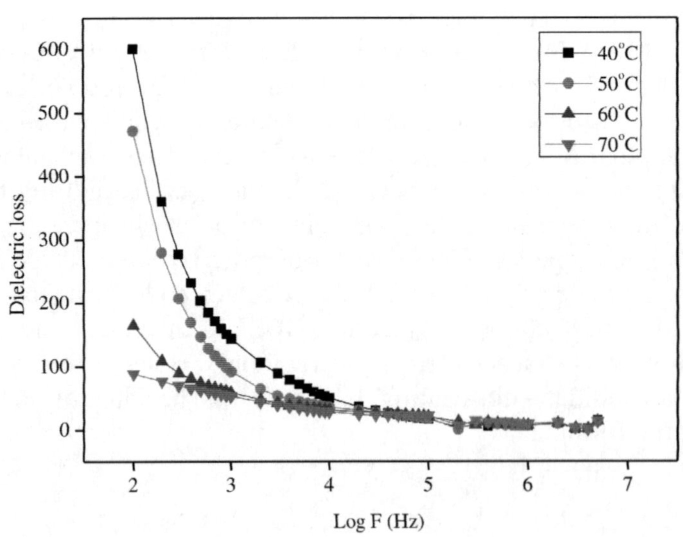

Figure 6. Plot between the dielectric loss and log frequency of 8HQF

4. Conclusion

A new organic material 8HQF was synthesized from solution using ethanol as solvent. The structure of the grown crystal were proved by the powder XRD data and compared with the literature data. Various functional groups present in the grown crystal confirmed from the FTIR spectrum. The UV cut off wavelength was found to 442 nm and the UV–Vis spectrum showed good transparency (43%) in the Uv -Vis region showing that 8HQF is a good candidate for NLO applications. The dielectric studies revealed that the grown crystal has low dielectric constant with less power dissipation and hence this crystal can be used for electro-optic applications.

Acknowledgment

The authors are grateful to centre for Sophisticated Analytical Instruments Facility (SAIF), Indian Institute of Technology- Madras, Chennai for Single Crystal X-Ray Diffraction studies and FT-IR studies.

References and Notes

[1] D.F. Eaton, *Science*281(1991) 25.

[2] C. Bosshard, K. Sutter, P. Pretre, J. Hulliger, M. Flörsheimer, P. Kaatz, P. Günter, *Organic Nonlinear Optical Materials*, Gordon and Breach, Basel(1995).

[3] P.N. Prasad, D.J. Williams, *Introduction to Nonlinear Optical Effect in Molecules and Polymers*, John Wiley & Sons Inc (1991).

[4] R.E. Bambury, in: M.E, Wolff (Ed.), *Burger's Medicinal Chemistry Part II*, John Wiley, New York, 41(1949).

[5] S.Yamaguchi, M.Goto, H. Takayanagi, H. Ogura, *Bull. Chem. Soc.* 1026, 61 (1998).

[6] I.M. Khan, A.Ahmad, *Spectrochim. Acta Part A* 966, 63 (2009).

[7] S. Franklin, T.Balasubramanian, *Acta Crystallographica* C65 (2009) o58.

[8] A.Philominal, S.Dhanuskodi, *J.Philp.Mater.Chem.Phys.* 139 (2013)1.

[9] N.Sudarsana, V.Krishnakumar, R.Nagalakshmi, *Mater. Res. Bull.*61 (2014) 136.

[10] V.Krishnakumar, R.Nagalakshmi, P.Janaki, *Spectrochim. Acta Part A*, 66 (2005) 1097.

[11] E.Selva Kumar, G.Anantha Babu, P.Ramasamy, Rajinikanth, B.Murugesan, A.Chandramohan, *Spectrochim. Acta Part A: Mol. Biomol. Spectrosc.* 117 (2014) 259.

[12] N. Sudarshana, V. Krishnakumar, R. Nagalakshmi,*J. Cryst. Growth.*398 (2014) *45.*

[13] N.Tigaur, V.Ciupinaa, G.Prodana, G.J.Rusub, C,Gherghies, E.Vasilec, *J.Optoelectronics Adv.Mater.* 6 (2006) 211.

[14] A.K.Chawla, D.Kaur, R.Chandra, *Optical Mater.* 29 (2007) 995.

[15]Mohd. Shkir B. Riscob, Mohd. Hasmuddin, Preeti Singh, V. Ganesh, M.A. Wahab, Ernesto Dieguez, G. Bhagavannarayana, *Opt. Mater.* 36 (2014) 675.

[16] A.M. Badr, H.A.Elshaikh, I.M. Ashraf, *J. Eng. Technol. Res.* 3 (2011) 62.

[17]S.Anandhi, T.S. Shyju, R. Gopalakrishnan, *Optik* 124 (2013) 5553.

[18] R.Sankar, C.M. Raghavan, M.Balaji, R.Mohan Kumar, R.Jayavel, *Cryst.Growth Des.* 7 (2007) 348.

[19] H.Ishihara, S.Dou, A. Weiss, Ber Bunsenges. *Phys.Chem* 95 (1991) 659.

[20] K.B.R. Varma, K.V.Rao, *Bull.Mater.Sci.* 5 (1983) 39.

Synthesis, Growth and Physicochemical Properties of Seminorganic NLO Crystal Bis(Thiourea) Ammonium Nitrate

S.M. Ravi Kumar [a]*, A.Anbarasi[b], S. Selvakumar[c], and G.J.Shanmuga Sundar[d]

[a]*Department of Physics, Government Arts College, Tiruvannamalai-606 603, Tamil Nadu,India*

[b]*Department of Physics, Periyar Government Arts College Cuddalore – 607 001, Tamil Nadu,India*

[c]*Department of Physics, Government Arts & Science College, Nandanam, Chennai-600 035, Tamil Nadu,India*

[d]*Department of Physics, Arignar Anna Government Arts College, Cheyyar-604 407, Tamil Nadu,India*

*Corresponding Author: ravism23@gmail.com, smravi78@rediffmail.com

Abstract

Nowadays, NLO materials are of current research interest due to their variety of applications such as second harmonic generation, optical bi-stability, laser remote sensing, optical disk data storage, laser driver fusion, medical and spectroscopic laser, photonic integrated circuitry, optical parametric oscillations and THz wave generation. This work concentrated on (bis)thiourea ammonium nitrate (BTAN) crystal was grown from aqueous solution by slow evaporation technique. The grown crystal has been subjected to single crystal X-ray diffraction to determine the unit cell dimensions. The grown crystal was also characterized by recording the powder X-ray diffraction patterns and by identifying the diffracting planes. The Fourier Transform Infrared (FT-IR) spectra have been recorded in the range 400–4500 cm^{-1}. Second harmonic generation (SHG) for the materials of this work was confirmed using Nd:YAG laser. The UV–visible spectra show that the grown crystals have wide optical transparency in the entire visible region. The Thermo gravimetric/Differential Thermal Analyses reveal that the materials have good thermal stability.

Keywords: Slow evaporation, Semi organic, NLO crystal, UV-vis_NIR study, TGA and DTA

1 Introduction

Photonics and optoelectronics are fast developing fields and are mainly focusing on nonlinear optical (NLO) materials. NLO materials are widely used in various fields of optics and optoelectronics such as second harmonic generation, optical bi-stability, laser remote sensing, optical disk data storage, laser driven fusion, medical and spectroscopic laser, photonic integrated circuitry, optical parametric oscillations and THz wave generation [1-4]. Materials exhibiting large non linearity remains very active in research both basic and applied sciences [5]. In recent years, to achieve large charge transfer and the optical transparency with fewer dislocations density extensive efforts have been made to develop new inorganic, organic and semi-organic NLO crystals [6-8]. The selection of material may depend not only on laser conditions but also on the physiochemical properties such as molecular nonlinearity, transparency, and conversion efficiency and laser damage threshold [9, 10]. The strong delocalization of π electron in the organic back bone of semi-organic establishes higher molecular polarizability and hence electron

IOP Conf. Series: Materials Science and Engineering **360** (2018) 012030 doi:10.1088/1757-899X/360/1/012030

density takes place. This functionalizes both the ends of the π bond system with suitable electron donor and acceptor groups and enhances the asymmetric electronic distribution in both ground and excited states of semi-organics. This enables an increased optical nonlinearity by means of such as π delocalization length and donor acceptor groups with nonlinear optical function bonds [11]. Hence, the synthesis of novel efficient frequency conversion materials has resulted in the development of semi organic materials, which possess efficient physicochemical properties, such as large nonlinearity over a broad frequency range, high optical damage threshold, low angular sensitivity, low dielectric constant, mechanical stability and inherent synthetic flexibility [12-14]. Semi-organic crystals possess both the good qualities of organic and inorganic, it is used in device fabrication technology due to their enhanced physiochemical properties such as thermal stability, wide range of transparency and excellent NLO coefficient [15, 16].

The NLO properties of semi organic crystals have attracted significant attention in the last few years, because both organic and inorganic components in it contribute specifically to process of second harmonic generation (SHG). Particularly, thiourea complex semi organic crystals are shows good second harmonic generation (SHG) efficiency. Thiourea molecules are an interesting in organic matrix modifier due to its large dipole moment and its ability to form and extensive network of hydrogen bonds [17]. The Centro symmetric thiourea molecule, when combined with inorganic salt yield non- centro symmetric complex, which has the nonlinear optical properties [18]. Hence, in the present study concentrated on synthesis, growth, structural, optical, SHG and thermal properties of thiourea complex semi organic crystal namely bis(thiourea) ammonium nitrate (BTAN) and the results are reporting for the first time.

2 EXPERIMENTAL PROCEDURES

Thiourea and ammonium nitrate supplied by Merck, India, were used without further purification. Thiourea mixed with ammonium nitrate in molar ratio 2:1. The saturated solution of ammonium nitrate was slowly added to the saturated solution of thiourea at room temperature and it has been stirred well for nearly 8 hours to obtain the homogenous solution. The solution of title compound was synthesis by the following chemical reaction.

$$\mathbf{2[CH_4N_2S] + NH_4NO_3 \rightarrow NH_4(CH_4N_2S)_2\ NO_3}$$

The saturated solution was filtered using Wattman filter paper. The filtered solution was taken in a beaker and covered with good quality perforated polythene cover to restrict the fast evaporation and it's kept at room temperature in a dust free compartment for slow evaporation. After the period of 40-60 days, colorless crystals with dimension 10mm x 6mm x 3mm were harvested. The grown crystals are non-hygroscopic and optically good transparent in nature. Well-faced good quality crystals of BTAN are shown in the figure 1.

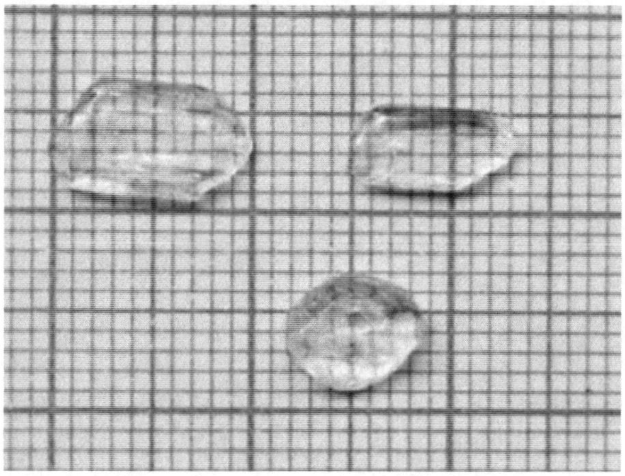

Figure 1 Photograph of as grown crystals of BTAN

3. Results and Discussion

The title compound was analyzed by single crystal XRD by ENRAF NONIUS CAD4-F single X-ray diffractometer with M_OK_α(λ=0.717Å) radiation. It is observed that the grown crystal crystallizes in orthorhombic system with space group P_{21}. The calculated lattice parameter for BTAN was found to be a = 8.73Å, b = 9.90Å, c=12.64Å and volume, V=1092Å^3.

The powder sample of grown crystal has subjected to Powder X-ray diffraction by using BRUCKER, Germany (model D8 Advance) X-ray diffractometer with CuK_{alpha} (wavelength=1.5405A^0) radiation. The Powder X-ray diffraction pattern of BTAN crystal was recorded at specific 2θ angles and also powder sample was scanned over the range 10° to 80° at a scan rate of 1° / min. The well defined peaks are reveals that the grown crystal has good quality and high crystalline nature. The observed powder XRD pattern of BTAN crystal is shown in the Figure 2.

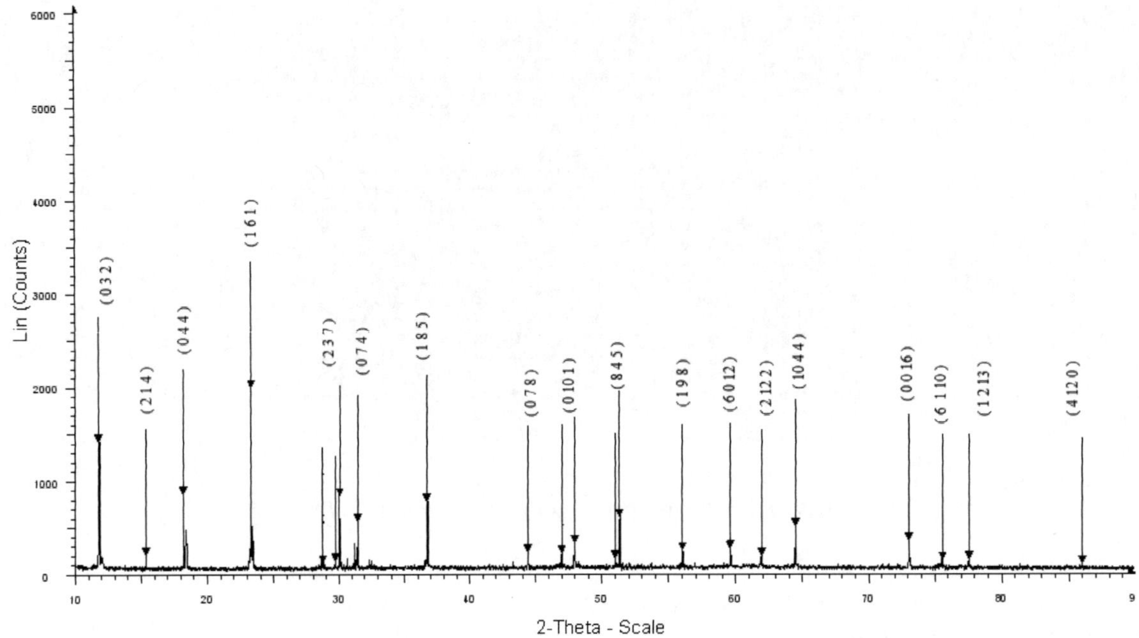

Figure 2 Powder X-ray diffraction (PXRD) Analysis of BTAN

The FTIR spectroscopy study effectively used to identify the presence of various functional groups presence in the materials. Figure 3 shows the FTIR spectrum of BTAN.

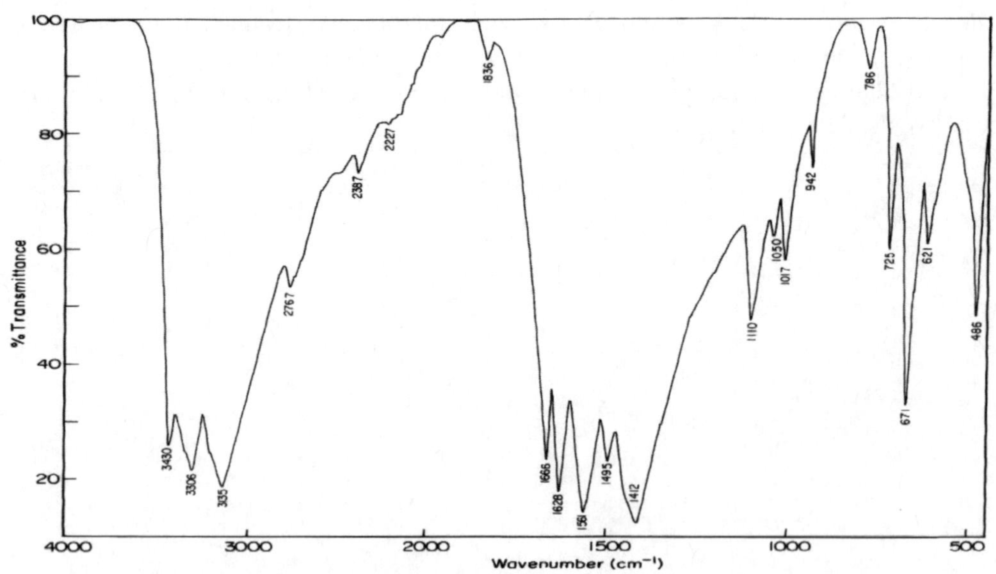

Figure 3 FTIR Spectrum of BTAN Crystal

IOP Conf. Series: Materials Science and Engineering **360** (2018) 012030 doi:10.1088/1757-899X/360/1/012030

The observed peak at 3410 cm^{-1} and 3183 cm^{-1} may be due to NH_2 stretching and N-H symmetric stretching vibrations respectively. The above mentioned peaks are the characteristic vibrations of thiourea and hence confirm the thiourea molecule in the grown crystal. The peaks at 2977 cm^{-1} is due to C-H stretching vibration. The NH_2 bending vibration is observed around 1620 cm^{-1}. The peaks at 1383 cm^{-1} confirm the C=S stretching vibration. The C-H deformation was observed at 666 cm^{-1}. The peak at 588 cm^{-1} is due to N-C-N stretching. The observed wave number and corresponding assignments of BTAN crystal is compare with thiourea is given in Table 1.

Wavelnumber (cm^{-1})		**Assignment**
Thiourea	**BTAN**	
3380	3410	NH_2 asymmetric stretching
3179	3183	N-H symmetric stretching
3090	2977	C-H stretching vibration
1627	1620	NH_2 bending
1412	1383	Symmetric C-N stretching
740	666	Symmetric C=S stretching
494	588	N-C-N Stretching

Table 1 Comparison of FTIR assignments of thiourea and BTAN crystal

The Suitable dimension and well polished crystal of BTAN was used for absorption study. A spectrum was recorded in the region 100-1000nm using VARIAN CARY 5E model spectrophotometer. The absorption spectrum obtained is shown in the Figure 4.

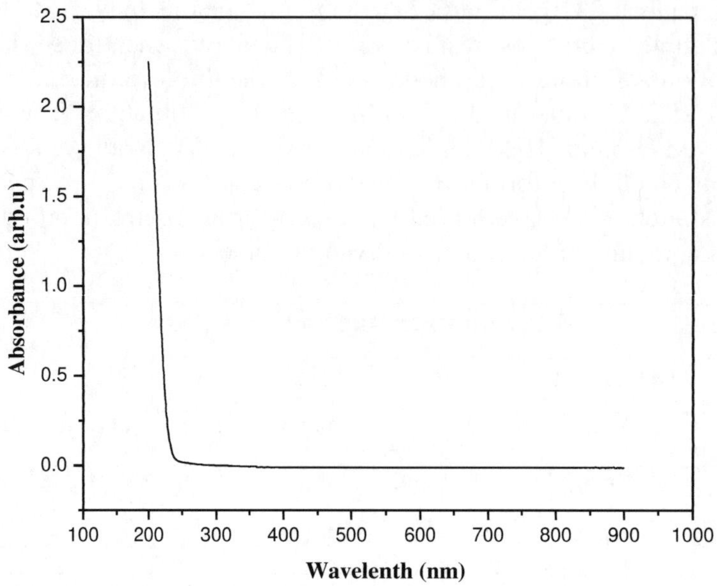

Figure 4. Optical absorption spectrum of BTAN

The absorbance is found to be low throughout the entire visible and near IR region due to the delocalization of electronic cloud through charge transfer axis. This is the most desirable property of the materials possessing NLO activity. The crystal shows good transmittance in the visible region and the lower cut off wavelength found to be 280nm. It is well known that the crystal has low cut off wavelength lies between 200 to 400 nm, suitable for device fabrication and more NLO applications. Hence, BTAN crystal is a suitable potential candidate for NLO applications.

Kurtz-Perry Powder second harmonic generation (SHG) measurements [25] was carried on grown crystal of BTAN using a a spectra-physics quanta-ray prolab 170 Nd:YAG laser with the first harmonics input at 1064nm and a pulse width of 10ns at repetition rate of 10Hz. This measurement offers possibility of assessing the nonlinearity of new materials. The grown crystal was identified as a NLO crystal due to the emission of high intense green radiation. Also the SHG efficiency was found to be 11.82 mJ. This value is compare with known NLO crystal of KDP which has the SHG efficiency 7.80 mJ. Hence, the SHG efficiency of BTAN is 1.52 times greater than that of KDP crystal. Therefore, BTAN crystal is a suitable candidate for device and various technological applications.

Information regarding phase transition and different stages of decomposition of the grown crystal BTAN crystal were identified by DTA and TGA thermal studies. BTAN crystal was weighed in an Al_2O_3 crucible with microprocessor temperature control system. DTA and TG curve of grown crystals were plotted in nitrogen atmospheres between ambient temperatures 100 to 1232 ºC shown in the figure 5.

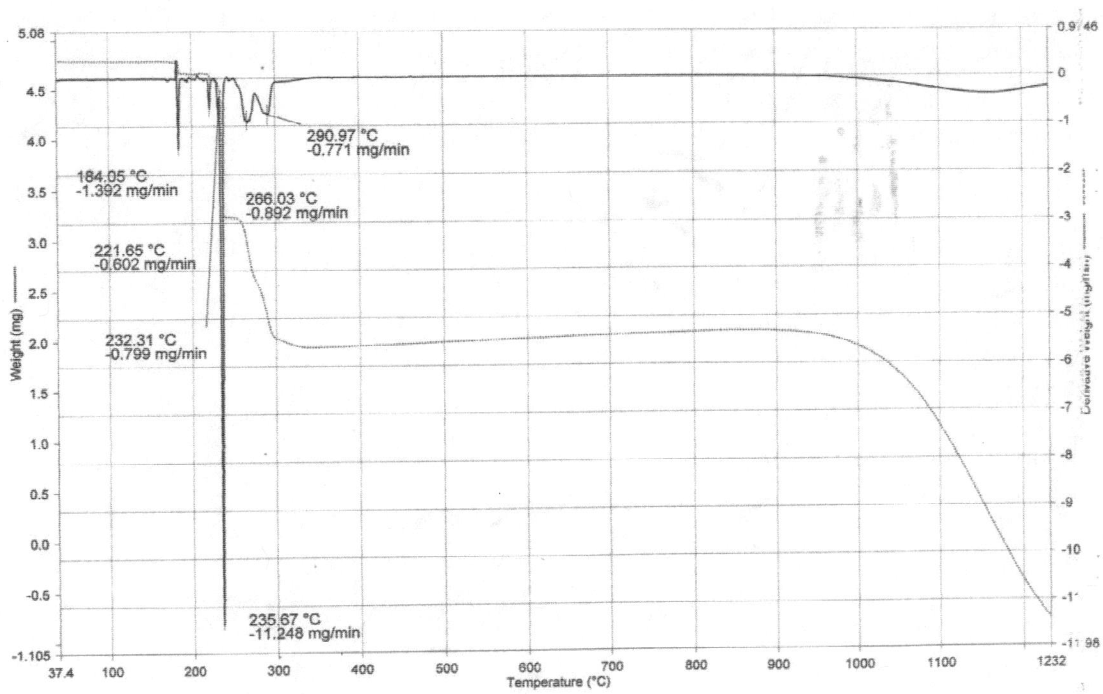

Figure 5. DTA and TG curve of BTAN crystal

There is no weight loss up to 184.05 ºC indicating that there is no inclusion of water in the crystal lattice. The recorded thermogram reveals that the major weight loss starts at 184 ºC and continues upto 1232 ºC. It is further observed that the DTA curve shown that first endo thermionic peak at 184 ºC is assigned to melting point of the BTAN compound and the remains exo thermionic peaks at 221 ºC, 232 ºC, 235 ºC, 266 ºC and 290 ºC are indicate the changes in the physical state of the crystal. The nature of the weight loss and sharpness of the peak ensure the decomposition and purity with crystalline nature of the material respectively. The observed decomposition point of BTAN crystal 184 ºC indicates that BTAN crystal is thermally stable up to 184 ºC.

4. Conclusion

Optically good crystals of (bis)thiourea ammonium nitrate has been grown from aqueous solution by slow evaporation technique under room temperature. Single crystal XRD analysis confirmed that the BTAN crystallize in orthorhombic system. Powder X-ray diffraction study revealed the crystalline nature of grown crystal. The optical quality of the grown crystal was justified by optical absorption studies. The sharp absorption on set at 280 nm and high absorption values of the grown crystal at wavelength above 380 nm, exhibit the optical quality. The powder SHG measurement shows the grown BTAN crystal having SHG efficiency of about 1.52 times that of KDP crystal. Thermal stability of grown sample was studied by TGA and DTA analyses which reveal that the BTAN crystal thermally stable up to 184 ºC.

Acknowledgement

The authors thank B. S. Abdur Rahman University, Vandalur, Chennai for providing instrument facility for characterization and also the corresponding author thankfully acknowledge the Science and Engineering Research Board (SERB), Department of Science and Technology (DST) for the financial support for Project **(FILE NO.EEQ/2016/000451).**

References

1. Santhanu Bhattacharya, Parthasarathi, Dustidar, T.N. Guru Row, *Chem. Mater.* 61, 531 **(1994)**.
2. T.Kaino, B.Cai, T.Takayama, *Adv. Funct. Mater.* 12, 599 **(2002)**.
3. M.Thakur, J.Xu, A.Bhowmilk, L.Zhou, *Appl. Phys. Lett.* 74, 635 **(1999)**.
4. A. Schneider, M.Neis, M.Stillhart, B.Ruiz, R.U.A.Khan, P.Gunter, *J.Opt. Soc. Am.* B23, 1822 **(2006)**.
5. R.J.Collins, D.F.Nelson, A.L.Schawlow, W.Bond, C.G.B.Garrett, W.Kaiser, *Phys. Rev. Lett.* 5, 303 **(1960)**.
6. Xiue Ren, Dongli Xu, Dongfeng Xue, *J.Crystal Growth,* 310, 2005 **(2008)**.
7. Dongfeng Xue, Henryk Ratajczak, *Chem. Phys. Lett.* 371, 601 **(2003)**.
8. Dongfeng Xue, Siyuan Zhang, *Chem. Phys. Lett.* 301, 449 **(1999)**.
9. F.Zernike, J.E.Midwinter, in:Applied nonlinear optics, *Wiley, Newyork*, **(1973)**.
10. P.N.Prasad, D.J.Willams, in:Introduction to nonlinear efforts in molecules and polymers, *Wiley, New York* **(1991)**.
11. Daqiu Yu, Dong feng Xue, Henryk Ratajczak, *J.Mol. Stuct.* 792 (2006) 280-285.
12. Y.R.Shen, The principles of nonlinear optics, *Wiley, New York* **(1984)**.
13. H.O. Marcy, L.F. Warren, M.S.Webb, C.A.Ebbers, S.P.Velsko, G.C.Kenndy, *Appl. Opt.* 31, 5051 **(1992).**
14. S.Ledoux, J.Zyss, *J. Int. Nonlinear opt. Phys.* 3, 287 **(1994)**.
15. S.M. Ravi Kumar, N. Melikechi, S. Selvakumar, P. Sagayaraj, *Physica B: Condensed Matter* 403, 4160 **(2008)**.
16. S. Selvakumar, S.M. Ravi Kumar, Ginson P. Joseph, K. Rajarajan, J. Madhavan, S.A. Rajasekar, P. Sagayaraj, *Materials Chemistry and Physics*, 103 (1), 153 **(2007).**

17. K.H. Hellwege, A.M. Hellwage, *Landolt-Bornstein Group II* 14, 584 **(1982)**.
18. M.Lawrence, J.Thomas Joseph Prakash, *Spectrochim. Acta Part A*, 91, 30 **(2012)**.

IOP Conf. Series: Materials Science and Engineering **360** (2018) 012031 doi:10.1088/1757-899X/360/1/012031

Synthesis, growth, structure, optical, mechanical, and electrical properties of an Inorganic new nonlinear optical crystal: sodium manganese tetra chloride (SMTC)

M. Packiya raj[a], S. M. Ravi Kumar[b*], R.Srineevasan[b] and R.Ravisankar[b]

[a]*Department of Physics, S.K.P. Engineering College, Tiruvannamalai 606 611,Tamil Nadu,India*
[b]*Department of Physics, Government Arts College , Tiruvannamalai 606 603, Tamil Nadu,India*

corresponding author:smravi78@rediffmail.com

Abstract

A new inorganic nonlinear optical single crystal of sodium manganese tetra chloride (SMTC) has been successfully grown form aqueous solution by the slow evaporation technique at room temperature. The crystals obtained by the above technique were subjected to different characterization analysis. The crystalline nature of the grown crystal of SMTC was analyzed by powder X-ray diffraction. Single crystal X-ray diffraction study reveals that the crystal belongs to orthorhombic system with non-centrosymmetric space group Pbam. Optical transmission study on SMTC crystal shows high transmittance in the entire UV–Vis region and the lower cutoff wavelength is found to be 240 nm. The mechanical strength of the grown crystal was estimated by Vicker'smicrohardness test. The second harmonic generation (SHG) efficiency of the crystal was measured by Kurtz's powder technique infers that the crystal has nonlinear optical (NLO) efficiency 1.32 times that of KDP.The dielectric constant and dielectric loss of the compound were measured at different temperature with varying frequencies. Photoconductivity study confirms that the title compound possesses a negative photoconducting nature. Growth mechanism and surface features of the as grown crystals were analyzed by chemical etching analyzing.
Key Words :sodium manganese tetra chloride, XRD, second harmonic generation

1 Introduction

The well known properties of laser radiation are important for a wide variety of applications. Laser radiation could be converted into one form of frequency to another through the nonlinear optics, hence the application of nonlinear optics is increased significantly in various fields in science and technology. Generally, nonlinear optical (NLO) interaction is made by one or two laser beam incident on a suitable material in which an output beam of the desired frequency is produced [1-2]. Harmonic generation, sum and difference frequency generation and parametric oscillation are included in the NLO interaction [3]. A Lower frequency pair of tunable output beam can be produced only by suitable material when it is interact (NLO) with high input laser beam. Mostly, NLO interaction imposes several demands on potential NLO materials. The field of nonlinear optics is one of the most attractive fields of current research because of its vital applications in various areas like optical switching, optical data storage for developing technologies in telecommunication and signal processing [4-6]. Since, the first demonstration was done in the year of 1961, which reveals that nonlinear frequency conversion is highly materials-limited field [7]. So materials should be optically transparent, quadratic susceptibility of sufficient

IOP Conf. Series: Materials Science and Engineering **360** (2018) 012031 doi:10.1088/1757-899X/360/1/012031

magnitude, allow for phase matching interaction and withstand the laser intensity without damaging. To date, the most important class of materials used in nonlinear optics is inorganic single crystals.

Inorganic materials, exhibiting second order nonlinear optical properties have attracted in the recent past due to their ability to process into crystals, wide optical transparency domain, large nonlinear figure of merit for frequency conversion, fast optical response time and wide phase matchable angle [8]. These ionic bonded inorganic crystals, easy to synthesis with high melting point and high degree of chemical inertness[9]. Highly polarisable, inorganic quality crystal and their efficient active second order harmonic generation (SHG) have been observed by Franken et al and co-workers in 1961[7].

Inorganic materials have advantages over organic materials, such as architectural flexibility for molecular design and morphology, high mechanical strength and good environmental stability with non toxicity andusability in high power applications. Molecular hyperpolarizability of inorganic nonlinear optical crystal are used in optical switching (modulation), frequency conversion (SHG, wave mixing) and electro-optic applications especially in EO modulation. Historically, inorganic NLO materials have been chronicled more extensively inorganic oxide crystal, $LiNbO_3$, $KNbO_3$, KDP and KTP, etc., have been studied for device application like piezoelectric, ferroelectric and Electro-optics [10]. This material has also been formed successful usage in commercial frequency doublers, mixers and paramedic generators to provide coherent laser radiation with high frequency conversion efficiency in the new region of the spectrum, inaccessible by other nonlinear crystal conventional sources.

The aim of this research work is to survey the processing and properties of inorganic nonlinear optical crystal sodium manganese tetrachloride (SMTC) with molecular formula Na_2MnCl_4 used in NLO frequency conversion.The structure of Na_2MnCl_4 was determined by Goodyear et al in the year of 1971[11]. The grown crystals of SMTC, chlorine ions coordinated octahedrally with Mn ions and form an infinite chain parallel to c axis and are held with sodium ions. Sodium ions are surrounded by four chloride ions at the corners of a trigonal prism. Binding of Mn-Cl and Na-Cl suggests that the structure is mainly ionic in character.Hence, an attempt hasbeen made on growth of sodium manganese tetra chloride (SMTC) single crystals by slow evaporation solution growth technique and its physical-chemical properties have been investigated for the first time.

2 Experimental Procedure

2.1. Synthesis

SMTC salt was synthesized by taking analytical reagent (AR) grade manganese chloride and sodium chloride in stoichiometric ratio 1:2 with double distilled water as a solvent. The synthesized SMTC salt has been obtained by the following chemical reaction.

$$MnCl_2 + 2NaCl \longrightarrow Na_2MnCl_4$$

Manganese chloride + sodium Chloride ⟶ Sodium manganese tetra chloride.

The scheme of the molecular structure of SMTC is as shown below.

IOP Conf. Series: Materials Science and Engineering **360** (2018) 012031 doi:10.1088/1757-899X/360/1/012031

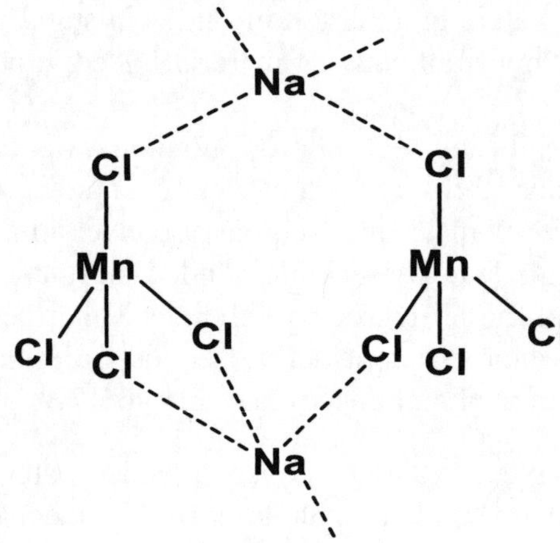

Scheme 1 Molecular arrangement of SMTC crystal

2.2 Solubility study

The growth rate of a crystal depends on its solubility and temperature. Solvent and solubility factors define super saturation, which is the driving force for the rate of crystal growth. The solubility study was carried out by taking an excess quantity of SMTC in the solvent and it was stirred continuously to achieve uniform concentration over the entire volume of the solution. The solubility of SMTC in the solvent of double distilled water was determined for various temperatures 30 - 55 ^{0}C in 5°C intervals using a constant temperature bath with an accuracy of ±0.01^{0}C attached to a cryostat facility. After attaining the saturation, the concentration of the solute was analyzed gravimetrically. The solubility curve for SMTC is shown in figure 1. It is observed that SMTC has a positive solubility (solubility increases linearly with temperatures) gradient in the water.

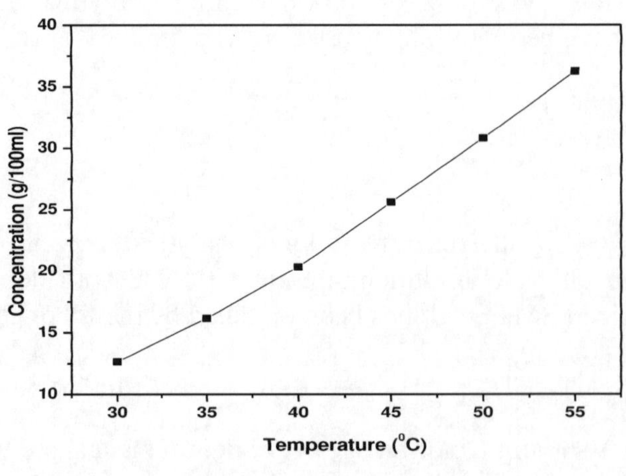

Figure 1 Solubility curve of SMTC

IOP Conf. Series: Materials Science and Engineering **360** (2018) 012031 doi:10.1088/1757-899X/360/1/012031

2.3 Crystal Growth

The prepared solutions were stirred vigorously at RT for 4 h. Continuous stirring with slightly rise in temperature ensures homogeneity and avoids co-precipitation of motives. Purification of synthesized salt was achieved by successive recrystallization process. The saturated mixture of solution was filtered two times with micron pore sizeWattmann filter paper. This synthesized clean solution was poured into a Petri dish and covered by polythene paper with pores, and allowed for slow evaporation of the water solvent. After a time span of 35 days, the solvent was evaporated and good quality SMTC crystal of dimensions $2\times2\times1$ mm^3 were harvested from the Petri dish. The growth rate was found to be 0.12mm per day. The grown crystal was defect less, optically transparent and with no inclusions. As-grown crystal of SMTC is shown in the figure 2.

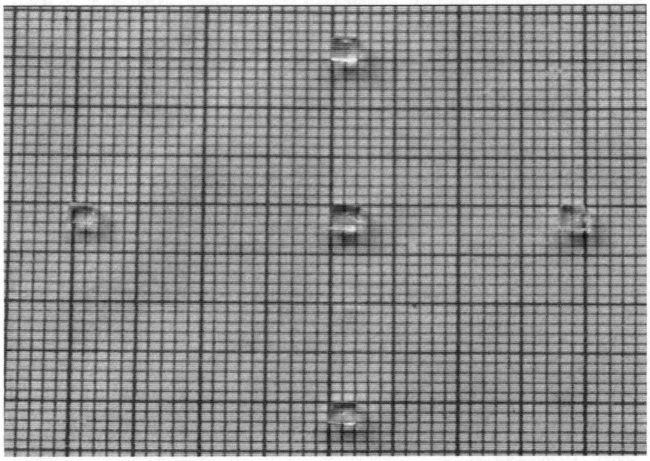

Figure 2 Photograph of as grown crystals of SMTC

3 Characterization of SMTC Single Crystal

The grown crystal of SMTC was subjected to single crystal and powder XRD analysis using ENRAF NONIOUS CAD4 X-ray diffraction meter and BRUKER, Germany (model D8 advance) X-ray diffractometer. Transmission behavior of the grown crystal was studied by using LAMBDA-35 UV-visible spectrophotometer. The NLO efficiency of the grown crystal was tested by KURTZ powder technique using ND:YAG laser of wavelength 1064 nm. Mechanical behaviour of the grown sample was investigated by Vicker's microhardness tester. Dielectric constant and dielectric loss studies were carried out by using HIOKI3532 HITESTER LCR meter. Keithley 485 PICOAMMETER was used to study the photoconductivity of the grown SMTC crystal.

IOP Conf. Series: Materials Science and Engineering **360** (2018) 012031 doi:10.1088/1757-899X/360/1/012031

3.1 Results and Discussion

3.1.1 Single crystal XRD studies.

The Single crystal XRD study confirms the unit cell parameters of as grown SMTC crystals a=6.93 Å, b=11. 82 Å, c=3. 86 Å,$\alpha=\beta=\gamma=90°$ and volume of the cell is found to be, 316.182 $Å^3$. Hence the SMTC crystal is Orthorhombic in structure and in thespace group Pbam. The lattice parameters are well coincide with a reported value [11].

3.1.2. Powder XRD studies

Powder XRD of as-grown crystal of SMTC has been studied and XRD pattern is shown in the figure 3. The powder sample was scanned over the range of 10-80° at a scan rate of 0.2°/s, using CuKα radiation of wavelength 1.545 Å. The obtained powder X-ray diffraction data were analyzed and compared with JCPDS data (JCPDS data card number 4137). A limited number of sharp peaks without any broadening in the XRD pattern confirms that the grown SMTC possesses good crystallinity.

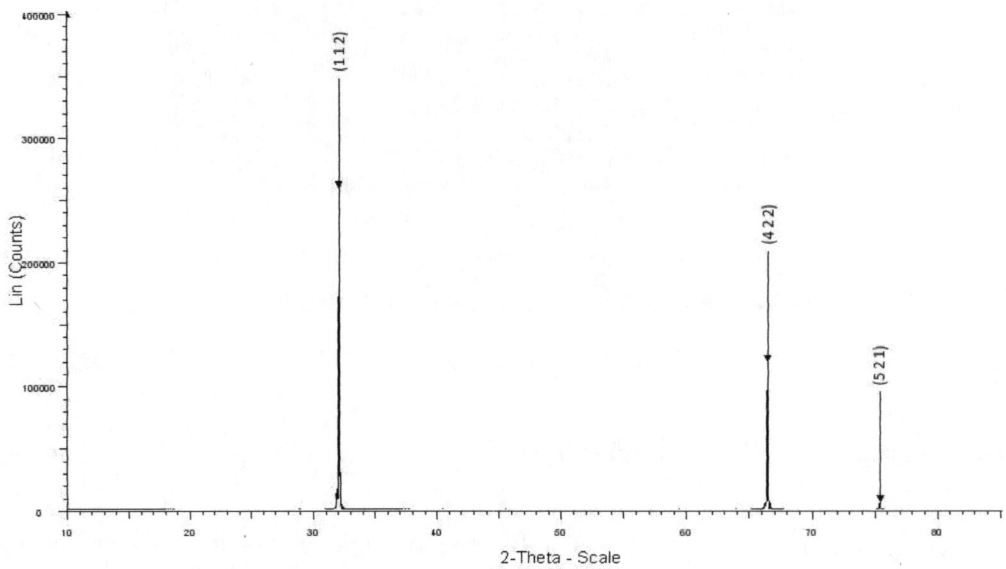

Figure 3 Powder XRD pattern of SMTC

3.1.3 Linear optical transmission studies

Since NLO crystals can be for practical use only when it has wide transparency window. The transmission range of SMTC crystal was determined by recording the optical transmission spectrum in the wavelength region of 200 - 900 nm. The optical transmission spectrum of SMTC crystal is shown in the figure 4. Optically polished single crystal of thickness 2mm was used to study the transmission behavior of SMTC. This recorded

IOP Conf. Series: Materials Science and Engineering **360** (2018) 012031 doi:10.1088/1757-899X/360/1/012031

spectrum, gives information about the structure of the molecule by absorption of UV and the visible light involves promotion of electrons in the σ and π orbital from ground state to a higher energy state [12]. The transmission spectrum shows that the grown crystal has a lower cutoff wavelength at 240 nm,which attributes the electronic transmission in the SMTC crystal. Absence of absorbance in the region between 240 nm and 900 nm is an essential property of the nonlinear optical crystals. Single crystals are mainly used in optical applications and hence an optical transmittance window and the transparency lower cutoff wavelength (200-400) is very important for the realization of the SHG output in the range for using lasers. Optical width of the as grown crystal SMTC was compared with NaCl and $Mncl_2$ complex inorganic crystals. The grown SMTC crystal has good transparency in UV-visible and IR region which ensures, that crystal can be used as sensor materials from UV, visible in the IR ranges and may be consider as a potential candidate for the photonic and optoelectronic applications [13]. The graph has been plotted to estimate the direct band gap values using Tauc's relation. The Tauc's plot has been drawn between $(\alpha h\nu)^2$ and hv as shown in the figure 5. The band gap value is obtained by extrapolating the straight portion of the graph to hv axis at $(\alpha h\nu)^2=0$. The estimated band gap value of grown sample SMTC is 5.4eV.

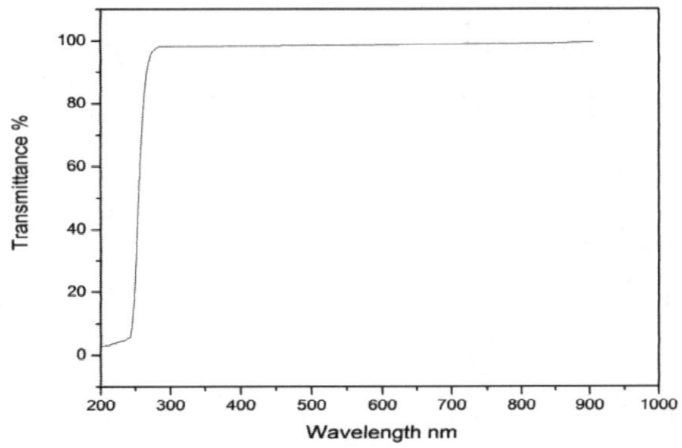

Figure 4 UV-Visible spectrum of SMTC crystal

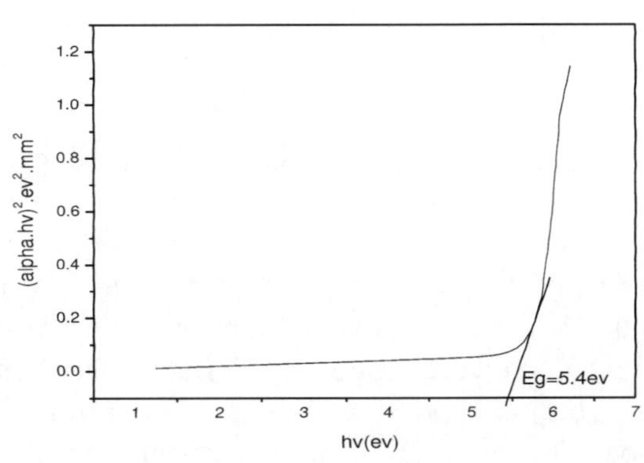

Figure 5 Tauc's plot of SMTC crystal

3.1.4 Second harmonic generation efficiency measurement

In order to confirm the nonlinear optical property, powdered sample of SMTC was subjected to Kurtz and Perry techniques, which remains a powerful tool for initial screening of materials for SHG [14]. The fundamental beam of wavelength 1064 nm from Q switched mode locked Nd:YAG laser was made to fall normally on the powder from of grind sample which was made available between two transparent glass slides. Pulse energy 2.9 mJ/pulse and pulse width 8 ns with a repetition rate of 10 Hz were used. The photo multiplier tube (Hamamatsu R2059) was used as a detector and 90 degree geometry was employed. The SHG signal generated in the sample was confirmed from emission of bright green (532 nm) radiation from the sample. The measured amplitude of second harmonic generation for SMTC crystal is 11.32 mJ and 8.8 mJ for KDP (KDP crystal was powdered to the identical size of SMTC and used as reference materials). It shows a powder SHG efficiency of SMTC crystal is about 1.3 times of KDP. The SHG efficiency of SMTC crystal is compared to few popular inorganic NLO crystals which are given in the table 1.

Compound	**SHG efficiency**
Potassium dihydrogen phosphate (KDP)	1
Potassium pentaborate	0.79
Ammonium pentaborate	0.2
SMTC(Present work)	1.32

Table 1 comparison of SMTC crystal with other NLO crystals

3.1.5 Microhardness study

Mechanical property of the grown sample was studied by measuring the hardness number with various applied loads. A well polished crystal was placed on the platform on the Vickers microhardness tester and the loads of different magnitude were applied over a fixed interval of time of 5 s. Here low loads are applied to measure the hardness of the samples. Microhardness analysis was carried out byusing Leitz Weitzler hardness tester fitted with a diamond indenter. The indentations were made on the flat surface with the load ranging from 25 to 100 g using Shimadzu HMV-2T fitted with Vicker's pyramidal indenter and attached to an incident light microscope. The indentation time has been kept 5 seconds for all the loads. Several indentations were made on the crystal surface with sufficient space for each load and the diagonal length (d) of the indented impressions is measured. The Vickers hardness number of the materials Hvhave determined by the relation, $Hv=1.8544\ P/d^2\ kg/mm^2$. Where

P is the applied load in kg and d is the diagonal length of indentation impression in mm. The hardness number was found to be increased with load and above 100 g, significant crack and inclusion are observed, which may be due to the release of internal stress generated locally by indentation [15]. The average value of the Vickers hardness number for the grown crystal is 100 kg/mm^2. The plot drawn between the corresponding loads and hardness values of SMTC is shown in figure 6. The work hardening coefficient (n) of the material is related to the load (p) by the relation $P=Ad^n$, Where A is an arbitrary constant. The work hardening coefficient 'n' of the sample has been determined from the slop of the log p vs log d (figure7) and is found to be 2.278 indicating that the crystal belongs to soft category. From careful observations on various materials Onitschand Hannemanpointed out that the value of n between 1.0 and 1.6 indicates a moderately hard material and n above 1.6 represents a soft material [16-17]. The value of n indicates that the SMTC crystal has higher mechanical strength and useful in device applications.

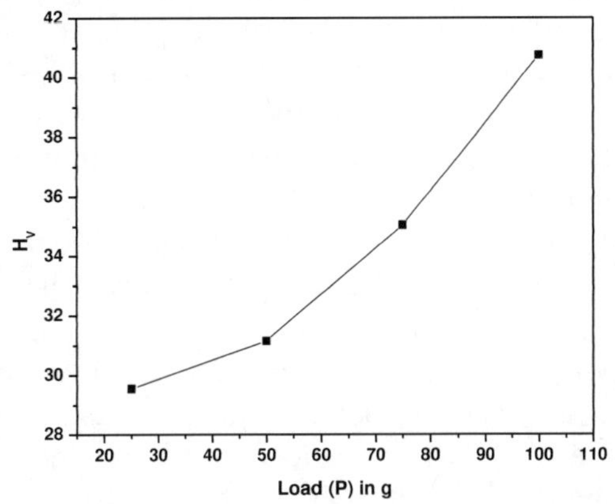

Figure 6 Vickers hardness number against load plot of SMTC crystal

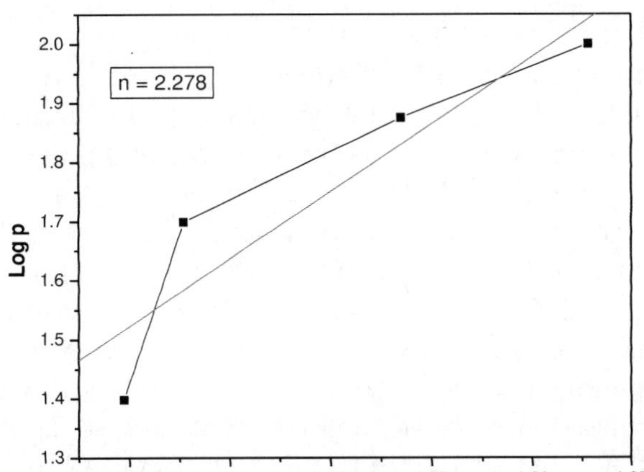

Figure 7 Log p vs log d plot of SMTC crystal

IOP Conf. Series: Materials Science and Engineering **360** (2018) 012031 doi:10.1088/1757-899X/360/1/012031

3.1.6 Dielectic studies.

The behavior of the crystal under the influence of electric field has good relation with the laser irradiation on the sample, hence the power dissipation factor can be studied from the dielectric studies. Dielectric studies for SMTC single crystal were carried out as a function of frequency for various temperatures. The SMTC crystal was coated with graphite on either side to form an electrode of a parallel plate capacitor. The dielectric constant was calculated using the formula

$$\varepsilon_r = Ct/\varepsilon_0 A$$

Where C is the capacitance; t is the thickness of the crystal,$\varepsilon_0$is the permittivity of the free space and A is the area of a cross section of the sample. The variations of dielectric constant and dielectric loss as a function of frequency (500 Hz to 5MHz) at different temperatures (308 -368 K) are shown in the figure 8 and 9. The dielectric constant as a function of frequency and temperatures are observed, which reveals that the dielectric constant decreases with increasing frequency. This behaviour of the sample referred as anomalous dielectric dispersion. High dielectric constant at low frequency is attributed to various polarization mechanisms of molecules. Generally, the polarization occurs as a function of time. It is obvious that at lower frequency, time taken for polarization is high. Hence, irrespective of the polarization mechanisms, measurement of dipole moment per unit volume would be high, resulting that the dielectric constant is independent of the frequency. At high frequency the change of polarization occurs even at very short time. Therefore, the polarization can occur only by the movement of electronic charge rather than the ions, results low dielectric constant. Electronic and space charge polarization predominantly in the low-frequency region suggest that the grown crystal possesses an enhanced optical quality with lesser defects and this phenomenon is an essential characteristic for nonlinear optical applications [18-20].

From the figure 9, the value of dielectric loss found to be high at low frequency region and dielectric loss found too low at higher frequency. The measure of low dielectric loss at higher frequencies is due to the dipole rotation. Also at high frequencies, the orientation polarization ceases and the energy need not be spent to rotate dipoles. It is also observed that both dielectric constant and dielectric loss depends on the temperature and increases slightly with an increase of temperature at a constant frequency. It is believed that materials with high dielectric constant lead to power dissipation.The Refractive index of the medium is related with dielectric constant according to the equation $n= (\varepsilon_r\mu_r)^{1/2}$. For non-magnetic materials, $\mu_r$is equal to one. This equation explains the electric field based refractive index of the medium could play vital role in nonlinear effect. SMTC crystals possess low dielectric constant and low dielectric loss and hence it will be suitable for electro-optic applications. Also the low value of dielectric loss at high frequency can be taken as a proof of the good optical quality of the crystal. The application part of nonlinear optics demand high quality crystals with lesser defects.

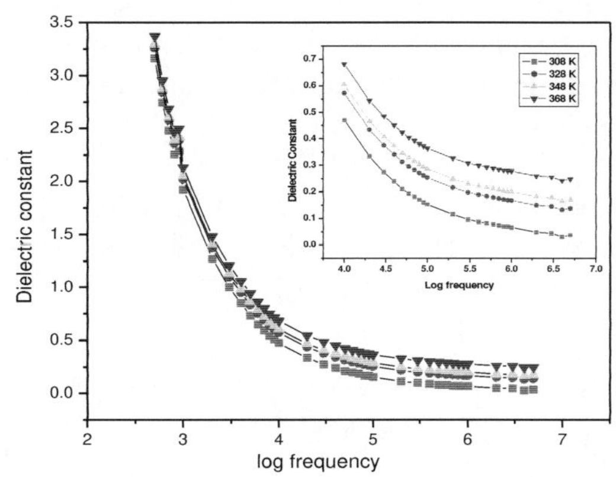

Figure 8 Variation of dielectric constant of SMTC single crystal with log frequency at different temperatures

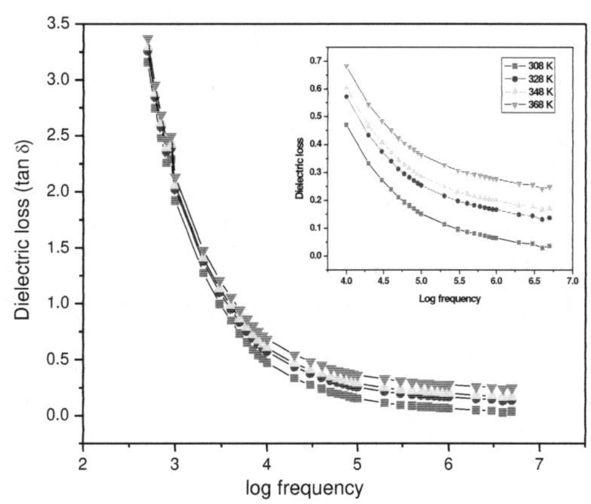

Figure 9 Variation of dielectric loss of SMTC single crystal with log frequency at different temperatures

3.1.7 Photoconductivity studies.

The photoconductivity study of SMTC crystal was carried out by connecting the sample in series with DC power supply and a picoammeter (Keithley 480) at room temperature. The details of the experimental setup are reported elsewhere [21]. By increasing the applied field from 10 to 150 V/cm and corresponding dark current without exposure of radiation was recorded. Photo current was recorded by exposing the crystal with halogen lamp of power 100 W containing iodine vapour for the same applied field. Dark current and Photo current against an applied field of same range were recorded in the same graph [Figure 10]. From the graph, it is observed that dark and photo current of the grown crystal increase linearly with applied field but photocurrent less than the dark current. This phenomenon is termed as negative photoconductivity.

Negative photo conductivity of being as grown crystal SMTC, may be due to decrease in either no number of free change carriers or their lifetime when subjected to radiation. Negative Photoconductivity of the grown crystal explained, according to Stockman model, the forbidden gap in the material contains two energy levels in which one is situated between the Fermi level and the conduction band while another is located close to the valence band. The second state has a high capture cross section for electrons and holes. As it captures electrons from the valence band the number of charge carriers in the conduction band gets reduced and the currentdecreases in the presence of radiation [22-23].

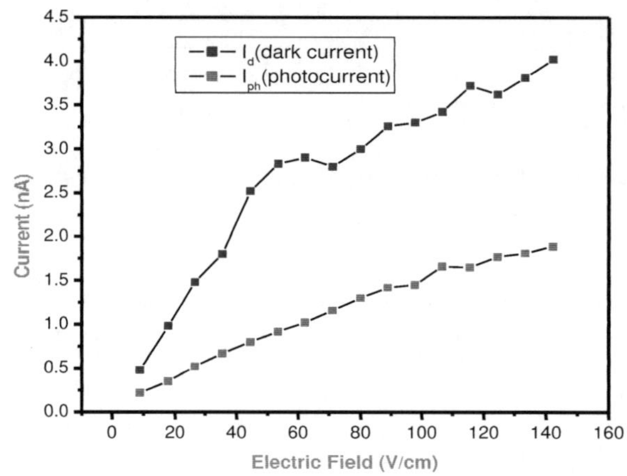

Figure 10 Field dependent conductivity of SMTC crystal

3.1.8 Chemical etching analysis

Chemical etching analyzing is the one of the most convenient and suitable technique for characterizing the growth defects. High quality or well define defect structure crystals are more important for device fabrication and technical applications now a day. Mostly, crystals grown from solution may suffer from imperfections like solvent inclusion, twins, grain boundaries and dislocation. Defects crystals are incident by laser beam, defects cause the scattering and absorption of laser beams. Hence, will affect the performance and use of crystal in laser applications. Generally, the etch patterns are captured at the dislocation sites by etching the crystal surface using suitable etchants [24]. The dissolution of a crystal in an etchant is an essential criterion for the etch pit formation. The region where the dislocations intersect the surface of a crystal is energetically different from the surrounding area. In the present study chemical etching was carried out using ethanol at room temperature. Once the damaged surface layer was removed by means of etching, a fresh surface appeared which in turn gave clear etch pits. The etched surfaces were dried using good quality filter paper and the surfaces were examined immediately. The etching study was carried for different time 5s, 10s and 15 s and observed etch patterns are shown in the figure 11(a, b, c) respectively. In the figure a, it is observed that whiskers type pits are noticed. Also in an etched surface shallow and deep etch pits are observed. By increasing the etching time there is no change of shapes, but the size of the pits slightly increased (figure11 b and c). The geometry of an etch pit also depends on the interatomic arrangement on the crystal. The observed etch pitch on the grown crystal surface implies that the crystal undergoes selective dissolution during the growth itself.

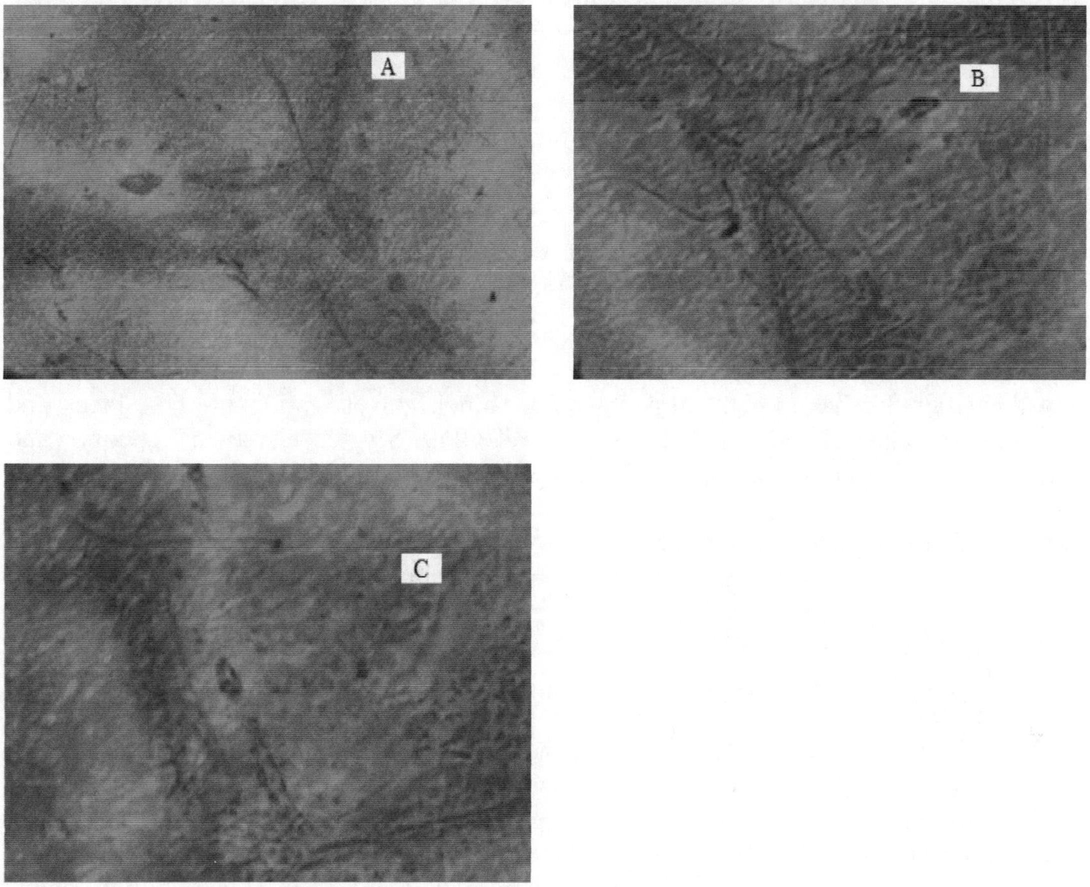

Fig. 11 Etch pattern on the SMTC crystal a) after etching for 5 s b) after etching for 10 s c) after etching for 15s

4 Conclusion

A potential inorganic nonlinear optical single crystal of sodium manganese tetra chloride was prepared at room temperature by slow evaporation of aqueous solutions. The well defined external appearance with bright, transparent and colourless crystals is obtained. The unit cell parameters and the space group were found using single crystal data. The good crystalline nature of the SMTC was confirmed by well defined peaks of powder X-ray diffraction studies. The FT-IR spectrum reveals the functional groups of the grown crystals. The grown crystal shows 99 % transmission with UV cut-off at 240 nm hence suitable for frequency conversion applications. The SHG efficiency of the SMTC was measured to be higher than that of KDP. From the mechanical measurements, it was observed that the hardness increases with increase of load. The dielectric constant and dielectric loss studies of SMTC establish the normal behavior. The above experimental results, viz., bulk size, extremely good crystalline perfection, optical transparency, SHG efficiency and mechanical strength may have possible NLO applications.

Reference

[1] A. H. Reshak, H. Kamarudin, and S. Auluck, Acentric Nonlinear Optical 2,4-Dihydroxyl Hydrazone Isomorphic Crystals with Large Linear, Nonlinear Optical Susceptibilities and Hyperpolarizability, J. Phys. Chem. 116, 4677(**2012**)

[2] Ali Hussain Reshak , I. V. Kityk , and S. Auluck, Investigation of the Linear and Nonlinear Optical Susceptibilities of $KTiOPO_4$ Single Crystals: Theory and Experiment, J.Phys.Chem.114,16705(**2010**)

[3] F. Peter, M.Bordui and Martin,Fejer.Annu. Rev. Mater. Sci. 23 ,321(**1993**).

[4] P.N. Prasad, D.J. Williams. Introduction to nonlinear optical effects in organic molecules and polymers, John Wiley & Sons, Inc., New York, USA (**1991**).

[5] H.O. Marcy, L.F. Warren, M.S. Webb, C.A. Ebbers, S.P. Velsko, G.C. Kennedy, G.C. Catella, Second harmonic generation in zinc tris(thiourea) sulfate, Appl. Opt. 31, 5051(**1992**).

[6] X.Q. Wang, D. Xu, D.R. Yuan, Y.P. Tian, W.T. Yu, S.Y. Sun, Z.H. Yang, Q. Fang, M.K. Lu, Y.X. Yan, F.Q. Meng, S.Y. Guo, G.H. Zhang, M.G. Jiang, Synthesis, structure and properties of a new nonlinear optical material: Zinc cadmium tetrathiocyanate, Mater. Res. Bull. 34 ,2003(**1999**).

[7] P.A. Franken, A.E. Hill, C.W. Peters, G.Weinreich, Generation of optical harmonics, Phys. Rev. Lett. 7 ,118(**1961**).

[8] H. Nalwa, Seizo Miyata.Nonlinear optics of organic molecules and polymers, CRC press, New York (**1996**).

[9] C.F. Dewey Jr, W.R. Cook Jr, R.T. Hodgson, J.J. Wynne,Frequency doubling in $KB_5O_8{\cdot}4H_2O$ and $NH_4B_5O_8{\cdot}4H_2O$ to 217.3 nm, Appl. Phys. Lett. 26 ,714(**1975**).

[10] D.S. Chemla, J. Zyss, Nonlinear optical properties of organic molecules and crystals. 01-02, Academic Press, Orlando, New York (**1987**).

[11] J. Goodyear, S.D.A. Ali, G.A. Steigmann, The crystal structure of Na_2MnCl_4, ActaCryst. B27 ,1672(**1971**).

[12] R. Sankar, C.M. Raghavan, M. Balaji, R. Mohan Kumar, R. Jayavel, Synthesis and Growth of Triaquaglycinesulfatozinc(II), $[Zn(SO_4)(C_2H_5NO_2)(H_2O)_3]$, a New Semiorganic Nonlinear Optical Crystal, Cryst. Growth Des. 7 ,348(**2007**).

[13] Y. Le Fur, R. Masse, M.Z. Cherkaoui, J.F. Nicoud, Z. Kristallogr. 856 (**1993**).

[14] S.K. Kurtz, New nonlinear optical materials, IEEE, J. Quantum Electron. 4 ,578(**1968**).

[15] T. Balakrishnan, K. Ramamurthi,Growth, structural, optical, thermal and mechanical properties of glycine zinc chloride single crystal, Mater. Lett. 62 ,65(**2008**).

[16] E.M. Onitsch, The present status of testing the hardness of materials. Mikroskopie. 95 ,12(**1956**).
[17] M. Hanneman,Metall. Manch. 23 ,135(**1941**).
[18] C. Balarew, R. Duhlew,Application of the Hard and Soft Acids and Bases Concept to Explain Ligand Coordination in Double Salt Structures, J. Solid State Chem. 55 ,01(**1984**).
[19] P.V. Dhanaraj, N.P. Rajesh,Growth and characterization of nonlinear opticalγ-glycine single crystal from lithium acetate as a solvent, Mater. Chem. Phys. 115 ,413(**2009**).
[20] K.V. Rao, A. Samakula, Dielectric properties of cobalt oxide, nickel oxide and their mixed crystals, J. Appl. Phys. 36 ,2031(**1965**).
[21] F.P. Xavier, A. Regis Indigo, G.J. Goldsmith, Role of metal phthalocyanine in redox complex conductivity of polyaniline and aniline black, J. PorphyrinsPhthaloeyanines. 3 ,679(**1999**).
[22] V.N. Joshi, Photoconductivity, Marcel Dekker, NewYork, **(1990)**.
[23] S. Abraham Rajasekar, K. Thamizhrasan, J.G.M. Jesudurai, D. Premanand, P. Sagayaraj, The role of metallic dopants on the optical and photoconductivity properties of pure and doped potassium pentaborate (KB5)single crystal. Materials Chemistry and Physics. 84(1) ,157(**2004**).
[24] Owczarek,K. Sangwal,Effect of impurities on the growth of KDP crystals: On the mechanism of adsorption on {100} faces from tapering data, J. Cryst. Growth. 99 ,827(**1990**).

IOP Conf. Series: Materials Science and Engineering **360** (2018) 012032 doi:10.1088/1757-899X/360/1/012032

Crystal Growth and Characterization of a Novel Non Linear Optical Single Crystal of Urea Phthalic Acid

N Indumathi[1], P Sanjay[1], K Deepa[2], J Madhavan[2] and S Senthil[1]

[1] Department of physics, Government Arts College For Men (Autonomous), Nandanam, Chennai – 35, Tamil Nadu, India.
[2] Department of Physics, Loyola College, Chennai – 34, Tamil Nadu, India.
Email: ssatoms@yahoo.co.in

Abstract. The Nonlinear optical (NLO) material of Urea Phthalic acid (UPA) was grown by slow evaporation method. The unit cell parameters were determined by single crystal XRD. The fundamental frequency vibrations of UPA were identified by FTIR analysis. The optical properties of UPA were determined by UV-Vis spectral studies. The Laser damage threshold value of UPA single crystal was found to be 5.3 GW/Cm^2. Z-scan techniques with He–Ne laser has measured the optical nonlinearity of UPA crystal.

1. Introduction

Developments of novel nonlinear optical (NLO) materials and crystal design technique for assembling the materials are used for many device applications in the domain of opto-electronics and photonics [1]. In particular, Organic NLO crystals are attracting a great deal of attention due to their high NLO coefficient, high damage threshold, high thermal stability and mechanical strength. The crystal engineering technology of the nonlinear optical material is based on the anchorage of organic molecules exhibiting a large NLO efficiency on the amino acid groups [2]. The unique optoelectronic properties of organics are due to the formation of weak Vander waals and hydrogen bonds hence, it shows high degree of localization. Urea is one of such organic NLO materials which have been explored as a first for NLO device for second to fifth harmonic generation [3]. In addition to this, the potentiality of urea as a first organic parametric oscillator is having extended transparency in the UV region, large birefringence and high optical damage threshold. The significant efforts were being made by many researchers to explore the potentialities of urea mixed systems and its derivatives through molecular engineering procedure. Such attempts involve the doping of other organic and inorganic NLO materials with urea. Many such efforts have become successful [4,5].With this reported background, we have ventured into growth and characterization of Urea Phthalic acid (UPA) single crystal and it is expected that this would be a non-centrosymmetric complex, but to our surprise it has been formed as a centrosymmetric structure in which third order nonlinearity which is more favourable and prominent. Similar results of third order nonlinear optical properties have been reported earlier [6]. In the present paper, the synthesis, crystal growth and characterizations by Single crystal XRD, FTIR, UV-Visible, LDT and Z-Scan of Urea Phthalic Acid (UPA) single crystal are reported in detail.

2. Experimental Details

The UPA crystal was synthesized by taking Urea and Phthalic acid in a 1:1 stiochiometric ratio in methanol and stirred separately for about half an hour using a magnetic stirrer. After that these solutions were mixed together and stirred well for about 1h and filtered. The filtered solution was allowed to dry at room temperature and the crystals were obtained by slow evaporation technique. The purity of the synthesized crystal was improved by successive recrystallization process thereby good optical quality colorless single crystals were obtained in 30 days which is shown in Figure 1.

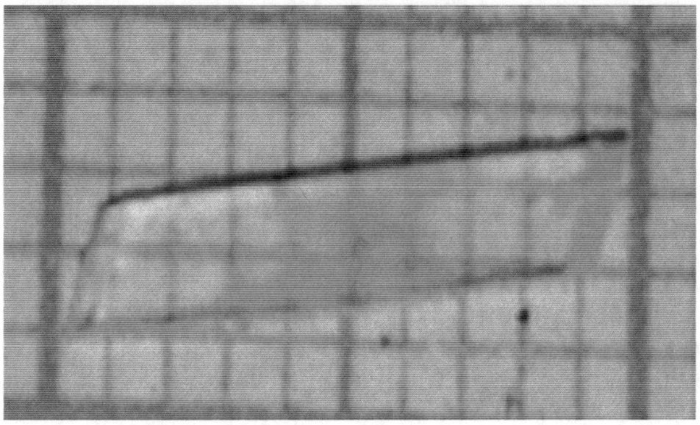

Figure 1. Photograph of as grown UPA single crystal

3. Results and Discussion

3.1. XRD analysis

The single crystal X-ray diffraction analysis for the grown UPA crystal has been carried out to identify the cell parameters using an ENRAF NONIUS CAD4 automatic X-ray diffractometer. The UPA crystal retained its triclinic structure with lattice parameters of a=7.402 Å, b=7.654 Å, c=10.052 Å and α=85.92°, β=81.94°, γ=65.20° and the volume of the crystal V=511.9 $Å^3$ with Centro symmetric group P1. The lattice parameters are in good agreement with the reported values [7].

3.2. UV-Vis Absorption spectroscopy

The UV-Vis spectral analysis of UPA was carried out using Varian Cary 5E spectrophotometer in the wavelength range 200 to 900 nm. The resulting spectrum is shown in figure. 2. The cut off wavelength as observed from the spectrum for UPA is 258 nm. It is due to the direct transition of an electron from a non - bonding 'n' orbital to an anti bonding 'π*' orbital (n →π*). It is important that for efficient NLO crystal, lower cut-off wavelength should be in the range of 200 to 400 nm. The steep decrease in the absorption of the compound around 276 nm may be assigned to the electronic excitation in coo^- group. As there is no appreciable change in the transmittance in the entire visible region upto 900 nm, the material can be useful as optical windows in the spectral instruments in the wavelength studied. The material is found to be transparent to all radiation in the wavelength range of 300-900 nm. The wide optical transmission window is an encouraging optical property seen in UPA and is of vital importance for NLO materials.

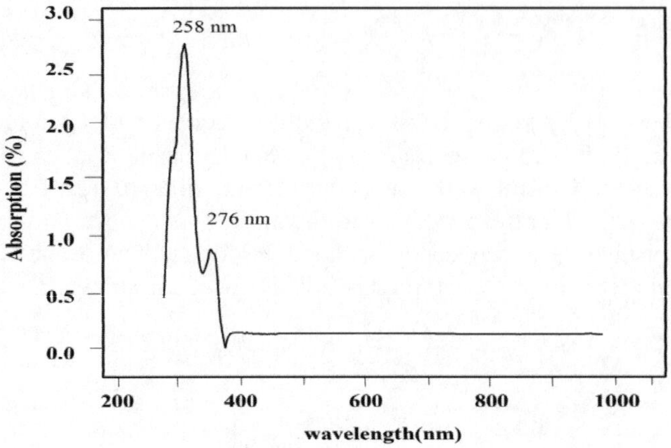

Figure 2. Absorption spectrum for UPA crystal

3.3. FT-IR analysis

The FTIR spectra of the grown UPA single crystal were observed on the powder sample is shown in Figure. 3. The spectrum of UPA crystal was recorded using the Brukers IFS66 spectrometer in the range 4000 – 400 cm^{-1} by KBr pellet to identify the functional group present in the crystal. In the high energy region the broad band in the range 3458 cm^{-1} is due to N-H stretching vibrations. The symmetric and asymmetric stretching vibrations of NH_2 are observed in range of 3233 cm^{-1} and 3337 cm^{-1} respectively. The absorption peaks at 3136 cm^{-1} and 1687 cm^{-1} are due to the symmetric and asymmetric stretching vibrations of C=O carboxylic group. The absorption peak at 1441 cm^{-1} is assigned for NH_2 bending which shows the presence of urea. The peaks observed at 1372 and 1014 cm^{-1} are corresponding to N-C-N stretching vibrations. The peak at 922 cm^{-1} is due to O-H out of plane bending vibration. All these observations clearly show the formation of the urea phthalic acid crystal.

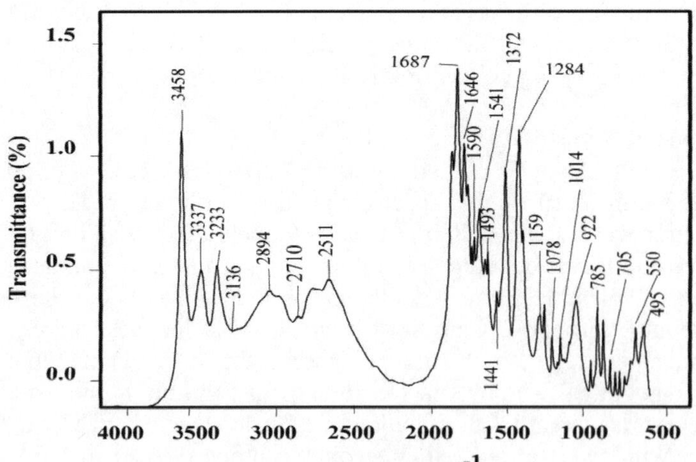

Figure 3. FT-IR vibrational spectrum of UPA crystal.

3.4. *Laser Damage Threshold Studies*

To measure laser-damage threshold for our UPA crystals, we used the laser with the pulse width 10 ns, the repetition rate 10 Hz and the fundamental wavelength 1064 nm. The laser beam with the diameter of 1 mm was focused on the sample. The sample was placed at the focus of a planoconvex lens (the focal length 30 cm). The energy of laser pulses was varied using attenuator that included a polarizer and a half-wave plate. The pulse energy of each laser shot was measured with a combination of a phototube and an oscilloscope. Finally, the surface damage threshold was calculated using the following relation for the power density (P_d), $P_d = E/\tau\pi r^2$. Here E is the input energy in mJ, τ the pulse width in ns and r the radius of the spot in mm. As a result, the laser damage threshold for the UPA crystals is found to be 5.3 GW/Cm^2, which is higher than that of Urea tartaric acid (4.6 GW/ Cm^2) [8].

3.5. Z-Scan analysis

The Z-scan method has gained rapid acceptance by the NLO community as a standard technique for determining the nonlinear changes in refractive index and nonlinear optical absorption. The nonlinear absorption and refractive index of UPA crystals were estimated using the single beam Z-scan method with laser beam intensity of 60mW and the wavelength of source used for the measurement was 632.8 nm. The study of nonlinear refraction by the Z-scan method depends on the position (Z) of the thin samples under the investigation along a focused Gaussian laser beam. The sample causes an additional focusing or defocusing, depending on whether nonlinear refraction is positive or negative. Such a scheme, referred to as an "Open aperture" Z-scan and it is suited for measuring nonlinear absorption in the UPA crystal. Results obtained from a typical closed aperture and open aperture of Z-scan study for the grown UPA crystals are represented by Figure. 4a and b. The nonlinear refractive index (n_2) of the crystal was calculated using the standard relations $\Delta Tp\text{-}v = 0.406(1\text{- }S)\ 0.25/\Delta\phi_o$. Where $S= 1\text{- }\exp(\text{-}ra^2/\omega a^2)$ is the aperture linear transmittance, $\Delta\phi_o$ is the on-axis phase shift. The on-axis phase shift is related to the third-order nonlinear refractive index by $|\Delta\phi_o| = kn_2L_{eff}\ I_o$ Where $k = 2\pi/\lambda$, $L_{eff} = [1\text{-}\exp(\text{-}\alpha L)]/\ \alpha$ is the effective thickness of the sample, α is the linear absorption coefficient, L the thickness of the sample, I_o is the on-axis irradiance at focus and n_2 is the third-order nonlinear refractive index. The calculated values of third order nonlinear refractive index (n_2) is 2.24 x 10^{-11} cm^2/w , nonlinear absorption coefficient (β) = 3.481 x 10^{-6} cm/w. the positive value of nonlinear refraction reveals the self-focusing nature and nonlinear absorption co-efficient (β) exhibits the two-photon absorption process. Real and imaginary value of third order susceptibility of UPA is 9.98 x 10^{-8} esu and 7.887 x 10^{-6} esu respectively. The third order susceptibility ($\chi^{(3)}$) is 7.88 x 10^{-6} esu and it is due to the π-electron cloud movement from the donor to acceptor which makes the molecule highly polarised.

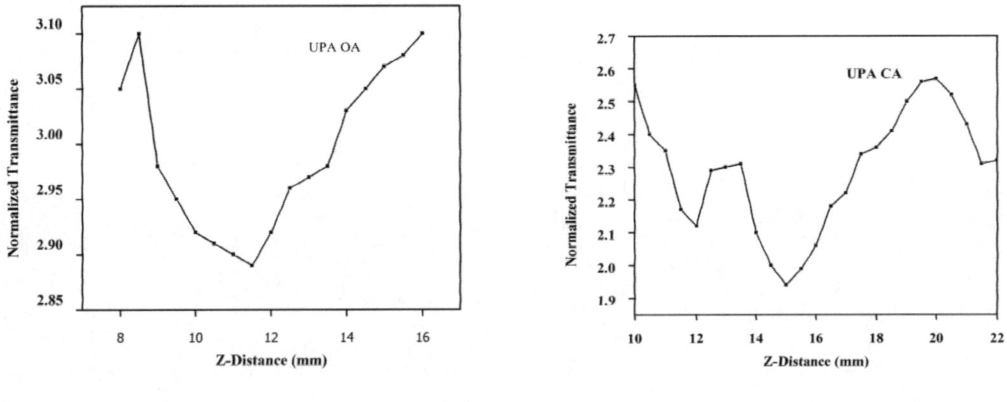

Figure 4. Z-scan spectrum of UPA (a) Open aperture (b) Closed aperture.

IOP Conf. Series: Materials Science and Engineering **360** (2018) 012032 doi:10.1088/1757-899X/360/1/012032

4. Conclusion

The novel single crystal of Urea Phthalic acid (UPA) was successfully grown from the methanol solvent by slow evaporation technique. The crystal structure has been confirmed and the lattice parameters have been calculated by single crystal X-ray diffraction studies. The absence of significant absorption in the entire visible region and lower cut-off wavelength indicates the suitability of UPA for optical applications. The vibration spectral analyses confirm the different functional groups in UPA. The laser damage threshold value for the grown crystal is 5.4 GW/cm^2. The third order nonlinear refractive index, absorption coefficient and optical susceptibility was calculated by the Z-scan techniques and it reveals that the UPA crystal possess self-focusing and two-photon absorption process.

Acknowledgments

The corresponding author gratefully acknowledges the support from the Department of Science and Technology (DST), Government of India for the research project (SB/EMEQ/248/2014).

References

[1] Fleck M, Petrosyan A M, 2010 Difficulties in the growth and characterization of nonlinear optical materials a case study of salts of amino acids *J. Cryst. Growth* .**312 (15),** 2284-2290

[2] Cyrac Peter A, Vimalan M, sagayaraj P, Madhavan J, 2010 Thermal, optical, mechanical, and electrical properties of a novel NLO active L-Phenylalanine phenylalaninium perchlorate single crystals *Physica B* **405,** 65-71

[3] Kato K, 1980 High efficiency, high power difference frequency generation at 2-4 μm in $LiNbO_3$ *J. Quant.Elect.* **16**, 810

[4] Chithambaram V, Jerome Das S, Arivudai Nambi R, Krishnan S, 2012 Synthesis growth and Characterization of nonlinear optical urea-ammonium chloride *Solid State Sci.* **14**, 216-218

[5] Selvarajan P, Glorium Arul Raj J, Perumal S, 2009 Characterization of Pure and urea doped γ-Glycine Single Crystals Grown by Solution Method *J. Cryst. Growth.* **311**, 15

[6] Shanthi A, Krishnan C, Selvarajan P, 2014, Growth and characterization of a single crystal of Urea adipic acid *Spectrochimica Acta A.* **122**, 521

[7] Videnova V, Adrabinska, 1996 The hydrogen bond as a design element in crystal engineering. Two- and three-dimensional building blocks of crystal architecture *J. Mol. Struct.* **374**, 199

[8] Vinothkumar P, Rajeswari K, Mohan Kumar R, Bhaskaran A, 2015 Structural, optical, thermal and mechanical properties of Urea tartaric acid single crystals, *Spectrochimica Acta Part A.* 1386

IOP Conf. Series: Materials Science and Engineering **360** (2018) 012033 doi:10.1088/1757-899X/360/1/012033

Ion irradiation induced effects on the optical properties of polymers: a possible correlation with the energy loss profile

V Kumar[1], P Chhokkar[2], B Rani[2], P K Goyal[3] and S Kumar[2]

[1]Department of Physics, University College, Kurukshetra, 136119, India

[2]Department of Physics, Kurukshetra University, Kurukshetra, 136119, India

[3]Department of Physics, YMCA University of Science and Technology, Faridabad, 121006, India

E-mail: vijphykaushik@gmail.com

Abstract. In the present study, the samples of CR-39 (monomer composition $C_{12}H_{18}O_7$) have been implanted with 100keV Ar^+ and Kr^+ ions to different fluencies ranging from $5x10^{14}$ to $1x10^{16}$ ions/cm^2. Further, these samples were subjected to UV-Visible, FTIR and Raman spectroscopic techniques to reveal the change in the optical behaviour of the samples in terms of induced structural changes in the polymer as a result of ion irradiation. The observed decrease in optical energy gap and increase in refractive index of the material under study after irradiation to Ar^+ and Kr^+ ions at the same energy have been tried to explain in terms of the energy loss profile along the trajectory of the respective ions. These studies are quite important towards the development of new materials with desired properties besides their fundamental importance.

1. Introduction

Polymeric materials have gained much importance in different fields of science and technology [1-8]. This is because of the good mechanical strength, high flexibility, good optical response etc. provided by these materials. It is well documented that the properties of the polymers can be modified to the desired extent through various means like chemical doping, irradiation, heat treatment, implantation etc. [8-15]. Among these, ion-implantation is quite promising towards the improvement of surface and near surface properties rather affecting the whole bulk. The ion implantation in polymers results in various effects such as ionization, displacing atoms, sputtering, carbonization, production of free radicals etc. [6-7, 14-22]. These structural modifications occurring in the ion implanted polymers may produce new electronic levels in the forbidden gap of the electronic band structure, thus, providing opportunities for the use of ion implantation in producing special materials for photonic applications. Especially, the ion implantation induced tailoring of optical energy gap and refractive index of transparent polymers is of great importance for the use of these materials in the fabrication of photonic device like optical waveguide, optical filters etc.[4-7, 21-22]. It is thus important to study optical properties of ion beam modified transparent polymers in a systematic manner for their utility in various optical devices.

IOP Conf. Series: Materials Science and Engineering **360** (2018) 012033 doi:10.1088/1757-899X/360/1/012033

Among the broad variety of available polymers, Poly(allyl diglycol carbonate), also known as CR-39 is considered an outstanding material. This polymer is highly transparent (over 90%) in the visible range and almost opaque in the ultraviolet region of electromagnetic spectrum. Further, it is very light weight polymer having refractive index comparable to that of crown glass. All these make it an excellent material for use in eye lenses, sunglasses, waveguides etc. [6-7, 23-26].

In the present work, the samples of CR-39 have been subjected to 100 keV Kr^+ and Ar^+ ion implantation. Some studies on the effect of ion implantation in CR-39 are available in the literature [6-7, 26-28], but from different point of view. We have carried out a systematic study on the effect of 100 keV Kr^+ and Ar^+ ion implantation on the optical energy gap and refractive index of CR-39 polymer and simultaneously tried to explain the observed changes in terms of the energy loss profile along the trajectory of the respective ions. Finally, FTIR and Raman spectroscopy have been used to correlate the observed results to the induced structural changes.

2. Experimental

The samples of area 15x15 mm^2 were cut from the bulk sheets of CR-39 (monomer composition $C_{12}H_{18}O_7$) with thickness 250 micron, procured from TASTRAK, Bristol, England. These samples were irradiated at room temperature and under high vacuum (~ 10^{-6} torr) with 100 keV Ar^+ and Kr+ ions to the fluences ranging from $1x10^{14}$ to $1.0x10^{16}$ ions/cm^{-2} utilizing the Low Energy Ion Beam Facility (LEIBF) available at Inter-University Accelerator Centre (IUAC), New Delhi, India.The beam current density was kept below 1 $\mu A/cm^2$ in order to avoid the thermal degradation of the samples. The energy loss profile along the trajectory of the implanted ions was found using SRIM code (version 2008.04) [29].

In order to study the optical properties, the virgin and the ion implanted samples were subjected to UV-Visible transmission and reflection studies by using Shimadzu Double beam Double Monochromator UV-VIS Spectrophotometer (UV-2550) in the wavelength range 190–900 nm with a resolution of 0.4 nm. The FTIR and Raman studies were carried out using the Shimadzu IR-Affinity-I, FTIR Spectrometer equipped with Pike, ATR-MAX-II accessory (Ge crystal) and Jobin-Yvon Raman spectrometer (He-Ne laser with λ=632.8nm) respectively.

3. Results and discussions

3.1. UV-visible studies

Before determining the optical energy gap and refractive index behaviour of the virgin- and ion-implanted samples, the recorded the UV–visible transmission (T) and reflection (R) spectra were taken into account the reflection losses at the surface using the method described in our earlier work [21].

3.1.1. UV-visible Absorption behaviour

The absorption behaviour of Ar^+ and Kr^+ implanted samples along with virgin CR-39 sample has been presented in figure 1. This figure clearly indicates the increase in absorption intensity and red shifting of the absorption edge with the increase in implantation dose. Further, the changes in absorption behaviour found to be more pronounced in case of Ar^+ implantation as compared to Kr^+ implantation in CR-39 polymer.

3.1.2. Determination of Optical Energy Gap

For the determination of optical energy gap (E_{OPT}) of virgin and ion implanted samples of CR-39 polymer, the absorption data corresponding to the fundamental absorption edge (Figure 1) as a function of wavelength (λ) have been considered [6, 12-14]. The data have been plotted in the form of $(\alpha h\nu)^{1/2}$ as a function of photon energy ($h\nu$) and presented in figure 2. From the intercepts of these plots on $h\nu$ axis, the values of E_{OPT} have been determined and are presented in table 1. This table depicts a clear-cut reduction in the value of E_{OPT} with the increase in ion fluence in case of both the ions. The value of E_{OPT} which was ~3.70 eV for virgin CR-39 has been reduced to ~2.54 eV and 1.90 eV at a dose of $1x10^{16}$ ions/cm^2 for Kr^+ and Ar^+ ions respectively.

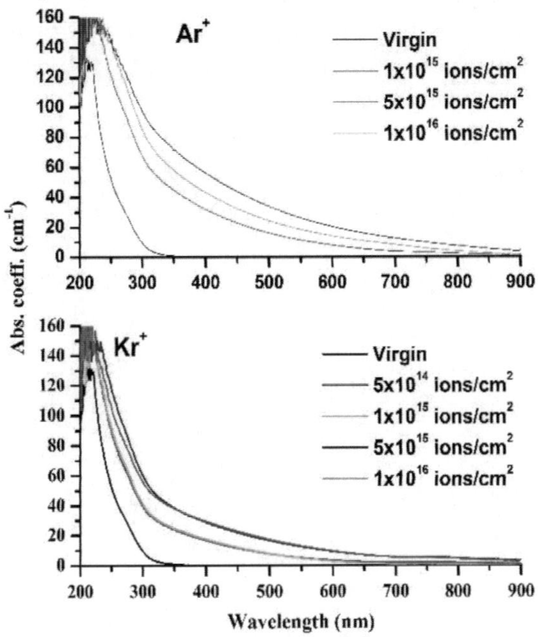

Figure 1. UV-Visible absorption spectra of Ar^+ and Kr^+ ion implanted CR-39 polymer.

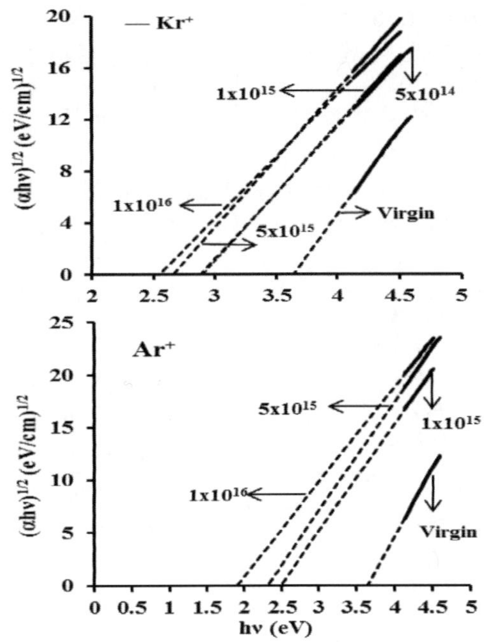

Figure 2. Plots for the determination of optical energy gap for virgin as well as Kr^+ and Ar^+ ions implanted CR-39 samples.

Table 1. The optical energy gap (E_{OPT}) values for virgin as well as Kr^+ and Ar^+ ion implanted CR-39 samples.

Ion fluence (ions/cm^2)	Optical energy gap, E_{OPT} (eV)	
	Kr^+ Implanted samples	Ar^+ Implanted samples
0	3.70 ± 0.01	3.70± 0.01
1x10^{15}	2.90 ± 0.02	2.48 ± 0.01
5x10^{15}	2.65 ± 0.01	2.32 ± 0.02
1x10^{16}	2.54 ± 0.01	1.90 ± 0.03

3.1.3. Determination of Refractive Index

The refractive index (n) values for virgin as well as Kr^+ and Ar^+ ion implanted CR-39 samples have been determined at different wavelengths using the relation [12, 21]

$$n=\left\{\left[\frac{4r}{(r-1)^2}-K^2\right]^{1/2}-\frac{r+1}{r-1}\right\}$$

where $K=\alpha\lambda/4\pi$, is the extinction coefficient with α as the absorption coefficient at wavelength λ and r is the reflection coefficient at the air-surface interface at the same wavelength.

The variation of the refractive index as a function of ion dose, at wavelengths 400 and 600 nm, has been presented in figure 3. From this figure, it is clear that the value of refractive index which remains almost ~1.56 for virgin sample has been found to be increased to ~2.00 and ~2.40 at the wavelengths 400 and 600 nm respectively, for the sample implanted at the dose of 1×10^{16} ions/cm^2 with Kr^+ ions. Further, for samples implanted at 1×10^{16} ions/cm^2 with Ar^+ ions, the value of refractive index has been found to increase to ~2.15 and 2.50 at the wavelengths 400 and 600 nm respectively.

The observed changes in optical parameters of Kr^+ and Ar^+ implanted CR-39 may be attributed to the formation of cross-linked and conjugated carbonaceous network in the near surface region of the implanted polymer [6-7, 14-17].

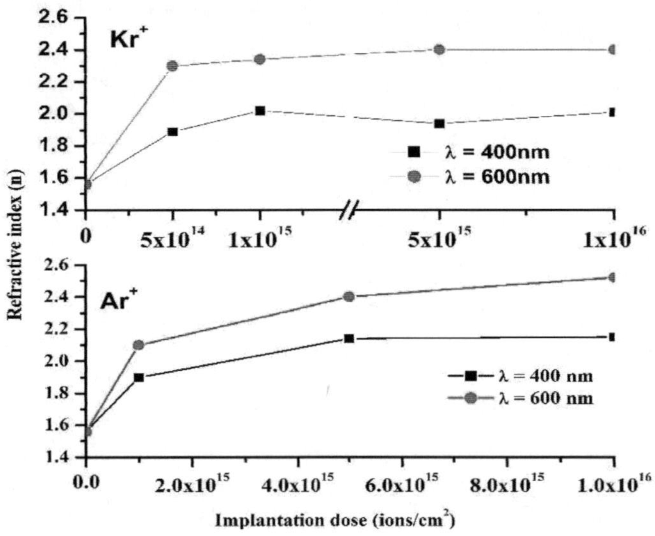

Figure 3. Variation of refractive index with ion dose, for virgin as well as Kr^+ and Ar^+ ion implanted CR-39 samples, at wavelengths 400 and 600 nm.

3.1.4. Relative comparison

A comparison between the observed values of various optical parameters (Table 1 & Figure 3) clearly indicates that the extent of changes produced in the case of Ar^+ implanted CR-39 is more pronounced as compared to Kr^+ implantation in CR-39. This may be explained on the basis of the spurs (a region of discrete energy loss by the ion along its trajectory) formed during the penetration of the implanted ions within the polymeric material [30].

For most of the polymers, the spur energy lies within an average value of 30 - 40 eV, which is sufficient to create the ion or radical pairs [7, 30]. The spur separation can be calculated after dividing the spur energy by the Linear Energy Transfer (LET). Therefore, a change in LET leads to a change in the spur separation. For low LET, the spurs are widely separated, while for increased LET the spurs are closely separated, resulting in the increased possibility of cross-linking. A separation of about 2 nm between the two adjoining spurs is sufficient for cross-linking. Further, the creation of closely spaced spurs is dominated in case of electronic energy loss in comparison to nuclear energy loss, by the penetrating ions in the implanted region [30].

Figure 4 presents the electronic LET plotted against penetration depth for 100 keV Kr^+ and Ar^+ ions in CR-39, as calculated through SRIM simulation for 1000 ions [29]. Figure 5 presents the profile of the separation between two adjoining spurs at different penetration depths corresponding to electronic LET (Figure 4). From this figure, it is clear that the spurs of separation ~0.03 – 2.0 nm and 0.05 – 2.0 nm are formed up to a penetration depth of ~140 and 200 nm in CR-39 after implantation to Kr^+ and Ar^+ ions respectively. Since, the spur separation up to ~2.0 nm favours cross-linking, therefore, it can be concluded that the effect of cross-linking is extended up to a larger thickness of the implanted sample in case of Ar^+ ions as compared to Kr^+ ion implantation.

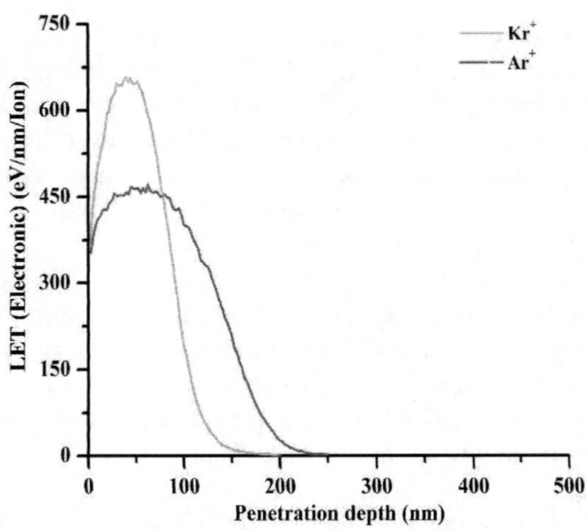

Figure 4. SRIM calculated electronic LET for 100 keV Kr^+ and Ar^+ ions in CR-39 polymer as a function of penetration depth.

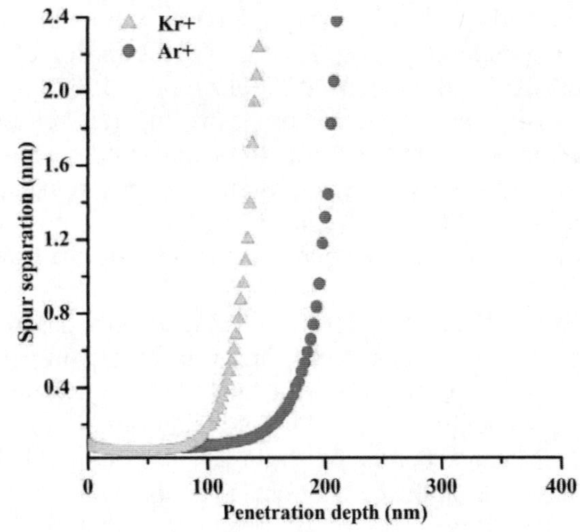

Figure 5. Separation between the Spurs formed as a result of 100 keV Kr^+ and Ar^+ ion implantation in CR-39 polymer, as a function of penetration depth.

3.2. Structural studies

3.2.1. FTIR

The FTIR-ATR spectra (recorded with Ge crystal) of virgin as well as Kr^+ and Ar^+ ion implanted sample of CR-39 have been presented in figure 6. In the spectrum for virgin sample, a number of peaks and bands at positions 783, 1007, 1255, 1394, 1526, 1742, 2308, 3615 and 3726 cm^{-1} have been observed. The identification of these bands has been made on the basis of data available in the literature and is in line with the monomer structure of CR-39 polymer [26, 31-32].

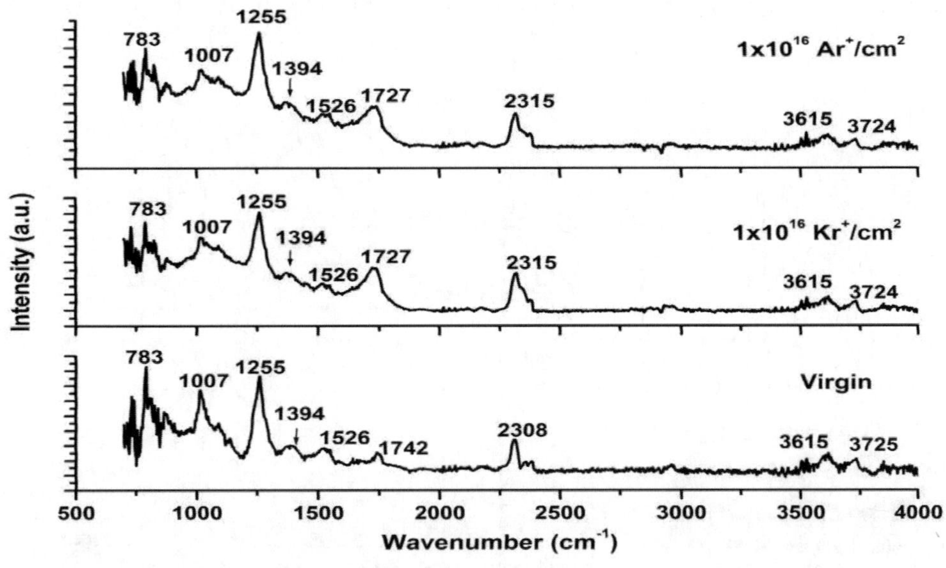

Figure 6. FTIR-ATR spectra of virgin- and ion-implanted CR-39 polymer.

Since, the present FTIR-ATR measurement was carried out by employing the Ge crystal that provides the structural information upto a depth of ~600 nm which is much larger than the thickness of the implanted layer (~140 nm for Kr^+ ions and ~200 nm for Ar^+ ions in CR-39 polymer), therefore the recorded spectrum is expected to reveal both the characteristic bands of virgin and the important features of implanted layer [7]. A careful investigation of the ATR spectrum (Figure 6) of the sample implanted with Kr^+ and Ar^+ ions at the dose of $1x10^{16}$ ions/cm^2, clearly indicates the reduction in the absorption intensity and broadening of all the important characteristic bands which were observed in case of virgin sample in the region ~700-4000 cm^{-1}. Such behaviour may be due to the formation of carbonized layer in the near surface region as an effect of implantation [7]. Further, the appearance of a new band in the region ~1600 – 1800 cm^{-1} and centered at position 1727 cm^{-1} can be attributed to the formation of a system of conjugated bonds in the thin surface layer of CR-39 after implantation [31-32].

3.2.2. Raman

The FTIR-ATR spectra discussed in previous section though give an indication of the formation of conjugated carbonaceous material in the implanted region but a more conclusive idea about the carbonization can be drawn on the basis of Raman analysis [7]. Figure 7 presents the Raman spectrum of virgin CR-39 polymer. Like IR spectrum, various characteristic peaks and bands have been observed in the Raman spectrum of virgin CR-39 polymer and again the identification of these peaks and bands has been done on the basis of data available in literature [32-34].

The Raman spectra for the sample implanted with Kr^+ and Ar^+ at the dose of $1x10^{16}$ ions/cm^2 (Figure 8), clearly shows the disappearance of characteristic features which were observed in virgin sample and the appearance of two new broad bands in the wavenumber region 1000-2000 cm^{-1}. For Kr^+ ion implantation, the Lorentzian fitting among these two bands depicts their centers at positions ~1357 cm^{-1} and 1580 cm^{-1}, which mark the presence of D and G bands of carbonaceous materials respectively [7, 14-15, 21]. For Ar^+ ion implantation, the Lorentzian fitting among these two bands depicts their centers at positions 1351 cm^{-1} and 1578 cm^{-1}, with their intensity ratio (I_D/I_G) ~1.2. The intensity ratio (I_D/I_G) in case of Kr^+ found to be ~1.3. Further, very small features of characteristic bands of virgin CR-39 found to appear along with the D and G bands in case of Kr^+ implantation. This may be due to the smaller penetration depth of Kr^+ ions in CR-39 as compared to the penetration depth of Ar^+ ions. Thus, Raman analysis supports the results that the effects produced by Ar^+ ion implantation are more prominent as compared to that produced by Kr^+ ion implantation. Similar trends have also been observed by our group in a study of N^+ and Ar^+ ions implantation in PMMA polymer [21], where the effects produced by N^+ ion implantation are found to more prominent as compared to that produced by Ar^+ ion implantation.

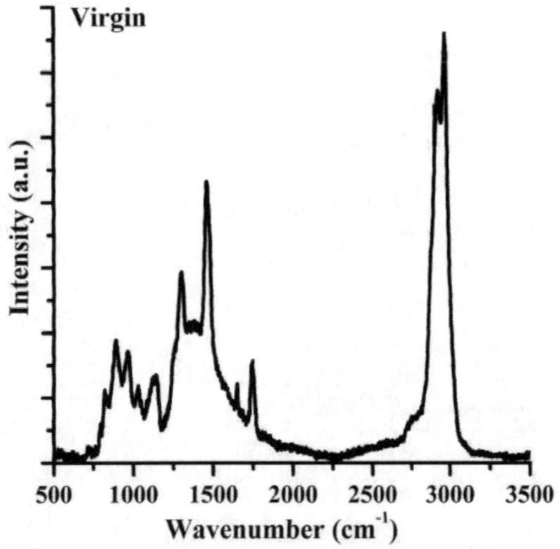

Figure 7. Raman spectra of virgin CR-39.

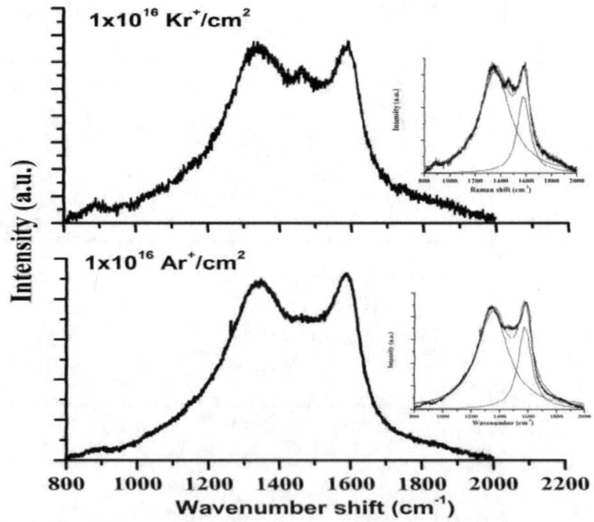

Figure 8. Raman spectra of Kr^+ and Ar^+ ion-implanted CR-39.

IOP Conf. Series: Materials Science and Engineering **360** (2018) 012033 doi:10.1088/1757-899X/360/1/012033

4. Conclusions

In the present work, the optical behaviour of CR-39 polymer exposed to 100keV Ar^+ and Kr^+ ions has been studied through UV-visible spectroscopic technique. From this study, it has been observed that the values of optical energy gap decreases while the values of refractive index increases with the increase in ion implantation dose. It has also been observed that the changes in optical parameters are more pronounced in case of Ar^+ implantation as compared to Kr^+ ions implantation in CR-39. This fact has been explained on the basis of energy loss profile along the trajectory of the respective ions and it has been inferred that the changes induced also depends upon the implanted ion and its depth of penetration. Further, all the observed changes have been well correlated with the structural changes through FTIR and Raman spectroscopy. Such studies are quite important towards the development of new materials with desired properties besides their fundamental importance.

5. References

[1] Cambre J N and Sumerlin B S 2011 *Polymer* **52** 4631
[2] Ma H, Jen A KY and L R Dalton 2002 *Advanced Materials* **14** 1339
[3] Long Y Z, Li M M, Gu C, Wan M, Duvail J L, Liu Z and Fan Z 2011 *Prog. Polym Sci.* **36** 1415
[4] Adhikari B and Majumdar S 2004 *Prog. Polym. Sci,* **29**, 699
[5] Billmeyer F W 2005 *Textbook of Polymer Science* (New York: Wiley-Interscience)
[6] Fink D 2004 *Fundamentals of ion-irradiated polymers* (Berlin: Springer-Verlag)
[7] Kondyurin A and Bilek M 2015 *Ion Beam Treatment of polymers* (UK: Elsevier)
[8] Radwan R M 2009 *J Phys D: Appl. Phys.* **42** 015419
[9] Mireault N and Ross G. G. 2008 *Appl. Surf. Sci.* **254** 6908
[10] Tomar A K, Mahendia S and Kumar S 2011 *Interna. J. Nanosci.* **10** 1107
[11] Valenza A, Visco A M, Torrisi L and Campo N 2004 *Polymer* **45**, 1707
[12] Migahed M D and Zidan H M 2006 *Current App. Phys.* **6** 91
[13] Shaban A M 1995 *Mater. Lett.* **25** 179
[14] Sharma T, Mahendia S, Aggarwal S, Kumar S and Kanjilal D 2011 *Opt. Mater.* **33** 1741
[15] Goyal P K, Kumar V, Gupta R, Mahendia S, Anita and Kumar S 2011 *Vacuum* **86**, 1087
[16] Shekhawat N, Aggarwal S, Sharma A and Nair K G M 2015 *J. Mater Sci* **50** 300
[17] Abdul-Kader A M and Radwan R M 2015 *J. Adhes. Sci. Technol.* **29**, 1607
[18] Shanthini MG, Catherine A M, Sakthivel N, Sarath C V, Elayaraja K, Lakshmi B S, Asokan K, Kanjilal D and Kalkura S N 215 *Applied Surface Science* **329**, 116
[19] Bumai Y A, Dolgikh N I, Kharchenko A A, Valeev V F, Nuzhdin V I, Khaibullin R I, Nagim F A, Lukashevich M G, and Odzhaev V B 2014 *J. Appl. Spect.* **81** 188
[20] Geworski A, Lazareva I, Gieb K, Koval Y and Müller P 2014 *J.Appl. Phys.* **116** 1
[21] Gupta R, Kumar V, Goyal P K and Kumar S 2012 *Applied surf. Sc.* **263** 334
[22] Jain I P and Aggarwal G 2011 *Surf. Sci. report* **66** 77
[23] Darraud C, Bennamane B, Gagnadre C, Decossas J L and Vareille J C 1994 *Polymer* **35** 2447
[24] Omichi H et al 1997 *Nucl. Instrum. Meth. Phys. Res. B* **131**, 350.
[25] Rafique M S, Bashir S, Ajami A and Husinsky W 2010 *Appl. Phys. A* **100** 1183
[26] Abdul-Kader A M, El-Badry B A, Zaki M F, Hegazy T M, Hashem H M 2010 *Philos. Mag.* **90**, 2543
[27] Shekhawat N, Aggarwal S, Sharma A, Sharma S K, Deshpande S K and Nair K G M 2011 *J Appl Phys.* **109**, 1
[28] Yap E, McCulloch D G, McKenzie D R, Swain M V, Wielunski L S and Clissold R A 1998 *J. Appl. Phys.* **83**, 3404
[29] Ziegler J F, Biersack J P and Ziegler M D 2008 SRIM - the stopping and range of ions in matter
[30] Lee E H 1999 *Nucl. Instrum. Meth. B* **151** 29
[31] Saad A F, Atwa S T, Yokota R and Fujii M 2005 *Radiat. Meas.* **40**, 780

[32] Yamauchi T, Takada S, Ichijo H and Oda K 2001 *Radiat. Meas.* **34** 69
[33] Sharma T, Aggarwal S, Kumar S, Mittal V K, Kalsi P C and Manchanda V K 2007 *J. Matter. Sci.* **42**, 1127
[34] Campbell D, Pethric R A and White J R 2000 *Polymer Characterization: Physical techniques* (United Kingdom: Stanley Thornes)

IOP Conf. Series: Materials Science and Engineering **360** (2018) 012034 doi:10.1088/1757-899X/360/1/012034

Retrapping probability of charge in traps: The parameter and its implication in luminescence

Ngangbam Chandrasekhar [1*], Lisham Paris Chanu[1], R.K. Gartia [1]
[1] *Department of Physics, Manipur University, Imphal-79500l, Manipur, INDIA*
*Corresponding author. Tel:(+91)8729870220; E-mail:ngchandrasekhar@manipuruniv.ac.in

ABSTRACT

The phenomenon of luminescence manifested by phosphors has many practical applications right from illumination, display to scintillators. However of late it has a totally new application in form of persistent luminescence wherein a phosphor emits enough photons to be visible for the whole night. These materials are known as Glow-in-the-Dark pigment. Invented in Japan, today it is a material that is used to illuminate the roads in some parts of the world. The heart of the phenomenon is the retrapping probability (A_n) of charge in traps, one of the intrinsic parameters that can be estimated by measuring the long period afterglow of phosphors. The parameter A_n and its sister parameter recombination probability (A_m) and their ratio have tremendous potentialities in designing Glow-in-the-Dark phosphors. This communication will evaluate the ratio A_m/A_n of some commercial phosphors to show the possible limit of optimization in development of these phosphors.

Keywords: Power law decay, Persistent Luminescence.

1 Introduction

Delayed luminescence (phosphorescence) is observed in many solids where there is a time lag between the absorptions of the exciting energy and the resulting luminescence. Becquerel [1] was one of the earliest pioneers who studied phosphorescence critically. The Becquerel type of decay is

$$I = I_0 / (t + t_0)^P \qquad (1)$$

Where I(t)= Intensity at time t

I_0=Intensity at t=0

t= time

t_0= Initial time

Where *P* depends upon the ratio of recombination to retrapping probability ($\gamma = A_m/A_n$).

In such situation a plot of log $I \sim$ log t gives a straight line when $t \gg t_0$. This has been serious implications in terms of providing physical significance to persistent luminescent materials where the materials exhibit enough light for visibility for unusually long time ($\approx$ 20 hours or more). The power *P* has been evaluated by many workers where it has been reported to be in the range (in organic solid) $0.5 \leq P \leq 2.0$ [2].

Adirovitch [3] provided a meaningful co-relation between *P* and $\gamma = A_m/A_n$, a graphical plot of which is presented in Fig.1 for ready reference and use in development of persistent luminescent materials.

In this paper we plot the log I ~ log t of some commercial glow- in-the- Dark Phosphors and discuss critically the use of Fig.1 in characterizing as well as developing new novel persistent luminescent materials. We would like to point out that while *P* seems to be a number A_m and A_m by themselves are intrinsic parameters that relate to the property of recombination centre and the trap that have critical role in luminescence including thermoluminescence, a technique non-exponential time dependence of afterglow in persistent luminescent materials are known for its many practical applications [4–7].

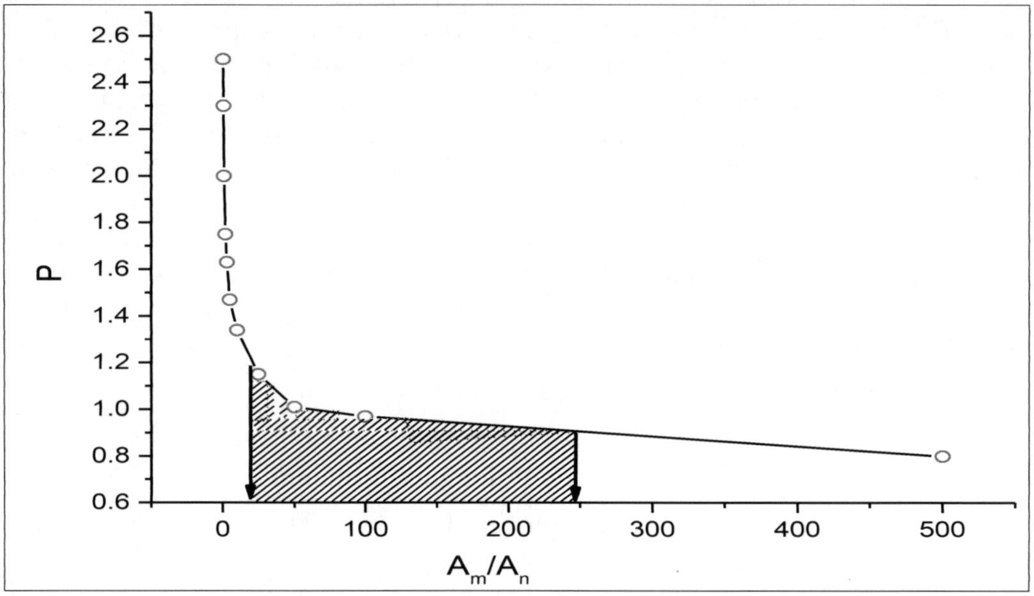

Fig.1
Plot between A_m/A_n and *P*. Shaded portion shows the range of A_m/A_n corresponding *P*.

2 Experimental Details

The persistent luminescent material used in the experiment is of commercial grade obtained from Jash Marketing, Hyderabad, India [8]. It is identified as $Ca_xSr_{1-x}S$:Eu. This is based on XRD and EDAX analysis. Its emission occurs at 647 nm. The excitation source used is an ordinary white LED. The duration of excitation is controlled with the help of EXPEYE Junior [9], an instrument developed by IUAC, New Delhi. The functioning of lighting LED is controlled by python program written by our group. The afterglow decays measurements are carried out with the help of Nucleonix TL Reader Type-TL 1009 (Nucleonix Systems Private Limited, Hyderabad).

3 Results and Discussions

It is quite common in many papers on persistent luminescence where the decay data are interpreted as sum of multiple exponential components [10–15]. We believe these workers need to check the possibility of fitting their data to Becquerel type of decay and evaluate the index P.

IOP Conf. Series: Materials Science and Engineering **360** (2018) 012034 doi:10.1088/1757-899X/360/1/012034

Incidentally, practically in all such papers one finds thermoluminescence (TL) data that also can be used to check the possibility of testing the presence or absence of the role of A_m/A_n.

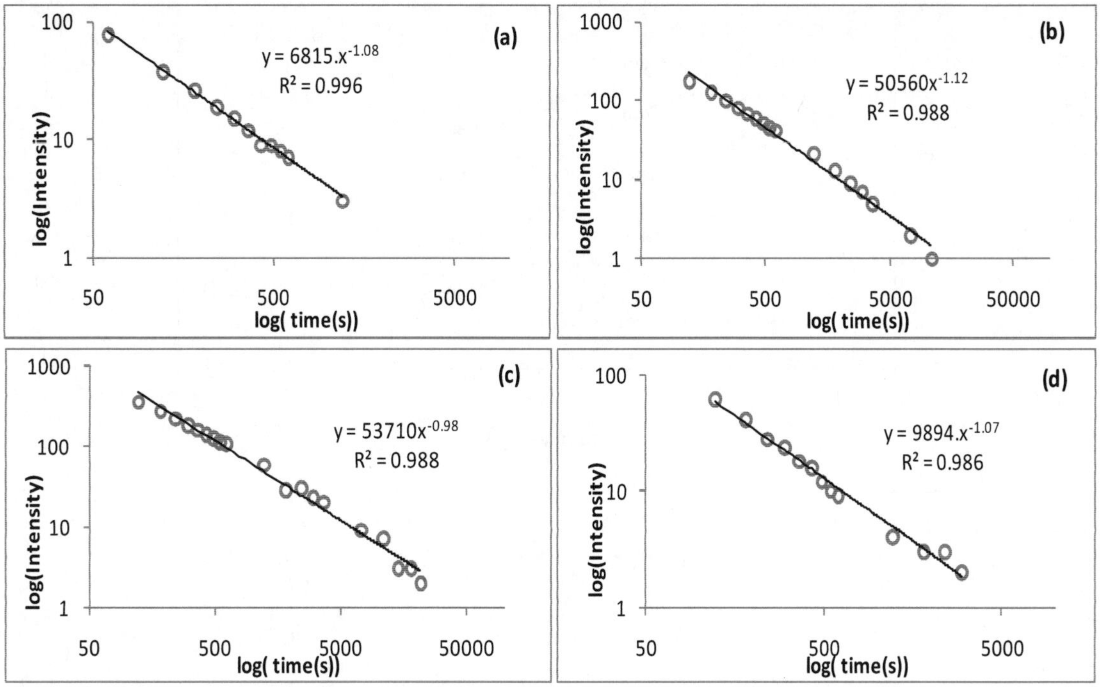

Fig.2 Luminescence decay of a commercial available persistent luminescent materials: (a) Zinc red glow in the Dark Powder; (b) Regular green glow paint; (c) Pure blue glow paint; (d) Zinc orange glow paint.

Table 1 Slopes of log *I*~log *t* plot corresponding Fig. 2.

Glow in dark Materials	Slope	R^2
(a) Zinc red glow in the Dark Powder	1.08	0.996
(b) Regular green glow paint	1.12	0.988
(c) Pure blue glow paint	0.98	0.988
(d) Zinc orange glow paint	1.07	0.986
(e) High Purple Glow paint	0.97	0.989

Some log *I*~ log *t* plots for data available for a commercial glow- in- the- Dark phosphors [16] are shown in Fig.2. In these the value of *P* is $0.98 \leq P \leq 1.12$ (Table 1).

Similar plots for $Ca_xSr_{1-x}S$:Eu as measured in our laboratory are shown in Fig. 3. The parameters are presented in Table 2.

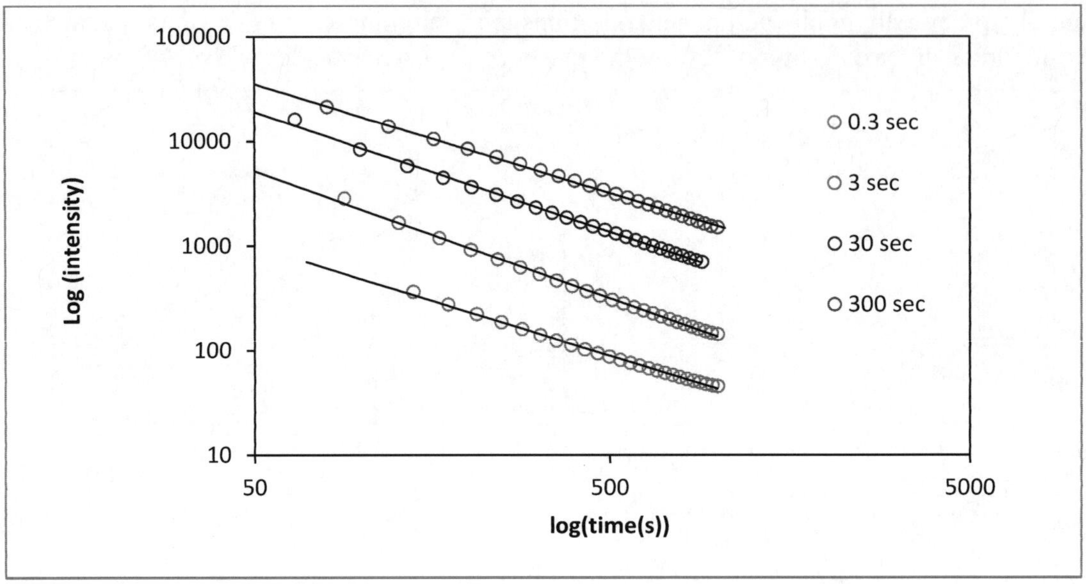

Fig. 3: Luminescence decay of red emitting (λ=647nm) commercial persistent phosphor $Ca_xSr_{1-x}S$:Eu, at different doses of excitation; (a) 300s, (b) 30s, (c) 3s and (d) 0.3s

Table 2 Slope of log I~ log t plot of $Ca_xSr_{1-x}S$:Eu .

Time of Irradiation(s)	Temperature(C) RT	slope	R^2
0.3	32.6	1.04	0.998
3	32.6	1.20	0.999
30	29.6	1.13	0.999
300	29.8	1.05	0.999

4 Conclusions

Estimation of the ratio of A_m/A_n done by us for commercial glow-in- the- Dark phosphors lie in the range $10 \leq A_m/A_n \leq 250$. Since A_m/A_n can be as low as 1 physically this investigation shows that there is scope of improving by appropriate combination of recombination and trapping centers.

Acknowledgements

Ngangbam Chandrasekhar is thankful to the authority of Manipur University for providing the UGC Non- Net fellowship.

References

[1] J. Becquerel, L. Lumiere, La lumière, ses causes et ses effets, F. Didot frères, 1867.
[2] A.K. Jonscher, A. de Polignac, The time dependence of luminescence in solids, J. Phys. C Solid State Phys. 17 (1984) 6493. doi:10.1088/0022-3719/17/35/018.
[3] E. Adirovitch, , J Phys Rad. 17 (1956) 705–707.
[4] R.K. Gartia, Defect Diffus. Forum. 357 (2014) 193–215.
[5] R.K. Gartia, N. Chandrasekhar, Thermoluminescence of Persistent Luminescent Materials, Defect Diffus. Forum. 357 (2014) 171–191.
[6] R.K. Gartia, L. Rey, T. Tejkumar Singh, T. Basanta Singh, , Nucl. Instrum. Methods Phys. Res. Sect. B Beam Interact. Mater. At. 274 (2012) 129–134.
[7] R.K. Gartia, Nucl. Instrum. Methods Phys. Res. Sect. B Beam Interact. Mater. At. 267 (2009) 2903–2907.
[8] JASH MARKETING SERVICES, (n.d.). http://www.glowpaint.in/.
[9] http:// www.expeyes.in
[10] R. Sakai, T. Katsumata, S. Komuro, T. Morikawa, J. Lumin. 85 (1999) 149–154.
[11] W. Pan, G. Ning, X. Zhang, J. Wang, Y. Lin, J. Ye, , J. Lumin. 128 (2008) 1975–1979.
[12] T. Katsumata, K. Sasajima, T. Nabae, S. Komuro, T. Morikawa, J. Am. Ceram. Soc. 81 (1998) 413–416.
[13] O. Arellano-Tánori, R. Meléndrez, M. Pedroza-Montero, B. Castañeda, V. Chernov, W.M. Yen, M. Barboza-Flores, J. Lumin. 128 (2008) 173–184.
[14] Y. Jin, Y. Hu, L. Chen, X. Wang, , J. Am. Ceram. Soc. 97 (2014) 2573–2579.
[15] Z. Mu, Y. Hu, Y. Wang, H. Wu, C. Fu, F. Kang, , J. Lumin. 131 (2011) 676–681.
[16] Glow Inc., http://glow.glowinc.com/

IOP Conf. Series: Materials Science and Engineering **360** (2018) 012035 doi:10.1088/1757-899X/360/1/012035

Transport of MMA through SBR for fabrication of IPN based on Cross-linked SBR and PMMA

Jose James[a,b,c*] George .V. Thomas [b], Sabu Thomas[b, c]

[a]St.Joseph's College, Moolamattom, Idukki , 685591,Kerala, India ,
[b]School of Chemical Sciences, Mahatma Gandhi University, Kottayam, 686560, Kerala, India,
[c] International and Interuniversity Centre for Nanosience and Nanotechonology, Kottayam, 686560, Kerala, India
**Corresponding author Email: josejameskadathala@gmail.com, cell +919447149547*

Abstract

The diffusion and transport of inhibitor-free Methyl methacrylate (MMA) monomers through cross-linked Styrene Butadiene Rubber (SBR) have been studied at various temperatures and is compared with its transport through Natural Rubber (NR). SBR has been vulcanized by conventional, efficient and peroxide vulcanization techniques. The dependence of diffusion rate on the crosslinking system, cross -link density, size of the penetrant, thickness of the matrix and type of polymer matrices have been studied for the systems. The influence of temperature on the sorption has been evaluated using the diffusion data. The peroxide system showed lowest uptake and the effective system showed highest uptake. Mechanism of transport of MMA through SBR system was also reported in this paper and it follows Fickian mode of transport.

Keywords:

Cross- linked SBR, NR, diffusion, sorption, Fickian mode of transport

1. Introduction

Transport of liquids through polymeric membranes is a key controlling factor in many of their applications. The transport of organic liquids through polymers is of technological importance in a variety of applications such as pervaporation, reverse osmosis and food packaging.[1] The transport properties of various engineering polymers like polyurethanes, cross-linked polystyrene, natural rubber and various synthetic rubbers have been extensively studied using organic solvents with various polarities and interacting groups and also using aqueous salt solutions of various acids by many researchers. The transport of small molecules through a polymer membrane happens due to random molecular motion of individual molecules.[2]

The diffusion process is a kinetic parameter and it depends on the free volume within the matrix, the nature of the polymer, crosslink density, the nature of fillers, the penetrant size, temperature and the degree of reinforcement.[3] In literature there are numerous studies on the transport properties of solvents through various types of polymers including elastomers[4,5], Interpenetrating Polymeric Networks (IPNs) and blends .[6,7] Unnikrishnan et al. [8] analyzed the role of the nature of crosslinks on the sorption and diffusion of aromatic

IOP Conf. Series: Materials Science and Engineering **360** (2018) 012035 doi:10.1088/1757-899X/360/1/012035

hydrocarbons through cross-linked natural rubber. Diffusion of aromatic solvents through SBR vulcanized by different vulcanizing agents and carbon black filled SBR were reported by George et al. [9]

The swelling and sorption behaviors of SBR in various liquids are of great interest[10], as SBR is a well-known polymer having extensive use in everyday life. SBR is a very important general purpose low cost synthetic rubber having wide range of applications in tyre and non-tyre fields. This is a random copolymer whose properties are determined by the amount of styrene in the copolymer. In the present study, SBR having a styrene content of 25% has been selected for the present study, which shows the best elastomeric properties. The swelling of SBR samples in the presence of monomers like styrene[11], methyl methacrylate, acrylonitrile, butyl acrylate, etc., is of great research interest, since it constitutes the preliminary step in the preparation of interpenetrating polymer networks (IPN).[12, 13]

IPN is a special type of polymer blend. An IPN is an intimate combination of two polymers, where at least one of which is synthesized or cross linked in the immediate presence of the other. Unlike usual polymer blends, by forming networks with chemical bonds or permanent entanglements, IPNs exhibit high interfacial strength between the two polymers and have fewer phase separation problems. [14]

To prepare a rubber-based IPN, the swelling characteristics of the rubber in monomers are to be studied in detail. A clear and accurate knowledge about the swelling rate of the polymer in the monomers helps in controlling the composition, morphology and properties of the resulting IPN. However, no detailed investigation has been made so far on the kinetics of diffusion and sorption of methacrylate monomers through SBR cross linked by different curing systems such as conventional (CV), efficient (EV) and peroxide (DCP) and systems.

In this paper the swelling of SBR in methyl methacrylate monomer has been studied in the temperature range of 30-50°C. This will be the primary step for the development of IPN based on SBR and poly methyl methacrylate (PMMA) having various morphological features and property profiles. The SBR was vulcanized by different techniques, viz. conventional, efficient and peroxide vulcanization systems. The influence of different vulcanizing systems on diffusion properties is analyzed. The effect of crosslink density on sorption has been investigated. Finally the temperature dependence of diffusion is determined.

2. Experimental Details

2.1 Materials

Styrene butadiene rubber (Synaprene 1502) with 25% styrene content was used for this study. The natural rubber (NR) used was of ISNR-5 grade and was supplied by Rubber Research Institute of India (RRII). The MMA monomer (from Sigma Aldrich)is made free of inhibitor by washing with 0.1% NaOH and dried using anhydrous calcium chloride before using for diffusion experiments. All other compounding ingredients were of analytical grade. Important properties of SBR and MMA are summarized in the **Table: 1.**

Table1 Description of Materials used

(A) Description of material used

Material	Molecular weight (g/ mol)	Density (g/cm^3)	Solubility Parameter $(Cal\ cm^{-3})^{1/2}$	Source
SBR	$M_n = 4 X10^5$	0.930	8.1	ISRL
MMA	100.12	0.936	9.3	Sigma Aldrich

(B) Description of solvents used

Material	Molecular weight (g/mol)	Molar volume	Density (g/cm^3)	Boiling point (°C)	Solubility Parameter $(Cal\ cm^{-3})^{1/2}$
Benzene	78.80	88.70	0.870	96.47	9.2
Toluene	92.14	105.9	0.867	110.6	8.9
Xylene	106.17	123.45	0.860	138.4	8.8

2.2 Preparation of rubber sheet

Mixing of SBR and other ingredients was done in a laboratory scale two roll mixing mill. The formulations of different mixes are given in **Table 2**. Cure properties of the compounded samples were determined using a Moving Die Rheometer at a temperature of 150 °C. Samples were then moulded by using an electrically heated hydraulic press under a pressure of 150 bar at150 °C. Only peroxide vulcanized 1phr NR (Natural rubber) sample (100gm.NR+1gm.DCP) is prepared as for the comparison purpose.

Table 2: Formulation of the SBR mixes

Ingredients	Efficient System	Conventional System	Peroxide System (1/2/4 phr)
SBR	100 gm.	100 gm.	100 gm.
Zinc oxide	5 gm.	5 gm.	-
Stearic acid	1.5gm.	1.5 gm.	-
CBS	1 gm.	1.5 gm.	-
TMTD	1.5gm.	-	-
MBTS	1.5gm.	-	-
MBS	-	1.5 gm.	-
Sulphur	0.6gm.	2 gm.	-
DCP	-	-	1/2/4 gm.

*phr -parts per hundred rubber

2.3 Diffusion experiments.

Circular samples were cut from the vulcanized sheets using a sharp-edged steel die. Thickness of the sample was measured at several points using a screw gauge. These circular samples were weighed and then dipped in 30 ml of methacrylate taken in diffusion bottles. Samples were taken out of the bottles at constant time intervals and the adhering solvent was

IOP Conf. Series: Materials Science and Engineering **360** (2018) 012035 doi:10.1088/1757-899X/360/1/012035

blotted off from the surface gently and then immediately weighed on a highly sensitive electronic balance. Weighed samples were immediately immersed into the test bottle. The weighing was continued until equilibrium swelling was obtained. In order to avoid the error due to the evaporation of the solvent weighing must be completed within 30 seconds. Transport properties were studied at 3 different temperatures, namely 30°C, 40°C, 50°C using MMA.

Diffusion of solvent through different rubber samples was determined. Mole% uptake is determined using the formula [15],

$$Q_t\,(mol\%) = \frac{Mass\ of\ solvent\ sorbed\ by\ the\ polymer}{Molecular\ mass\ of\ the\ solvent\,/\,initial\ mass\ of\ polymer}\ 100 \qquad (1)$$

Diffusion curves are obtained by plotting the mole percentage uptake (%Qt) for the solvent against the square root of time. At equilibrium swelling Q_t becomes Q_α.

3. Results and Discussion

3.1 Effect of nature of crosslinks

The diffusion curves of cross-linked SBR using different vulcanizing systems under a common torque are plotted in Figure1.

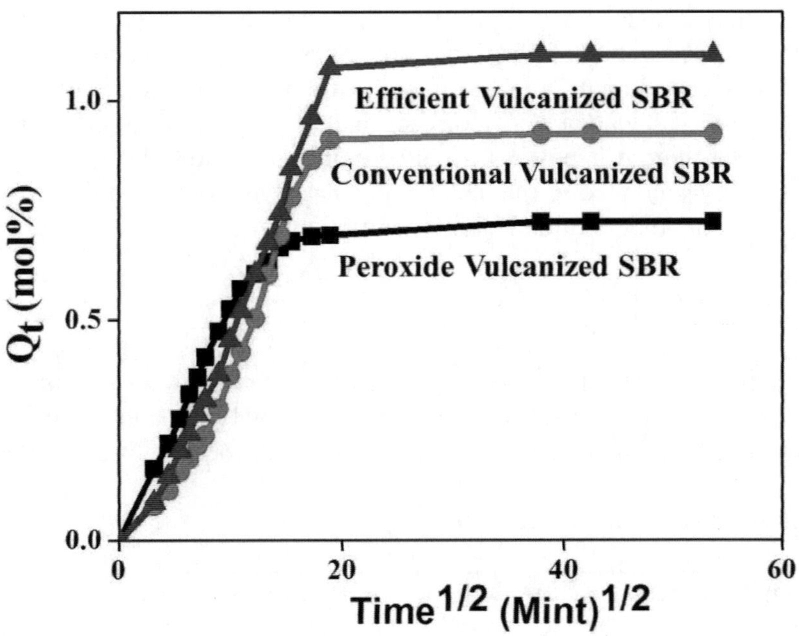

Figure1. Sorption curve showing the mole per cent uptake of optimum cured Efficient Vulcanized, Conventional vulcanized and Peroxide vulcanized SBR samples at 30°C.

From the figure, it is clear that Qt (mol%) values decreases in the order efficient vulcanized system > conventional vulcanized system > Dicumyl peroxide vulcanized systems (EV> CV> DCP). The difference in the solvent uptake value for SBR vulcanized with different vulcanizing systems may be due to the formation of different types of bonds between rubber chains during vulcanization. The Figure 2 depicts the general structural features of vulcanized networks. [16]

IOP Conf. Series: Materials Science and Engineering **360** (2018) 012035 doi:10.1088/1757-899X/360/1/012035

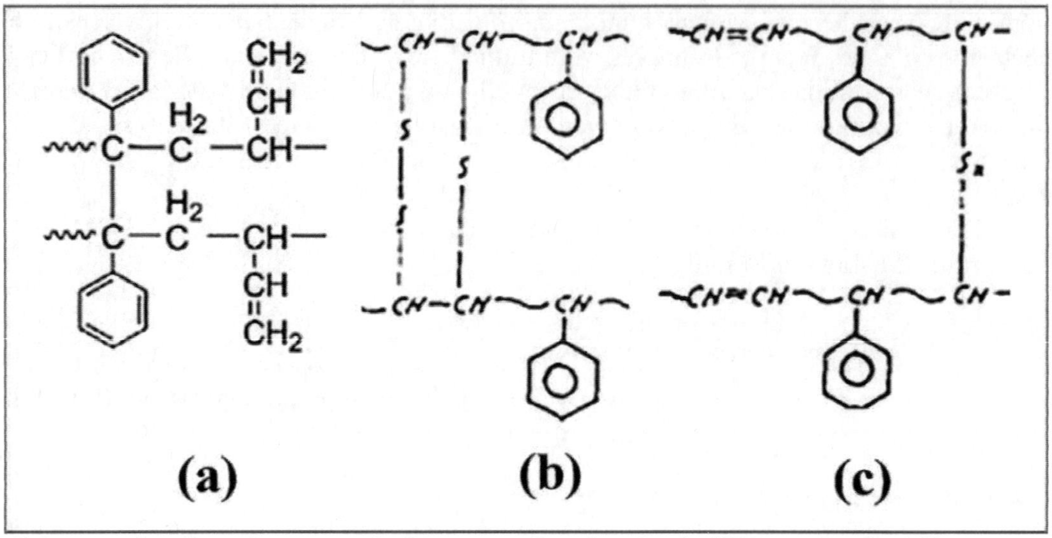

Figure2.Scheme showing the general features of vulcanized SBR network (all systems included).

The peroxide vulcanized sample contains only C-C linkages (a) with lowest solvent uptake, since they contain most rigid cross links which are resistant to permanent mobility. Contrary to our expectations, effectively vulcanized samples showed the highest amount of solvent uptake even though it contains mostly monosulphidic and disulphidic linkages (b),whereas SBR vulcanized with conventional system showed lower solvent uptake than EV systems even though it possess highly flexible polysulphidic linkages (c). However the observed anomaly could be explained on the basis of volume fraction of solvent swollen rubber (Vr) and can be determined using the relation.[17]

$$V_r = 1/\{1 + \left({\rho_p}/{\rho_s}\right)\left({M_\alpha}/{M_0}\right) - \left({\rho_p}/{\rho_s}\right)\} \qquad (2)$$

M_∞ is the mass of solvent taken up at equilibrium and M_0 is the initial mass of polymeric sample. ρ_p and ρ_s represent the densities of polymer and solvent respectively. The **table3** shows the value of volume fraction of swollen rubber and in the order DCP> CV> EV.

Table3. Volume fraction of swollen SBR

Solvent	EV system	CV system	DCP system
MMA	0.4769	0.5213	0.5813

Volume fraction of swollen rubber is proportional to crosslink density. It can be concluded that as the volume fraction value decreases the crosslink density also decreases. As a result solvent uptake behavior follows the order DCP< CV< EV.

3.2. Effect of crosslink density

The nature of interlocking is the one of the major controlling factors in the physico–mechanical and thermal properties of polymers. The assessment of the interaction of polymer with solvent is very much essential to understand the extent of interlocking. The cross-linking content in polymer can act as one of the controlling factor in its transport behavior. Here we consider the sorption behavior of 1phr, 2phr and 4phr SBR (peroxide vulcanized) samples in MMA. Result can be depicted in the Fig .3. As the crosslink density increases, penetration of solvent is hindered by excess crosslinks in the system. This conclusion is crystal clear from the Figure3.

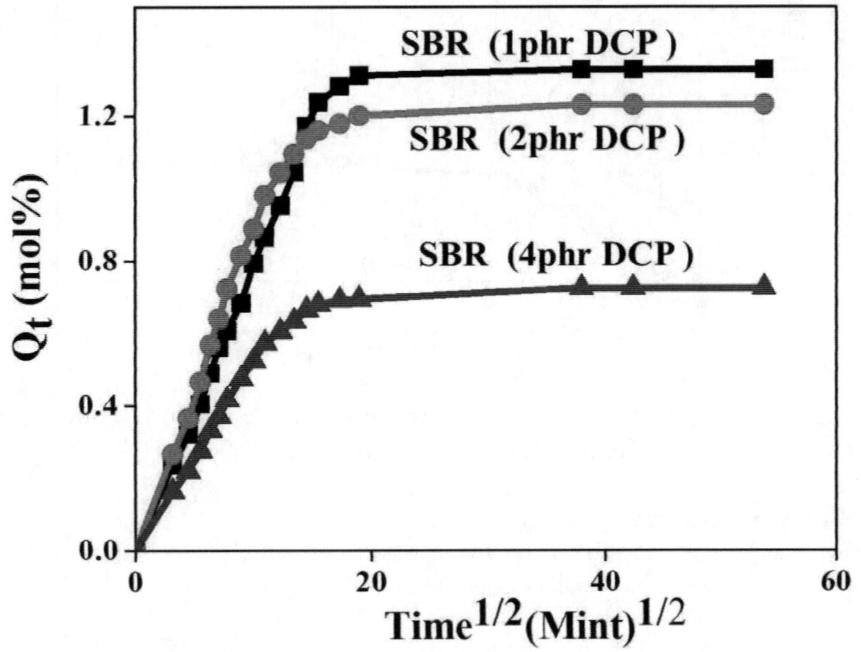

Figure3. Sorption curve showing the mole per cent uptake of optimum cured 1phr, 2phr and 4phr peroxide Vulcanized SBR samples at 30°C.

3.3 Effect of Penetrant Size

There is a systematic trend of sorption behavior of liquids of different molecular sizes.[18] With an increasing size of the solvent molecules (Methyl methacrylate [MMA] to Glycidyl methacrylate [GMA]), there is a decrease in the value of Qt mol%, when these liquids flowing through the 1phr peroxide SBR sample. As the penetrant size increases, corresponding Q_∞-value decreases. By increasing the size of penetrant, the resistance offered by the polymer to the transport of the solvents through the matrix is enhanced. Therefore, the uptake was low for penetrants of higher size. In other words, the high molecular weight solvent shows the lowest uptake and the low molecular weight shows the highest uptake. The increase in uptake with decrease in penetrant size may be due to the lower activation energy required for the activation of the diffusion process. It is clear from the Fig:4.

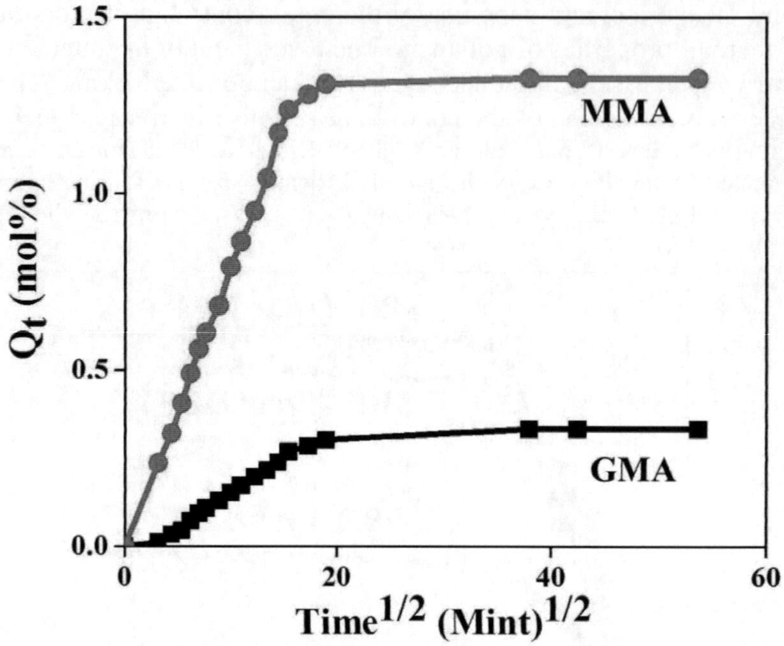

Figure4. Sorption curve showing the mole per cent uptake of MMA and GMA through optimum cured SBR (1phr peroxide Vulcanized) sample at 30°C.

3.4 .Effect of polymer matrix

Here we compare the sorption behavior of MMA through 1phr NR and 1phr SBR (both are peroxide vulcanized) samples. The results are depicted in the Figure 5.

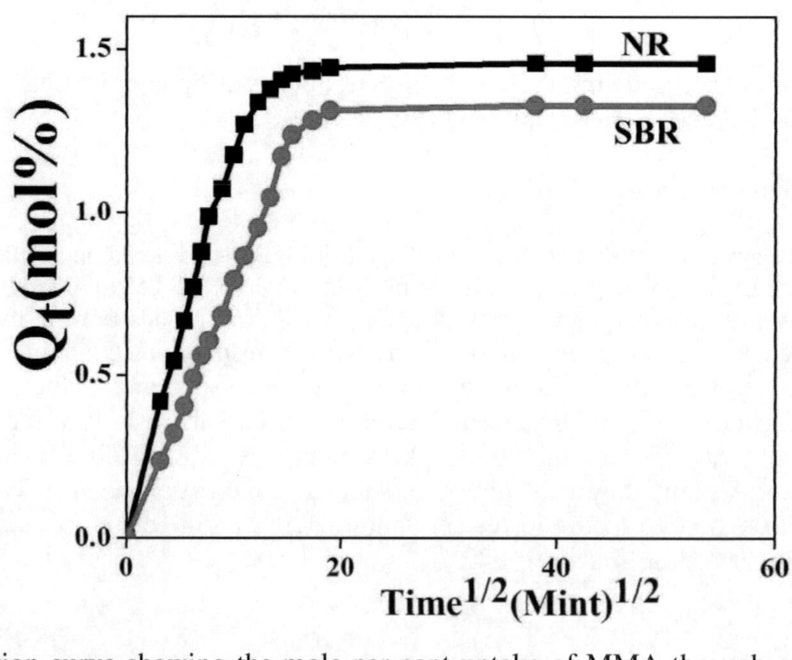

Figure5. Sorption curve showing the mole per cent uptake of MMA through optimum cured 1phr NR and 1phr SBR (peroxide vulcanized) samples at 30°C.

NR sample possess high Q_t value than SBR sample because bulkier benzene side rings in SBR hinders the solvent penetration. Shape and orientation of functional groups in polymer matrix have deciding role in its transport behavior.

3.5. Effect of thickness of polymer Matrix

Here we compare the sorption behavior of 1phr peroxide SBR samples with variable thickness (0.8mm,2mm and 6mm) and are depicted in Fig:6 .As the thickness of the sample decreases rate of sorption increases .This tendency follows Fick's laws of diffusion.[19]

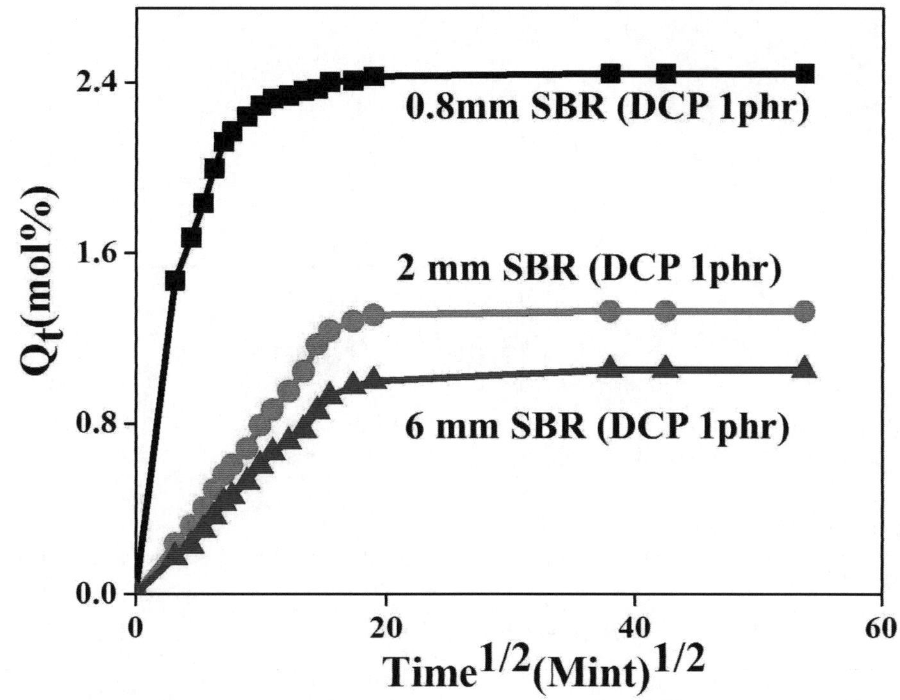

Fig:6. Sorption curve showing the mole per cent uptake of MMA through optimum cured 1phr SBR (peroxide vulcanized) samples with variable thickness at 30°C.

3.6. Effect of temperature

Diffusion of MMA through SBR sample was conducted at 30, 40 and 50°C in order to study the effect of temperature on the diffusion phenomenon in SBR.As the temperature increases the rate of diffusion increases. The Q_∞ values show increased solvent uptake with increasing temperature. This is mainly due to two reasons. As the temperature enhances, there is an increase in free volume of the SBR system leading to an increase in segmental motion of the polymer matrix. The gain in kinetic energy by the solvent molecules is resulted from the increased number of collisions. It is also found that the slope of the linear portion increases with temperature showing that the transport properties are temperature dependent. The effect of temperature on diffusion is clearly proved in Fig:7.

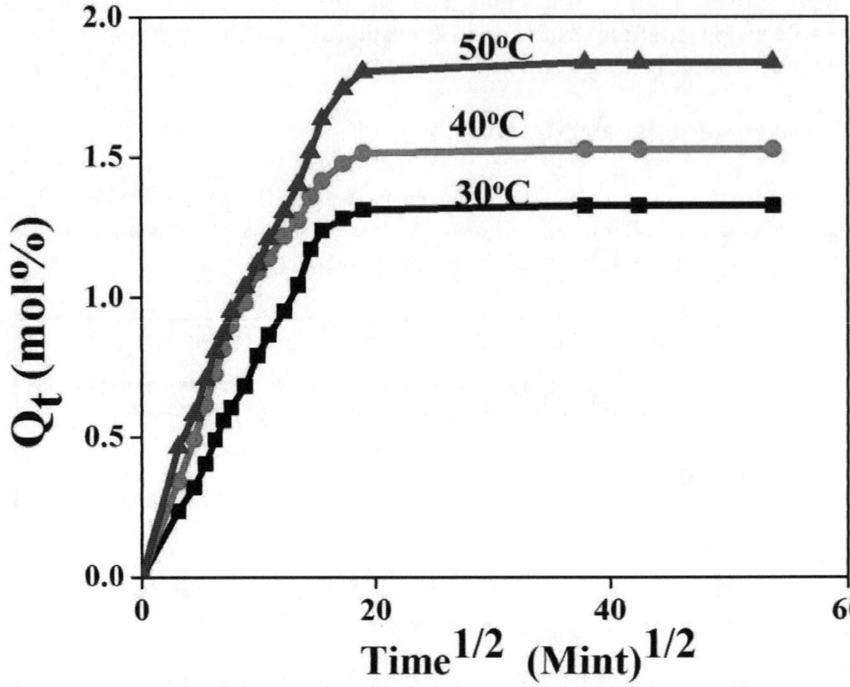

Fig:7. Sorption curve showing the mole per cent uptake of MMA through optimum cured 1phr SBR (peroxide vulcanized) samples at different temperatures.

3.7 Transport Mechanism

The transport properties of polymeric membranes can be followed by the empirical equation [20],

$$log\left({}^{Q_t}/_{Q_\infty}\right) = logk + nlogt \qquad (3)$$

Where Q_t and Q_∞ are the mol % increase in uptake at time t and at equilibrium respectively. k is a constant. The value of n and k are determined by linear regression analysis. In all cases the correlation coefficients were found to be 0.99

The slope of the plot of $log\left({}^{Q_t}/_{Q_\infty}\right)$ versus $logt$ gives the value of **n** which indicates the mechanism of transport and its Y intercept gives the value of k which depends upon the structural significance of polymer as well as its interaction with solvent.

The value of "n" gives an insight into the mode of transport and its relations with relaxation of polymer chains.[21] This information can be summarized as follows. When value of n = 0.5, the rate of diffusion of solvent molecules (r_1) is less than that of the rate of relaxation of polymer chains (r_2) and is called Fickian mode of transport. When r_1 is greater than r_2, it adopts non-Fickian mode of transport with n =1. If n varies from 0.5 to1, the system adopts anomalous mode of transport where $r_1$is approximately equal to r_2.

'n' and 'k' values at different temperature for SBR(peroxide vulcanized) samples are summarized in the table 4.

Table4. Values of n and k for MMA through SBR at different temperatures

System	Solvent	Temperature	n values	k values
SBR (peroxide-1phr)	MMA	30°C	0.5126	-1.292

SBR (peroxide-1phr)	MMA	40°C	0.4981	-0.2661
SBR(peroxide-1phr)	MMA	50°C	0.5081	-0.3821

In all the conditions, transport of MMA through peroxide vulcanized SBR system follows Fickian mode of transport. The Fick's laws are usually followed by rubbery polymers. Here SBR does not show deviation from accepted law. There is no regular trend in 'K' values.

4. Conclusions

In conclusion, this study on the diffusion of MMA monomer through SBR shows following interesting results. Among different vulcanization systems of SBR, the sorption rate is in the order efficient vulcanized system > conventional vulcanized system > Dicumyl peroxide vulcanized systems (EV> CV> DCP). As the crosslink density of a particular vulcanized system increases, the rate of diffusion decreases. The diffusion rate of solvent through SBR system decreases with enhancement in the size of penetran . For the same vulcanized system, NR poses high sorption rate than SBR system. As the thickness of the SBR sample decreases, its sorption rate increases. The diffusion of MMA through SBR is a temperature dependent process and as the temperature increases, rate of diffusion increases. Diffusion of MMA through peroxide vulcanized SBR samples follow Fickian mode of transport. From the above results, we can conclude that sorption of MMA in SBR can take place successfully and are governed by existing laws of diffusion. This study could be act as a pretest for the synthesis of IPN based on SBR and PMMA.

Acknowledgements

The authors would like to thank University Grant Commission for granting Faculty Development Scheme in 12th UGC plan .They also wish to express their sincere thanks to Apollo Tyres, Challakudy, India for providing SBR.

References

1. J.Abraham, H.J. Maria, S. C. George, N. Kalarikkal and S. Thomas,J. *Phys. Chem.Chem. Phys.* 17, 11217 (2015).
2. A.P.Mathew, S. Packirisamy, M. G. Kumaran, and S. Thomas, J.*Polymer* 36, 4935 (1995).
3. S. C. George and S. Thomas, J.Prog. Polym. Sci., 26, 985 (2001).
4. R. Stephen, S. Varghese, K. Joseph, Z. Oommen and S. Thomas, J. Membr. Sci., 282 , 162 (2006).
5. A. E. Mathai, R. P. Singh and S. Thomas, J. Membr. Sci. 202, 35 (2002).
6. J. Wang and W. Wu, Eur. Polym. J., 41, 1143 (2005).
7. T. Johnson and S. Thomas, J. Mater. Sci., 34, 3221 (1999).
8. G. Unnikrishnan and S. Thomas, J.Polymer,35, 5504 (1994).
9. S. C. George, S. Thomas and K. N. Ninan, J.Polymer, 37, 5839 (1996).
10. S. C. George and S. Thomas, J. Macromol. Sci., Part B: Phys., 39(2) , 175 (2000).
11. B.Das. and T.Gangopadhyay, Eur. Polym. J., 8, 867 (1992).

12. B.Das. and T.Gangopadhyay, Eur. Polym. J., 29, 57 (1993).

13. A.A.Donatelli, L.H. Sperling. and D.A.Thomas, J. Macromolecules,9,676 (1976).

14. J. James, G.V. Thomas, H. Akhina and S. Thomas, in Micro-and Nano-Structured Interpenetrating Polymer Networks: From Design to Applications, Edited S.Thomas,D.Grande,U.Cvelbar,K.V.S.N.Raju,R.Narayan,S.P.thomasand H. Akhina, John Wiley & sons Publishers, Hoboken, New Jersey(2016),pp.1-27.

15. J. Wang, W. Wu and Z. Lin, J. Appl. Polym. Sci., 109, 3018 (2008).
16. R. M. Barrer, J. A. Barrie and M. G. Rogers, J. Polym. Sci., PartA: Gen. Pap., 8, 2565 (1963).
17. T. M. Aminabhavi and H. T. S. Phayde, J.Polymer, 36, 1023 (1995).
18. P. V. Thomas, S. Thomas, S. Bandyopadhyay, A. Wurm and C. Schick, J. Compos. Sci. Technol., 68, 3220 (2008).
19. S. B. Harogoppad and T. M. Aminabhavi, J. Macromolecules, 24, 2598 (1991).
20. L. Mathew, K. U. Joseph and R. Joseph, J.Bull. Mater. Sci.,29, 91 (2006).
21. J. James, G.V.Thomas, K.P.Pramoda and S.Thomas, *Polymer*,116, 76(2017).

Advanced Ablative composites for Aerospace applications

Mohan Kumar.L, [*a,b] Usha. KM.,[b] Chakravarthy.P.[a]

[a]*Department of Aerospace Engineering, Indian Institute of Space Sciences and Technology (IIST), ISRO, Valiamala, Thiruvananthapuram- 695 58,Kerala, India*

[b]*Vikram Sarabhai Space Centre (VSSC), ISRO, Thiruvananthapuram- 695013. Kerala, India.*

Corresponding author Email:* ***mohankumarvssc@gmail.com*

Abstract- The external surface of a re-entry space vehicle experiences high heat flux and temperatures of the order of 2500°C during re-entry into the atmosphere and the internal contour of a rocket motor nozzle has to encounter temperatures above 2000°C during its operation. Special Thermal Protection Systems (TPS) are employed to protect the rocket hardware and re-entry capsule from such extreme thermal environments. Ablative composites are cost-effective and well-proven candidates for such applications. However, processing, characterisation and qualification of ablatives are quite complex and involves long cycle times.

Ablation is a heat and mass transfer process in which a large amount of heat is dissipated in a very short period of time with sacrificial loss of material. Ablative composites are polymeric composites where the reinforcement is generally a fibre with high melting point like Carbon or Silica and the matrix is a high char yielding resin like phenolics. In this paper, the process of ablation, the physical and chemical changes and the process of synthesis of ablatives are clearly elaborated. The recent advances in design, development, characterisation and testing of carbon and silica phenolics are explained with examples.

The process sequence and customised equipments used for the synthesis and characterisation are described. The raw materials, critical process parameters, quality control mechanisms, non-destructive testing methods etc are covered. The defects which were observed during the realisation and the corrective measures taken to solve them are explained.

Different laminate and subscale level tests have been done to qualify the process and verify the design margins. Finally a full scale rocket engine test is carried out for qualification before induction into an operational launch vehicle.

Keywords: *ablatives, composites, carbon, silica, phenolics, hydroclave, curing, tape-winding, moulding, rocket nozzles.*

IOP Conf. Series: Materials Science and Engineering **360** (2018) 012036 doi:10.1088/1757-899X/360/1/012036

1. Introduction

Ablative composites are an elite class of composites made of high melting point fibers and polymeric resins with very high char yield. Commonly used reinforcements include carbon, graphite, silica, glass, asbestos etc and resins include phenolics and furfuryl alcohol. An ideal ablative composite should possess high heat of ablation, high enthalpy of phase change, sufficient strength, high specific heat, high thermal shock resistance etc. At the same time, it should have low thermal conductivity, medium density, low molecular weight for the volatiles evolved during pyrolysis and as low an erosion rate as possible.

Ablation is an orderly heat and mass transfer process in which a large quantity of heat energy is dissipated in a very short period of time by sacrificial loss of material at a rate which can be estimated. It is a very complex process including many physical and chemical transformations including phase changes like melting, vapourisation, sublimation and pyrolysis. Many endothermic reactions occur during the process.

The ablative composites encounter extremely hostile conditions during the motor operation. The high temperature, pressure, velocity of hot exhaust gases, heat flux, particle impingement of the solid propellant particles etc contribute to harsh environment inside the nozzle. High temperature resistant metals or alloys alone cannot survive the operating conditions in a solid rocket motor. While the metallic structure provides the necessary structural capability, high performance composite materials are required for thermal protection. Ablative composites generally use Carbon or Silica as the reinforcement and phenolic resin as the matrix resin.

When the ablative is subjected to a very high heat flux as the hot exhaust gases pass through the nozzle, the surface temperature increases rapidly. Due to the low thermal conductivity of the material, the temperature builds up on the surface rather than the heat getting conducted to the backup structure. As the temperature near the surface reaches the pyrolysis temperature of the resin in the composite, decomposition of resin takes place. This leads to the formation of char on the surface of the ablative. As time progresses, the extent of char increases or the char front advances into the thickness of the material. Then the surface material starts eroding; erosion of the material can be due to the thermal degradation of the material and/or due to the mechanical erosion caused by metallic particle impingement due to the Aluminium particles of the solid propellant flowing along with the hot exhaust gases. As the layer on the surface erodes, the next layer gets exposed

IOP Conf. Series: Materials Science and Engineering **360** (2018) 012036 doi:10.1088/1757-899X/360/1/012036

and the process continues.

During pyrolysis, the volatile gases evolving at the reaction zone finds its way through the charred zone taking away significant amount of heat. Similarly, melting of the resin as well as the fibers also consume some heat. With all these processes, a large quantity of heat is expended with sacrificial loss of material, thereby, protecting the metallic substrate from thermal degradation.

As the ablative composite is exposed to heat, it acts as a heat sink and its temperature starts rising. At the pyrolysis temperature, the decomposition of the composite starts, evolving gases, leaving the rigid porous carbonaceous substance called passive char. During the further advancement of the heating zone, the pyrolysis gases developed, absorb the heat from the passive char layer and dissociate itself into solid substrates which fill the porous char and builds pressure in the porous cavities. As the pressure exceeds a threshold limit, char spallation takes place. All the above reactions are of endothermic in nature.

2. Experimental Details

Ablative composites were synthesised from phenolic resin as matrix and carbon/silica fibres. Phenolic resin is synthesised from phenol and formaldehyde. The typical properties of the resin matrix are given below:

TABLE I : PHENOLIC RESIN PROPERTIES

Sl. No	**Important properties**	
	Parameters	**Typical values**
1	Specific gravity at 30°C	1.2
2	Solid content, %	63.5
3	Viscosity at 30°C, cps	250-300
4	Degree of advancement, ml	13.5
5	Free phenol content, %	5
6	Free formalin content, %	2

Carbon/Silica fibers are used as the reinforcement. Rayon based Carbon fiber is preferred for ablative applications as it provides lower thermal conductivity and higher interlaminar Shear strength because of crenulated cross-section. Carbon fabric is made by weaving Carbon fibres with carbon content greater than 94%. This is made by successive carbonization of rayon. Polyacrylonitrile (PAN) and Pitch based carbon fabrics can also be used. 8 Harness Satin weave was chosen as the weave pattern considering the drapability for ease of processing. Typical parameter values of the carbon fabric used is listed below:

TABLE II: IMPORTANT PARAMETERS OF CARBON FIBERS

Sl. No	**Important properties**	
	Parameters	**Typical values**
1	Carbon content, %	94-96
2	Sodium content, ppm	600
3	Ash content, %	0.20
4	pH	8
5	Breaking strength, kg/inch width	80-100
6	Areal density, gm/sq.m	250-300
7	Thickness, mm	0.3-0.4
8	Specific gravity	1.75
9	Thread count, ends/inch	45-55

The carbon/silica fabric is impregnated with phenolic resin in an impregnation plant which can be either vertical or horizontal. The block schematic of a typical impregnation plant is given in figure 1.

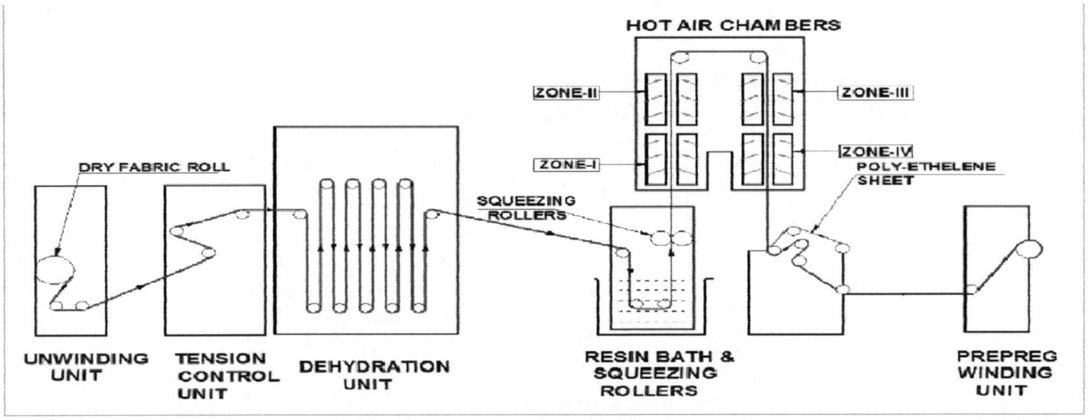

FIGURE 1: **BLOCK DIAGRAM OF AN IMPREGNATION PLANT**

Initially, the fabric is dehydrated above 100°C to drive out the moisture. Then it is passed through a resin tank to absorb the resin. The tension in the fabric is to be closely controlled to ensure uniform wetting. The squeeze rollers can be adjusted to control the resin content in the prepreg. It is then passed through heating zones to advance the resin. The resultant material is called Carbon/Silica phenolic prepreg. The important parameters of the prepreg are,

TABLE III: IMPORTANT PARAMETERS OF CARBON PHENOLIC PREPREG

Sl. No	**Important properties**	
	Parameters	**Achieved values**
1	Volatile content, (%)	4-10
2	Dry Resin content, (%)	35-40
3	Wet Resin content, (%)	40-50
4	Degree of advancement, Chang's index, (ml of water)	22-28

The prepreg is cut into the form of plies using a template and are either stacked together or wound on a metallic mandrel. The prepreg layup is compacted either giving vacuum or in a hydraulic press to get good as-wrapped density.

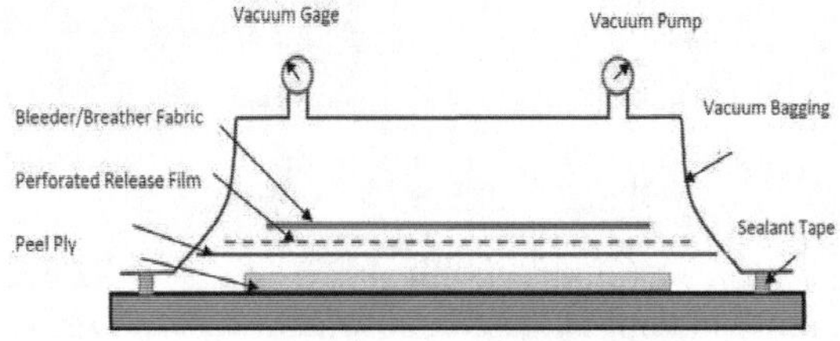

FIG. 2: SCHEMATIC OF LAYUP OF PREPREG

Fig.2 shows a schematic of the layup with bleeder/breather film, perforated release film and the vacuum bag in position. Perforated release film serves dual purpose; perforations allow the excess resin and volatiles to freely flow out of the liner and the release film prevents the unwanted adhesion of the liner to the mould. The bleeder material shall absorb the excess resin squeezed out of the liner. The vacuum bag is made from special grade polymer films capable of withstanding the curing temperature and pressure. For curing in Hydroclaves, impermeable high temperature resistant rubber bags with about 600% elongation are used. Alternatively, the prepreg is wound onto a metallic mandrel under pressure from a Teflon-coated metallic roller. The prepreg is preheated using a hot air blower and cooled using liquid nitrogen to ensure adequate flow of resin. This operation is done on a customised center-lathe modified with provisions for prepreg heating, cooling, very accurate pitch control, tension control devices, tape alignment systems and features to tilt the bed to accommodate winding mandrels of different angles and sizes.

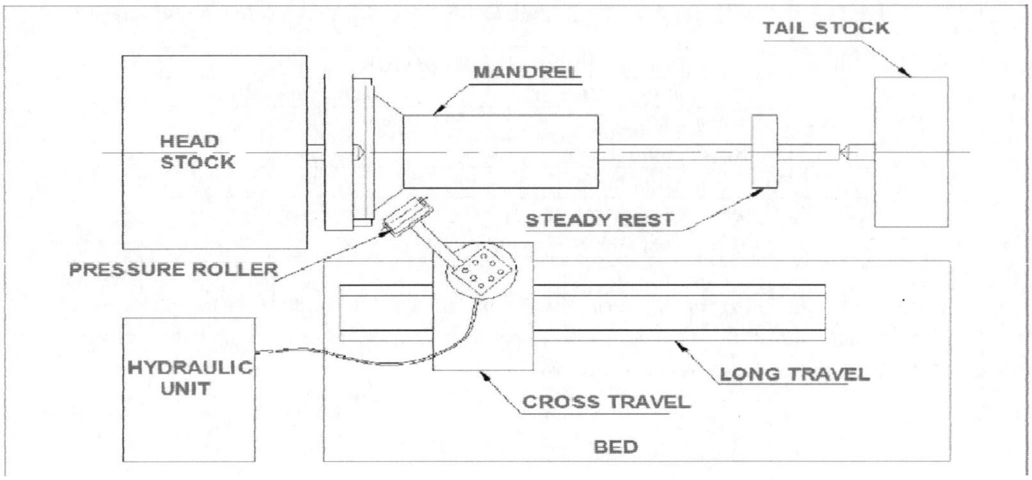

FIGURE 3: TAPE WINDING MACHINE

Curing or polymerisation is the most critical operation in the processing sequence of an ablative composite. It is done in sophisticated equipment like a hydroclave or autoclave under high temperature and pressure. After layup, the ablative liner is put in a vacuum bag and cured at 150 °C under pressure of 40 to 60 bar. Cross-linking of molecules of the polymer or polymerisation is achieved by curing under high temperature and pressure. In an oven, only heating is possible, whereas in an Autoclave/Hydroclave pressurisation is also done. The pressurising medium is air in an Autoclave which can go upto 10 bar pressure, while in a Hydroclave, pressure upto 70 bar can be applied since water is the pressurisation medium.

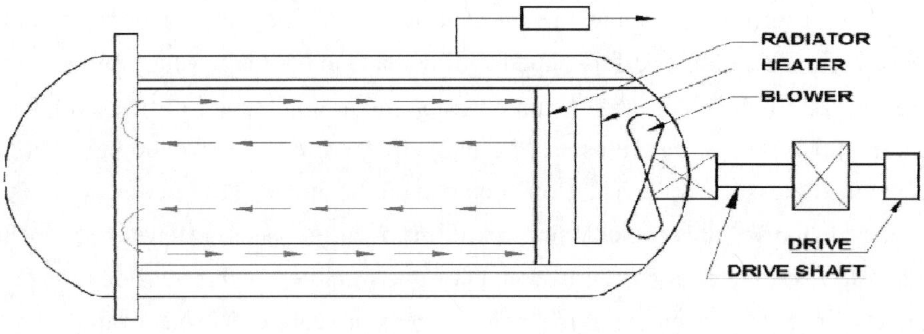

AUTOCLAVE

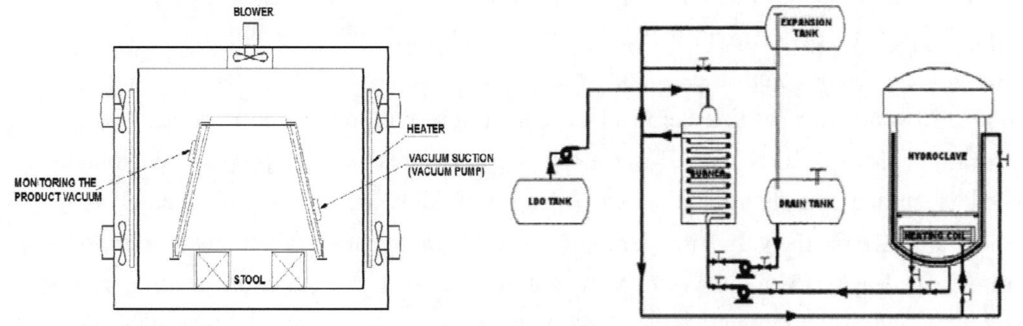

FIGURE 4: OVEN, AUTOCLAVE & HYDROCLAVE

The product with the vacuum connections is leak tested to ensure a minimum vacuum level. The product further undergoes a series of cure steps in a phased manner such as heating, pressurization, soaking, depressurization and cooling in the Hydroclave to get a defect free cured product. The heating rate, pressurization steps, maximum compaction pressure, cooling rate and the depressurization sequence are to be selected, considering the density of the final product, type and quantity of prepreg, tool design and previous experience of curing similar quality products. The product temperature at identified locations is monitored as the product temperature gradient during the entire curing cycle is to be minimized to avoid thermal stresses and to achieve the desired thermal and mechanical properties.

As the phenolic resin undergoes condensation polymerization reactions during the heating and soaking phase, the volatile compounds evolved such as alcohol, water vapour, free phenol etc. are to be completely sucked and positively driven out by giving sufficient compaction pressure and vacuum. Insufficient pressure and inadequate vacuum for curing will lead to delaminations and defects in the cured product. The pressurization is done based on the product temperature and maximum pressure is applied at a specified product temperature at which the resin changes from the visco-elastic to the viscous stage. When the phenolic resin is heated, viscosity of the resin decreases initially as the temperature rises, reaches a minimum value and then starts rising on further heating. If the maximum pressure is applied at low viscosity then more resin will be squeezed out resulting in insufficient resin and delaminations in the cured product, and on the other hand if the pressure is applied at a high viscosity, the compaction pressure will not get transmitted to the inner regions of the product. The importance of pressure application based on the product temperature is to apply the maximum pressure at a reasonable viscosity level to avoid excess resin squeeze out as well as to enable uniform compaction by transmitting

IOP Conf. Series: Materials Science and Engineering **360** (2018) 012036 doi:10.1088/1757-899X/360/1/012036

the pressure as a viscous medium rather than at a solidified state which does not permit the matrix penetration between the reinforcement. Another peculiar phenomenon in the curing of the ablative composites is the "resin exotherm phase". During the initial phase of heating, it is quite normal that the product temperature will always be lagging the water temperature in the Hydroclave. However at a particular stage the product temperature especially in the middle regions starts rising fast and goes above the vessel water temperature. Even if the vessel temperature is not increased, the product temperature rises continuously superseding the vessel temperature due to a phenomenon peculiar to the phenolic resins. At this juncture the vessel temperature also needs to be brought along with the rise in temperature of the product to minimize the product temperature gradient. The maximum pressure is applied during the exotherm phase. The exotherm regime and the pressure application at the required product temperature are unique and very important to achieve defect free and quality products with better properties. A typical cure cycle is given below.

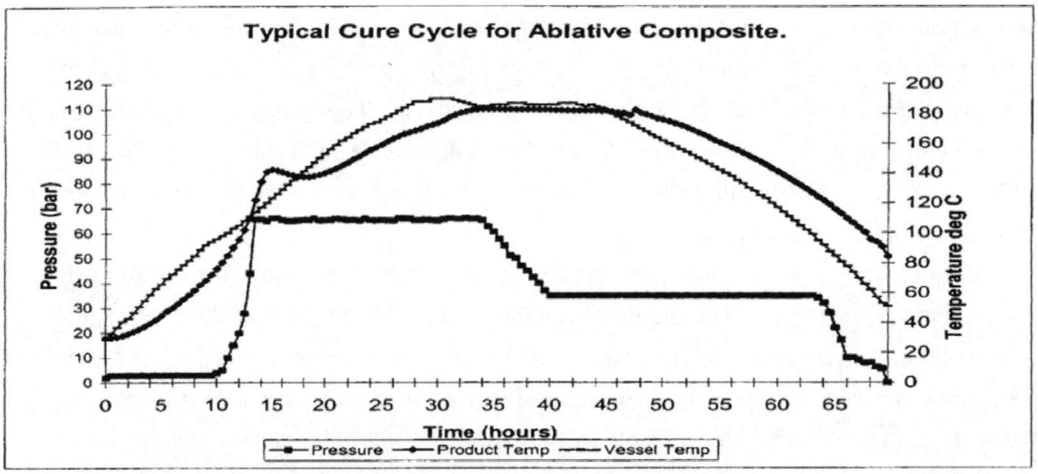

FIGURE 5: A TYPICAL CURE CYCLE FOR ABLATIVES

The product is further heated to the specified temperature (keeping the maximum pressure) and soaked at this temperature to complete the polymerization reactions. After soaking, the product is slowly cooled to the room temperature keeping the product temperature gradients to a reasonable minimum value to avoid thermal stresses, which are critical in the cooling regime. Parallelly, pressure is also reduced in a phased manner and is totally released as the product reaches a temperature less than 90°C. Thereafter product is allowed to cool naturally to room temperature before extraction from the tool.

The vacuum pumps are connected to the vacuum ports of the hydroclave vessel through valves and resin traps. The vacuum bagged product is connected to the vacuum ports in the vessel through flexible SS vacuum hoses. The volatile compounds formed during the curing and the excess resin is sucked by the vacuum pump and the volatiIes are periodically removed from the resin trap. Vacuum bagging is very important since any bag failure or vacuum leak will lead to water seepage and can result in product delaminations.

Hydroclaves use thermic fluid heating system. The light diesel oil (LDO) stored in the LDO tank is pumped, injected and fired in the burner to heat the thermic fluid which is circulated from the burner to the hydroclave through piping. The hot thermic fluid heats the water in the hydroclave through heating coils and jacketed vessel areas of the hydroclave. The thermic fluid heating system enables a controlled heating for the composite product. The thermic fluid circulation pump has to withstand the high temperatures of the order of 200°C and shall work continuously for about 70 percent of the cure cycle duration. The capacity of the burner system depends on the vessel volume and the maximum heating rate required. Positive displacement plunger pumps and control valves enable pressurization and depressurization. The pressure inside the pressure vessel needs to be positively maintained above the critical pressure of water at that temperature to avoid steam formation and associated safety problems. The pressure relief valves set above the maximum operating pressure relieve excess pressure in case of overshoot. High rate of pressurization is possible in hydroclave and this helps in immediate pressure rise required during the exotherm phase. The pressure vessel, rubber gasket, control valves and the pressure pipe joints in the pressurization and depressurization lines are very critical as any leak in any of these lines is not permissible from the safety point of view as the pressure drop to below the critical pressure value forms steam. If the joints are perfect and there is no leak, the load on the pressure pump will be so less that during the heating phase the requirements of switching on the pump is rare, as the increase in temperature itself will be sufficient to provide the required pressure increment. For redundancy in the pressurization two pressure pumps are provided in the system.

The water temperature inside the pressure vessel needs to be maintained uniform to enable uniform heating of the product. The water circulation system maintains the spatial distribution of temperature within reasonable limits. A high capacity water circulation pump circulates water in the vessel at all the working pressures and temperatures. Interlocks are made in such a way that pressurization is possible only after switching the

IOP Conf. Series: Materials Science and Engineering **360** (2018) 012036 doi:10.1088/1757-899X/360/1/012036

circulation ON. The circulation pump is rugged and highly reliable, as it has to work continuously for the entire cure cycle.

The cooling medium is water, circulated through cooling coils housed inside the pressure vessel. Since the cooling starts at high temperatures of the order of 150°C and above depending on the type of phenolic resin system, water entering the cooling coils immediately attains 150°C causing sudden steam formation resulting in very high back pressure and knocking in the cooling pipes. It is very important to note that cooling water circulation needs to be at extremely low and controlled flow rate to minimize the pressure drop due to sudden temperature fall inside the Hydroclave. Special control valves having very low to medium and high flow rate are essential for controlled cooling without pressure drop.

Rust preventive coating / painting over the interior surface of the Hydroclave is a mandatory requirement as the rusting will reduce the effective wall thickness and thereby decreasing the available margin.

After curing, the ablative composite is machined to the required configuration using special tools. Polycrystalline diamond (PCD) or Tungsten Carbide tools are used due to the abrasive nature of the material. Machining of ablatives is totally different from conventional materials like metals; here the material removal is in the form of carbon or silica powder and not in the form of chips. This necessitates dust collection suction systems at the cutting location which has to move along with the tool post. Coolant is not used during the machining of ablatives as it would cause deterioration in the properties of the composite.

3. Results and Discussion

Dimensions of the required part are measured and Non-destructive testing is done to confirm that no defects were present in the liner. Common defects likely in ablative composites include delaminations, cracks, voids, porosity, resin lean lines, resin rich lines, resin starvation, resin patches, non-uniform resin distribution, waviness, wrinkles etc. Visual inspection, tap test and alcohol wipe test was done initially. This was followed by Ultrasonic inspection by Pulse echo and through transmission methods. Wherever the signal strength was less or suspected delaminations were reported, tangential radiography was done to rule out the presence of delaminations.

Co-cured specimens are tested for mechanical and thermal properties. Specimens are fabricated from the end rings of the liners as per ASTM standards and tested.

IOP Conf. Series: Materials Science and Engineering **360** (2018) 012036 doi:10.1088/1757-899X/360/1/012036

4. Conclusion

The indigenous manufacturing processes for the development of high performance thermal protection ablative composites have been discussed in detail. The experimental details of synthesis and processing of carbon phenolic ablative composites are explained. The processed ablatives were characterized and all the critical properties have been evaluated. The various state-of-the art process equipments for ablative composite manufacturing have been explained. Indigenous process developments and equipment design has been discussed. These ablative composites are being used in the launch vehicles satisfactorily.

Acknowledgments

The authors gratefully acknowledge the permission, support, guidance and advice of Dy. Director, CMSE and Director, VSSC for allowing to publish this paper. We wish to thank the support rendered by all the divisions of Composites Entity for their active support and encouragement in this activity. The significant contribution of the Indian industry in developing the process equipments indigenously is gratefully acknowledged. Our sincere thanks to all the staff members of the processing sections of Ablatives Division for their whole-hearted support in this activity.

References:

1. M. Balasubramaniam, *Composite materials and Processing,* CRC Press, USA. **(2013)**.
2. Fitzer, E; Manocha, L M *Carbon reinforcements and Carbon/Carbon components***,** Springer Verlag, Berh **(1998)**.
3. Sutton, K., *An Experimental study of a Carbon-phenolic ablation material*, NASA Technical Note NASA-TN-D 5930. (**1970**)
4. T.L.Elegange and R.R. Bowman, *Nozzle fabrication for space shuttle solid rocket motor,* AIAA/SAE 14th Joint Propulsion Conference.
5. H.R. Clements and G.T. Ward, *Development of fabrication techniques for large solid booster nozzles*, AIAA 6th Solid Propellant Rocket Conference, Washington DC.
6. Mohan Kumar. L, Nikhil K Mawari, et al, *Development of Bias wrapping technology for Ablative liners in S139 Solid rocket motor,* INCCOM13 Proceedings, November 2014.
7. Gutowski Timothy G, *Advanced Composite Manufacturing*, Wiley, 1997.

IOP Conf. Series: Materials Science and Engineering **360** (2018) 012037 doi:10.1088/1757-899X/360/1/012037

Cure Characteristics and Rheological Properties of Organically Modified Lanthanum Oxide-Natural Rubber Composites

Sreevalsa S[a], Sruthi Sreekumar[a], Sreedha Sambhudevan[a], Balakrishnan Shankar [b]

[a] *Department of Chemistry, Amrita University, Amritapuri, India.*
[b] *Department of Mechanical Engineering, Amrita University, Amritapuri, India.*

Email: sreedha@am.amrita.edu

Abstract: The present work emphasizes on the cure features and rheological behaviour of organically modified gadolinium oxide-Natural rubber vulcanizates. The main aim of incorporating gadolinium oxide is to impart X-ray shielding property to rubber composites. Earlier lead based materials were used for this purpose, but due to high toxicity of lead, lanthanum based materials can be an excellent alternative. Gadolinium oxide was organically modified with methacrylic acid and is used as filler in rubber vulcanizates at different loadings from 0-40 parts per hundred rubbers. Modification was done to increase the adhesion between an inorganic gadolinium oxide and natural rubber matrix. Characterization of modified gadolinium oxide using X-ray diffraction shows that the particle size got reduced to nano regime which highly influences the vulcanizates properties. The increase in torque minimum and torque maximum with filler loading is in covenant with mechanical property studies. The swellingindex and swelling co-efficient decreaseswhile cross link densityenhances with filler percentage which reveals that a better interaction in molecular level is occurring between the modified filler and natural rubber than the unmodified one.

1. Introduction

Incorporation of nano- sized fillers into rubber composites is a well-practised technology to improve the performance of materials without much expensive procedures [1-4]. The rubber composites containing lanthanides are found to be a new system for over all and efficient X-ray shielding materials. Composite materials are used for many reasons such as low cost, light weight, durability, shielding protection etc. Besides, organic materials with higher numbers of hydrogen atoms in a given volume such as poly isoprene act as an outstanding support for X-Ray shielding. Due to the very fine size of particles, rare earths possesweakinteraction with rubber matrix and hence the dispersion of these inorganic particle in organic rubber matrix is a very crucial step [5-6], particularly for a upper loading which eventually leads to the lessening of the chance of inteferencesamong the dispersed phase and an X-ray beam. The main aim of rare earth modification organically is to attain an assured amount of enhancement in distribution and hence compatibility.

The aim of the present work is to prepare X-ray shielding natural rubber composites where gadolinium oxide in its modified form is used as the radiation absorbing material. Usually high-denseparticles are stronglytangible than low-dense alternatives for lessening the intensity of gamma and X radiation which possess higher energy levels. Due to its high atomic number, lead is the most accepted material for reducing the impact of gamma rays and x-rays. Atomic number means thenumber of protons present in an atom with a corresponding number of electrons. Lead with higher number of elecronswill hinder the gamma and x-ray particles that may pass through a lead shield and the level of shielding can be enhanced with thicker shielding barriers. The present work discusses the use of gadolinium in its oxide form as a promising radiation shielding material instead of lead.

2. Materials and methods

Gd_2O_3, methacrylic acid (MAA) and Dicumyl peroxide (DCP) were purchased from Sigma Aldrich. Natural rubber (ISNR 5) was, obtained from the Rubber Research Institute of India. Gadolinium oxide prior to mixing with rubber was modified using methacrylic acid. Gd_2O_3, MAA and water in the molar ratio 1:12:12 is taken in a 50 ml RB and the reaction was carried out by refluxing at 75°C for 90 minutes. The solution after filtration was thickened by distilling water for 3 h. A white precipitate was obtained after the addition of excess methanol to the liquor and leaving it overnight. The resultant compound was characterized using FTIR. The modified Gd_2O_3 was incorporated into the rubber matrix according the following recipe for sulphur vulacnization.

Ingredient	Composition(pphr)
NR	100
Filler	0 to 40
ZnO	5
Stearic acid	1
TDQ	1
CBS	1
Sulphur	2.5

Table 1 : Recipe for sulphur based cure system

The system was also cured with DCP (Dicumyl peroxide) to compare the effect of curing system in vulcanizate properties.

The XRD pattern of the samples were collected at room temperature on a Bruker AXS D8 advance diffractometer, equipped with X-ray source Cu of wavelength 1.5406 Å with step size of 0.020 2è and step time 32.8 sec. The surface morphology was studied using FEI Quanta 200 scanning electron microscope. The cure characterestics was studied using an oscillating Disc Rheometer (ODR) model 4308, Zwick, Germany. The parameters measured includes Torque minimum (M_L) and torque maximum (M_H).

3. Results and discussion

3.1 FTIR Spectroscopy :

The modified gadolinium oxide was characterized using FTIR as shown in figure 2. The stretching frequency of ester carbonyl group is observed at 1696 cm^{-1} and vinyl stretching frequency at 1635cm^{-1}. The Gd-O stretching band is retained at around 947 cm^{-1}. The stretching frequency of ester group as seen in the FTIR spectrum confirms the fact that modification has been carried out successfully.

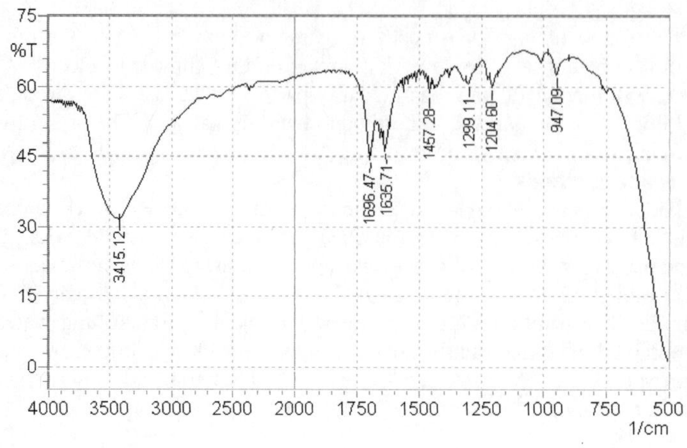

Figure 1: FTIR spectrum of modified Gd_2O_3

3.2 X-ray diffraction

Figure 2 depicts the XRD profile of gadolinium oxide before and after organic modification. A broad peak at around 29^0 is identified as the 222 reflection of cubic Gd_2O_3. The broadening of peaks and decrease in peak intensity may be attributed to the lowering of particle size to nano regime [7]. The peak positions and 2θ values are well agreeing with the early reported one for Gd_2O_3 nano particles [8]. In general, the atomic scattering factor is high for atoms with high atomic number (Gd : Z=64), but the low peak intensities observed for the Gd_2O_3 sample in Figure 2 is probably also a consequence of gadolinium fluoresce [9]. X-ray analysis reveals the crystalline nature of $Gd(MAA)_3$.. The well - defined diffraction peaks establish the crystalline nature of the sample after modification. The basic structure of Gd_2O_3 is retained with an increase in the crystalline nature after modification.

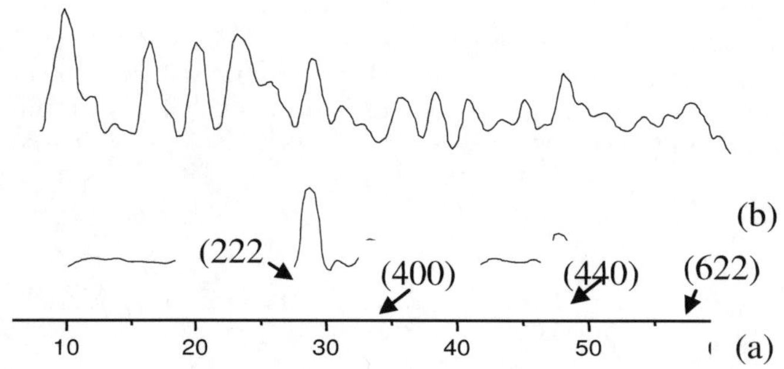

Figure 2 : XRD spectrum of (a) before modification and (b) after modification

3.3 Cure characterestics

The cure characterestics of natural rubber vulanizates are shown in figure. It was found that the torque minimum increases apparently with rise in filler loading for all vulcanizates. As the torque minimum is a direct measure of stock viscosity, the presence of nano sized fillers imparts additional resistance to flow owing to the restricted molecular mobility of the nano composites [10]. Or in other words presence of modified filler increases the viscosity of the samples. According to Einstein, Guth and Gold equation [11]

$$\eta_f = \eta_u(1 + 2.5\,c + 14.1\,c^2)$$

Where η_f and η_u are the viscosities of modified and un modified natural rubber composites and c is the volume fraction of the filler. The torque maximum increases with increase in filler loading which clearly shows that incorporation of modified filler particles apparently enhance the stiffness of nano composites. Torque difference which is an indirect mesure of cross link density of nano composites also rises with filler loading reveals that there is a increasedcollaborationamong the rubber matrix and filler.

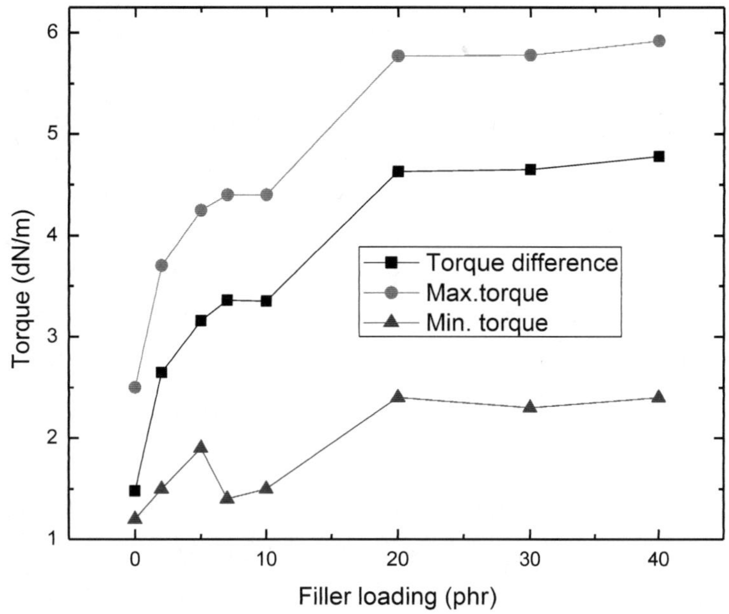

Figure 3: Cure Characterestics of modified Gd_2O_3 based natural rubber composites

3.4 Swelling measurements

The swelling experiments of the samples were carried out according to ASTM D471. Anisotropic swelling studies directly relates to the amount of distribution of fillers in the rubber matrix. Swelling index is estimated using the equation

$$Swellingindex\% = \frac{(W_2 - W_1)}{W_1} \times 100$$

where W_1 and W_2 are the initial and final swollen weights of the sample.

The swelling nature of the composites can also be studiedusing the swelling coefficient values and is givenby the equation

$$Swellingcoefficient\alpha = \{A_s/m\} \times [1/d]$$

where A_s is the weight of the solvent absorbed at equilibrium swelling, m is the mass of the sample before swelling and d is the density of the solvent used.

The cross link density of vulcanizates v can be calculated from the Flory-Rehner equation [12-13].

$$v = \frac{-Ln[(1 - v_r) + v_r + \chi v_r^2]}{v_s \times \left(v_r^{\frac{1}{3}} - \frac{v_r}{2}\right)}$$

where v_r is the volume fraction of the polymer in the swollen gel, χ is the polymer – solvent interaction parameter, v_s is the molar volume of the solvent.

Filler loading (phr)	*Swelling Index (%)*	*Swelling Coefficient*	*Cross link density*

		($\times10^{-3}cm^3/g$)	($\times10^{-5}g/cm^3$)
0	329.10	3.79	18.8
10	310.92	3.59	19.55
20	283.46	3.25	21.05
30	274.62	3.17	23.5
40	240	2.77	26.68

Table 2 : Swelling characterestics of vulcanizates.

The values clearly shows that rise in filler percentage enhances the cross linking between the particles and rubber matrix. Larger cross Link density values is a direct evidence of enhanced restriction on the network and thus leads to a lowering of swelling because of theexistence of fillers.

4. Conclusions

The present study deals with the cure characterestics and swelling measurements of modified Gd_2O_3-Natural rubber nano composites. It is concluded that the presence of fillers increases all the properties and there is a dramatic increase with filler loading also.

References

1. M.Z Rong, M. Q Zang and W. H Ruan, *Materials Science and Technology*, 22, 7, **(2006)**.
2. Sabu Thomas and Ranimol Stephen, *Rubber Nanocomposites: Preparation, Properties and Applications*, John Wiley and Sons, **(2010)**.
3. G Sui, W. H Zhong, X. P Yang, Y. H Yu and S. H Zhao,*Polymers advanced technologies*, 19, 11, **(2008)**.
4. R. Sengupta, S. Chakraborty, S. Bandyopadhyay, S. Dasgupta, Mukhopadhyay, K. Auddy and A.S. Deuri, *Polymer Engineering & Science*, 47, 11, **(2007)**.
5. Brian P. Kirkmeyer, Robert A. Weiss, Kareni. Winey, *Journal of polymer science: Part B: polymer physics*, 39, **(2001)**.
6. S.Yochiomi, N.Koichi, A. Missko, J. Soc. *Rubber Ind*, 67,**(1994)**.
7. H. Jensen, J. H. Pedersen, J. E. Jorgensen, J. Skov Pedersen, K.D Joensen, S.B. Iversen, E.G. Sogaard, *J. of Experimental Nanoscience*, 1,**(2006).**
8. The XRD, SAED patterns and d vaules were indexed with reference to the crystal structures from the PCPDF files (PCPDF card no.00-012-0797).
9. Maria Ahrén, Linnéa Selegård, Fredrik Söderlind, Mathieu Linares, Joanna Kauczor, Patrick Norman, Per-Olov Käll and Kajsa Uvdal,*Journal of nanoparticle research*, 8, doi:10.1007/s. 11051-012-1006-2, **(2012).**
10. Ishak ZAM, Bakar AA, *European Polymer Journal*, 31, **(1995).**
11. Blow, Rubber Technology and Manufacture, Institution of the Rubber Industry; Butterworth, London; **1971**.
12. Walter Chasse, Michael Lang, Jens-Uwe Sommer, Kay Saalwachter, *Macromolecules,*45, 2, **(2012)**.
13. P. J. Flory, J. Rehner, *J. Chem. Phy*, 11,**(1943)**.

IOP Conf. Series: Materials Science and Engineering **360** (2018) 012038 doi:10.1088/1757-899X/360/1/012038

Study the Structural and Optical behaviour of Conducting Polymer based nanocomposites: ZrO_2-Polypyrrole Nanocomposites

Gaganpreet Kaur Sidhu and Rajesh Kumar*
Department of Physics, Panjab University, Chandigarh- 160014, India

**Corresponding author Email:* rajeshbaboria@gmail.com

Abstract. Conducting polymer provides an intriguing type of environment for various inorganic structures resulting in the generation of exciting properties in the composite owing to the mutual interactions between the two and this provides interesting dimensions for the development of new multifunctional devices. ZrO_2 ceramic is one of the most investigated materials for its outstanding mechanical properties ionic, conduction properties and due to its high oxygen ion conduction. In order to achieve novel properties of ZrO_2 nanoparticles, nanocomposites of ZrO_2 with Polypyrrole (PPy) were prepared through in-situ method using ZrO_2/pyrrole dispersions. Firstly, ZrO_2 nanoparticles were synthesized by co-precipitation method and then embedded in pyrrole solution for polymerization. The structural and optical properties of ZrO_2/PPy nanocomposites were investigated comprehensively by X-Ray diffraction (XRD), UV-visible spectroscopy and Fourier-transform infrared spectroscopy (FTIR). XRD and FTIR highlight the presence of tetragonal phase of ZrO_2 nanoparticles in nanocomposites, whereas UV-Visible and PL spectra indicate the presence of different electronic levels.

1. Introduction

During the recent years there has been an explosive growth in the field of nanocomposites. It represents a new class of materials having improved properties as compared to pure conducting polymers and metal oxides[1]. These nanocomposites possess interesting electrical and potential applications in the fields like chemical sensors, biosensors, electronic devices, etc [2-4]. Such materials are of great interest both in the field of physics and chemistry due to their novel properties, which are entirely different from that of constituents[5,6]. There are number of examples regarding polymer/metal oxide hybrid materials in which polymers such as polystyrene[7] and poly (vinyl chloride)[8] are combined with metal oxides or salts such as CuO, SiO_2[9], ZrO_2[10], $BaSO_4$[11]. Polymeric materials can be obtained in different form depending on the requirement for various applications like in powder form, thin films, nanorods, etc. Several methods have been used by researchers to improve the properties of polymers, among them one of the best methods is the synthesis of polymer/metal oxide nanoparticles. Polymer/metal oxides acquired an important part in the present technology. Among various conducting polymers, Polypyrrole (PPy) has attracted much attention due to its unique electrical properties like electrical conductivity and is environmental friendly nature[12]. PPy can be easily prepared either by oxidative chemical or electrochemical polymerization of pyrrole. PPy has been used in biosensors,[13,14] gas sensors,[15,16] wires,[17] microactuators,[18] and anti-electrostatic coatings[19]. Metal oxides materials provide good mechanical strength and stability to the polymers. However, among different metal oxides, we use ZrO_2. It is an

IOP Conf. Series: Materials Science and Engineering **360** (2018) 012038 doi:10.1088/1757-899X/360/1/012038

excellent oxide material with wide band gap (5 - 7 eV). It exists in three different forms Monoclinic (Room temperature-1170 ˚C), Tetragonal (1170-2370˚C) and Cubic (above 2370˚C. The most important phase of ZrO_2 for application point of view is tetragonal phase which can be stabilized at low temperature by number of methods[20]. It possesses various applications such as fuel sensors, oxygen sensors[20], catalyst[21] etc. All these properties are mainly due to high specific surface area and presence of oxygen vacancies[22]. Here, we report the synthesis of polymer/metal oxide nanocomposites comprising PPy as the polymer and ZrO_2 as the metal oxide component. PPy/ZrO_2 nanocomposites were synthesized by embedding synthesized ZrO_2 nanoparticles into a PPy matrix during the polymerization of pyrrole and studied their structural and optical properties.

2. Experimental Details

Pyrrole, ferric chloride hexahydrate and distilled water all were obtained as AR grade reagents commercially are used for the synthesis of PPy. PPy was synthesized by chemical oxidative polymerization using pyrrole as the monomer and ferric chloride hexahydrate as the oxidizing agent. Polymerization was carried out in a beaker by mixing 1 ml of vacuum distilled pyrrole with 100 ml of distilled water containing 9.74 gm of ferric chloride hexahydrate at room temperature. The optimal ratio of Fe (III)/ monomer is 2.5. The solution was stirred for 24 hrs on magnetic stirrer giving rise to the formation of black precipitates. These precipitates were then filtered with the help of vacuum and washed with the copious amount of deionised water until the solution becomes clear. Precipitates were dried overnight in air and then again dried in oven at 25˚C.

PPy/ZrO_2 nanocomposites has been synthesized by adding 1 ml of vacuum distilled pyrrole to 100 ml of distilled water containing 9.74 g of ferric chloride hexahydrate and ZrO_2 nanoparticles at room temperature. The solution was stirred for 24 hrs on magnetic stirrer which gave rise to the formation of black precipitates followed by the same procedure as mentioned above.

X-Ray diffraction (XRD) patterns were recorded on a Philips PW-3710 diffractometer using Cu Kα radiation (λ = 1.5406 Å) in the 2θ range 10˚– 80˚ to determine the crystal structure and estimate the average crystallite size. Fourier-transform IR (FTIR) spectra (4000–400 cm^{-1}) were recorded for the prepared powder to confirm the formation of composite. UV-Visible and PL spectra of PPy, and PPy/ZrO_2 nanocomposites were recorded to understand the absorption and emission spectra of polymer and nanocomposites.

3. Results and Discussion

Figure 1 shows the XRD patterns of pure PPy and ZrO_2/PPy nanocomposites. A broad peak at around 2θ = 25.12˚ indicating the short range ordering of PPy powders, which assigned to the repeated units of pyrrole ring, implying that polymer chain is highly oriented[23]. However, the XRD pattern of ZrO_2/PPy nanocomposites consist some additional peaks that correspond to the tetragonal phase of ZrO_2 nanoparticles. Peaks at 2θ = 30.69˚, 50.74˚, and 60.27˚ can be indexed as (1 1 1), (2 0 0) and (2 1 1) crystal planes of ZrO_2 nanoparticles, respectively. The peak positions of nanoparticles are almost at same position as is observed in pure nanoparticles which imply that even in composite, nanoparticles retain their identical structure. However, nanoparticles peaks are not so sharp indicating the amorphous nature of nanocomposites. The average crystallite size has been calculated using basic Scherrer's formula given as: $D=\frac{k\lambda}{\beta Cos\theta}$, in which β is the full width at half maximum (FWHM), λ is the X-ray wavelength, θ is the diffraction angle and k is the Scherrer's constant of the order of unity. The average crystallite size of nanoparticles present in nanocomposites has been calculated using basic Scherrer's formula as mentioned above. The average crystallite size has been calculated by fitted all the curves using Gaussian fit. When PPy is polymerized using $FeCl_3$, then average crystallite size of globules was found to be 5.8 Å. However, the average crystallite size of nanoparticles in nanocomposites comes out to be ~ 8 nm. The full width half maxima and the average crystallite size corresponding to all peaks are listed in Table 1.

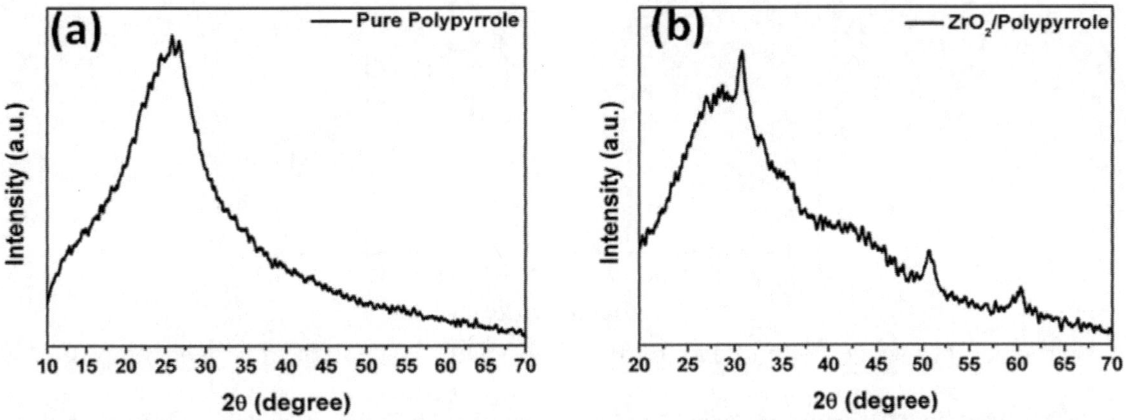

Figure 1: XRD spectra of (a) pure Polypyrrole (PPy) and (b) of ZrO_2/PPy nanocomposites.

Table 1: Average crystallite size calculated for Pure PPy and ZrO_2 nanoparticles in nanocomposites.

Material	2θ	FWHM (β)	Crystallite Size d (size) (nm)
Polypyrrole	28.56°	14.18	0.58
	30.76°	0.902	8.45
ZrO_2	50.74°	1.171	7.37
	60.29°	2.096	4.45

Figure 2 shows the Fourier transform infrared spectra of pure PPy powder which was recorded using KBR pellets in wavelength range of 4000 to 400 cm^{-1}. The band at 1681 cm^{-1},1533 cm^{-1} and weak band at 1447 cm^{-1} are assigned to C–N in plane bending vibration, stretching vibration of C = C and C – C stretching vibration respectively, which may be due to the over oxidization of PPy. PPy shows characteristic peaks of C–H in-plane bending vibration and C–O–C stretching vibration of pyrrole at 1159 cm^{-1} and 1016 cm^{-1}, respectively in the IR spectrum. The band at 1283 cm^{-1} corresponds to the C=N bond and the adsorption band at 858 cm^{-1} corresponds to the out-of plane vibration of the C-H bond[24]. All peaks correspond to PPy are present in ZrO_2/PPy nanocomposites FTIR spectra, but with little shifts in peak positions. Peak at 1681 cm^{-1} in PPy shifts to 1672 cm^{-1}, 1533 cm^{-1} shifts to 1525 cm^{-1},

1447 cm^{-1} to 1439 cm^{-1}, 1283 cm^{-1} to 1279 cm^{-1}, 1159 cm^{-1} to 1146 cm^{-1}, 1016 cm^{-1} to 1018 cm^{-1} and 858 cm^{-1} to 852 cm^{-1}. All peaks corresponding to their respective bands are listed in Table 2. These shifts in peak positions indicate the presence of interaction between the chains of PPy and nanoparticles. In addition to these peaks, another peak at 752 cm^{-1} is also observed which corresponds to Zr-O bond. ZrO_2 nanoparticles show its characteristics FTIR peak in the range from 500-800 cm^{-1}.

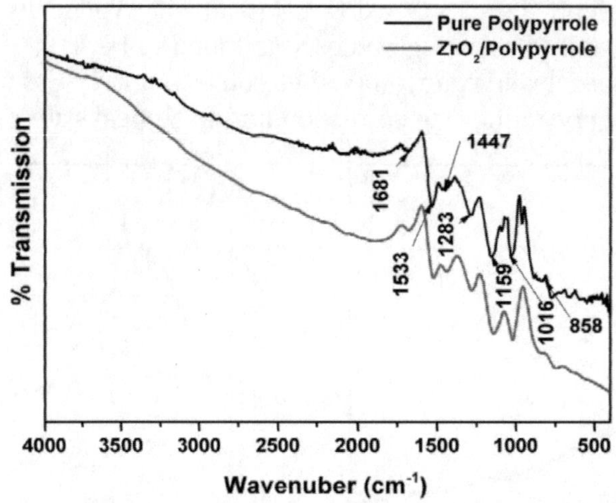

Figure 2: FTIR spectra of pure Polypyrrole (PPy) and ZrO_2/Polypyrrole nanocomposite.

Table 2: Various FTIR peaks observed in pure PPy and ZrO_2/Polypyrrole nanocomposite.

Characteristic Bands	Peak positions (cm^{-1})
C–N in plane bending vibration	1681
C = C stretching vibration	1533
C – C stretching vibration	1447
C–H in-plane bending vibration	1159
C–O–C stretching vibration	1016
C=N stretching vibration	1283
C-H bond out-of plane vibration	858
Zr-O vibration	752
O-H vibrations	3500-3000

Figure 3, shows the absorption spectra of pure PPy powder with absorption peak at around 232 nm, whereas ZrO_2/PPy nanocomposites shows absorption peak at around 234 nm. In literature, UV-Vis peak of PPy was reported at around 340 nm and at 460 nm[8]. In our case only single peak around 232 nm is observed which is due to the transition of electron from the highest occupied molecular orbital (HOMO) to the lowest unoccupied molecular orbital (LUMO) and is related to the π^*-π electronic transition[25]. However, in ZrO_2/PPy nanocomposites a small shift in peak is observed towards the higher wavelength. This relocation in absorption peak reveals a change in electronic structure and transpose of polaron levels in PPy due to the interaction with ZrO_2 nanoparticles. A sizeable overlay of optical transition takes place between PPy and extrinsic states of ZrO_2 nanoparticles[26,27].

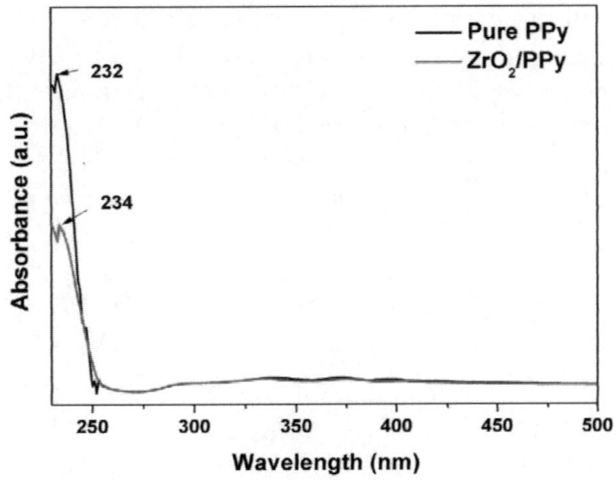

Figure 3: Absorption spectra of Pure PPy and ZrO_2/PPy nanocomposites.

PPy-nanocomposites are amorphous materials and obey Tauc relation. So the band gap of samples has been calculated using UV-Vis spectra as shown in Figure 4. The Tauc relation is given by: $\alpha h\nu = A(h\nu - E_g)^n$, in which α is known as absorption coefficient, hv known as photon energy, Eg is the optical bandgap, and A is constant. The indices *n* attained different value depending on the type of transitions as 1/2, 3/2, 2, 3 for direct allowed and forbidden or indirect allowed and forbidden transitions, respectively. As shown in Figure 4, the absorbance is plotted against the photon energy for PPy and its composites, a satisfactory fit is obtained for *n* = 1/2, which corresponds to the existence of a direct allowed bandgap. The intercept is drawn to the curve on the photon energy axis gives the bandgap of all the samples. A decrease in bandgap of nanocomposites is observed with the addition of nanoparticles which is due to the alteration of electronic structure. The band gap calculated for pure PPy and ZrO_2/Ppy nanocomposites is 6.1 and 5.07 eV, respectively.

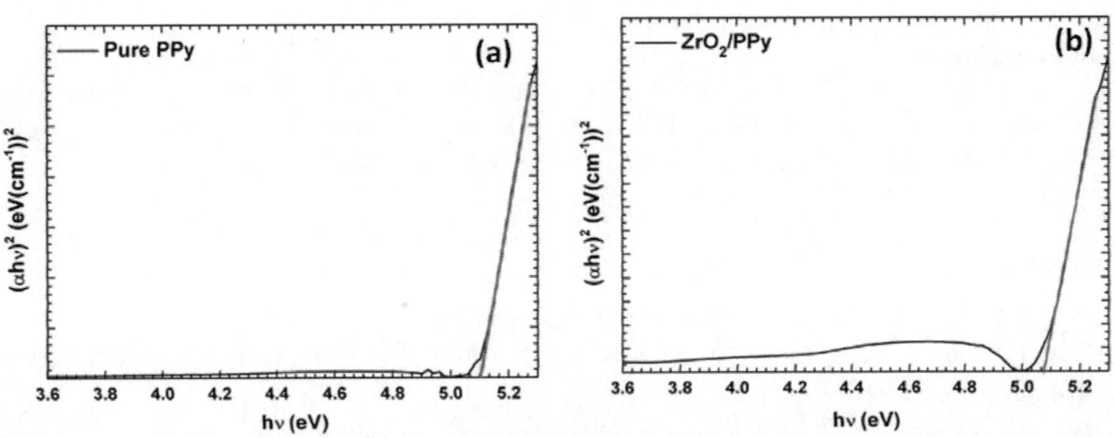

Figure 4: Variation of absorption coefficient with photon energy of (a) Pure PPy and (b) ZrO_2/PPy nanocomposite.

IOP Conf. Series: Materials Science and Engineering **360** (2018) 012038 doi:10.1088/1757-899X/360/1/012038

The photoluminescence spectrum of the PPy and its composite were tested and shown in Figure 5. A broad peak centered at 508 nm was collected in pure PPy. ZrO_2/PPy also shows much broader emission peak with lesser intensity as compared to pure PPy which is due to the presence of ZrO_2 nanoparticles[28]. At 430 nm excitation in the near UV region, both polymer and composite exhibited one strong green emission band at around 508 nm. The presence of PL peak at same position in composite indicates the dominating role of polymer as compared to nanoparticles.

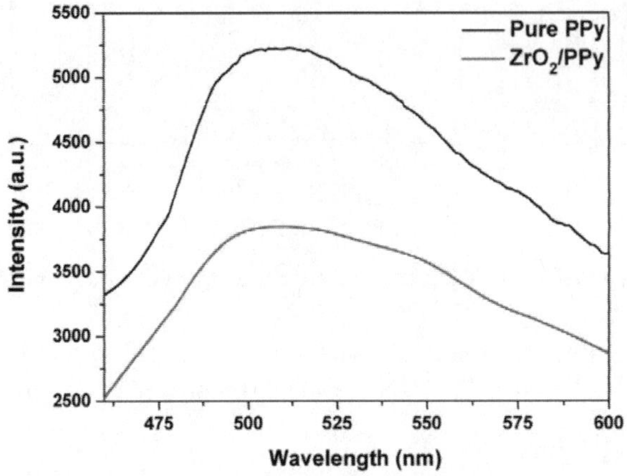

Figure 5: PL spectra of pure PPy and ZrO_2/PPy nanocomposite.

4. Conclusion

ZrO_2/PPy nanocomposites were synthesized by chemically oxidative polymerization. The prepared nanocomposites was analysed for structural and optical properties. XRD reveals the combined broad peak of PPy and various peaks of ZrO_2 nanoparticles in ZrO_2/PPy nanocomposite indicating the presence of tetragonal phase of nanoparticles is still present in nanocomposites. FTIR peaks confirm the formation of composite, whereas optical properties shows absorbance and emission peak of PPy at 232 and 508 nm, respectively with a small shift in both peaks in its composite.

Acknowledgments

The authors are thankful to CIL Department of Panjab University, Chandigarh for various characterizations and IISER, Mohali for PL measurements. One of the authors Ms. Gaganpreet Kaur Sidhu is thankful to UGC for research fellowship.

References

1. Gangopadhyay R and De A 2000 *Chem. Mater.* **12** 608
2. Jiua T, Liua H, Gana H, Li Y, Xiao S, Li H, Liu Y, Lu F, Jiang L and Zhu D 2005 *Synthetic Met.* **148** 313
3. Rein H, Ruckpaul K and Haberditzl W 1973 *Chem. Phys. Lett.* **20** 71
4. Halls J J M, Walsh C A, Greenham N C, Marseglia E A, Friend R H, Moratti S C and Holmes A B 2002 *Nature* **376** 498
5. Sui X, Chu Y, Xing S and Liu C 2004 *Mater. Lett.* **58** 1255

6. Murugendrappa M, Parveen A, and Prasad M A 2007 *Materials Science and Engineering: A* **459** 371
7. Lascelles S F and Armes S P 1995 *Adv. Mater.* **7** 864
8. Banerjee P and Mandal B M *Synthetic Met.* 1995 **74** 257
9. Maeda S and Armes S 1995 *Synthetic Met.* **73** 151
10. Bhattaeharya A, Ganguly K M, De A and Sarkar S 1996 *Mater. Res. Bull.* **31** 527
11. Gan L M, Zhang L H, Ghan H S O and Chew C H 1995 *Mater. Chem. Phys.* **40** 94
12. Yao T, Lin Q, Zhang K, Zhao D, Lv H, Zhang J, Yang B 2007 *J. Colloid Interface Sci.* **315** 434
13. Vidal J C, García E and Castillo J R 1999 *Anal. Chim. Acta* **385** 213
14. Gao M, Dai L and Wallace G G 2003 *Electroanal.* **15** 1089
15. Cherenack K and Pieterson L 2012 *J. of Appl. Phys.* **112** 091301
16. Kemp N T, Flanagan G U, Kaiser A B, Trodahl H J, Chapman B, Partridgeh A C and Buckley R G 1999 *Synthetic Met.* **101** 434
17. Jerome C, Labaye D, Bodart I and Jerome R 1999 *Synthetic Met.* **101** 3
18. Jager E W, Smela E and Inganas O 2000 *Science* **290** 1540
19. Navale S T, Khuspe G D, Chougule M A and Pati V B 2014 *J. Phys. Chem. Solids.* **75** 236
20. Srinivasan R, De Angelis J, Ice G and Davis B H 1991 *Journal of Materials Research* **6** 1287
21. Su C, Li J, He D, Cheng Z and Zhu Q 2000 *Appl. Catal. A* **202** 81
22. Kumar A M and Rajendran N 2012 *Surf. Coat. Tech.* **213** 155
23. Orinakova R, Fedorkova A and Orinak A 2013 *Chem. Pap.* **67** 860
24. Huyen D N, Tung N T, Vinh T D and Thien N D 2012 *Sensors* **12** 7965
25. Abdulla H S and Abbo A I 2012 *Int. J. Electrochem. Sci.* **7** 10666
26. Dey A and De S 2006 *J. Phys. D. Appl. Phys.* **39** 4077
27. Singh S K and Shukla R 2015 *Mat. Sci. Semicon. Proc.* **31** 245
28. Wang W, Yu D and Tian F 2010 *J. Lumin.* **130** 494

IOP Conf. Series: Materials Science and Engineering **360** (2018) 012039 doi:10.1088/1757-899X/360/1/012039

Artificial lipid membrane: surface modification and effect in taste sensing

Saurav Kumar[1, 2], Amol P Bhondekar[1, 2*], Prateek Jain[3], Sudeshna Bagchi[1, 2] Anupma Sharma[1, 2], Ritesh Kumar[1, 2], Sunita Mishra[1, 2]

[1]CSIR-Central Scientific Instruments Organisation, Sector-30 C, Chandigarh
[2]Academy of Scientific and Innovative Research (ACSIR), New Delhi-110025, India
[3] National Agri-Food Biotechnology Institute (NABI) , Mohali, Punjab-140308

E-mail: amolbhondekar@csio.res.in; amol.bhondekar@gmail.com; Tel.: +91-0172-2657190(489)

Abstract. All articles *must* contain an abstract. The abstract text should be formatted using 10 point Times (or Times Roman, or Times New Roman) and indented 25 mm from the left margin. Leave 10 mm space after the abstract before you begin the main text of your article. The text of your article should start on the same page as the abstract. The abstract follows the addresses and should give readers concise information about the content of the article and indicate the main results obtained and conclusions drawn. As the abstract is not part of the text it should be complete in itself; no table numbers, figure numbers, references or displayed mathematical expressions should be included. It should be suitable for direct inclusion in abstracting services and should not normally exceed 200 words. The abstract should generally be restricted to a single paragraph. Since contemporary information-retrieval systems rely heavily on the content of titles and abstracts to identify relevant articles in literature searches, great care should be taken in constructing both.

1. Introduction

Taste is one of the five fundamental sensory perception identified in humans. The sensory apparatus by which we can identify the signal i.e. tongue, can distinguish five fundamental tastants which are saltiness, sourness, umami, sweetness and bitterness [1, 2]. Among these tastes, sweetness is a signal of presence of carbohydrates that serve as an energy for living organisms. Biologically speaking, taste substances interact with the bio-membrane on the taste buds and sweetness is received by T1R2 and T1R3 taste receptors, which are G-Protein coupled receptors (GPCRs) [3–5]. Sweetness is a pleasant taste for most of the people and it is the major attribute for the value of food and its demand. Therefore, evaluation of sweetness is an important task for most of the food industry. Sweet taste substances are commonly categorized as sugars, which is of a different kind as per their chemical structure and degree of sweetness. Major sugars -Sucrose, Fructose, and Glucose are carbohydrates with aldehyde or ketone group or their derivatives [6]5. Since sugars are principally nonionic and in the aqueous phase exhibit their molecular structure without charge, hence potentiometric measurement with lipid/polymer membrane shows low sensitivity to sugars compared to other basic taste substances like NaCl, HCl, MgCl2, etc [7].

The taste sensor has the characteristic of using in common partially-selective sensor electrodes and measuring the analytes. The development of taste sensor need to fulfill following aspects: (a) the response of the taste sensor should be comparable to that of human taste; this fact is called global selectivity; (b) the taste sensor threshold for each taste should accord with the human sense; (c) the unit of taste information from the taste sensor should be clearly defined; (d) the taste sensor can detect interactions between taste substances or taste qualities such as a suppression effect that appears between sweetness and bitterness [8]. The taste sensor can discriminate the samples into the five basic taste qualities [1,9]. The taste sensor may utilizes lipid/polymer membranes as the taste-sensing part and outputs a change in the membrane potential caused by the interaction between the lipid/polymer membrane and taste substances. Lipid/polymer membranes are composed of a polymer support, a

IOP Conf. Series: Materials Science and Engineering **360** (2018) 012039 doi:10.1088/1757-899X/360/1/012039

lipid, and a plasticizer [10]. The lipid and plasticizer are responsible for controlling hydrophobic and electrical properties of the membrane surface, whereas the polymer support plays the role in encapsulating and immobilizing the lipid. Lipid/polymer membranes are designed and developed so that they can selectively respond to each taste quality. The lipid/polymer membrane responds to taste substances in sample solutions (i.e. said as analytes), and this response is called relative value, which is the difference between the membrane potential for the sample solution and that for the reference solution as the source. The properties of the lipid membrane formed from hydrocarbon solvents are identical in all the aspect to those of biological membrane resistance, which is generally much lower in cell membrane- and is there attributed to the presence of special translocators that locally modify the bilayers. However, the resistance of the experimentally prepared membrane can also be lowered to that of cell membrane by the insertion of translocator and protein or similar structure which could bring similar change. These molecules, actually disturb the fluidity and permeability of the membrane and further bring change to its surface charge density [11]. The same can also be achieved by the application of nanoparticles which is biocompatible and better catalytic activity.

Gold nanoparticles (AuNPs) have become increasingly attractive for nano-biotechnological, biomedical and pharmaceutical applications due to their unique optical, electronic and thermal properties as well as their straightforward synthesis, stability and the simple incorporation with lipid membrane and proteins [12][13]. The nanoparticles show similar dimensions to those of biomolecules, and the integration of nanoparticles with biomaterials exhibits unique recognition, catalytic, and inhibition properties, yields novel hybrid nano-bio materials for synergetic properties and functions [14]. The use of nanostructured thin films of conducting polymers and a lipid-like material coating gold microelectrodes, it is possible to mimic the human perception of taste. An electronic tongue composed of nanostructured films of lipid and plasticizer, these material acts as a transducers as it provide signals on interaction with electrolytes [15–17].

It is important to determine the necessary and sufficient conditions for fabricating a sensor, which may provide possible understanding for the reaction mechanism. This may be a basis for creating sensors that can respond to natural substance based on their type. The studies on the interaction between metal cations and bio-membranes have attracted considerable interest in recent years. Up to now, studies have been made on the interaction of Ca+2, Cu+2 [14, 16]. Researchers have found that charged particle penetration into the lipid membrane (like AuNP) had effect on the fluidity of the membrane lipid and its charge density. Due to the physicochemical interaction, AuNP have considerable influence on the electrical properties of artificial lipid membranes [15, 18]. The interaction of nanoparticles (here, AuNP) and artificial lipid membrane depends upon the surface charge. When positively charged AuNPs are attached to the negatively charged membrane, then the membrane would try to maintain the original charge distribution [19].

During the process, membrane probably would lose its rigidity, and its morphology would be changed and this statement suggest that the fluidity and permeability of the membrane might increase. Neutral and negatively charged nanoparticles hardly adsorb onto negatively charged cell surface due to low adhesion [20,21]. Conversely, TDAB offers positively charge on the membrane surface, and citrate coated AuNP holds negative charge which might offer different interaction behavior as explained above and hence it definitely offer higher adhesion and stronger interaction.

In this experiment, we fabricated sweet sensitive AuNP modified artificial lipid membrane to sense the sweet taste analyte. The developed sweet-sensitive lipid/polymer membrane composed of tetra-dodecylammonium bromide, di-n-octyl-phenylphosphonate and polyvinylchloride [2,10,22,23]. Modification of lipid membrane was done with AuNP synthesized by citrate method. The taste sensor was studied for different taste substance from low concentration of 0.01mM (i.e. 10μM) to 1000mM. The modified lipid membrane's response has been compared with non-modified lipid membrane at different concentration of taste substances on gold electrode. It was tried during the experiment to monitor how the interaction of nanoparticles (AuNP) affects the sensitivity and selectivity of artificial lipid membrane. Further investigation towards the interaction of AuNP with artificial lipid membrane has been made from UV-Vis spectroscopic studies and the explanation was provided in its mechanism.

IOP Conf. Series: Materials Science and Engineering **360** (2018) 012039 doi:10.1088/1757-899X/360/1/012039

2. Materials and Methods

2.1 Reagents

Tetradodecylammonium bromide (TDAB), di-octyl-phenylphosphonate (DOPP), polyvinyl chloride (PVC), tetrahydrofuran (THF) and trisodium citrate and tetrachloroauric acid (HAuCl4) was purchased from Sigma-Aldrich, Inc. The taste substance, i.e. analytes sucrose, D+-glucose, fructose were purchased from Himedia laboratory Pvt Ltd, Mumbai; Magnesium chloride (MgCl2) purchased from Loba Chemie Pvt Ltd, Mumbai. Chemicals like sodium chloride (NaCl), hydrogen chloride (acid) (HCL) and sodium-L-glutamate (MSG) obtained from Chem Lab NV, Industrieterrein "De Arend" 2 B-8210 Zedelgem (provided by Alpha MOS in the testing kit). The glassware used for synthesizing GNPs was thoroughly cleaned and siliconized using 1% 3-glycidoxypropyltrimethoxysilane solution (GOPS) (Sigma, India). Water was purified with a Milli-Q ultra-pure system (Millipore, India) having a resistivity >18 MΩ/cm.

2.2 Lipid/polymer membrane

The lipid/polymer membrane is composed of tetradodecylammonium bromide (TDAB) as the lipid, di-octylphenylphosphonate (DOPP) as the plasticizer and Polyvinyl chloride (PVC) as the polymer. The concentration used herein in the experiment has already been used in the commercial application of taste sensor and other earlier work. The reported work showed that the CPA value of the sensor is highly sensitive, a potential drop due to absorption onto membrane was reported smaller than -3mV for all the major taste analyte [7]. The procedure followed for fabricating the membrane was in accordance with previously reported work (reference). The membrane carries a positive charge because TDAB is positively charged in aqueous solution. The membrane was designed to contain lipid concentrations in the predefined ration of 1:2:3 with plasticizer and PVC represented in table 1. Because of the ratio, the properties of the membrane, i.e. the surface charge density, its fluidity, and hydrophobicity, which affect the potential value across the membrane [10]. The amount of the analyte substance adsorbed onto the lipid /polymer membrane depends upon lipid concentration, because it gives the site in the form of the charge layer to enable interaction (hydrophobic).

2.3 Synthesis of aqueous gold nanoparticles

An aqueous solution of GNPs was prepared by reducing tetrachloroauric acid (HAuCl4) with glutamic acid under refluxing to yield colloidal gold particles. For this, 50 ml of 0.01% tetrachloroauric acid in Milli-Q water was taken in a reaction flask and refluxed for 10 min with 50 ml aqueous solution of trisodium citrate of 5 mM molarities was used [24]. The reduction of the gold metal ions (Au3+) to yield GNPs (AuNP) was confirmed by the appearance of a dark cherry–red to blue color in the different colloidal solutions.

2.4 Modification of Lipid membrane

The lipid/polymer membrane modified to study the response behavior of sweet analyte comprised of the same discussed in section 2.2. The chemical species of AuNP are introduced at the time of preparation of the lipid/polymer membrane and concentrations of lipid and plasticizer kept same. The modification of lipid/polymer membrane with AuNP brings about charges in the surface charge density, fluidity and hydrophobicity of the membrane surface.

2.5 Measurement using Taste Sensor

The relative value of the variation of membrane potential was measured using an electrochemical work station, (Zahner electrik electrochemical workstation IM6) composed of a gold working electrode and a common reference electrode. The sensing electrode to which lipid/polymer membrane (modified and

IOP Conf. Series: Materials Science and Engineering **360** (2018) 012039 doi:10.1088/1757-899X/360/1/012039

non-modified) is attached measures the changes in the membrane potential that is generated when the electrode is immersed in a sample. The membrane prepared deposits on the electrode. The details of the sample used in the study are summarized in table 1. The measurement procedure is as follows (Figure 2). The sensor electrode was immersed in the reference solution, and the membrane potential is measured (Vr), this process is followed by membrane potential in the sample solution (V¬s). After this once again, the sensor electrode is treated with the reference solution. The highest value achieved during the sample solution (90% of the highest value) and the initial value of reference solution (Vr), the difference is taken as relative change due to the sample solution interaction with the membrane. In the end, the rinsing solution, which consists of 100mM KCl, 10mM KOH, 30% vol of Ethanol was used to rinse the sensor electrode. The process was repeated for five times and mean value of the consistent response is taken into consideration to avoid possible human error during the experiment.

Table 1. Analytes used during the taste sensing

Taste Sample	Analyte	Composition
Control	PBS	137mM NaCl, 2.7mM KCl, 10mM Na_2HPO_4, 1.8mM KH_2PO_4
Sweet	Glucose	0.01mM -1000mM + control
	Sucrose	
	Fructose	
Salty	NaCl	
Sour	HCL	
Bitter	$MgCl_2$	
Umami	MSG	

2.6 UV-Vis Spectroscopic Investigation

The UV-Visible absorption spectra were obtained using Lambda 25, Perkin Elmar, UV/VIS Spectrometer with the slit width of 10 nm and the measurement were performed in the range of 200 to 700 nm.

3. RESULT AND DISCUSSION

This section may be divided by subheadings. It should provide a concise and precise description of the experimental results, their interpretation as well as the experimental conclusions that can be drawn.

3.1. Chemiscal characteristics of modificed artificial lipd membrane

The taste sensor based on artificial lipid membrane, was synthesized by using hydrocarbon compound was analyzed for all basic tastes. The artificial lipid membrane was modified by using gold nanoparticles (AuNPs) which may to affect the surface property of the lipid membrane. The synthesis process was a chemical process comprised of lipid part: plasticizer part and supporting polymer part in the ratio of 1:3:2 [1–3]. Figure 1 shows the chemical structure of the lipid molecule (tetradodecylammonium bromide i.e. TDAB), plasticizer molecule (dioctyl phenyl phosphonate i.e. DOPP) and supporting polymer (polyvinyl chloride i.e. PVC). The concentration of TDAB on the membrane surface is a factor causing the surface of the lipid/polymer membrane to be positively charged because TDAB is a positively charged lipid [1,4,5]. During the reaction, lipid part which acts as ammonium salt provide bromide ion to plasticizer and give rise to the appearance of positive surface charge over the membrane surface of TDAB.The AuNP which was prepared by the process mentioned in the section 2.3, consists of the process of refluxing with citrate solution. This process gives rise to the appearance of partial negative charge on the AuNP surface [6]. The modified artificial lipid which is achieved by the doping of AuNP during the synthesis of artificial lipid membrane gives rise to a membrane which is more conjugated. The interaction between the AuNP and artificial lipid membrane brought charges rearrangement within the membrane, this eventually affects the fluidity, charge density and permeability of the membrane. The physico-chemical behavior of the lipid membrane depends upon the surface charge of the membrane [6–8], here hydrophilic part which is outside comprised of ammonium ion which is partially positive and the natural adherence is achieved to the citrate coated AuNP which is partially negative. In this situation the membrane probably would lose its rigidity, and its morphology would be changed and this reason suggests that the fluidity and permeability of the membrane might increase [9–11].

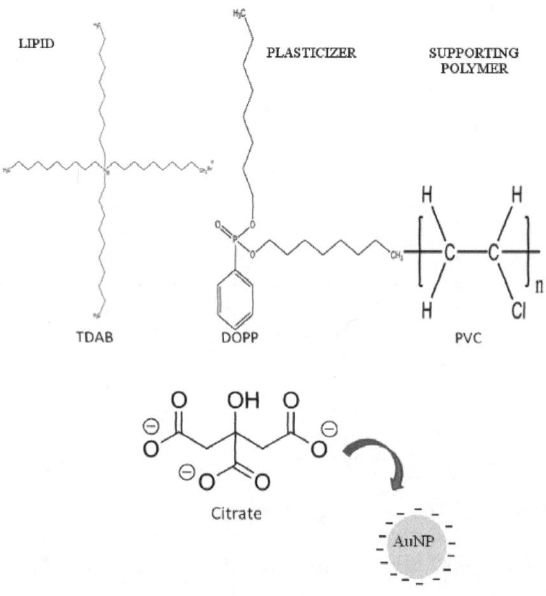

Figure 1. Structure of the lipid, plasticizer and supporting polymer used for the lipid-polymer membranes. Gold nanoparticles (AuNPs) with negative charge on the surface.

IOP Conf. Series: Materials Science and Engineering **360** (2018) 012039 doi:10.1088/1757-899X/360/1/012039

3.2. Sensing characteristics of modificed artificial lipd membrane

In the Figure2, the process of sensing mechanism is illustrated, where the relative values were taken as the electronic response. Baseline measurement were carried by using the control solution (mentioned in the table 1). The values of open circuit potential changes in the presence of different analytes. It was observed that the baseline shift towards higher electronic response with in the presence of sugar molecules in the solution while the decrement was obtained in case of salt, sour and bitter taste sample in the solution.

Table 2. List of chemicals used during the synthesis of lipid membranes

Chemical used with abbreiviation	Weight of chemicals used
tetradodecylammonium bromide/TDAB	100mg
dioctyl phenyl phosphonate/DOPP	310μL
polyvinyl chloride/PVC	200mg

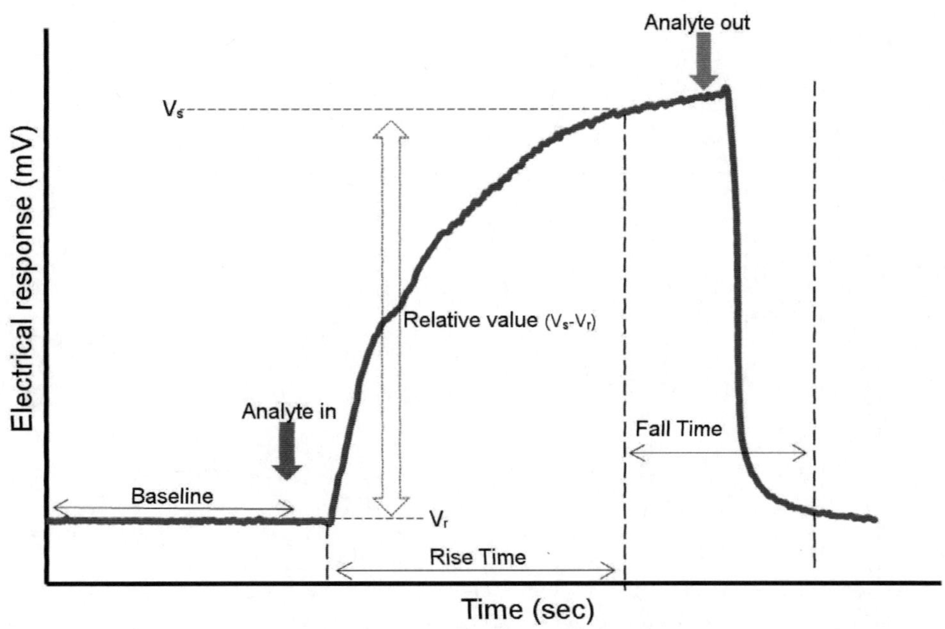

Figure 2. Schematic representation of taste sensing measurement process

IOP Conf. Series: Materials Science and Engineering **360** (2018) 012039 doi:10.1088/1757-899X/360/1/012039

3.3. Spectroscopic characteristics of the modificed artificial lipd membrane

In Figure 3, the UV-Vis graph shows the change in the absorption spectra of artificial lipid membranes due to its modification with AuNP. Inset figure depicts the absorption peak of the AuNP at 520 nm. This AuNP was doped during the synthesis process of the artificial lipid membrane. Nanoparticles, which increase the electron density of lipid membranes, it consequently increases the interaction between part of the lipid membrane based on its polarity [10,12,13]. THF which was used during the UV-Vis measurement appears to be a good solvent in which the chains are relatively extended, so the light can be absorbed without much mutual interference. The absorption peak at 300 nm ± 5 nm may arise due to n-->π* transition of C=O, the similar transition effect might occur in the complex of TDAB-DOPP with PVC. The peak wavelength shows the tendency to shift towards long wavelength as the size of the conjugated system increases [14]. In Figure 3, the absorption peak of 300 nm shift towards longer wavelengths shows the presence of the larger size of the conjugated system in the modified artificial lipid membrane. The additional band appeared in the modified structure likely to be attributed due to the covalent interaction of AuNP to the polymers 25. The strong absorption band centered at 520 nm arises due to the presence of AuNP in a dielectric medium, localized surface plasmon give rise to such peaks. During the lipid membrane synthesis this dielectric medium changed and hence peak got suppressed and a small hump is visible with the small red shift. This supression in the intensity may also suggest the decrease in the free AuNP in the solvent after self assembly with artificial lipid membrane.

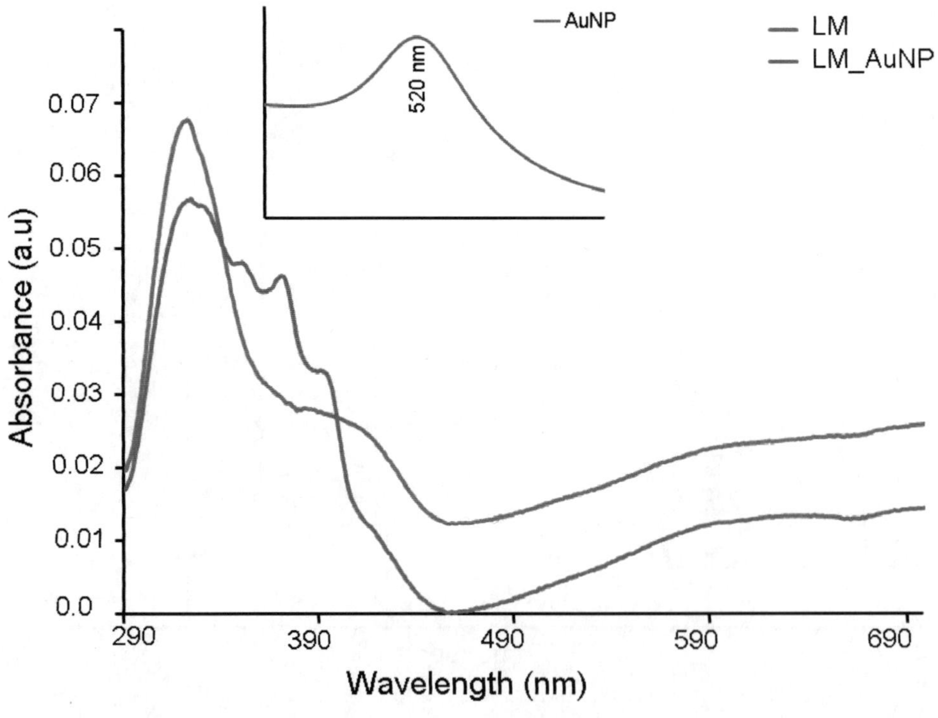

Figure 3. UV-Vis spectra of modified artificial lipid membrane (LM_AUNP) and TDAB-DOPP based artificial lipid membrane (LM). The inset shows the UV-Vis graph of AuNP.

3.4. Taste sensing

The response study behavior was studied subsequently using developed open circuit potential within the synthesized artificial lipid membrane deposited on the electrode. Figure 4 shows the response graph of AuNP modified artificial lipid membrane (LM_AuNP) and non-modified artificial lipid membrane (LM), in the different concenrtrations, range between 10 μM to 1000 mM. dipicted in (a) to (f). Relative values were considered during the response measured, where it was found that sugar molecules showed enhanced response behavior towards the modified artificial lipid membrane which might be due to dissociation state of -COOH of the AuNP incorporated with the lipid membrane [5,6,15,16]. It was observed earlier that the carboxyl or phospahte group in the lipid/polymer membranes is essential for the sugar (surose) response of the sweetness sensor, and the hydroxyl group may not hold the necessary position for such event. Although, carboxylate groups of citrate molecule may react with the -OH group of sugars [17]. Among the sugars, the response behavior for modified lipid membrane was more prominent for the fructose sugar rather than conventional sucrose, this may be due to the presence of 'keto' group of fructose and it is also known to us, The response behavior for the compounds like NaCl, HCl and $MgCl_2$ showed consistent decrease in the increment in potential across the artificial membrane. The relative value of the NaCl increases negatively with the lipid membrane due to the screening effect of electrolyte Cl^- anions [18]. The similar kind of behavior can be suggested for HCl and $MgCl_2$.

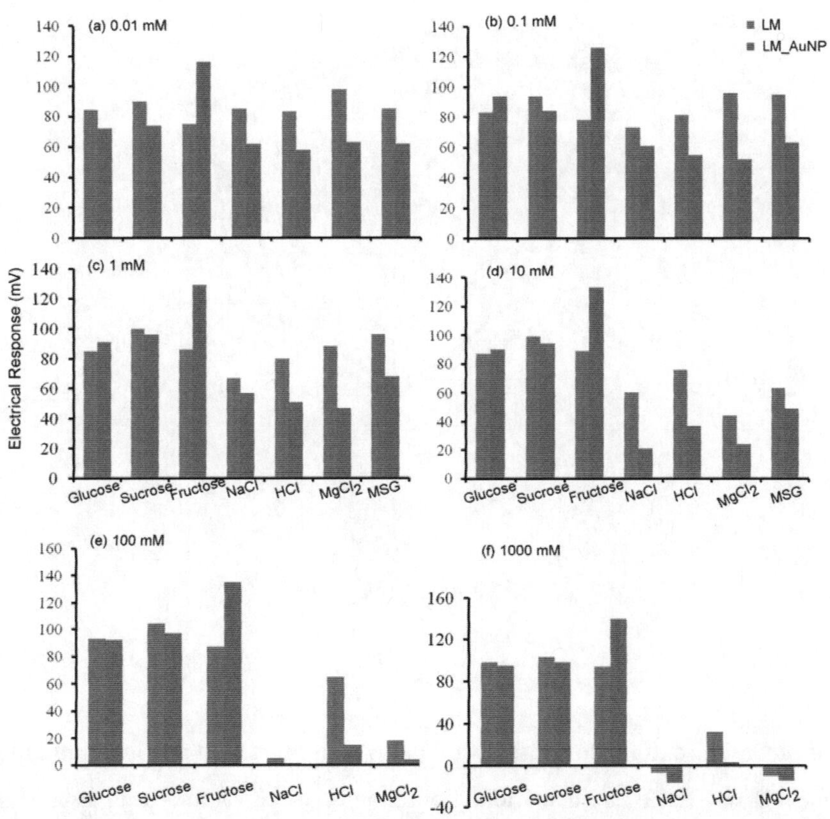

Figure 4. Taste response analysis of AuNP modified artificial lipid membrane (LM_AUNP) and TDAB-DOPP based artificial lipid membrane(LM) on basic taste analytes for sweet (glucose, sucrose and fructose), salty (NaCl), sour (HCl), bitter (MgCl2) and umami taste (MSG). (a) 0.01mM; (b) 0.1 mM; (c) 1 mM; (d) 10mM; (e) and (f) shows 100 mM and 1000 mM respectively for sweet (glucose, sucrose and fructose), salty (NaCl), sour (HCl), and bitter (MgCl2) taste.

In the results, it was also observed that the steady relative response between modified and unmodified lipid membrane was found in the MSG compound, which might be due to its neutrality in an aqueous medium [19,20].

Figure 5 showed the radar plot for the alaysis of sensitivity towards sweet analytes, and in the result, it was observed that the LM_AuNP artificial lipid membrane showed enhanced sensitivity towards fructose molecule. Based on the response graph (Figure4) and radar plot (Figure5), we can suggest that, the behaviorial difference in sensitivity of modified lipid membrane (LM_AuNP) and non-modified lipid membrane (LM). The increased response to fructose, the behavior of NaCl towards Au doped TDAB-DOPP (LM_AUNP) based artificial lipid membrane electrode is a matter of investigation and we are conducting further experiments to ascertain these facts.

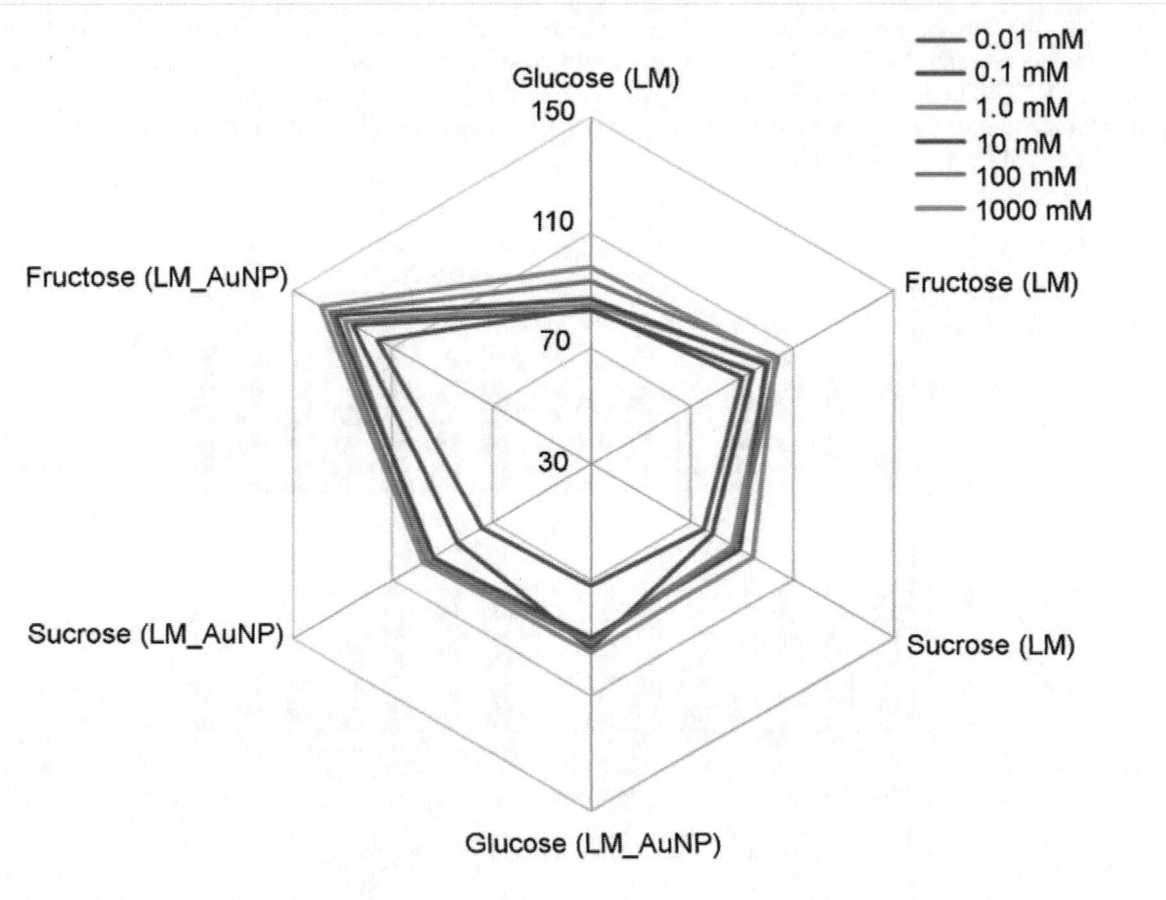

Figure 5. Radar plot representing the relative values in AuNP modified artificial lipid membrane (LM_AuNP) and TDAB-DOPP based artificial lipid membrane (LM) for sweet taste analytes (glucose, sucrose, and fructose).

4. Conclusions

In this study, a lipid membrane with AuNP modification was developed, which shows high sensitivity towards sugar molecules. The UV/Vis graph clearly showed the presence of conjugation due to the AuNP, and decrease in the free AuNP into the system. The interaction between hydrophilic ammonium part and the AuNP might be positive which leads to the change in the physicochemical behavior of the modified lipid membrane. The comparison of the response of TDAB-DOPP (artificial lipid membrane) with AuNP with unmodified TDAB-DOPP, sowed higher sensitivity to fructose molecules. The results from the other analyte's response suggested that the surface charge changed due to the doping, which results in the variation of response for different taste analytes like salty, sour and bitter. Comprehensively, we can state that the observed response in the presence of AuNP, which is the citrate coated, showed enhanced sensitivity to fructose. We are conducting a further experiment to clarify the relationship between modified the artificial lipid membrane and change in its sensitivity towards the same taste analytes groups.

Acknowledgments:

The authors extend their sincere thanks to the Director, "CSIR-Central Scientific Instruments Organisation", to laboratory members of FPIL group of Agrionics Division.

Author Contributions:

Saurav Kumar and Amol P. Bhondekar conceived and designed the experiments; Saurav Kumar, Amupma Sharma performed the experiments; Ritesh Kumar, Sudeshna Bagchi analyzed the data; Prateek Jain contributed to characterization; Saurav Kumar wrote the paper; Amol P.Bhondekar, Sunita Mishra edited the paper. All authors have read and approved the final manuscript.

Conflicts of Interest

The authors declare no conflict of interest

5. References

[1] Toko K 1996 Taste sensor with global selectivity *Mater. Sci. Eng. C* **4** 69–82
[2] Yamafuji K, Toko K and Hayashi K 1996 Taste sensing system using artificial lipid membranes
[3] Krull U J and Thompson M 1987 Chemo-receptive lipid based membrane transducers
[4] Toko K 2000 Taste sensor *Sensors Actuators B Chem.* **64** 205–15
[5] Toyota K, Cui H, Abe K, Habara M, Toko K and Ikezaki H 2011 Sweetness sensor with lipid/polymer membranes: sweet-responsive substances *Sens. Mater* **23** 465–74
[6] Cho E C, Xie J, Wurm P A and Xia Y 2009 Understanding the role of surface charges in cellular adsorption versus internalization by selectively removing gold nanoparticles on the cell surface with a I2/KI etchant *Nano Lett.* **9** 1080–4
[7] Verma A and Stellacci F 2010 Effect of surface properties on nanoparticle–cell interactions *Small* **6** 12–21
[8] Tatur S, Maccarini M, Barker R, Nelson A and Fragneto G 2013 Effect of functionalized gold nanoparticles on floating lipid bilayers *Langmuir* **29** 6606–14
[9] Brockman H 1994 Dipole potential of lipid membranes *Chem. Phys. Lipids* **73** 57–79
[10] Broda J, Setzler J, Leifert A, Steitz J, Benz R, Simon U and Wenzel W 2016 Ligand-Lipid and Ligand-Core Affinity control the Interaction of Gold Nanoparticles with Artificial Lipid Bilayers and Cell Membranes *Nanomedicine Nanotechnology, Biol. Med.*
[11] Hanemann T and Szabó D V 2010 Polymer-nanoparticle composites: from synthesis to modern applications *Materials (Basel).* **3** 3468–517
[12] Behera M and Ram S 2013 Spectroscopy-based study on the interaction between gold nanoparticle and poly (vinylpyrrolidone) molecules in a non-hydrocolloid *Int. Nano Lett.* **3** 1–7

[13] Chitte H K, Bhat N V, Karmakar N S, Kothari D C and Shinde G N 2012 Synthesis and characterization of polymeric composites embeded with silver nanoparticles *World J. Nano Sci. Eng.* **2** 19

[14] Ionescu D, Popescu A, DRĂGUŞIN M, DIMA M, IFTIME A and GANEA C 2007 Modulation by quercetin of the effect of certain Hofmeister anions on artificial lipid bilayers *Rom J Biophys* **17** 85–90

[15] Yasuura M, Okazaki H, Tahara Y, Ikezaki H and Toko K 2014 Development of sweetness sensor with selectivity to negatively charged high-potency sweeteners *Sensors Actuators, B Chem.* **201** 329–35

[16] Habara M, Beppu D, Cui H, Ikezaki H and Toko K 2007 Detection of sugars using lipid/polymer membranes *Sens. Mater* **19** 325–31

[17] Wittmann V 2006 The Organic Chemistry of Sugars. Edited by Daniel E. Levy and Péter Fügedi.

[18] Kobayashi Y, Habara M, Ikezazki H, Chen R, Naito Y and Toko K 2010 Advanced taste sensors based on artificial lipids with global selectivity to basic taste qualities and high correlation to sensory scores *Sensors* **10** 3411–43

[19] Corradini D 2010 *Handbook of HPLC* (CRC Press)

[20] Daub C D, Leung K and Luzar A 2009 Structure of Aqueous Solutions of Monosodium Glutamate *J. Phys. Chem. B* **113** 7687–700

IOP Conf. Series: Materials Science and Engineering **360** (2018) 012040 doi:10.1088/1757-899X/360/1/012040

EPR studies on VO^{2+}doped sodium –borate glasses synthesized by microwave method

Asha Rajiv[1], Jayagopal Uchil[2],C. Narayana Reddy[3*]
[1,2,]Department of Physics, School of Graduate Studies, Jain University, Bangalore 560027, India.
[3*] Department of Physics, Sri Kongadiyappa College, Bangalore University, Bangalore -56, India.

*Corresponding author: Email: nivetejareddy@gmail.com
ashaarajiv@gmail.com, Mob: +91 9945617403

ABSTRACT
Electron paramagnetic resonance (EPR) of VO^{2+}ions in sodium borate glass system has been studied over a wide range of composition. The EPR spectra exhibited hyper fine splitting (hfs) due to considerable weakening of interaction between vanadium ion concentrations. The hfs is seen between 10 and 35 mol% of vanadium concentration. The appearance of hfs is due to the interaction of $3d^1$ electron spin of ^{51}V ion with its nuclear spin, which increases with increasing concentration till 35 mol%. The spin -Hamiltonian parameters (g and A), the dipolar hyperfine coupling parameter (P) and Fermi contact interaction parameter (k) have also been evaluated. The values of spin-Hamiltonian parameters indicate that VO^{2+} ions are in octahedral coordination with a tetragonal compression which can be attributed to the structural changes taking place with the increasing concentration of V_2O_5

Key words: Glasses, Electron paramagnetic resonance, hyper fine splitting, octahedral coordination.

1. INTRODUCTION
Glasses are a class of amorphous materials exhibiting a variety of useful proprties which can be used for technological applications. These materials were extensively synthesized by conventional melt quenching method. In this method thermal energy is delivered to the surface of the material, thus thermal gradient will develop in the long heating process and material loss may occur. However, in micrwave heating uniform and volumetric heating of the material takes place. Since microwave heating is faster and energy efficient, we adopted this method to prepare the samples. Glasses containing transition metal ions have attracted researchers due to their physical applications in electronic and electro-optic devices [1,2].Vanadium pentoxide based materials have been used as cathode materials in alkali batteries [3,4].These materials are suitable for electrochemical propertiesbecause the electrochemical redox potential of V^{5+}/ V^{4+} couple as compared to the Li^+/Li couple is high as lithium glasses offers sufficiently high Li^+ diffusion coefficient.Glasses containing $V_2O_5$are of interest for studying the correlation between electrical properties and changes in the glass network because V_2O_5 is unique, being itself, a conditional glass former. In amorphous V_2O_5, the closest site spacing between V – V is about 3.5 Å, a value which is almost equal to twice the value of V – O spacing [1.75 Å] in the basic structural unit and which corresponds to the O –V – O distance of the basal plane. Thus, the glass structure optimizes the packing of such units. The basic boro-vanadate glasses consists of trigonal $[BO_{3/2}]^0$ and tetrahedral $[VOO_{3/2}]^0$ units. During modification, it is important to decide which of $[BO_{3/2}]^0$and $[VOO_{3/2}]^0$ units modify preferentially. The preference of modification is decided by the Sandersons electronegativity principle [5]. Since both the structural units $[BO_{3/2}]^0$and $[VOO_{3/2}]^0$ have the same electronegativity, they have equal a ptiori probability to attract the oxide ion.

Electron Paramagentic Resonance (EPR) spectroscopy is one of the sensitive and important spectroscopic technique to study paramagnetic centers of transition metal ions. There are classic review reports on EPR technique [6,7]. The applicability of the EPR technique to V_2O_5 containing glass system is mainly due to the presence of V^{4+} paramagnetic centers, whose outer shell structures, $3p^6$, $3d^1$ enables unpaired magnetic moment of $3d^1$ electron to interact with the applied field. Since a

V^{4+} ion has an electronic spin s =1/2, a nuclear spin I = 7/2 leading to the interactions between corresponding magnetic moments resulting in a hyperfine structure [8, 9]. At lower concentrations of V_2O_5, hyperfine structure is seen in all the spectra typical of V^{4+} (isolated) ions. With increasing V_2O_5 concentration, hyperfine structure was superposed by a broad signal without showing any hyperfine structure characteristic of clustered ions. Gupta *et al*[10] pointed out that, long range spin-spin interactions between V^{4+} ions and the spin orbit coupling cause an anisotropy of the g-factor and the broadening of the individual lines [11]. In glasses only orientation averaged spectra can be observed, that leads to additional reduction of hyperfine structure lines. It was seen in V_2O_5-TeO_2 glasses that, the disappearance of hyperfine structure lines at higher contents of V_2O_5 is due to superexchange interaction of V^{4+}- O - V^{5+}linkages [12]. In this paper we attempted to study EPR spectra of microwave synthesized glasses doped with vanadyl ions.

2. EXPERIMENTAL

Glasses were prepared by using the general formula x V_2O_5 . (100 – x) [$0.5Na_2O : B_2O_3$] (where x = 10, 15, 20, 25, 30, 35 mol %) by microwave melting technique using analytical grade sodium carbonate, orthoboric acid, and vanadium pentoxide as starting materials. Appropriate quantity of weighed chemicals were mixed and thoroughly ground to homogenize the mixture and kept in a silica crucible inside a domestic kitchen microwave oven operating at 2.45 GHz, 850 W for about 3 to 5 minutes. Glass was obtained by rapidly quenching the melt between two copper blocks preheated to 100^0C to avoid cracking of the samples. Glasses were annealed in a muffle furnace for about 2 hours at 200^0C to remove thermal strains that could have developed during fast quenching.

Electron paramagnetic resonance spectra have been recorded on an EPR spectrometer (JEOL FE1X) operating in the X-band frequency (9.205 GHz) with a modulation frequency of 100 kHz. The magnetic field was scanned from 2200 to 4200 G with a scan speed of 625 G /min and the microwave power used was 5 mW. A powdered glass specimen of 100 mg was taken in a quartz tube for EPR measurements. The EPR spectrum of $CuSO_4 \cdot 5H_2O$ powdered sample was also recorded as a reference to calculate the number of spins. Polycrystalline DPPH with an effective *g* value of 2.0023 was used as a standard field marker.

3. RESULTS AND DISCUSSION

3.1. EPR spectroscopic studies

EPR is a technique which provides information comprising the valence states of transition metal ions (TMI), local environment and nature of interaction between them. The EPR spectra exhibited hyper fine structure (hfs) due to considerable weakening of interaction between vanadium ion concentration. The EPR spectra of vanadyl ions [VO^{2+}] in glasses are very interesting because the valance state of vanadyl ions in glasses changes with the glass composition. The presence or absence of hfs in the EPR spectra is due to the modification of the glass structure caused by the modifier oxide (Na_2O). Fig.1 illustrates EPR spectra of sodium boro-vanadate glasses. EPR spectrum consisting of eight lines corresponding to eight values of nuclear magnetic quantum numbers m_I =-7/2, -5/2, . . . , +7/2. The hfs is seen between 10 and 35 mol% of V_2O_5. The appearance of hfs is due to the interaction of $3d^1$ electron spin of ^{51}V ion with its nuclear spin, which increases with increasing concentration till 35mol% which is evident from Fig .1. The disappearance of hfs for V_2O_5concentrations above 40 mol% can be attributed to an enhanced electron hopping between aliovalent vanadium centers along V^{4+} – O – V^{5+}linkages. Short life time of $3d^1$ electron on V^{4+} center causes the broadening of the energy levels of hfs and consequently leads to its disappearance.

It can be seen from Fig.1, a strong absorption line arises from the fact that at high V_2O_5 content, most of the vanadium ions are in the V^{4+} state. The absence of hyperfine structure (hfs) points to the interaction between vanadium centers via V^{4+}-O-V^{5+}superexchange mechanism. Generally, such glasses exhibit electronic conductivity [13]. At 35 mol% of vanadium, the EPR spectra consists of V^{4+} line with a weak but visible superimposed hfs, which indicates the cross-over from hfs regime to non-hfs regime. However, the mechanism of conduction in glasses with high concentration of V_2O_5 has been suggested as the transfer of an electron from V^{4+} site toV^{5+} site. Structural groups formed in V_2O_5 rich glasses, provide the path for hopping of electrons. The increase in V_2O_5 rich glasses can be explained by considering the decrease in average distance between the TMI site and redox ratio.

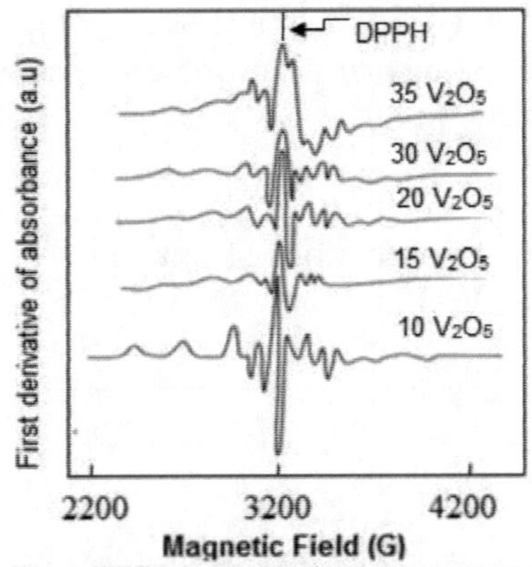

Fig. 1: EPR spectra obtained at room temperature for NBV glass system with varying mol% of V_2O_5 and identical mol% of Na_2O and B_2O_3.

The EPR spectra is well resolved from x = 10 to 35mol% due to ^{51}V and interpreted using axially symmetric spin Hamiltonian [14]. The solutions of the spin Hamiltonian are given by the following equations:

$$B_{\parallel}(m) = B_{\parallel}(0) - mA_{\parallel} - \left[\frac{63}{4} - m^2\right]\frac{A_{\perp}^2}{2B_{\parallel}(0)} \quad (1)$$

$$B_{\perp}(m) = B_{\perp}(0) - mA_{\perp} - \left[\frac{63}{4} - m^2\right]\frac{A_{\parallel}^2 + A_{\perp}^2}{4B_{\perp}(0)} \quad (2)$$

where m is the magnetic quantum number, ν is the frequency of the spectrometer, h is the Planck's constant, β is the Bohr magneton, B is the magnetic field, g and A are the spin-Hamiltonian parameters.

Also, $B_{\parallel}(0) = \frac{h\nu}{g_{\parallel\beta}}$ and $B_{\perp}(0) = \frac{h\nu}{g_{\perp\beta}}$

$A_{\parallel}$and$A_{\perp}$have been computed using the relations

$$A_{\parallel} = \frac{1}{7}\,[B_{\parallel}(-\frac{7}{2}) - B_{\parallel}(\frac{7}{2})] \quad (3)$$

$$A_{\perp} = \frac{1}{7}\,[B_{\perp}(-\frac{7}{2}) - B_{\perp}(\frac{7}{2})] \quad (4)$$

The spin Hamiltonian parameters $g_{\parallel}$,$g_{\perp}$, $A_{\parallel}$and $A_{\perp}$ have been calculated and the values are listed in Table 1. Fig.2 show the variation of $A_{\parallel}$and $A_{\perp}$withV_2O_5mol%. As can be seen from Fig.2 that both $A_{\parallel}$and $A_{\perp}$gradually decreases with increasing V_2O_5 content, which suggests that the covalence between the central atom and the surrounding oxygen ligands.

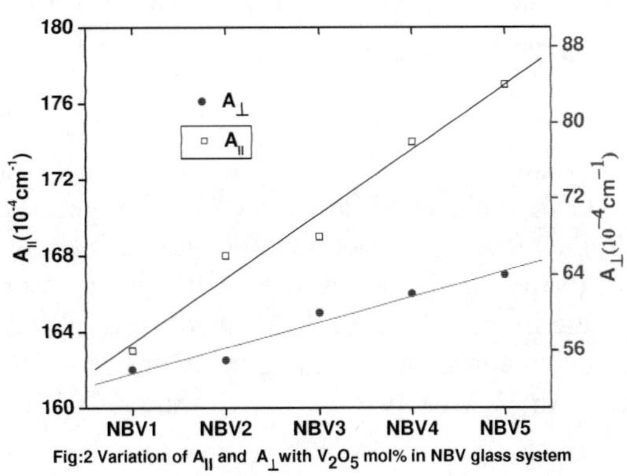

Fig:2 Variation of $A_{\parallel}$ and $A_{\perp}$with V_2O_5 mol% in NBV glass system

IOP Conf. Series: Materials Science and Engineering **360** (2018) 012040 doi:10.1088/1757-899X/360/1/012040

It is clear from Table 1,that the values of $g_{\perp}$increases while $g_{\parallel}$decreases with increasing V_2O_5concentration suggesting a greater tetragonal distortion of vanadyl complex. Further, the value of $g_{\perp}>g_{\parallel}$suggesting that vanadium present in glass as vanadylion[VO^{2+}] at a tetragonally distorted octahedral site [15].
It has been qualitatively shown from molecular orbital theory that $A_{\parallel}$ and $A_{\perp}$consists of contribution from $3d_{xy}$ electrons and anamolus contributions from the s- electrons. The variation of anamolouscontributions $A'_{\parallel}$and $A'_{\perp}$with V_2O_5mol% are shown in Fig. 3.

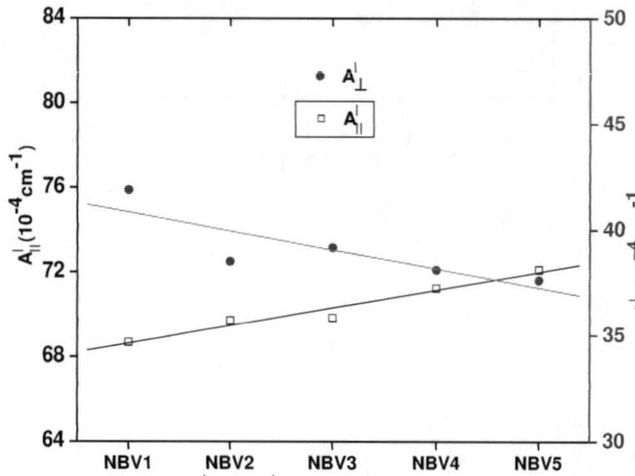

Fig:3 Variation of $A'_{\parallel}$ and $A'_{\perp}$ with V_2O_5mol% in NBV glass system

$A'_{\parallel}$and $A'_{\perp}$were calculated using the values of $A_{\parallel}$, $A_{\perp}$, P and k . The dipolar coupling constant (P) and fermi – contact coupling parameter (k) are evaluated using the expression given by Kivelson and Lee [16].

$$A_{\parallel} = -P[\left(k+\frac{4}{7}\right)-\Delta g_{\parallel}-\left(\frac{3}{7}\right)\Delta g_{\perp}] \quad (5)$$

$$A_{\perp} = -P\left[\left(k-\frac{2}{7}\right)-\frac{11}{14}\Delta g_{\perp}\right] \quad (6)$$

where g_e= 2.0023, $\Delta g_{\parallel} = g_{\parallel} - g_e$and $\Delta g_{\perp} = g_{\perp} - g_e$. The values of P and k evaluated using the relations (5) and (6) are presented in Table 1.
Equation (5) and (6) can be written as

$$A_{\parallel} = A'_{\parallel} - P\beta^2 k \quad (7)$$

$$A_{\perp} = A'_{\perp} - P\beta^2 k \quad (8)$$

The hyperfine components of $A_{\parallel}$ and $A_{\perp}$ consists of the contributions of $A'_{\parallel}$and $A'_{\perp}$ of the $3d_{xy}$electrons to the hyperfine structures. It is evident from Fig. 3 that both $A'_{\parallel}$increases and $A'_{\perp}$decreases with increasing concentration of V_2O_5 , indicating that as V_2O_5 content increases the V – O bond becomes more covalent and $3d_{xy}$orbitals are dilated. Assuming the free electron g- value (2.0023), $\Delta g_{\parallel}$ and $\Delta g_{\perp}$ are calculated. It can be seen from Table .1 that the ratio $\frac{\Delta g_{\parallel}}{\Delta g_{\perp}}$ increases with increasing V_2O_5 concentration. Besides, the ratio $\frac{\Delta g_{\parallel}}{\Delta g_{\perp}}$ is a measure of the degree of teragonality. The values of g -tensor depends on distortion type. In general, the axial distortion can be either three fold or four fold symmetries to describe the crystal field of V^{4+}ions in glasses. The bonding of the vanadyl ion in ligand field of octahedral symmetry has been well characterized by earlier workers [17-19].
The V^{4+} site in the VO^{2+} ion forms planar bonds with each of the four oxygen ligands. VO^{2+} is an octahedral environment of oxygen with tetrahedral distortion (C_{4V}) where the vanadyl oxygen forms the apex V– O bond. Pk /γ is evaluated using 'γ' the gyromagnetic ratio which is equal to 1.468 for ^{51}V. The variation of P and k with V_2O_5 concentration is shown in Fig. 4 while that of Pk/γ versus V_2O_5mol % is shown in Fig. 5.

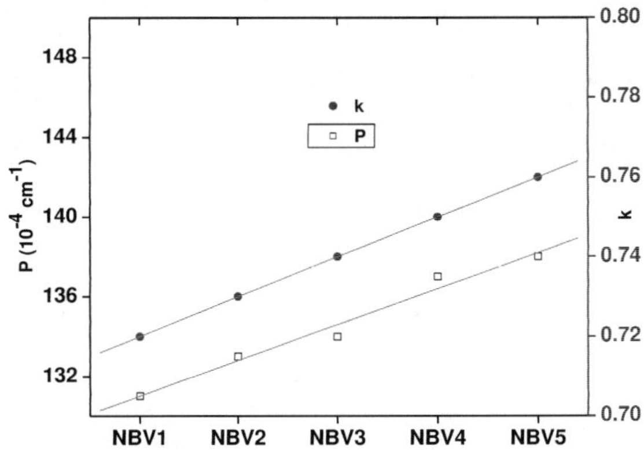

Fig:4 Variation of P and k with V_2O_5 mol% in NBV glass system

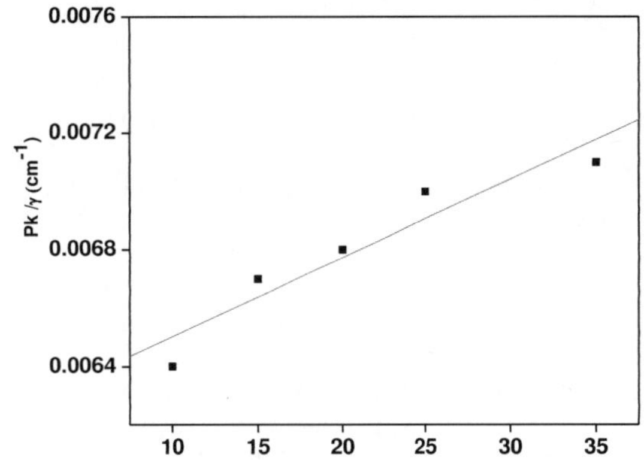

Fig: 5 Variation of Pk /γ with V_2O_5 mol% in NBV glass system.

The variations seen in these parameters indicate the s-character of the magnetic spin of V_2O_5, which arises due to the partial unpairing or polarization of s-electrons as a result of an interaction with unpaired d- electron [20]. Pk/γ is found to increase with increasing concentration of V_2O_5indicating an increase in tetragonal distortion and is also a measure of 's' character of the spin of the vanadium. The bonding coefficient α^2 can be evaluated using the relation [21].

$$\alpha^2 = \left[1 - \frac{g_{\parallel}}{g_e}\right] \frac{E}{4\lambda\beta^2} m \qquad (9)$$

where E is the energy of the transition [$^2B_2$ - $^2B_1$] and β^2 is a measure of the in-plane π-bonding with the equilateral ligand and λ is the spin orbit coupling constant and is equal to 170 cm^{-1}. The value of α^2 calculated using the relation (9) is given in Table 2.

4. CONCLUSIONS

The spin Hamiltonian parameters extracted from the spectra shows that vanadium ions are present in the matrix of glass as a vanadyl [VO^{2+}] complex. A – tensor, g- tensor, dipolar hyperfine coupling parameter (P) and Fermi – contact coupling parameter (k) exhibit composition dependent trend. Increase in V_2O_5 concentration greatly affects the tetragonal distortion of the vanadyl complex. From EPR studies, it can be concluded that the tetrahedral vanadium ions in the investigated glasses exist in vanadyl complexes with the tetragonally distorted octahedral sites.

Table:1 Spin Hamiltonian parameters of VO^{2+} ion in glass system.

Code	$g_{\parallel}$	$g_{\perp}$	$\Delta g_{\parallel}$	$\Delta g_{\perp}$	$\frac{\Delta g_{\parallel}}{\Delta g_{\perp}}$	$A_{\parallel}$ 10^{-4} cm^{-1}	$A_{\perp}$ 10^{-4} cm^{-1}	P 10^{-4} cm^{-1}	k	Pk/γ cm^{-1}
NBV1(10V)	1.962	1.970	0.4030	0.0320	1.26	163	54	131	0.72	0.0064
NBV2(15V)	1.959	1.978	0.0433	0.0243	1.78	168	55	133	0.73	0.0067
NBV3(20V)	1.956	1.980	0.0463	0.0223	2.07	169	60	134	0.74	0.0068
NBV4(25V)	1.955	1.982	0.0473	0.0203	2.33	174	62	137	0.75	0.0070
NBV5(35V)	1.954	1.985	0.0483	0.0173	2.79	177	64	138	0.76	0.0071

Table: 2 Molecular orbital coefficients and crystal field splitting parameters.

Code	λ	$\Delta_{\parallel}$	β^2	α^2
NBV3(30V)	482	20746	0.7367	0.7207
NBV2(20V)	477	20964	0.7137	0.7128
NBV1(10V)	473	21141	0.6264	0.6257

REFERENCES

[1] T Minami T, K Imazawa, M Tanaka, "Formation region and characterization of superionic conducting glasses in the systems AgI-Ag_2O-M_xO_y",1980 *J.Non-cryst. Solids*. **42** pp.469-476.

IOP Conf. Series: Materials Science and Engineering **360** (2018) 012040 doi:10.1088/1757-899X/360/1/012040

[2] U Selvaraj, K J Rao, "Characterization studies of molybdophosphate glasses and a model of structuraldefects", 1985 *J.Non-Cryst. Solids*. **72** pp.315-334.

[3] M Levy, J L Souquet, "Amorphous and vitreous materials as electrodes in electrochemical cells", 1989 *Mater. Chem. Phys*. **23** pp. 171-188.

[4] M C Ungureanu, M Levy, J LSouquet, "Mixed conductivity of glasses in the P_2O_5-V_2O_5-Na_2O system", 1998 *Ionics*. **4** pp.200-206.

[5] R T Sanderson, `*Polar Covelance'* 1983 Academic Press, New York

[6] C P Poole Jr , "Electron Spin Resonance- A Comprehensive Treatise on Experimental Techniques", Edition, 1967, 1983 *Wiley-Interscience, New York; Second edition, Wiley, New York*.

[7] A Abragam, Bleancy, "EPR of Transition Metal Ions", 1970 *Clarendon Press Oxward*. .pp.270-425.

[8] V Dimitrov, "Structural changes in vitreous vanadate systems",1995 *J . Non- Cryst. Solids*., pp.183-186.

[9] J E Garbarzyk, Tykarski, P Machowski , M Wasiucionek, "EPR studies of mixed-conductive glasses in the AgI–Ag_2O–V_2O_5 –P_2O_5 system", 2001 *Solid State Ionics*.**140** pp.141-148.

[10] S Gupta, N Khanijo, A Mansingh, "The influence of V^{4+} ion concentration on the EPR spectra of vanadate glasses", 1995 *J Non -Cryst Solids*. **181** pp.58-63.

[11] R Muncaster, M Parke, "ESR spectra of glasses in the system V_2O_5-TeO_2",1977 *J Non-Cryst Solids*. **24** pp.399-412.

[12] S Sindhu, S Sanghi , A Agarwal, V P Seth, N Kishore, "Effect of Bi_2O_3 content on the optical band gap, density and electrical conductivity of $MO{\cdot}Bi_2O_3{\cdot}B_2O_3$ (M=Ba, Sr) glasses", 2005 *Mater.Chem.Phy*., **90**(1) pp. 83-89.

[13] J E Garbarczyk, P Machowski, M Wasiucionek, L Tykarski , R Bacewicz, A Aleksiejuk, "Studies of silver-vanadate-phosphate glasses by Raman,EPR and impedance spectroscopy methods", 2001 *Solid State Ionics*. 136-137 pp.1077-1083.

[14] C Narayana Reddy, R Damle, RV Anavekar, " Spectroscopic and structural studies on calcium borate glasses containing V_2O_5", 2006 *Physics and Chemistry of Glasses-European Journal of Glass Science and TechnologyPart B*. **47** (1) pp. 34-40.

[15] A Yadav, V P Seth, " Electron paramagnetic resonance of Cu^{2+} and VO^{2+} ions in phosphate glasses", 1982 *J. Mater. Sci.* **22** pp. 239-243.

[16] D Kivelson, S K Lee S K "ESR studies and the electronic structure of Vanadyl ion complexes,"1964 *J. Chem. Phys.,* **41** pp. 1896-1903.

[17] B Sreedhar , J Lakshmana Rao, S V J Lakshman, "Electron spin resonance spectra of VO^{2+} ions in alkali zinc borosulphate glasses", 1990 *J. Non-Cryst. Solids*.**116** pp. 111–114.

[18] V P Seth, S Gupta, A Jindal, S K Gupta, 1993 *J. Non-Cryst. Solids*. **162** pp.263.

[19] R V Anavekar, N Devaraj, K P Ramesh, J Ramakrishna, "Mössbauer effect and electron paramagnetic resonance studies on zinc borate glasses containing transition metal oxides" 1992 *Phys.Chem.Glasses*. **33** pp. 116-121.

[20] R P Sreekanth Chakradhar, K P Ramesh , J L Rao, J Ramakrishna, "Mixed alkali effect in borate glasses—EPR and optical absorption studies in xNa_2O–$(30-x)K_2O$–$70B_2O_3$ glasses doped with Mn^{2+}", 2003 *J. Phys.Chem. Solids*. **64** pp. 641–650.

[21] A K Bandyopadhyay , "Optical and ESR Investigation of borate glasses containing single and mixed Transition Metal Oxides,"1981 *J. Mater. Sci.*, **16** pp.189-203.

A comparative study on the postbuckling behaviour of circular plates

Anju V Nair[1], Abdul Rahman Mohd Kasim[1*], Mohd Zuki Salleh[1]

[1]Faculty of Industrial Sciences and Technology, Universiti Malaysia Pahang

Lebuhraya Tun Razak, 26300, Kuantan, Malaysia

Corresponding author: rahmanmohd@ump.edu.my

Abstract. Materials/metals, polymers, reinforced plastics, composites, functionally graded materials, ceramics and nano materials find significant place in many engineering and scientific applications. The material structural members such as plates and beams are subjected to structural deformation owing to several external and internal factors such as temperature, hysteresis and so on. Therefore, to analyze the structural deformation of these material structures, its nonlinear behavior modeling is relevant. The rapid progress in high speed computing devices resolve research problems involving material non-linearity behavior such as post-buckling behavior in varied structures. The post-buckling behavior of commonly used structural elements especially plates under thermal and mechanical loads are of great practically important research topic. Therefore, a comparative study on the thermal post buckling behavior of plates by various numerical methods such as finite element formulation, shear deformation theory, Rayleigh – Ritz method, iteration methods and so on will be helpful to obtain a cutting edge on numerical solutions and their applications in varied engineering applications. This paper reports a novel mathematical formulation involving substitution method to evaluate and compare the thermal post buckling load carrying capacity of circular plates with minimal errors.

1. Introduction

Study of deformation of structural members was based on high margins of safety. The effects due to nonlinearities involved in either strain – displacement relations termed as geometric nonlinearity, stress-strain relations termed as material nonlinearity is gaining research importance in field of structural mechanics. The nonlinearity in the behavior of the material is termed as material nonlinearity or physical nonlinearity. The analysis of structural members with physical nonlinearity is gaining research importance because of introduction of new materials such as metals, polymers, reinforced plastics, composites, functionally graded materials, ceramics and nano materials in aerospace and other industries. Application of these materials in the structural members under severe thermo-mechanical loads is of prime concern for researchers.

Beams, rectangular plates and circular plates (thin and thick) are commonly used structural elements in idealizing large aerospace structures. During their service conditions, these members/elements are subjected to severe thermo mechanical loads owing to aerodynamic/solar heating. Therefore, the effect

IOP Conf. Series: Materials Science and Engineering **360** (2018) 012041 doi:10.1088/1757-899X/360/1/012041

of nonlinearities (geometric and material) in structural members when subjected to the loading conditions as close to reality as possible has to be analysed mathematically. Numerical formulations is proved to be a simple and powerful mathematical technique for designers, practicing engineers as well as for researchers who wish to have accurate engineering results with less mathematical intervention and computational effort. Most of the earlier studies have been using the classical methods/energy methods/ variational methods and the versatile finite element method. However, the solution procedure in all the above mentioned studies involves solving of higher mathematical terms and solving higher order differential equations or involve in cumbersome calculations.

In the present study, the prime focus is on the post-buckling analysis of circular plates considering a wide range of loading conditions. Similar to other plates, the post-buckling of circular plate also occurs owing to in-plane loading. The precise knowledge of critical buckling loads, mode shapes and the post buckling behaviour is important for a reliable circular plate design. The simply supported and clamped boundary conditions are used in most of the post-buckling studies of circular plates. Different numerical methods such as finite element approach, iteration methods, and shear deformation theory has been developed owing to the variation in the thickness of circular plates.

Friedrichs and stoker [1] explained the post buckling behaviour of a simply supported circular plate at large deflection, by solving the coupled nonlinear von Karman equations, using power series and perturbation method. The method is extended to clamped circular plates by Bodner [2]. The thermal post buckling behaviour of isotropic circular plates with simply supported and clamped boundary conditions using finite element analysis has been presented in [3]. The numerical results are compared with the Rayleigh Ritz solutions to check the accuracy of the results. Table 1 compares the linear bucking load value of circular plates under simply supported and clamped boundary conditions through different numerical analysis techniques. This linear buckling load is used to evaluate the post-buckling load of circular plates under mentioned boundary conditions and further explained in the mathematical formulation part of the manuscript. This manuscript reports a novel numerical technique to determine the thermal post buckling load of circular plates and to best of our knowledge the error percentage for the calculated value is minimal to other reported post-bucking values for circular plates.

Table 1: Tabulation of numerical values of linear buckling load with simply supported and clamped boundary conditions.

Ref.	**Contents**	**Linear buckling load for simply supported circular plates**	**Linear buckling load for clamped circular plates**
[3]	The thermal post buckling behavior of isotropic circular plates has been studied using finite element formulation and compared with Rayleigh Ritz method	4.1978	14.6896
[4]	The post buckling behavior of linearly tapered circular plates for different taper parameters having orthotropic properties has been investigated using finite element formulation	4.1978	14.6826
[5]	A simple finite element formulation for the post buckling analysis of circular plates has been presented. The approximate solutions are evaluated by using (i) iterative method (ii) step by step method and (iii) linear vector method.	4.198	14.69
[6]	The post buckling behaviour of elastic circular plates are reinvestigated using a modified finite element analysis.	4.1978	14.6894

[7]	A finite element method is used to analyse the post buckling behaviour of circular plates with elastically restrained edges under thermal loads.	4.1978	14.6826
[8]	A simple formulation to predict the thermal post buckling load of circular plates by evaluating the radial edge tensile load has been presented.	4.1978	14.6896
[9]	Based on radial edge tensile load a novel formulation to study the post buckling behaviour of circular plates elastically restrained edges against rotation has been presented.	4.1977	14.6986
[10]	The large amplitude vibrations of circular plates has been investigated by using a modified Berger's approximation. The governing equations are considered based on Von Karman nonlinearities.	-	-
[11]	The post buckling behaviour of orthotropic circular plates using a simple finite element has been presented. The numerical results are described in the form of radial load ratios with their corresponding empirical formulae.	4.1979	14.6896

2. Mathematical formulation

The objective of this section is to develop a mathematical formulation to evaluate the thermal post buckling load carrying capacity of the circular plates. By considering a circular plate of radius 'a' and of uniform thickness 't' under a radial uniform compressive load 'N_r' per unit length at the boundary, the von Karman type nonlinear strain – displacement relations of the circular plate for large lateral axisymmetric displacements are

$$\varepsilon_r = \frac{du}{dr} + \frac{1}{2}\left(\frac{dw}{dr}\right)^2 \quad (1); \qquad \varepsilon_\theta = \frac{u}{r} \quad (2); \qquad \chi_r = -\frac{d^2w}{dr^2} \quad (3); \qquad \chi_\theta = -\frac{1}{r}\left(\frac{dw}{dr}\right) \quad (4)$$

where r and θ are the radial and circumferential coordinates, $\varepsilon_r, \varepsilon_\theta$ are the strains and χ_r, χ_θ are the curvatures.

The strain energy U of the plate with orthotropic material properties, is given by

$$U = \frac{1}{2}\int_0^{2\pi}\int_{r_1}^{r_2}\left[C_1\varepsilon_r^{\,2} + C_2\varepsilon_\theta^{\,2} + C_{12}\varepsilon_r\varepsilon_\theta + D_1\chi_r^{\,2} + D_2\chi_\theta^{\,2} + D_{12}\chi_r\chi_\theta\right] r\,dr\,d\theta \qquad (5)$$

where $C_1 = \frac{E_r h}{1-\nu^2}$, $C_2 = \frac{E_\theta h}{1-\nu^2}$, $C_{12} = \nu_r C_2 = \nu_\theta C_1$, $D_1 = \frac{E_r h^3}{12(1-\nu^2)}$, $D_2 = \frac{E_\theta h^3}{12(1-\nu^2)}$, and $D_{12} = \nu_r D_2 = \nu_\theta D_1$

Using appropriate substitutions, eqn. (5) can be written as

$$U = \frac{1}{2}\int_0^1\left[\left(\frac{d^2w}{dr^2}\right)^2 + \beta\frac{1}{r^2}\left(\frac{dw}{dr}\right)^2 + 2\nu\frac{\beta}{r^2}\left(\frac{d^2w}{dr^2}\right)\left(\frac{dw}{dr}\right)\right]dr \qquad (6)$$

for the orthotropic parameter $\beta = E_\theta / E_r$. If $E_\theta = E_r$ (β = 1), then the plates are isotropic and otherwise the plates are orthotropic ($E_\theta \neq E_r$).

The work done, W by the external load N_r per unit length at the boundary of the element is given by

$$W = \frac{1}{2}\int_0^{2\pi}\int_{r_1}^{r_2}\overline{N}_r\left(\frac{dw}{dr}\right)^2 r\,dr\,d\theta \qquad (7)$$

By using substitution, equation (7) can be reduced to

$$W = \frac{\lambda\beta}{2}\int_0^1 \frac{d^2w}{dr^2}dr \tag{8}$$

The total potential energy, is the function of in – plane displacement fields and lateral displacement fields of the plate can be stated as $\Pi = U - W$ (9)

By assuming suitable admissible function for the lateral displacement 'w', which satisfies the geometric boundary conditions, the uniform radial edge tensile load N_{r_T} is obtained. The value of the Poisson's ratio ν is taken as 0.3 and both simply supported and clamped boundary conditions are considered in this study.

The following admissible function F taken from [12] for the lateral displacement 'w' is used to study the thermal post buckling behavior of circular plates.

$$F = b_0\left[1+\alpha_1\left(\frac{r}{a}\right)^2+\alpha_2\left(\frac{r}{a}\right)^4\right]+b_1\left[\alpha_3\left(\frac{r}{a}\right)^2+\alpha_4\left(\frac{r}{a}\right)^4+\left(\frac{r}{a}\right)^6\right]+b_2\left[\alpha_5\left(\frac{r}{a}\right)^4+\alpha_6\left(\frac{r}{a}\right)^6+\left(\frac{r}{a}\right)^8\right] \tag{10}$$

The values of α_1, α_2, α_3, α_4, α_5 and α_6, are given in terms of ν and β represent the simply supported and clamped boundary conditions for the function F. By solving equation (9), one can get the eigen values and eigen vectors with appropriate boundary conditions and can evaluate the linear buckling load parameters. The total equivalent compressive uniform radial edge load carrying capacity of the circular plate $N_{r_{NL}}$ can be mathematically represented in the non-dimensional form, as

$$\overline{N}_{r_{NL}} = \overline{N}_{r_{Cr}} + \overline{N}_{r_T} \tag{11}$$

where each term in equation (11) is non – dimensionalised as $\overline{N}_r = \frac{N_r a^2}{D}$ (12)

and D is the plate flexural rigidity.

From equations (1) and (2), the radial edge tensile load per unit length at any point r on the radius is evaluated as

$$N_r = \frac{\beta t}{1-\nu^2}\left[\frac{du}{dr}+\frac{1}{2}\left(\frac{dw}{dr}\right)^2+\nu\frac{u}{r}\right] \tag{13}$$

By using the approximation proposed by Berger [13] which states that the second invariant of the strains are neglected or $\varepsilon_r << \varepsilon_\theta$, N_r can be written as

$$N_r = \frac{\beta t}{1-\nu^2}\left[\frac{du}{dr}+\frac{1}{2}\left(\frac{dw}{dr}\right)^2\right]. \tag{14}$$

The uniform radial tensile load developed in the circular plate due to large deflections, can be obtained in the non – dimensional form as

$$N_{r_T} = \frac{12}{\beta(1-\nu^2)}\int_0^1\left(\frac{dw}{dr}\right)^2 dr \tag{15}$$

The post buckling load carrying capacity (γ), is obtained from equation (11) using the values of radial edge tensile load parameters evaluated from the assumed admissible functions for the lateral displacement w and the linear buckling load parameters as

$$\gamma = \frac{N_{r_{NL}}}{N_{r_{cr}}} = 1 + c\left(\frac{b_0}{t}\right)^2 \tag{16}$$

The values of γ are evaluated for different values of b_0/t for $\beta = 1$ which represents isotropic circular plates.

3. Numerical results and discussions

The radial edge tensile load and the thermal post buckling load carrying capacity is evaluated by assuming suitable admissible function for the lateral displacement which satisfies both simply supported and clamped boundary conditions and the numerical results obtained with varying values of b_0/t for $\beta = 1.0$. Tables 2 and 3 represent the post buckling load carrying capacity of isotropic circular plates with both simply supported and clamped boundary conditions. The numerical results from the known literatures are compared including the results of the present formulation. It can be shown that the values of the post buckling load carrying capacity of the isotropic plates are in good agreement. The percentage errors from the reference values are also evaluated and it can be shown that the highest error percentage is 1.41% for simply supported and 1.66% for clamped boundary conditions. The error is occurred in the values of γ due to the edge boundary conditions and depends on the admissible function for the lateral displacement.

Table 2: Representing the values of post buckling load carrying capacity of simply supported isotropic circular plates.

b_0/t	Post buckling load carrying capacity (γ)						
	PRESENT	REF [3]	REF [14]	REF [15]	Percentage error from REF [3]	Percentage error from REF [14]	Percentage error from REF [15]
0.0	1.0000	1.0000	1.0000	1.0000	0	0	0
0.2	1.0820	1.0730	1.0706	1.0730	0.83	1.06	0.83
0.4	1.3042	1.2931	1.3043	1.2930	0.85	0.01	0.86
0.6	1.6873	1.6637	1.6847	1.6637	1.41	0.15	1.41
0.8	2.2041	2.1907	2.2174	2.1907	0.61	0.59	0.61
1.0	2.9086	2.8818	2.9022	2.8816	0.92	0.22	0.93

Table 3: Representing the values of post buckling load carrying capacity of clamped isotropic circular plates.

b_0/t	Post buckling load carrying capacity (γ)						
	PRESENT	REF [3]	REF [14]	REF [15]	Percentage error from REF [3]	Percentage error from REF [14]	Percentage error from REF [15]
0.0	1.0000	1.0000	1.0000	1.0000	0	0	0
0.2	1.0201	1.0210	1.0208	1.0205	0.08	0.06	0.03
0.4	1.0805	1.0840	1.0832	1.0835	0.32	0.24	0.27
0.6	1.1812	1.1893	1.1873	1.1887	0.68	0.51	0.63
0.8	1.3221	1.3373	1.3330	1.3366	1.13	0.81	1.08

1.0	1.5032	1.5286	1.5203	1.5278	1.66	1.12	1.61

The following graphs demonstrate a comparative study of the thermal post buckling load carrying capacity of circular plates for various b_0/t values with simply supported and clamped boundary conditions. From the figures, it is clear that the numerical results in the present study, along with those obtained from the references give more or less the same values of γ for simply supported and clamped boundary conditions.

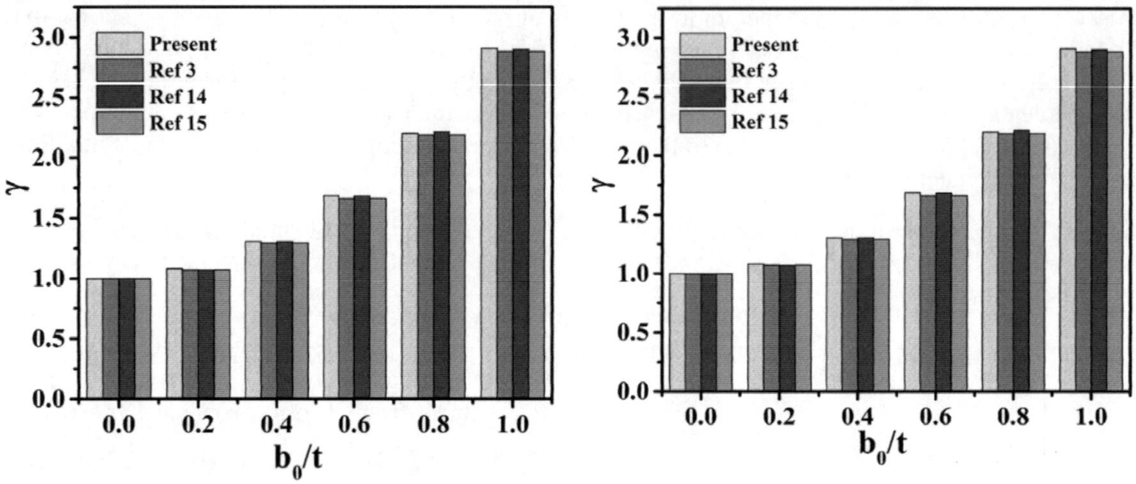

Figure 1: The bar graph comparing the present numerical results with the standard reference literature values for simply supported and clamped boundary conditions.

4. Conclusions

A comparative study on the thermal post buckling behavior of circular plates with both simply supported and clamped boundary conditions are compiled. The radial edge tensile load and hence the thermal post buckling load carrying capacity of the circular plates are evaluated by using a mathematical formulation based on Von Karman nonlinearities; the results obtained from the different known literatures are compared. The numerical results obtained from the present formulation minimizes the error reported in the previous studies and the calculated post buckling load carrying capacity of circular plates fits well with the values from the literatures. The present mathematical formulation can be used as a simple numerical method in calculating the post-buckling load of circular plates.

Acknowledgement

The authors would like to thank the Universiti Malaysia Pahang and Ministry of Higher Education, Malaysia for the provision in terms of financial form from the research grants (RDU160330).

Nomenclature

a = radius of the circular plate
N_r = uniform radial edge compressive load per unit length
$N_{r_{cr}}$ = linear buckling load
$N_{r_{NL}}$ = total uniform radial edge compressive load per unit length
N_{r_T} = uniform radial edge tensile load per unit length due to large lateral displacements

T	= thickness of the plate
$\varepsilon_r, \varepsilon_\theta$	= in – plane strains
v	= Poisson's ratio
U	= strain energy
r, θ	= radial and circumferential coordinates
W	= work done
χ_r, χ_θ	= curvatures
r_1, r_2	= internal and external radii
α_1 - α_6	= generalized displacements
β	= orthotropic parameter
E_r, E_θ	= Young's moduli in radial and circumferential directions
λ	= post buckling load parameter
b_0	= central (maximum) lateral displacement of the circular plate

References

1. Friedrichs, K.O. and J. Stoker, *Buckling of the circular plate beyond the critical thrust*, 1941.
2. Bodner, S.R., *The post buckling behavior of a clamped circular plate,* Quarterly of Applied Mathematics, 1955, **12**(4): p. 397-401.
3. Raju, K.K. and G.V. Rao, *Thermal post-buckling of circular plates,* Computers & Structures, 1984, **18**(6): p. 1179-1182.
4. Raju, K.K. and G.V. Rao, *Post-buckling of cylindrically orthotropic linearly tapered circular plates by finite element method,* Computers & Structures, 1985, **21**(5): p. 969-972.
5. Rao, G.V. and K.K. Raju, *Post-buckling behaviour of elastic circular plates using a simple finite element formulation,* Computers & Structures, 1979, **10**(6): p. 911-913.
6. Rao, G.V. and K.K. Raju, *A reinvestigation of post-buckling behaviour of elastic circular plates using a simple finite element formulation,* Computers & Structures, 1983, **17**(2): p. 233-235.
7. Rao, G.V., K.K. Raju, and N.R. Naidu, *Post-buckling behaviour of circular plates with elastically restrained edges subjected to thermal loads,* Computers & Structures, 1992, **45**(1): p. 209-210.
8. Rao, G.V. and R.R. Varma, *Simple formulation to predict thermal postbuckling load of circular plates,* AIAA Journal, 2007, **45**(7): p. 1784-1786.
9. Varma, R.R. and G.V. Rao, *Novel Formulation to Study Thermal Postbuckling of Circular Plates with Edges Elastically Restrained against Rotation,* Journal of Engineering Mechanics, 2011, **137**(10): p. 708-711.
10. Venkateswara Rao, G., et al., *Solution for large amplitude vibrations of circular plates via modified Berger's approximation,* The IES Journal Part A: Civil & Structural Engineering, 2012, **5**(1): p. 56-61.
11. Kanaka Raju, K. and G. Venkatewara Rao, *Finite-element analysis of post-buckling behaviour of cylindrically orthotropic circular plates,* Fibre Science and Technology, 1983, **19**(2): p. 145-154.
12. Yamaki, N., *Influence of large amplitudes on flexural vibrations of elastic plates,* ZAMM-Journal of Applied Mathematics and Mechanics/Zeitschrift für Angewandte Mathematik und Mechanik, 1961, **41**(12): p. 501-510.
13. Berger, H.M., *A new approach to the analysis of large deflections of plates,* 1954.
14. Ramaraju, R.V. and V.R. Gundabathula, *Reinvestigation of intuitive approach for thermal postbuckling of circular plates,* AIAA Journal, 2009, **47**(10): p. 2493-2495.
15. Li, S.-R., R.C. Batra, and L.-S. Ma, *Vibration of Thermally Post-Buckled Orthotropic Circular Plates.* Journal of Thermal Stresses, 2007. **30**(1): p. 43-57.

Effect of deposition parameters on Structural and Optical Properties of ZnS Thin Films

Priya K[1], Gowrish Rao K[1], Ganesh Sanjeev[2]
[1]Department of Physics, Manipal Institute of Technology, Manipal University, Udupi, Karnataka-576104, India
[2]Microtron Centre, Department of Studies in Physics, Mangalore University, Mangalore, Karnataka-574199

Corresponding Author: priyakmb1991@gmail.com

Abstract: Thin films of ZnS have been vacuum coated on glass substrates at room temperature. The crystal structure and optical properties of the films were studied using X-ray diffractometer (XRD and UV visible spectrophotometer, respectively. The X-ray diffraction patterns showed that films were polycrystalline in nature with cubic structure. The optical band gap was measured close to 3.7 eV. The influence of rate of deposition on the grain size of the films were analyzed.
Keywords: ZnS, Thermal evaporation

1. Introduction

The synthesis of binary metal chalcogenides in a crystalline form has been a rapidly growing area of research due to their important physical and chemical properties [1]. ZnS, as a typical II–VI semiconductor compound with band gap energy of 3.7 eV at room temperature, has attracted considerable attention [2]. ZnS has been used as a wide band gap material in electroluminescent and optoelectronic devices such as blue light-emitting diodes, Mn-doped electroluminescent devices, electro- optic modulator and n-window layers of solar cells [3, 4]. ZnS is also emerging as photodetector in the ultraviolet region which has application in both civil and military fields [5].

Subsequently, several techniques have been used to produce ZnS thin film such as spray pyrolysis [6], chemical bath deposition (CBD) [7], SILAR [8], molecular beam epitaxy [9], sputtering [10] and thermal evaporation [11]. Among the above mentioned methods, thermal coating is a simplest physical method. In this research ZnS thin films are coated on glass substrates by thermal evaporation method.

2. Experimental

In the present work, thin films of ZnS have been coated using thermal evaporation method. ZnS pellets of 99.9% purity were procured from Sigma Aldrich for this purpose. Films were deposited on glass substrates using molybdenum boat under 10^{-6} torr. The glass substrates were cleaned with chromic acid, distilled water and acetone. Stainless steel shield was used to ensure constant rate of evaporation in each trail.

The deposited films were characterized using X-ray diffractometer (XRD) and UV visible spectrophotometer. X-ray pattern were obtained using Rigaku Minflex 50 X-ray diffractometer using CuKα radiation (λ = 1.54 Å). Transmittance and absorbance data was obtained from Shimadzu UV 1800 spectrophotometer.

3. Results and discussions

3.1. Thickness measurement

Optical thickness of the films were found using Swanepoel method [12]. For uniform thickness, the interference fringes obtained are as shown below in Figure 1. The fringes clearly are in the region of medium and weak absorption. In this region, the refractive index can be calculated from the equation

$$n = [N + (N^2 - s^2)^{1/2}]^{1/2}$$

where

$$N = 2s\frac{T_M - T_m}{T_M T_m} + \frac{s^2+1}{2}$$

Where s is the refractive index of the substrate (for glass s = 1.5), T_M is the maxima of the interference fringe and T_m is the corresponding minima.

Thickness of the film can be obtained from the formula

$$t = \frac{\lambda_1\lambda_2}{2(\lambda_1 n_2 - \lambda_2 n_1)}$$

where n_1 and n_2 are the refractive indices at two adjacent maxima at λ_1 and λ_2 wavelengths. The thickness was calculated using the above formulas for different transmission plots. The thickness values, hence, calculated were about 292 nm, 370 nm and 441 nm. Below 290 nm the films lacked uniform substrate coverage and above 441 nm they showed poor adhesion. From the above thickness values, the rate of deposition was found by dividing the thickness values by the time of deposition. Corresponding to 292 nm, 370 nm and 441 nm thickness values, 3.2 Ås^{-1}, 4.1 Ås^{-1}, 4.9 Ås^{-1} rates of deposition were calculated.

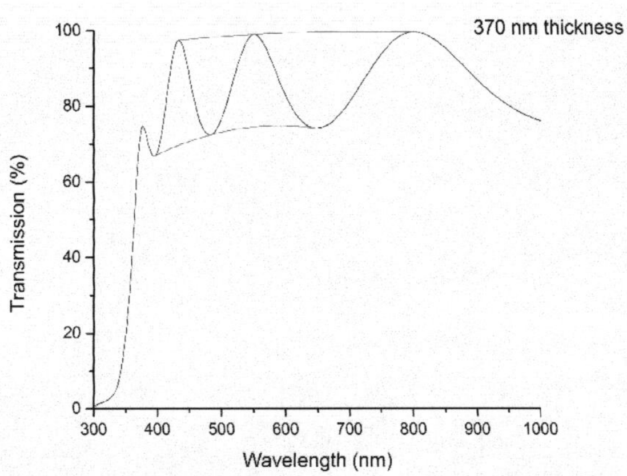

Figure 1. Transmission spectrum of ZnS film

3.2. XRD analysis

Figure 2 shows the XRD pattern of ZnS thin films deposited at three rates of deposition namely, 3.2 $Ås^{-1}$, 4.1 $Ås^{-1}$, 4.9 $Ås^{-1}$. The deposited films were polycrystalline in nature. The peaks were sharp at $2\theta \approx 29°$ indicating cubic structure with preferred orientation along (111) plane. Grain size was calculated from the XRD pattern using the equation

$$D = \frac{0.9\,\lambda}{\beta\, cos\theta}$$

Where λ is 1.54 Å, wavelength of CuKα radiation used, **β** is the full width at half maximum (in radian) and **θ** corresponds to the maximum height of the peak.

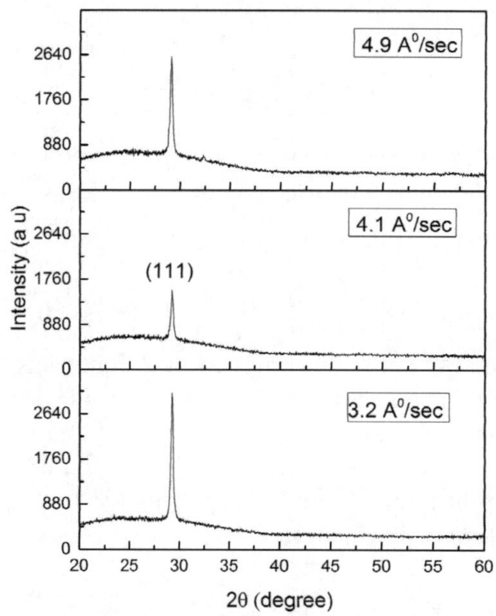

Figure 2. XRD pattern of ZnS for different rate of deposition

Table 1. The grain size values for different rates of deposition

Thickness **t** (nm)	Deposition Rate **R** (Å/sec)	**2θ** (degree)	FWHM **β** (x 10^{-3} radian)	Grain size **D** (nm)	Interplanar spacing **d** (Å)	Lattice parameter **a** (Å)
292	3.2	29.24	5.70	25	3.05	5.28
370	4.1	29.2	9.01	15	3.05	5.28
441	4.9	29.12	6.68	21	3.05	5.28

From the above table, it is observed that, at low rate of deposition the grain size is high and it reduces with high deposition rate from 25 nm to 15 nm for 3.2 $Ås^{-1}$ to 4.1 $Ås^{-1}$. This is in accordance with literature [13] which says that, lower the rate of deposition, higher is the

IOP Conf. Series: Materials Science and Engineering **360** (2018) 012042 doi:10.1088/1757-899X/360/1/012042

deposition time and hence, large clusters can be formed, which finally leads to higher grain size. But from the deposition rate of 4.1 Ås^{-1} to 4.9 Ås^{-1} the grain has increased from 15 nm to 21 nm. This can be attributed to increase in thickness of the films with which crystallinity improves [13]. The lattice parameter of the films was found to be 5.28 Å.

3.3. Optical properties

The absorption spectra for three different deposition rates is as shown in Figure 3. The spectra are measured in the range of 300 to 800 nm. A strong absorption in the ultraviolet region in contrast with visible region was observed for all three rates of deposition.

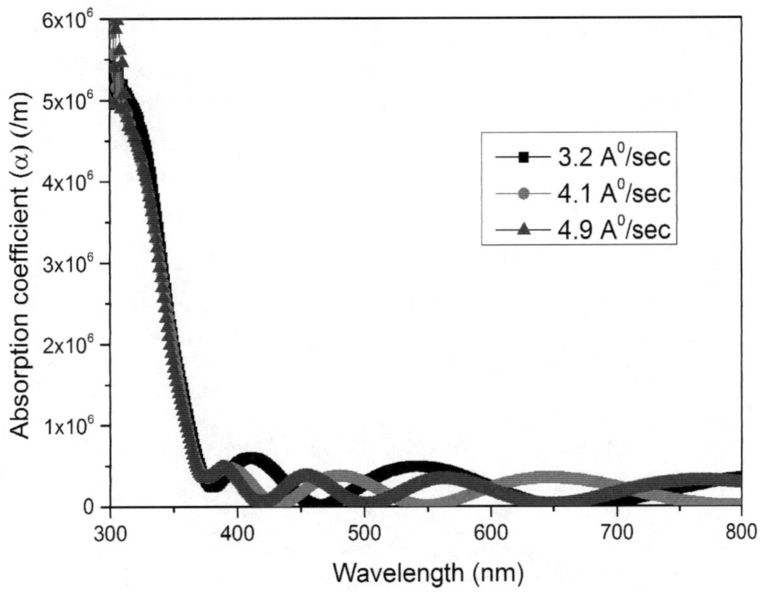

Figure 3. Absorption spectra at various rate of deposition

The absorption coefficient (α) was analyzed using the following expression

$$\alpha h\nu = K\,(h\nu - Eg)^{1/2}$$

where K is a constant.

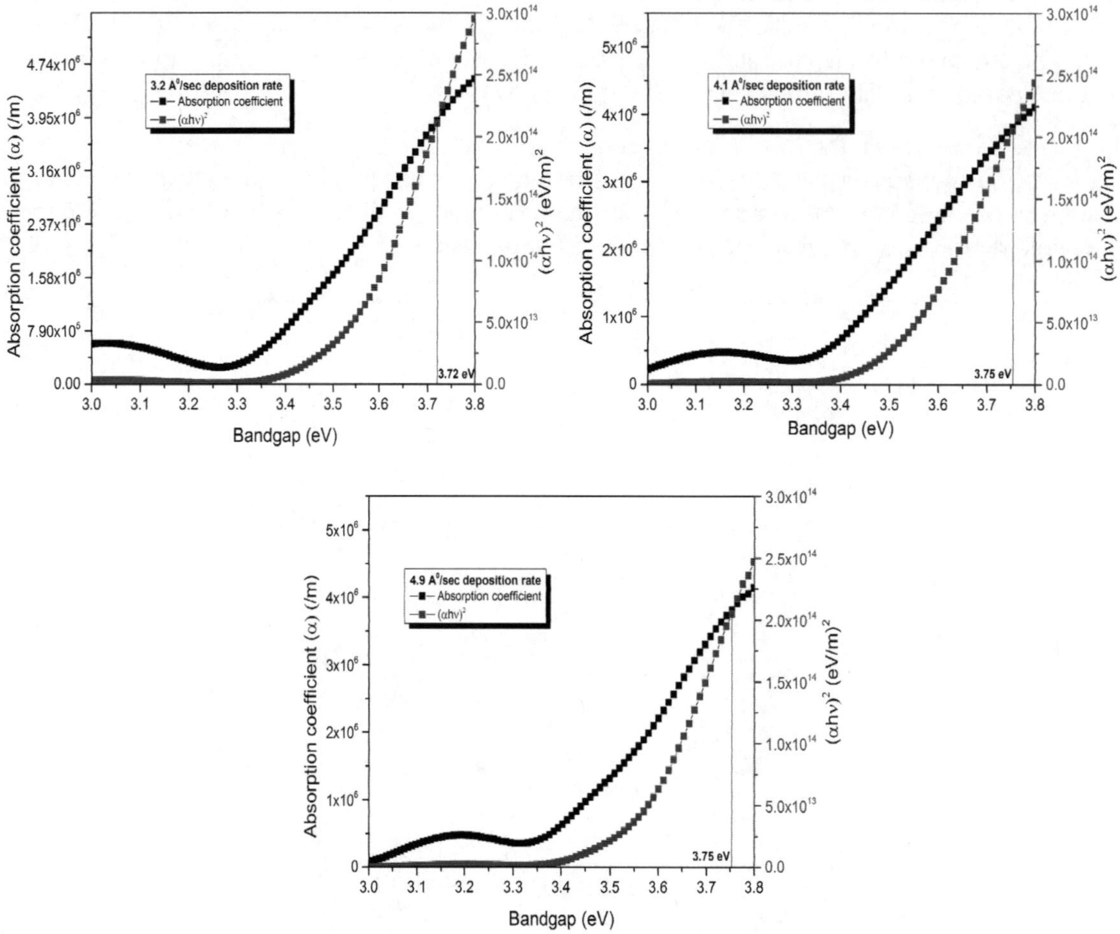

Figure 4. Bandgap vs absorption coefficient and tauc plot for different rate of deposition

The value of absorption coefficient (α) increased from 370 nm to 320 nm which can also be confirmed from bandgap values obtained from intersection α and tauc plot [14], as shown in Figure 4.

The bandgap of the films were close to 3.7 eV. It increased marginally from 3.72 to 3.75 eV with increase in rate of deposition as well as with thickness.

4. Conclusion

ZnS thin films have been successfully coated on glass substrate at various rates of deposition. The deposited polycrystalline films show cubic phase. As rate of deposition increases grain size reduces and with increase in thickness value, crystallinity improves. The deposited films have bandgap around 3.7 eV and are ideal for ultraviolet radiation detection.

Acknowledgement
The authors are thankful to Manipal Institute of Technology, Manipal and DAE-BRNS, Govt of India (sanction no: 34(1)/14/37/2014 dated: 27/11/2014).

References

[1] Anuar Kassim, Saravanan Nagalingam, Ho Soon Min, Noraini Karrim, Arabian Journal of Chemistry, 3, 243, 2010.

[2] G. F. Garlick, A. F. Gibson, Journal of Optical Society of America, 39, 935, 1945.

[3] I. Nducwe, Solar Energy Materials and Solar Cells, 40, 123, 1996.

[4] R. Ortega Borges, D. Lincot, J. Vedel. In: Proceedings of the 11th E.C. photovoltaic Solar Energy. Conference, 862, 1992.

[5] E. Monroy, F. Omnes, F. Calle, Semiconductor Science and Technology, 18, 33, 2003.

[6] A. N. Yazici, M. Oztas, M. Bedir, Journal of Luminescence, 104, 115, 2003.

[7] Fei Long, Wei-Min Wanga, Zhan-kui Cui, Li-Zhen Fan, Zheng-guang Zou, Tie-kun Jia, Chemical Physics Letters, 462, 84, 2008.

[8] Y.F. Nicolau, J. C. Menard, Journal of Applied Electrochemistry, 20, 2915, 1990.

[9] M. Yoneta, M. Ohishi, H. Saito, Journal of Crystal Growth, 127, 314, 1993.

[10] L. Z. Shao, K.H. Chang, H. L. Hwang, Applied Surface Science, 212, 305, 2003.

[11] V. Dimitrova, J. Tate, Thin Solid Films, 365, 134, 2000.

[12] R. Swanepoel, Journal of Physics E: Scientific Instruments, 16, 1214, 1983.

[13] R. E. Thun, in Physics of Thin Films, Edited G. Hass and R. E. Thun, Academic Press, New York, 1, 187, 1963.

[14] W. E. Spicer and F. Abeles, Optical Properties of Solids, North-Holland publishers Amsterdam, 1972.

IOP Conf. Series: Materials Science and Engineering **360** (2018) 012043 doi:10.1088/1757-899X/360/1/012043

A Report on Growth And Spectroscopic Studies on The L-Tartaric Acid Doped Potassium Dihydrogen Phosphate Non-Linear Optical Single Crystals

S Arulmani[1], P Sanjay[1], E Chinnasamy[1], K Deepa[2], J Madhavan[2] and S Senthil[1]
[1]Department of Physics, Government Arts College for Men (Autonomous), Nandanam, Chennai - 600035
[2]Department of Physics, Loyola College, Chennai -600034
Email: ssatoms@yahoo.co.in

Abstract. In recent years study of non-linear optical (NLO) materials is gaining rapid momentum due to their needs in several device applications. Potassium Dihydrogen Phosphate (KDP) is a well-known NLO material, whose non-linearity is enhanced by doping with amino acids. In this present study, single crystal ofL- Tartaric acid doped potassium Dihydrogen phosphate (LTKDP) have been grown by slow evaporation method at room temperature. Good quality transparent crystals were obtained with 27-30 days. The crystallinity and cell parameters were characterized by X-ray diffraction analysis (XRD), the shifts in frequency assignments of various functional groups of KDP due to the addition of L-Tartaric acid was analysed by Fourier transform infrared (FTIR) and Fourier transform Raman spectroscopic studies. The dielectric constant and the dielectric loss of the single crystals were studied as a function of frequency. Kurtz- Perry power technique was employed to determine the Second Harmonic Generation (SHG) efficiency of LTKDP single crystal.

1. INTRODUCTION

The rapid development in the field of science and technology necessitates searching for newer and efficient non–linear optical materials. Potassium dihydrogen phosphate (KDP) crystals have created interest because of its Piezo-electric, electro-optic, non-linear optical properties and its extensive application in x-ray monochromators [1, 2].The electro-optic effect in KDP is used to obtain phase and amplitude modulations[3]. KDP crystal based world's largest laser to generate UV beams has been demonstrated [4]. In order to get good quality crystals, many techniques have been introduced by several workers. The modern world is witnessing revolutionary advancements in the various aspects of science and technology. Organic non-linear materials are attracting a great deal of attention as they have large optical susceptibilities inherent ultra-fast response times and high optical thresholds for laser power as compared with inorganic materials [5]. A number of such materials have been reported in Literature for this potential application [6]. Among organic class of NLO materials, amino acids exhibits some specific features such as molecular chirality weak vanderwaals hydrogen bonds, absence of strongly conjugated bonds wide transparency range in the visible and ultraviolet regions, amenability for synthesis multi-functional substitution, higher resistance to optical damage and manoeuvrability for device application etc [7,8]. Among amino acids molecular are usually linked through the hydrogen bond. The aim of the present study is to investigate the effect of L-Tartaric acid in KDP crystal by slow evaporation method at room temperature. The doping of L-Tartaric acid into KDP was achieved by adding 0.4wt %, this solution was prepared in double distilled water, and the solution was stirred for 3-4 hours to achieve homogenization. The solution was filtered and sealed with porous lid and placed in a dust free atmosphere for slow evaporation. The optically good quality and transparent crystals were harvested in 27-30 days. The crystals were characterized using powder x-ray diffraction studies, Fourier transform infrared (FT-IR) and FT- RAMAN analysis, Second harmonic generation (SHG), Dielectric studies etc.

2. EXPREMENTALMETHODS

2.1. Crystal Growth

Single crystal of L-Tartaric acid doped KDP was grown by slow evaporation technique at room temperature. Higher grade KDP salt was dissolved in double distilled water and and 0.4% of L-Tartaric Acid was doped with KDP, the saturated solution of LTKDP was prepared in accordance with the solubility data. The solution was thoroughly stirred for homogenization and then filtered into a Borosil beaker using whatmann filter paper. The pH of the solution was noted. The saturated solution was closed with perforated cover and kept in a dust free atmosphere to allow slow evaporation of the solvent. Transparent and good quality LTKDP crystals were harvested. The photograph of as grown doped LTKDP crystal is shown in figure.1.

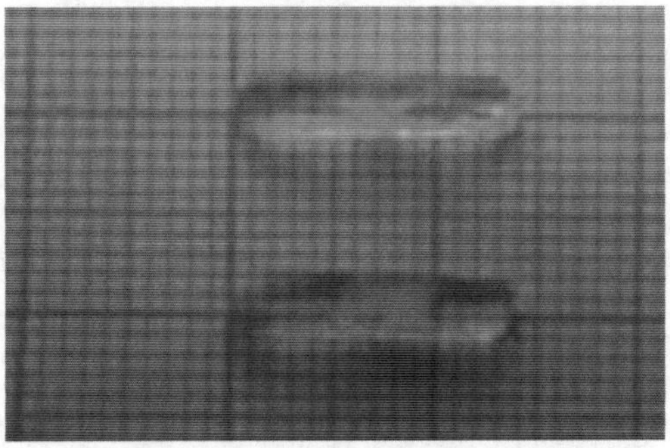

Figure 1. Photograph of as grown LTKDP crystal

3. RESULT AND DISCUSSION

3.1. Powder X-Ray Diffraction Studies (XRD)

The powder X-ray diffraction studies have been carried out to confirm the crystallinity and to determine the lattice parameter of the grown crystals. The powder x-ray diffraction (PXRD) analysis was carried out using XPERT–PRO Diffractometer with CuKα radiation of wavelength 1.5406 Å with the scanning rate 2degree per minute over the range of 10 - 70°. The PXRD studies for the grown crystal show the appearance of sharp and strong peaks which confirms the good crystallinity of the grown crystals, there are no other phases emerging besides the tetragonal system. The cell parameter value of the grown crystal is a = b = 7.433Å, **c** = 6.932Å with the angles $\alpha = \beta = \gamma = 90^{\circ}$ having space group 1.42 d and the calculated lattice parameter values are in agreement with the reported values. Powder x-ray diffraction pattern for LTKDP crystal is shown in figure (1).

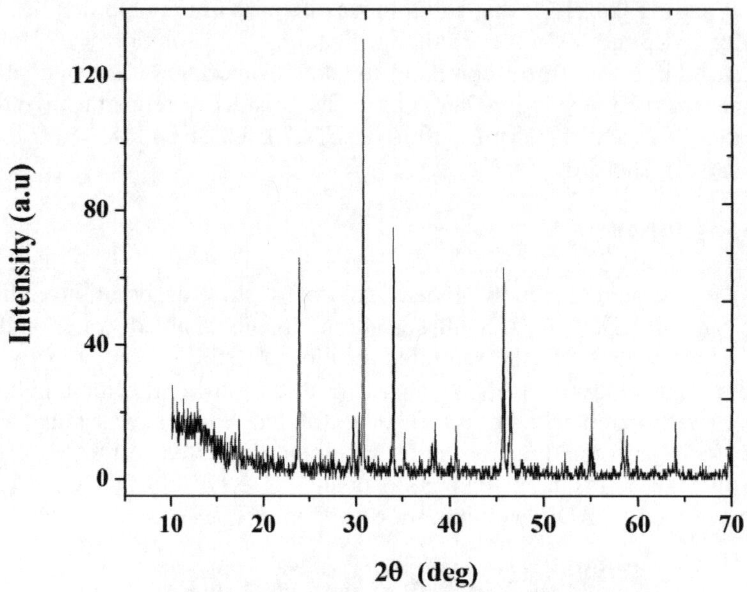

Figure 2. Powder XRD pattern of LTKDP

3.2.FT-IR and FT- RAMAN spectral analysis

FT-IR and FT-RAMAN studies are very useful techniques for qualitative analysis and identification of compounds. The recorded FT-IR and FT-RAMAN spectra of LTKDP are depicted in Figure 3 and Figure 4. The spectral bands are interpreted and compared with the standard spectra of the functional groups[9]. The bands which appear in the spectral region 450-4000 cm^{-1}arise from the internal vibration of LTA , and water molecules , the vibration of O-H and O=H types of hydrogen bonds. In the high frequency region of IR spectra, the sharp peaks observed at 536cm^{-1} are ascribed to PO_4 stretching vibration in the crystal. Their Raman equivalents are observed in the region of 514 cm^{-1}. In the IR spectra the peak will be observed at 3223 cm^{-1} are in O-H stretching hydrogen bonded of KDP. Their Raman equivalents are not observed due to insufficient detector sensitivity in the region above 3091 cm^{1}. An intensive IR bands assigned to the in -plane band vibration is observed at 1096 $cm\text{-}^{1}$ with equivalent Raman band at 1295cm^{-1}. It is observed that some of the bands originate from both LT and KDP simultaneously. The IR bands at 2738 cm^{-1} is assigned to P-O-H asymmetric stretching vibration. In Raman it cannot be observed at this region. In IR bands the frequency of 2376cm^{-1} in the doped crystal corresponds to P-O=H stretching [10]. The broad absorption peak at 1694 cm-[1] is due to O=P-OH bending vibration. The O-H deformation and P=O stretching will be observed in the frequency of 1297 cm^{-1}. In IR the absorption band at 912 cm-[1] is due to P-O-H bending vibration in the doped LTKDP. In Raman the sharp peak will be observed at 1079 cm^{-1} are ascribed to P-O-H stretching molecule in the doped LTKDP crystal. In Raman spectra, the peaks will be observed at 681 cm^{-1} for O-H bending (out of plane). In the IR band 550 cm^{-1} and in Raman 514 cm-[1] frequencies will be observed due to the P=O symmetric bending. The frequencies of 481 cm^{-1} in IR band and 402 cm^{-1} in Raman band.

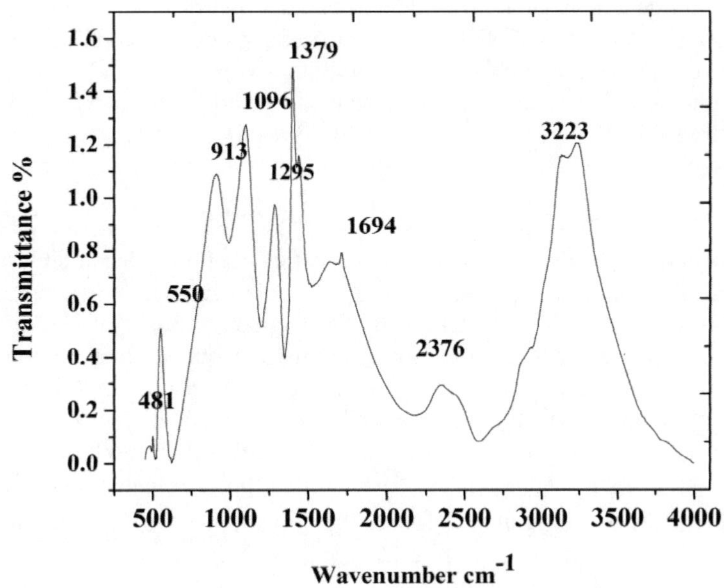

Figure 3.FT-IR Spectrum of LTKDP

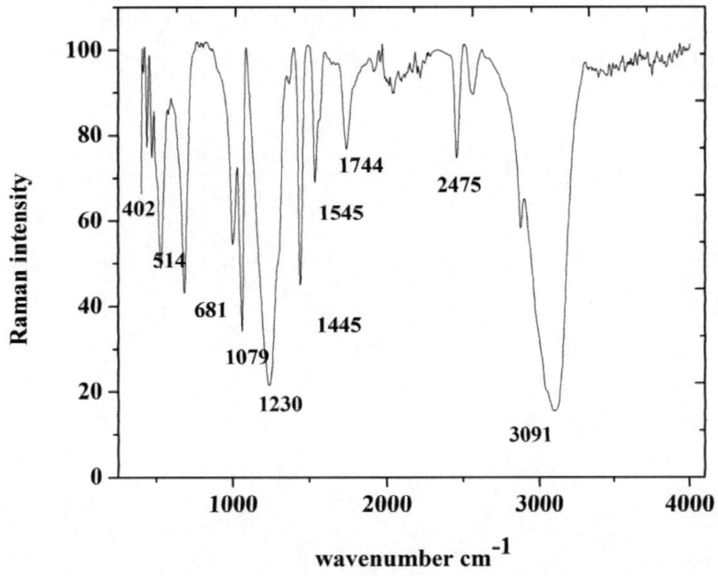

Figure 4.FT-Raman spectrum of LTKDP

3.3. Second Harmonic Generation efficiency studies

Kurtz and Perry second harmonic generation efficiency test (SHG) is performed for the comprehensive analysis of second order nonlinearity. Nonlinear optic measurements were carried out

IOP Conf. Series: Materials Science and Engineering **360** (2018) 012043 doi:10.1088/1757-899X/360/1/012043

by using Kurtz powder technique. A Q-switched Nd: YAG laser beam of 1064nm wavelength with 1.9 mj/ pulse input power, 8ns pulse width and repetition rate 10Hz was used to estimate SHG efficiency of the grown crystals. The grown crystals of LTKDP were grounded into a fine powder and then packed in a micro-capillary of uniform bore and exposed to laser radiation. The fundamental input radiation (1064nm) was separated or filtered by a monochromator and the output was measured. Second harmonic radiation generated by the randomly oriented micro- crystals was focused by a lens and detected by a photo multiplier tube. SHG was confirmed by the emission of green light. Using the pure potassium dihydrogen phosphate (KDP) crystalline powder as reference material, the output of SHG range was compared and found the SHG conversion efficiency of LTKDP is 1.09 times greater than that of reference KDP.

3.4. Dielectric Studies

Every material has a unique set of electrical characteristics that are dependent on its dielectric properties. The variation of dielectric constant of the sample with frequency is studied at room temperature using H10K1 3532 LCR HITESTER in the frequency range 300HZ to 3MHZ and is shown in figure 5. The high value of dielectric constant at high frequency may be due to the presence of all the four polarizations namely space charge, orientation, electronic and ionic. The dielectric characteristic of a crystal is determined by the dielectric constant. The dielectric constant determines the share of the electric stress which is absorbed by the material. It is the ratio between the permittivity of the medium and the permittivity of free space. If E_0 is the applied electric field, Ep is the electric field produced due to polarization of the material, then the dielectric, $\dot{\varepsilon} = \frac{E0}{E0-Ep}$. The dielectric constant $\dot{\varepsilon}$ has no units and it is a measure of the polarization in the dielectric material. Any small change in dielectric constant with respect to temperature can be attributed to the transition of the material. When the temperature raises, the orientation of the dipoles is increased and this increases the dielectric constant $\dot{\varepsilon}$. Potassium dihydrogen phosphate (KH_2PO_4) has very good photorefractive performance and electro-optic properties due to its high dielectric constant. Figure 6 shows the variation of dielectric loss with frequency. The crystal possesses enhanced optical quality with lesser defects and this parameter play a vital role for the fabrication of nonlinear optical device because of low dielectric loss with high frequency for the samples.

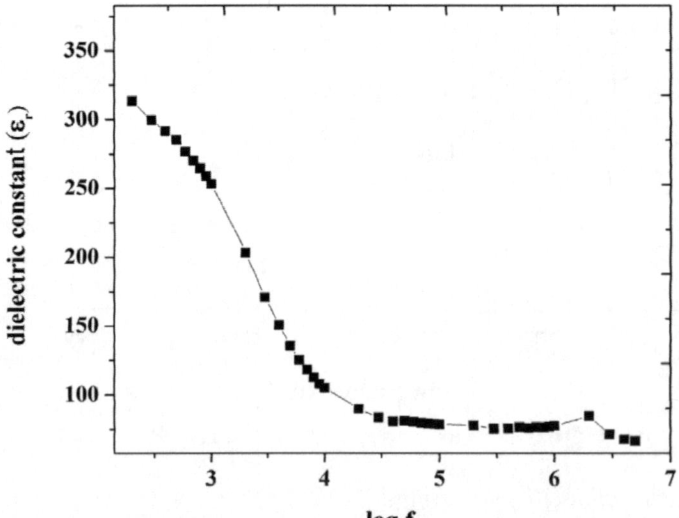

Figure 5. Variation of dielectric constant of LTKDP

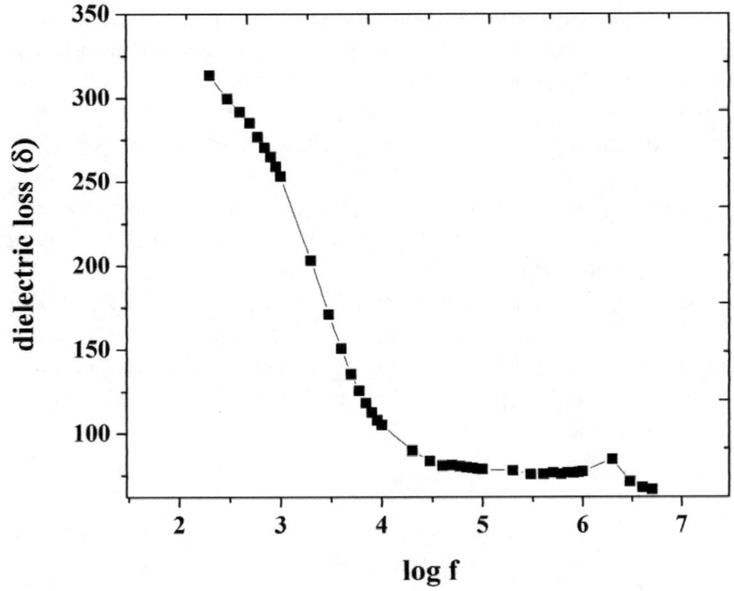

Figure6. Dielectric constant of LTKDP

4. Conclusions

High quality and transparent LTKDP single crystal has been grown by slow evaporation method at room temperature. Powder XRD studies shows that the grown crystals are tetragonal in structure. The transmission spectrum reveals that the crystal has sufficient transmission in the entire visible and IR region. The vibrational frequencies were assigned from FT-IR and FT-RAMAN spectral analysis which confirms the presence of functional groups. The SHG efficiency of the grown crystal was measured by Kurtz and Perry power method and its efficiency was found to be 1.09 times that of the pure KDP crystal. All the above results strongly attributes LTKDP is a novel material for non-linear optical applications.

Acknowledgment

The authors acknowledge University Grand Commission (UGC), India for funding this research project- F.No.4-4/2015-16 (MRP/UGC-SERO) – 2914.

References

[1] Dhumane N R, Hussaini S SDongre V G, Ghughare P, Shirsat M D, 2009Growth and characterization of glycine doped bis thiourea cadmium chloride single crystal Research *and Technology* **44**269–27

[2] Suresh Kumar B, .RajendraBabu K,2008 Effects of L-Arginine, L-Histidine and Glycine on the growth of KDP single crystal and their characterization *Indian Journal of Pure & Applied Physics, (IJRST)*.**46**123-12**6.**

[3] Parikh K D, Parekh B B, .Dave D J, Joshi M J, 2007 Thermal, FT-IR AND SHG efficiency studies of L-Argine doped KDP Crystals *Bulletin of Materials science* **30** 105-112

[4] Ferdousi Akhtar and Jiban Podder, 2011A Study on Structural, Optical, Electrical and Etching Characteristics of Pure and L-Alanine Doped Potassium Dihydrogen Phosphate Crystals *Journal of Crystallization Process and Technology.* **1** 55-62 .

[5] Jagadish P and Rajesh N P, 2011J.Effect of L-proline on the growth and NLO properties of KDP crystal*Journal Of Optoelectronics And Advanced Materials.* **13** 962.

[6] Muley G G, 2012 , Study on Growth Habit, SHG and Thermal properties of urea Phosphate Doped KDP Crystal *Journal of Science and Technology*.**2**109-113

[7] Dhumane N R, Hussaini S S, Dongre V G, Ghugare P, Shirsat M D 2009 Growth and Characterisation of L-Alanine doped Zinc Thiourea Chloride Single Crystal (ZTC) *Applied Physica A* **95** 727-732.

[8] Dave d J, Parikh K D, Parekh B B,JoshiM J, 2009Growth and spectroscopic, thermal, dielectric and SHG studies of L- threonine doped KDP crystals *Journal Of Optoelectronics And Advanced Materials,* **11**602 – 609.

[9] Dhanaraj P V, Rajesh N P, Mahadevan C K and G. Bhagavannarayana, 2009Growth and Characterization of Non Linear Optical Crystal: L-Alanine Doped KDP *Physica B.* **40** 404.

[10] KumeresanP, MoorthyBabu S, Anbarasan P M, *2008* Thermal, dielectric studies on pure and amino acid L-glutamic acid,L-histidine, L-valine doped KDP single crystals *Journal of. Crystal. Growth.***30** 1361–1368.

IOP Conf. Series: Materials Science and Engineering **360** (2018) 012044 doi:10.1088/1757-899X/360/1/012044

Growth and Characterization of L-Leucenium Hydrogen Maleate Single Crystals for Nonlinear Optical Applications

A Hemalatha[1,2], S Arulmani[2],P Sanjay[2], K Deepa[3], J Madhavan[3] and S Senthil[2]

[1]Department of Physics, Quaid-E-Millath Government College for Women, Anna salai, Chennai – 02,Tamil Nadu, India.
[2]Department of Physics ,Government Arts College For Men (Autonomous), Nandanam, Chennai – 35, Tamil Nadu, India.
[3]Department of Physics, Loyola College, Chennai – 34, Tamil Nadu, India.
E mail – ssatoms@yahoo.co.in

Abstract. Good quality single crystals of amino acid based L-Leucenium maleate (LLM) has been grown from aqueous solution by slow evaporation method at room temperature. The crystalline nature of the crystal has been confirmed by the X-ray diffraction analysis Using Cuk$_\alpha$ radiation of wavelength of 1.54 A°. The presence of various functional groups in the LLM is confirmed by FT-IR using the Bruker IFS66 spectrometer. The functional groups present in the sample were also confirmed by FT-RAMAN spectral analysis. The thermal stability of LLM crystal has been analyzed by TGA and DTA studies. Second harmonic generation test was done on the LLM sample using Kurtz and Perry technique. The NLO efficiency of the grown crystal is nearly 0.6 times that of the standard KDP crystal. The transparent nature of the sample has been conformed from the UV-Vis-NIR spectrum and optical band gap of the LLM crystal also calculated.

1. Introduction

Nonlinear optical materials have been extensively studied in the recent years due to their potential applications in various fields like optical data storage, optical switching images, processing and manipulation. Therefore, there is a need to produce high efficiency NLO materials. Among the class of NLO materials, organic NLO materials are generally preferred to be more efficient than the inorganic counterpart due to their favorable nonlinear responds [1]. Amino acid crystals have subjected to extensive investigation by several researchers for their excellent characteristics. Amino acids are interesting useful organic materials for NLO applications as they contain a proton donor carboxyl acid (-COOH) group and proton acceptor amino (NH_2) group in them, known as zwitterions which produce hydrogen bonds. Due to this dipolar nature, amino acids have physical properties which make them ideal candidates for application [2]. Among the various amino acids, L-Leucine has

IOP Conf. Series: Materials Science and Engineering **360** (2018) 012044 doi:10.1088/1757-899X/360/1/012044

formed several complexes, which are promising materials for second harmonic generation. Maleic acid, basically dicarboxylic acid with large π conjugation has attracted a great deal of attention [3]. The intermolecular hydrogen bond in maleic acid is very strong. Maleic acid forms crystalline maleate of various organic molecules through hydrogen bonding and π -π interaction [4]. On the basis of earlier reports on L-Leucine salt [5], we have successfully grown a good quality crystal of L-Leucinium hydrogen maleate, a new organic compound. In this paper, the crystal growth and its characterization by powder XRD, FT-IR, FT-RAMAN, UV-Vis-NIR, NLO and Thermal analysis are reported for the grown compound.

2. ExperimentalProcedure

L-Leucinium hydrogen maleate (LLM) crystal was synthesized from highly pure L-Leucine and L-Maleic acid and they were taken in the equimolar ratio of 1:1. The required quantity of L-Leucine and L-Maleic acid was thoroughly dissolved by adding double distilled water according to their solubility data and stirred well for about five hours using a magnetic stirrer to obtain a homogenous mixture. The solution was filtered to remove insoluble impurities using Whatman filter paper of pore size ten micrometers. Then, saturated solution was taken in a beaker and tightly covered with a polythene paper of perforated sheet in order to control the evaporation rate and kept undisturbed at room temperature for crystallization. Finally, a well-defined single crystal was obtained after 35 days by slow evaporation method. The photo graph of the as grown crystal of LLM is shown in Figure. 1.

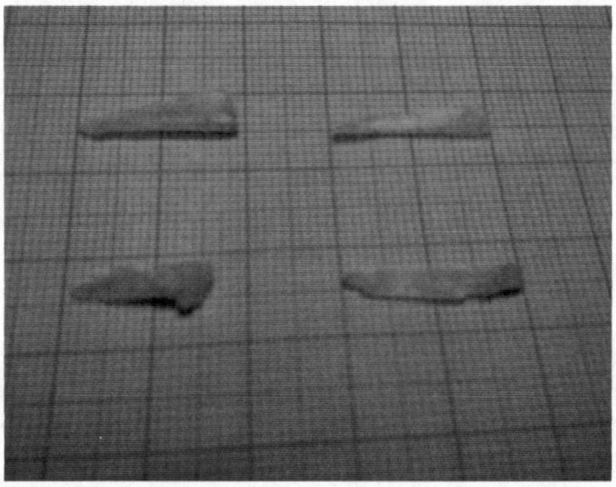

Figure1. Photograph of as grown single crystal of LLM

3.Results and Discussion

3.1. .X-ray diffraction analysis

The crystalline nature of the grown crystal was checked by taking the X-ray diffraction pattern of powder samples of L-Leucinium hydrogen maleate (LLM) with Cuk$_\alpha$ (λ = 1.5406 A°) radiation. The sample was scanned over a 2θ in the range of 10 -30° at the rate of 2°per min. From this measurement we found the lattice dimensions as a=21.132 A°, b=5.226 A°, c=31.452 A°, β=98.439° having the space group C2, it is crystallize in monoclinic system and is well matched with the reported literature [5]. The presence of a sharp and well defined peak confirms the good crystalline nature of the L-Leucinium hydrogen maleate (LLM) crystal. The differences in the peak amplitude can be attributed to the different sizes and orientation of the powdered grains. The recorded powder XRD pattern is shown in Figure 2.

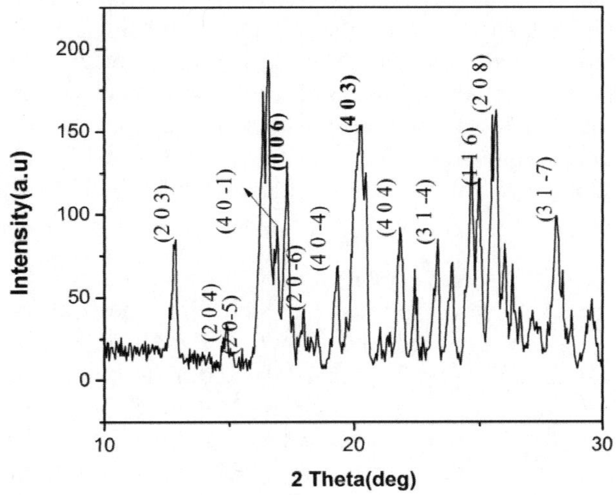

Figure 2. Powder X-ray diffraction pattern of the LLM crystals

3.2. UV-Vis Absorption spectroscopy

The optical absorption study is an important tool in identifying the usefulness of a NLO material in the visible regions. The UV-Visible absorption spectrum was recorded for the grown compound in wavelength range of 200 - 900nm using Varian carry 5E model UV-Vis spectrometer. There is no appreciable absorption of light in the entire visible range as in the case for all amino acids [6]. No absorption was found in the entire visible region of the UV-Vis-NIR spectra. UV-Vis-NIR studies also give important structural information because absorption of UV and visible light involves promotion of the electron in π and n orbital from the ground state to higher energy state [7]. The optical absorption spectrum of LLM is shown in Figure. 3a. The lower cut of wave length is around 212nm. There is no sufficient absorption from 200nm to 900nm. This is an advantage of the use of grown crystal, where the absence of strongly conjugated bonds leads to wide transparency ranges in the visible and UV spectral regions. The optical absorption coefficient of photon energy helps to study the band structure and explains the type of transition of electrons.

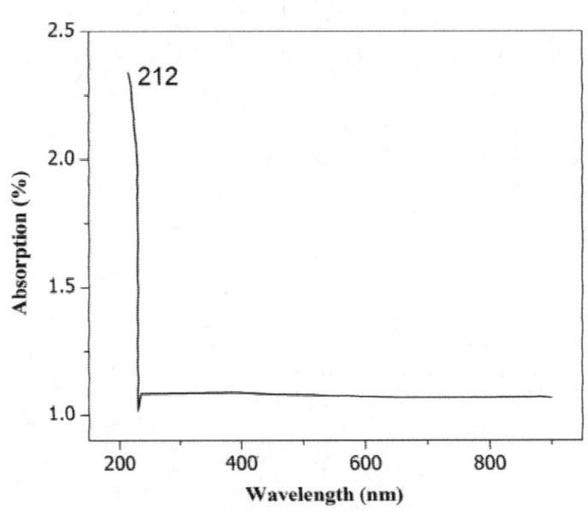

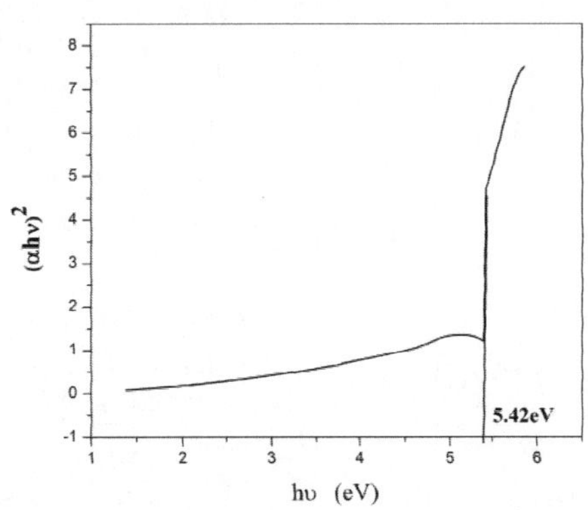

Figure 3.(a) UV-Vis absorption spectrum of LLM **Figure 3. (b)** Plot of $(\alpha h\nu)^2$ versus hν of LLM

The optical absorption coefficient α is calculated using the following relation α = Abs/t where Abs is the absorbance and t is the thickness of the crystal [8]. The optical energy band gap of the crystal is determined from the transmittance spectra using the relation $\alpha h\nu = A(h\nu - E_g)^{1/2}$ where A is a constant which varies with transitions, h is the Plank's constant, ν is the frequency of the photon, Eg is the band gap of the material. The band gap of LLM crystal was estimated by plotting $(\alpha h\nu)^2$ versus hν as shown in Figure. 3b. As per Tauc's idea, the optical band gap has been calculated from the extrapolation of linear part at absorption edge. The value of band gap was found to be 5.42 eV. As a consequence of wide band gap, the grown crystal has large absorption in the visible region.

3.3. FT-IR analysis

The FTIR spectra of the grown LLM single crystal were observed on the powder sample is shown in Figure. 4. The FTIR spectrum of LLM crystal was recorded using the Brukers IFS66 spectrometer in the range 4000 – 400 cm^{-1} by KBr pellet to identify the functional group present in the crystal. The medium intensity band at 3131 cm^{-1} appearing in IR spectrum is due to the NH_3^+ asymmetric stretching mode [9]. The stretching vibrations of the CH_2 group have been observed in the region 2960–2840 cm^{-1}. The position and intensity of the peaks are at 2959 cm^{-1} [10,11] and 2873 cm^{-1} is due to the asymmetric and symmetric stretching of CH_2 group of LLM [4,9]. The combination and overtone vibrations overlap in the range 2653 – 1880 cm^{-1} [4]. The strong carboxyl absorption at 1730 cm^{-1} is assigned to C=O stretching confirms the –COOH and $COO^{-1}$of compound [1,2]. The presence of N-H bending observed at 1572 cm^{-1} confirms presence of amino acid in the crystal lattice [9]. The absorption at 1390 cm^{-1} is due to the COO- ions. CH_2 wagging vibration is absorbed due to the peak at 1321 cm^{-1}[8,12]. The symmetric C-O stretching frequency is observed at 1241 cm^{-1} in the spectrum[13,4]. The broad band of CH_2 wagging vibration absorbed at 1035 cm^{-1} and CH_2 rocking vibration between 839 – 894 cm^{-1}. The strong band of the C-C stretching is observed at 867 cm^{-1} [3,4]. The COO^{-1} bending, scissoring, wagging vibrations observed at 749, 662, 589 cm^{-1}. The vibrational study confirms the LLM exists as zwitterions in which the carboxyl group is present as COO^- Carboxylate ion and the amino group exists as NH_3^+ ammonium ion.

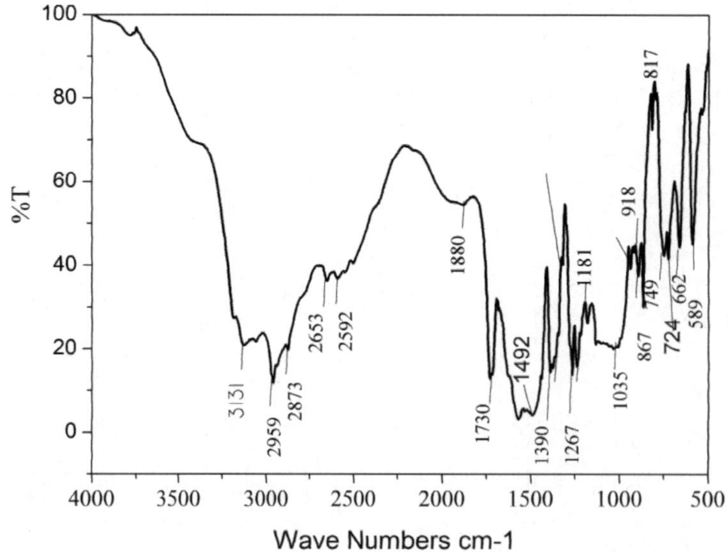

Figure 4 FT-IR spectrum of the grown LLM crystal

3.4. FT-RAMAN analysis

Figure 5 shows the Raman spectra of LLM crystals recorded in the range of 500 – 3500cm^{-1} at room temperature using Bruker IFS27 Raman spectrograph. The peaks at 3058.47 cm^{-1} confirm NH_3^+ stretching of amino group. The strong band at 1695 cm^{-1} and a shoulder at 1623 cm^{-1} are attributed to NH_3^+ asymmetric deformation mode. The peak against 2961 cm^{-1} and 2936 cm^{-1} are corresponding to the CH_2 asymmetric stretching vibrations. NH_3^+ rocking is observed at 1179 cm^{-1}. Symmetric stretching of CH band in the Raman spectrum is at 2871 cm^{-1}. C-C stretching revealed due to the peak at 1460 cm^{-1} 897 cm^{-1}. The corresponding COO^{-1} symmetric and COO^- wagging are confirmed due to the peaks at 1390 cm^{-1} in FT Raman spectrum. NH_3^+ rocking is observed at 1179 cm^{-1} and 962 cm^{-1}. Therefore the functional groups identified by FT-IR spectrum are also confirmed by FT-RAMAN spectrum.

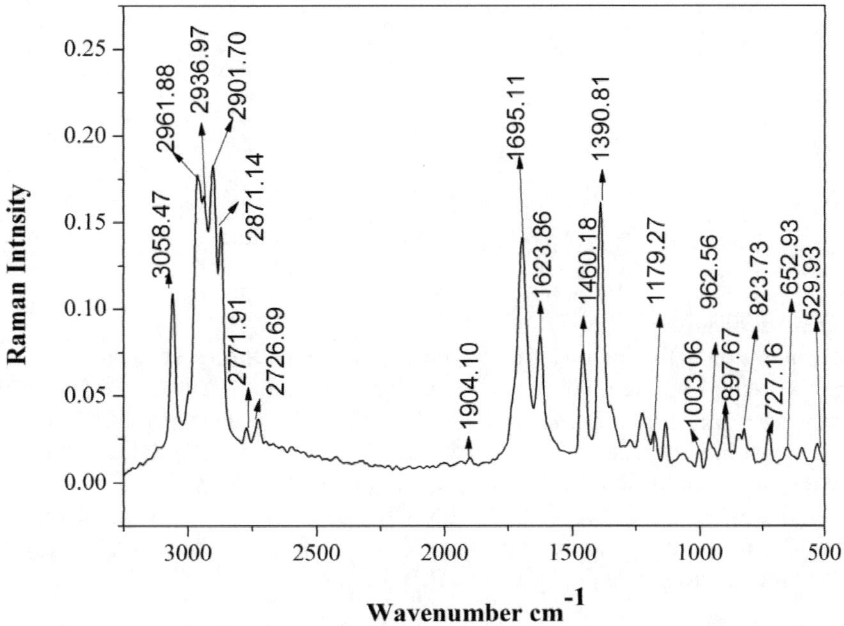

Figure 5. FT –Raman spectrum of the grown LLM crystal.

3.5. Thermal analysis

Thermo gravimetric and Differential thermal analysis gives information regarding phase transition, water of crystallization and different stages of decomposition of the crystal. The Thermo gravimetric analysis (TGA) and Differential Thermal analysis (DTA) of the crystal were carried out between 25°C to 650°C at a heating rate of 10K/min in the nitrogen atmosphere. The resulting spectrum is shown in Figure 6. There was no weight loss up to 140°C, hence the material is free of any solvent entrapped in the crystal lattice. The TGA curve shows a weight loss in one stage and the same near 150°C .is assigned to loss of water. The weight loss due to water is also associated with melting of the sample, which is clearly observed in the DTA curve as an endothermic transition occurred at 142°C. At this temperature, LLM and water molecule were dissociated from the compound and the LLM started to melt as its melting point is nearly 150°C, an exothermic transition as observed at 478°C and 554°C in the DTA curve. From the results of DTA it is established that no transformation in structure was observed before melting. Hence the material can be exploited for any suitable application up to its melting.

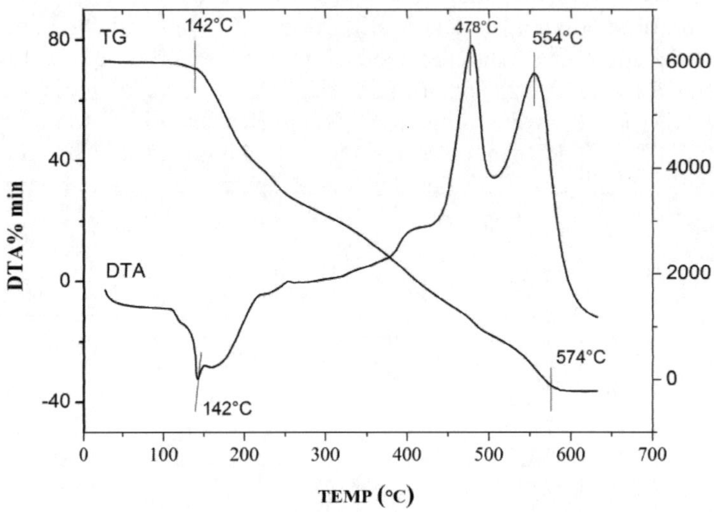

Figure 6.TG/DTA curve for LLM crystal

3.6. Nonlinear optical studies

Second harmonic generation efficiency of the grown crystal was confirmed using Kurtz and Perry powder technique [14]. A Q-switched Nd: YAG laser emitting a fundamental wavelength of 1064nm and a pulse width of 8ns with a repetition rate of 10 Hz was used. The incident input energy of 0.7J/s was incident on the crystalline powder packed between two transparent glass slides. The emission of green light from the sample confirmed the frequency doubling of LLM crystal. The second harmonic generation output signal of 2.98mW obtained for LLM crystal whereas the standard KDP crystal gave an SHG signal of 5.03mW for the same input energy. Hence, SHG efficiency of LLM crystal was 0.6 times that of potassium dehydrogenate phosphate (KDP) crystal.

4. Conclusion

Single crystals of L-Leucenium maleate (LLM) were successfully grown by slow evaporation technique at room temperature. Powder X-ray diffraction analysis reveals the monoclinic crystalline nature of the LLM crystal. The UV-Vis-NIR studies shows that these crystals are transparent in the entire visible region with a lower cut-off wavelength 212 nm and band gap value were calculated for LLM crystal. The presence of various functional groups was identified by FT-IR analysis and is confirmed by FT-Raman. The TG/DTA studies established that the compound undergoes no phase transition and is stable up to 150°C. The SHG efficiency studies show the suitability of the crystals for NLO application.

Acknowledgments

The corresponding author gratefully acknowledges the support from the Department of Science and Technology (DST), Government of India for the research project (SB/EMEQ/248/2014).

IOP Conf. Series: Materials Science and Engineering **360** (2018) 012044 doi:10.1088/1757-899X/360/1/012044

References

[1] Victor Antony Raj M and Madhavan J, 2011 Linear and nonlinear optical properties of pure and doped L-alaninium maleate single crystals *J. Archives of Physics Research (Scholars Research Library)* **2(1)** 160-168.

[2] Fuchs B A,Chaisyn K, StephanVelsko P,1989 Electrical Studies on Organic Nonlinear Optical Single Crystal: L-leucine Nitrate *J. Appl Opt* **28** 113-119.

[3] Natarajan S, Brito S A M, and Ramachandiran E S, 2006 Growth Thermal Spectroscopic and optical Studies of L-Alaninium Maleate a New Organic Nonlinear Optical Material Optical Material *J. Crystal Growth and design* **6** 137-140.

[4] Baraniraj.T., PhilominathanP.2010, Growth and Characterization of NLO based L-arginine maleate dihydrate single crystal *J. Spectrochem Acta.***75A** 74-76.

[5] Sergey G Arkhipov, Denis A Rychkov, Alexey M Pugachev and Elena V Boldyreva, 2015 New hydrophobic L-amino acid salts: maleates of L-Leucine L-Isoleucine and L- Norvaline *J. Acta Crystal Structural Chemistry* **C17** 1-9**.**

[6] Rodrigues Jr J J, Misoguti L, Nunes F D, Mendonca C R, Zilio S C, 2003 Optical properties of L-threonine crystals, *J. Opt Mater* **22** 235-240.

[7] Prakash.M,.Geetha D, Lydia Caroline M, 2011 Crystal growth and characterization of L-phenylalaninium trichloroacetate-A new organic nonlinear optical material, *J.Physica B.* **406(13)** 2621-25.

[8] MohdShkir, Haider Abbas, 2014 Physico chemical properties of L-asparagine L-tartaric acid single crystals: A new nonlinear optical material *J. Spectrochemica Acta Part A:Molecular and biomolecular.Spectroscopy* **118** 172-176.

[9] Soma Adhikari, TanusreeKar, 2012 Bulk single crystals growth and characterization of L-Luecine- A nonlinear optical material ,*J. Material Chemistry and physics* **133** 1055-1059.

[10] Bhagavananarayan G, Riscop B, Mohemed Shakir, 2011 Growth and characterization of L-Luecine L-Leucinium picrate single crystal: A new nonlinear optical material, *J. Materials Chemistry and Physics* **126** 20-23.

[11] Madhavan J, Aruna S, Anuradha A, Premanand D,VethaPotheher I,Thamizharasan K, Sagayaraj P, 2007 Growth and characterization of a new nonlinear optical L-histedine acetate single crystal *J. Optical Materials* **29** 1211-16.

[12] Brigt Mary.M,Sasirekha.V,Ramakrishnan.V,2006,Vibrational spectral analysis of DL-valine DL-valinium and DL-methionine DL-methioninium picrates *J. Spectrochemica Acta Part A* **65** 955-963.

[13] Praveen Kumar P, Manivannan V, Tamilselvan S, Senthil S, Victor Antony Raj, Sagayaraj P Madhavan J, 2008 Growth and characterization of a pure and doped nonlinear optical L-histidine acetate single crystals *Optics Communications* **281** 2989-95.

[14] Kurtz S K. Perry T T, 1968 A Powder Technique for the Evaluation of Nonlinear Optical Materials *J. Appl. Phys.* **39** 3798-13**.**

IOP Conf. Series: Materials Science and Engineering **360** (2018) 012045 doi:10.1088/1757-899X/360/1/012045

Crystal Structure, Vibrational Spectra and Theoretical Studies on Glycine Lithium Sulphate

A Shiny Febena[2], A Chitra[1,2], R Subaranjani[2] and J Madhavan[2]*
[1]Department of Physics, S.I.V.E.T College,Gowrivakkam,Chennai-73,Tamilnadu,India

[2]Department of Physics, Loyola College, Chennai -34,Tamilnadu, India.

*Corresponding Author: jmadhavang@yahoo.com**

Abstract. In this work we present structural and FT-IR studies on the nonlinear semi organic molecular crystal Glycine Lithium Sulphate (GLS). Vibrational spectral measurements were made for GLS. Optimized geometrical structure and harmonic vibrational frequencies were computed by DFT (B3LYP) method using 6-31++G (d, p) basis set. Complete assignments of the observed spectra were proposed. To investigate microscopic second order nonlinear optical behavior of the examined complex the electric dipole, the polarizability and the first order Hyper polarizability were computed.

1. INTRODUCTION

Recently several studies dealing with organic, inorganic and semi organic materials possessing nonlinear properties are being reported due to the increasing need for photonic applications [1]. In this work, the growth and characterization of Glycine Lithium Sulphate by slow evaporation method from aqua solution are reported. Also Literature survey shows that computation studies have not been done so far on the title compound. The aim of this study is to give a complete description of the molecular geometry and molecular vibrations of the GLS.

2. EXPERIMENTAL DETAILS

Glycine Lithium Sulphate crystal were synthesized by taking Analytical grade of Glycine and Lithium Sulphate in the molar ratio and dissolved in double distilled water.

$$C_2H_5NO_2 + Li_2(SO_4) \longrightarrow Li_2(SO_4)(C_2H_5NO_2)$$

Based on the solubility data as shown in Figure.1 supersaturated solution of GLS was prepared and stirred well for 3 hours. The beaker containing solution was kept in the constant temperature bath and tiny crystals of GLS were observed at the bottom of the beaker. The transparent tiny crystals were chosen as the seeds for growing bulk crystals. Crystals of size 10 x 11 x 12 mm^3 were harvested and is as shown in Figure.2. The grown sample was subjected to powder XRD analysis and its lattice parameters were calculated.

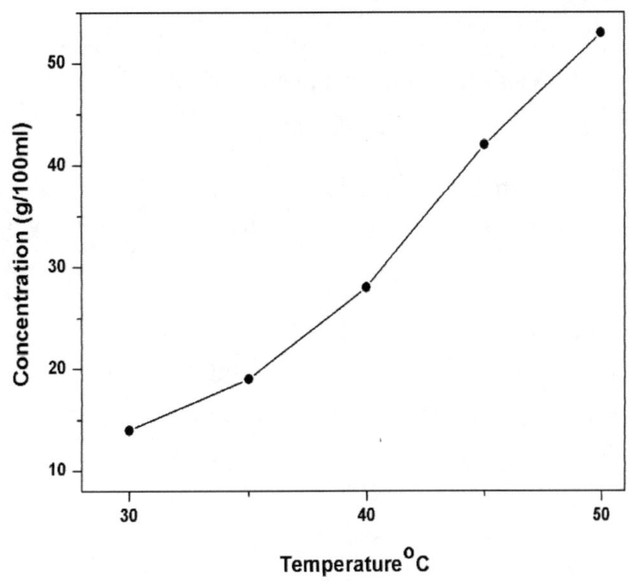

Figure 1.The solubility curve of GLS crystal

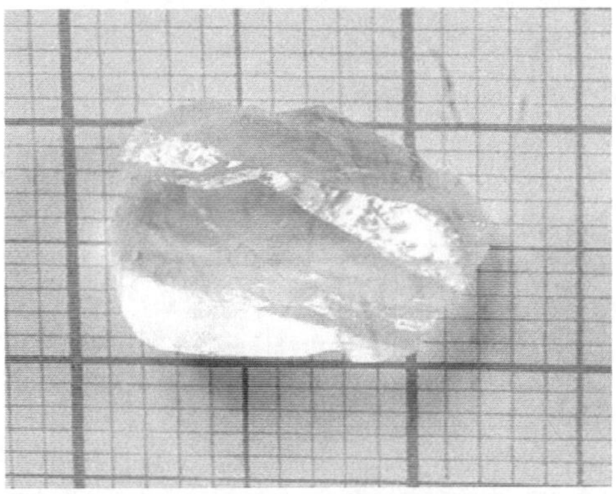

Figure 2.The grown GLS crystal

3. RESULTS AND DISCUSSION

3.1 X-ray diffraction analysis

From the powder x-ray diffraction analysis, it was observed that the crystal belongs to orthorhombic crystal system having non-centro symmetric with $Pna2_1$ space group. The crystal data are given in Table 1. The recorded powder XRD pattern is given in Figure 3.

3.2 Computational Details

All calculations are performed using Gaussian03 program package. Vibration frequencies are calculated using B3LYP/631(d,p) [2,3]. Vibration mode assignments were also compared with the

IOP Conf. Series: Materials Science and Engineering **360** (2018) 012045 doi:10.1088/1757-899X/360/1/012045

experimental results. The optimized geometrical parameters, fundamental vibrational frequencies, IR intensity, reduced mass, and force constant were calculated.

Table 1. Crystal parameters of GLS

Crystal Data	
Empirical formula	$C_2 H_5 Li_2 N O_6 S$
Crystal System	Orthorhombic
Space Group	$Pna2_1$
a(Å)	16.423
b(Å)	5.0050
c(Å)	7.654
β(°)	629.1

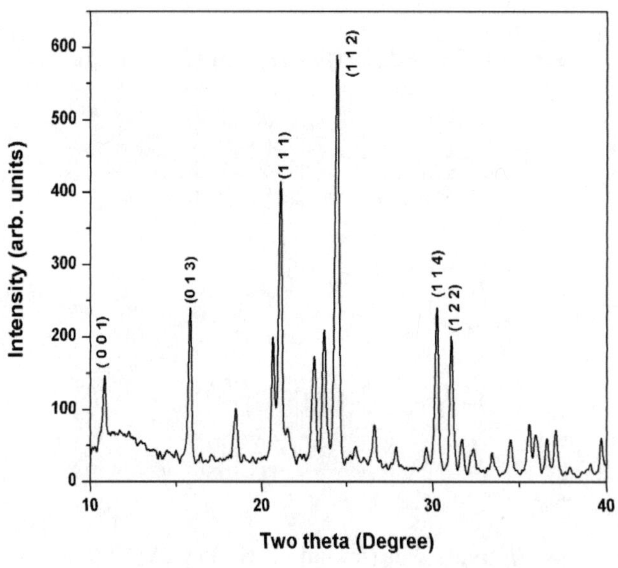

Figure 3. Powder XRD spectrum of GLS crystal

3.3 Molecular Geometry

The atomic numbering scheme for GLS is shown in Figure 4.. The most optimized geometric parameters (bond length and bond angles) were calculated and were depicted in Table 2. From the theoretical values, the most optimized bond lengths are slightly shorter as well as longer than the experimental values. Many researchers explained the changes in bond length of the C-H bond on substitution due to a change in the charge distribution on the carbon atom of the benzene ring [5]. The experimental values of C_4-H_5 bond length is longer than the theoretical values. A notable difference among the methods occurs in the computed N_7-H_9, H_{10}-Li_{11}, N_7-Li_{11} and O_{15}-O_{17}bond lengths. The DFT gives bond angle O_2-C_1-O_3 = 120, H_8-N_7-H_9=109.47 and N_7-O_{12}-Li_{11}=135.2 whereas the corresponding XRD values are 124.267,108.58 and 134.645 respectively. The comparison of Table 2. reveals the DFT calculation describes the geometry of the crystal with good quality.

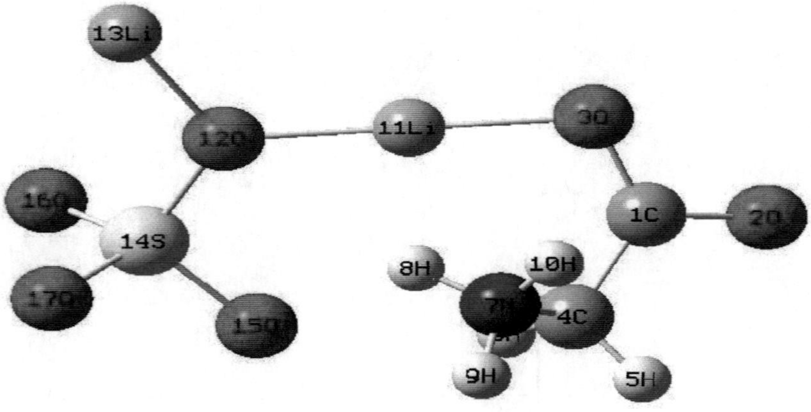

Figure 4. Atomic numbering system adapted for *ab* initio computations of GLS molecule

Table 2. Selected bond lengths and bond angles of GLS molecule

S.No	Molecule	Bond length (Å)	Molecule	Bond angles (°)
1	C_1-O_2	1.430000	O_2-C_1-O_3	120
2.	O_2-O_3	2.476833	C_1-C_4-H_6	109.4712
3.	O_3-Li_{11}	1.89	H_9-N_7-O_{12}	118.5122
4.	C_4-H_5	1.07	H_8-N_7-H_9	109.4712
5.	H_5-N_7	2.086720	O_3-C_1-C_4	120

3.4 Vibrational Assignments

The Fourier Transform Infra Red spectra of GLS was recorded on IFS 66V FT-IR SPECTROMETER using KBr pellet in the range 4000 cm^{-1} to 500 cm^{-1}. The observed FT-IR is shown in Figure 5. The GLS molecule has 17 moieties and is in stable conformation with C_1 symmetry then exhibits 45 normal modes of vibrations. The normal modes of GLS is distributed amongst the symmetry species as

$$\Gamma_{3N-6} = 31A' \text{ (in-plane)} + 14A'' \text{ (out-of-plane) respectively}$$

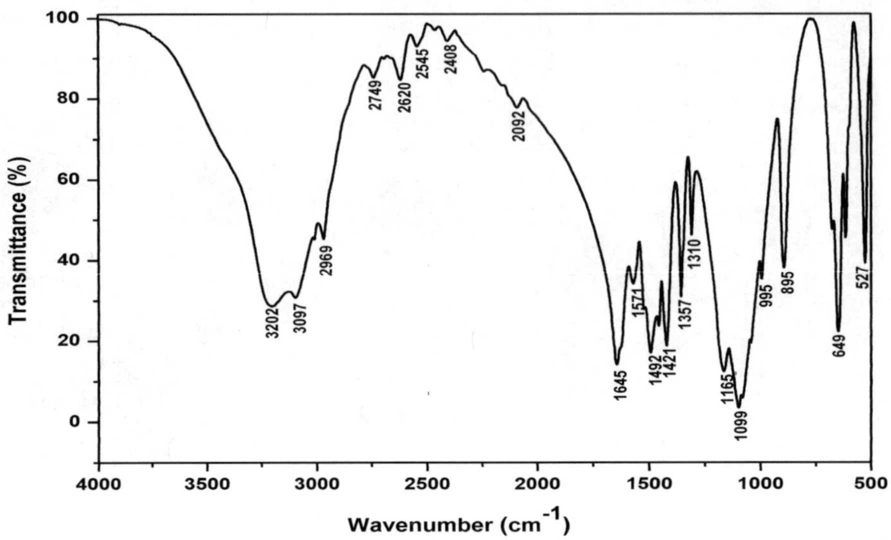

Figure 5. Experimental FT-IR Spectrum of GLS

Nitrogroup vibrations

Aromatic nitro compounds have strong absorptions due to asymmetric and symmetric vibrations of the nitro group at 1570-1485 cm^{-1} and 1370-1320 cm^{-1} [4,5]. The theoretical asymmetric stretching vibrations occur at 1503 cm^{-1} and the corresponding experimental peak at 1571 cm^{-1}.

C=O vibrations

The stretching vibration C=O is commonly expected in the region1690±30 cm^{-1}. The C=O in-plane deformation is expected in the region 645±55 cm^{-1}. The C=O out-of-plane deformation is in the range 540±80 cm^{-1}. The theoretical stretching vibration appears at 1679 cm^{-1} and the in-plane deformation at 653 cm^{-1}(theoretical) and 649 cm^{-1} (experimental).

C=N vibrations

According to Henry et al (2008), IR spectrum exhibits strong and broad bands due to C=N at 1635 cm^{-1}. In the present work the theoretical and experimental strong and broad bands due to C=N appears at 1626 cm^{-1} and 1645 cm^{-1} respectively.

The selected vibrational assignments of fundamental modes of GLS by DFT methods are reported in Table 3.

Table 3. Vibrational frequencies of GLS

No.	Wavenumber cm^{-1}		Spectroscopic assignment	Force constant	Reduced mass
	B3LYP	Expt.			
1.	457.3058	457	COO^- opd	1.7983	14.5946
2.	520.3388	527	COO^- d	1.8572	11.6421
3.	653.1643	649	COO^- roc ipd	1.7693	7.0388
4.	897.9603	895	CC st	2.4910	5.2433

5.	970.3730	995	CN st	0.9178	1.6543
6.	1050.167	1099	I Ph	2.1834	3.3602
7.	1183.694	1165	CH_2 roc	1.3022	1.5774
8.	1449.395	1492	COO^- symst	1.4858	1.2004
9.	1503.990	1571	NH_3^+ sym d	1.7639	1.3236
10.	1626.886	1645	NH_3^+ asy d	1.6559	1.0619

Symst-symmetric stretching, asyst-asymmetric stretching, st(ph)-stretching of phenyl, roc-rocking, opb-out of plane bending, opd-out of plane deformation, wag-wagging, ipd-in plane deformation, d-deformation

3.5 Hyperpolarizability
First hyperpolarizability is a third rank tensor that can be described by a 3x3x3 matrix. The 27 components of this 3D matrix can be reduced to 10 components due to Kleinman symmetry [6]. In the presence of applied electric field, the energy of a system is a function of electric field. The components of β are defined as the coefficients in the Taylor series expansion of the energy in the external electric field[7]. The first static hyperpolarizability (β_0) and its related properties have been calculated using B3LYP/6-31G level based on finite field approach. The calculated dipole moment(μ), polarizability(α) and first order hyperpolarizability values for GLS molecule are given in Table 4.

Table 4. Dipole moment, Polarizability and Hyperpolarzability

Dipole Moment in Debye		Polarizability in esu		Hyperpolarizability in esu	
μ_x	1.7681	α_{xx}	67.4945318	β_{xxx}	339.8812369
μ_y	4.2611	α_{xy}	0.9920316	β_{xxy}	68.9682643
μ_z	13.2300	α_{yy}	58.9864308	β_{xyy}	130.2321786
μ_{tot}	14.0113	α_{xz}	-4.5362599	β_{yyy}	268.4342981
		α_{yz}	6.6848818	β_{xxz}	-78.79471
		α_{zz}	85.163995	β_{xyz}	-16.2330359
		α_o	5.9043×10^{-31}	β_{yyz}	5.503936
				β_{xzz}	140.9210373
				β_{yzz}	151.7801336
				β_{zzz}	-114.5418743
				β_{tot}	$6.7368\ 10^{-30}$

3.6 Hyperpolarizability
First hyperpolarizability is a third rank tensor that can be described by a 3x3x3 matrix. The 27 components of this 3D matrix can be reduced to 10 components due to Kleinman symmetry [6]. In the

IOP Conf. Series: Materials Science and Engineering **360** (2018) 012045 doi:10.1088/1757-899X/360/1/012045

presence of applied electric field, the energy of a system is a function of electric field. The components of β are defined as the coefficients in the Taylor series expansion of the energy in the external electric field[7]. The first static hyperpolarizability (β_0) and its related properties have been calculated using B3LYP/6-31G level based on finite field approach. The calculated dipole moment(μ), polarizability(α) and first order hyperpolarizability values for GLS molecule are given in Table 4.

3.7 HOMO-LUMO energy gap

HOMO and LUMO analysis have been used to elucidate the information regarding charge transfer within the molecule. The intermolecular charge transfer from the donor to acceptor group through a single-double bond conjugated path can induce large variations of both the molecular dipole moment and the molecular polarizability. The HOMO and LUMO energy calculated by B3LYP/6-31G (d,p) method as shown in Figure 5.

HOMO energy (B3LYP) = -0.336 a.u.
LUMO energy (B3LYP) = -0.157 a.u.
HOMO–LUMO energy gap (B3LYP) = 0.179 a.u.

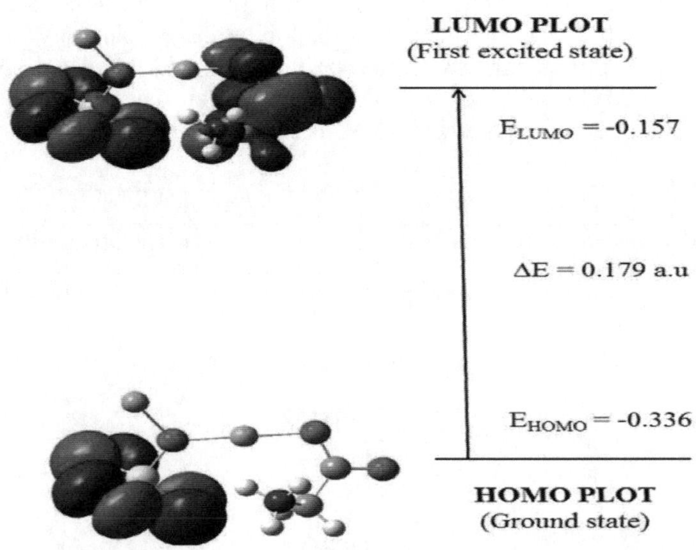

Figure 6. HOMO – LUMO plot of GLS at B3LYP/6-31 G (d, p)

4. CONCLUSION

Good optical quality bulk single crystal of GLS has been grown successfully by slow evaporation growth technique. The single crystal analysis shows that the crystal belongs to orthorhombic crystal system with $Pna2_1$ space group. DFT computations of GLS molecule calculated by DFT (B3LYP) level with 6-31G (d,p) basis set gives the optimized structure. Experimentally obtained bond lengths and bond angles are compared with theoretically calculated one. Theoretical and experimental IR spectroscopic analysis was carried out and the presence of functional groups in GLS molecule was qualitatively analyzed. Nonlinear optical NLO behaviour of the examined sample was investigated by the determination of the electric dipole moment μ, the polarizability α and the hyperpolarizability β using DFT method. HOMO-LUMO analysis reveals the molecular energy gap.

Acknowledgment
The authors acknowledge University Grants Commission (UGC), India, for funding this research project-F.No.4-4/2015-16(MRP/UGC-SERO) - 2914.

REFERENCES

[1] Sankar D, Vinay Raj Menon, Sagayaraj P and Madhavan *J.,Physica B.*, 405 (2010) 192.
[2] Borisenko, C.W. Boch, I.J. Hargittai, *J. Phys. Chem.* 98 (1994) 1442
[3] Frisch M.J, Trucks G.W Pope J.A, Gaussian 03, Revision C.02,Gaussian Inc. Wallingford, CT, 2004
[4] N.Sundaragganesan,G.Elango.S.Sebastian&P.Subramani,*Indian Journal of Pure& Applied Physics*,Vol.47,2009,pp.481-490
[5] Vijayakumar T, I. Hubertjoe, C.P.R. Nair, V.S. Jayakumar, *Chem. Phys.* 343 (2008) 83.
[6] Wojtkowaik.B, Chabanel.M, Spectrochimie Molecular Technique Documentn.Paris(1977)265.
[7] Henry, Lagrenee, Abraham, Inorg. *Chem. Commun*, 4 (2008) 1071
[8] Kleinman A, *J. Phys.* Rev. 126 (1962) 1977.
[9] Anoop M R,Binil P S, Suma S, Sudarsanakumar M R, Sheena Mary Y, Hema Tresa Varghese, YohannanPanicker C, *J.Mole.Structure* 969 (2010) 48-54.

IOP Conf. Series: Materials Science and Engineering **360** (2018) 012046 doi:10.1088/1757-899X/360/1/012046

Vibrational, Spectral Investigation of Nonlinear Optical L-Isocleucine D-Alanine: Experimental and Theoretical Confirmation

R Vincent Femilaa[1], P Geetha[1,2] and J Madhavan[1]*

[1]*Department of Physics, Loyola College, Chennai, Tamilnadu, India*
[2]*Department of Physics, Quaid-E-Millath Government College for Women, Chennai, TamilNadu, India.*

*E-mail: jmadhavang@yahoo.com

Abstract. Optically transparent amino acid based nonlinear crystal of L-Isoleucine D-Alanine (LIDA) was grown from its aqueous solution by slow solvent evaporation technique. The material was studied in-depth by using vibrational spectrum and density functional theory (DFT). The Fourier transform infrared (FT-IR) spectrum was recorded in the range 4000 cm^{-1} – 450 cm^{-1}. Meanwhile, the DFT computations are performed at B3LYP/6-31G+ (d, p) level to obtain equilibrium geometry, vibrational wave numbers and first hyperpolarizability. The scaled theoretical wave numbers are also shown to be in good agreement with experimental data. From the UV-Vis spectrum the transmittance abilities of the crystal were studied. The band gap of the material was determined from the Tauc plot.

1. INTRODUCTION

In the modern world, the development of science in many areas has been achieved through the growth of single crystals. Crystal growth of organic materials has been recently attracting scientific attention in the search of new nonlinear optical materials. Many organic materials have been found to have greatly nonlinear or optoelectrical properties than inorganic substances [1]. Some organic compounds exhibit large NLO response, in many cases, order of magnitude larger than widely known inorganic materials. They also offer the flexibility of molecular design and the promise of virtually an unlimited number of crystalline structures. Amino acids are the famous organic materials which can play a role in nonlinear optics as they contain proton carboxylic acid (COOH) group and the proton acceptor amino (NH_2) group. Considerable efforts have been made on the amino acid mixed complex crystals in order to make them suitable for device fabrications. L-Isoleucine is one of the protein genic amino acids that aids in the production of protein. It is a branched chain amino acid (BCAA) that is classified as a hydrophobic amino acid. One of its main benefits is increased strength and rapid muscle repair when combined with other amino acids. It is also beneficial for children and teens that are in their growing years because it is vital for healthy growth. Alanine is an efficient organic NLO compound under the amino acid category. It is the simplest amino acids with an asymmetric carbon atom. Alanine and DL-Alanine are the usually available forms of alanine. In this chapter, we tried to report the growth and characterization of L-Isoleucine D-Alanine (LIDA) single crystal.

IOP Conf. Series: Materials Science and Engineering **360** (2018) 012046 doi:10.1088/1757-899X/360/1/012046

2. EXPERIMENTAL DETAILS

The LIDA compound was synthesized by the direct combination of L-Isoleucine and D-Alanine in distilled water at room temperature. The synthesized compound was further purified by repeated recrystallization. Purified salt of LIDA were collected and further used for bulk single crystal growth. The grown crystal is shown in the figure.1 of size measuring 8 x 5 x 7 mm^3. Finely crushed powder of LIDA crystal was subjected to powder X-ray diffraction analysis using CuK_α (λ = 1.5418 Å) radiation. LIDA crystal has a monoclinic structure and the lattice parameters are tabulated in Table.1.

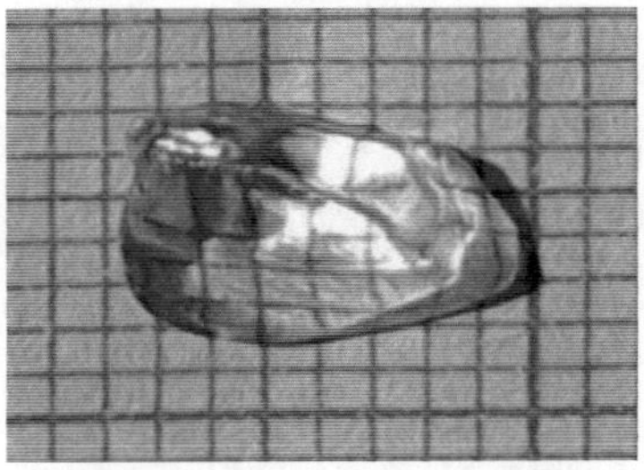

Figure.1 Photograph of as grown LIDA single crystal

Table 1. Crystal data

Crystal data	
Empirical formula	$C_9H_{20}N_2O_4$
Crystal System	Monoclinic
Space Group	$P2_1$
a(Å)	9.8944
b(Å)	4.7425
c(Å)	12.9045
β(°)	93.374

3. RESULTS AND DISCUSSION

3.1 Vibrational Spectral Analysis

The main objective of the vibrational analysis is to find vibrational modes connected with specific molecular structures of the compound. The best possible fitting was done between the calculated and experimental vibrational wavenumbers. The title molecule consists of 35 atoms, hence undergoes 99 normal modes of vibrations. These normal modes of vibrations are further classified into 67 in-plane vibrations and 32 out plane vibrations. The experimental FT-IR spectrum is as shown in Figure.2.

NH_3^+ group vibration

In zwitterionic molecules, theoretically the NH_3^+ symmetric stretching band appears at the value 3085.4737 cm^{-1} and experimentally appears at the value 3088 cm^{-1} respectively [2]. The NH_3^+ symmetric deformation theoretically appears at 1537.3389 cm^{-1}[3].

Carboxylate group vibration

The asymmetric and symmetric stretching modes of carboxylate ion vibrations theoretically occur at 1557.9300 cm^{-1} and 1451.4792 cm^{-1} respectively [4]. COO^- deformations theoretically occurs in the region 528.6117 cm^{-1}-582.6304 cm^{-1} and experimentally present at the value 539 cm^{-1}[5]. The COO^- deformation coupled with rocking mode appears theoretically at the value 675.0159 cm^{-1}- 740.3881 cm^{-1} in IR as a very strong band.

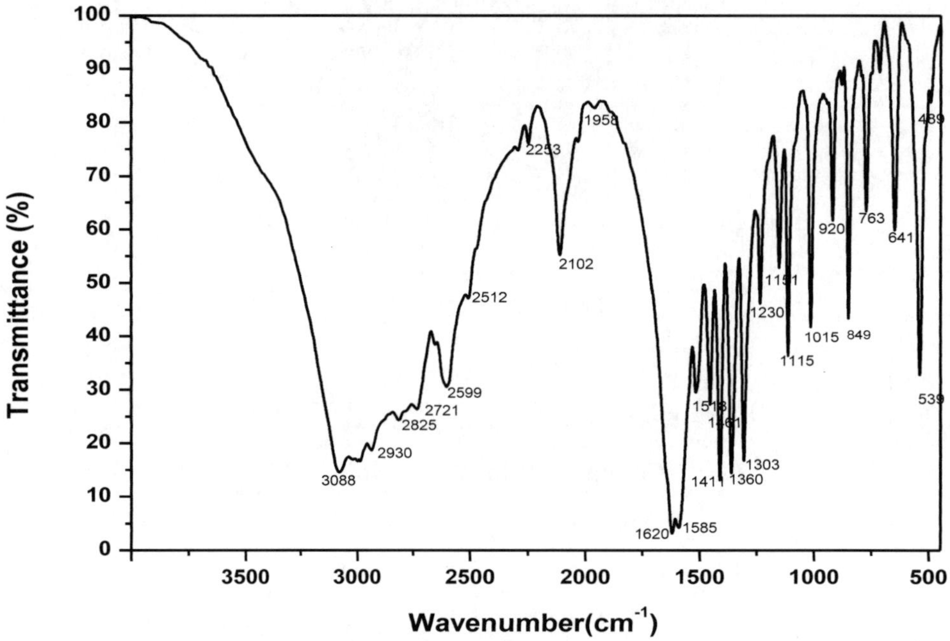

Figure 2. Experimentally obtained FT-IR spectrum of LIDA

C-N vibrations

C-N stretching absorptions are assigned theoretically in the region 1271.3646 cm^{-1} -1383.5730 cm^{-1} and experimentally at 1303 cm^{-1}, 1360 cm^{-1} for primary aromatic amine with nitrogen directly on the ring.

NH_2 vibrations

The NH_2 asymmetric stretching vibrations theoretically give rise to a strong band in the region 3404.9086 cm^{-1} - 3429.2699 cm^{-1} and the symmetric NH_2 stretching is observed as weak band in the region 3152.8348 cm^{-1} - 3160.8503 cm^{-1} theoretically.

O-H vibrations

The O-H out-of-plane bending mode for the free molecule lies in the region 5.0714 cm^{-1} - 289.9627cm^{-1} theoretically and it is beyond the infrared spectral range of the present investigation. However, for the associated molecule the O-H out-of-plane bending mode lies theoretically in the region 528.6117 cm^{-1} - 675.0159 cm^{-1} and experimentally at 539 cm^{-1} in both intramolecular and intermolecular associations, the frequency is at a higher value than in free O-H.

COOH vibrations

The carbonyl group is contained in many different classes of compounds, for which a strong absorption band due to C=O stretching vibration is observed theoretically in the range of 1557.9300 cm^{-1} - 2086.7851 cm^{-1}. The vibrational assignments are presented in Table.2.

Table 2. Vibrational Assignments of LIDA Molecule

No	Wavenumber (cm^{-1})		Spectroscopic assignment	Force constant	Reduced mass
	B3LYP	Expt.			
1.	528.6117	539	COO^- def	0.3631	2.2058
2.	843.3942	849	C-N abs	0.5149	1.2286
3.	932.3841	920	NH_3 roc	1.0113	1.9744
4.	1024.5942	1015	NH_2 t	0.7400	1.1965
5.	1109.4946	1115	OH ipb	1.1096	1.5299
6.	1138.4561	1151	CH_2 roc	1.4700	1.9250
7.	1233.3331	1230	COO^- st &b	1.3043	1.4553
8.	1309.8000	1303	COO^- st & b	1.6224	1.6051
9.	1362.7537	1360	COO^- st & b	1.8737	1.7125
10.	1421.2563	1411	C-H b+ C=C st	2.1903	1.8403
11.	1460.9092	1461	C-H b+ C=C st	4.0324	3.2067
12.	2929.0913	2930	CH_2 st+ C-H st	5.3970	1.0677
13.	3085.4737	3088	NH_3^+ sym st	5.9061	1.0529

def- deformation, st- stretching, abs- absorption, st &b-stretching &bending, t- twisting, sym st- symmetric stretching.

3.2 Molecular Geometry and Hyperpolarizability

Computational chemistry is an important tool to design and modeling the NLO materials. The geometry of the investigated compound was treated as an isolated gas molecule. Becke-Lee-Yang-Parr hybrid exchange correlation three parameter (B3LYP) level at 6-31G basis set is used to derive the optimized structure in Gaussian 03W software package. Optimized structure was confirmed to minimum energy conformations. The optimized molecular structure of the isolated LIDA is shown in Figure.3. The second-order polarizability or first hyperpolarizability was calculated by B3LYP using 6-31G (d, p) basis set on the basis of the finite-field approach. Theoretical calculations on molecular hyperpolarizability become one of the key factors in the second-order NLO materials design. Theoretical determination of hyperpolarizability provides a guideline to experimentalists for the design and synthesis of organic NLO materials. Nonlinearity in organic chromophores can be synthetically modulated by varying the composition or length of conjugated p-systems, and by evaluating the effects of various electron-donor and electron-acceptor groups. They determine not only the strength of molecular interactions (long-range interaction, dispersion force, etc.) and the cross sections of different scattering and collision process but also the NLO properties of the system. The hyperpolarizability values are listed in Table.3.

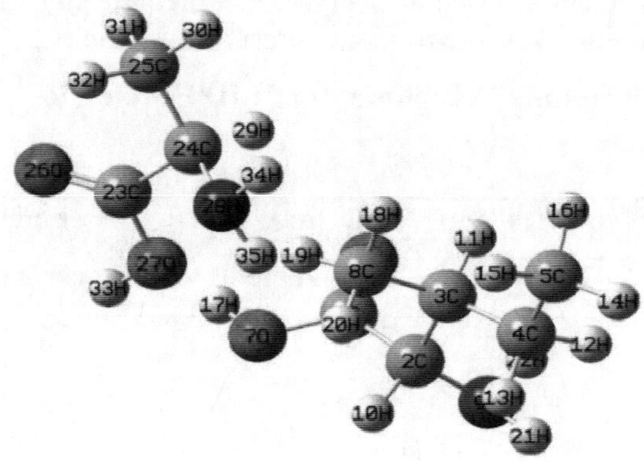

Figure 3. Atomic numbering for ab initio computations of LIDA molecule

Table 3. Hyperpolarizability of LIDA in esu

β_{xxx}	-275.705
β_{xxy}	-109.758
β_{xyy}	-187.947
β_{yyy}	-9.47754
β_{xxz}	-0.0259567
β_{xyz}	-0.00107375
β_{yyz}	-0.0120082
β_{xzz}	6.55646
β_{yzz}	-30.2164
β_{zzz}	0.0216665

3.3 Optical Absorption Spectrum

In order to determine the optical transmission characteristics of the grown crystal UV-Vis spectrum was recorded on the cut and polished grown sample. The UV-Vis spectral absorption was studied using a Varian Cary 5E UV-Vis-NIR spectrophotometer with a single crystal of 6mm thickness in the range of 200–800nm. The recorded spectrum is shown in Figure. 4 and 5. The crystal has sufficient transmission in the entire visible and IR region. The lower cut off wavelength is around 240 nm. From the absorbance energy band gap was evaluated. The E_g could be estimated from the plots of hν versus $(\alpha\ h\nu)^2$.

IOP Conf. Series: Materials Science and Engineering **360** (2018) 012046 doi:10.1088/1757-899X/360/1/012046

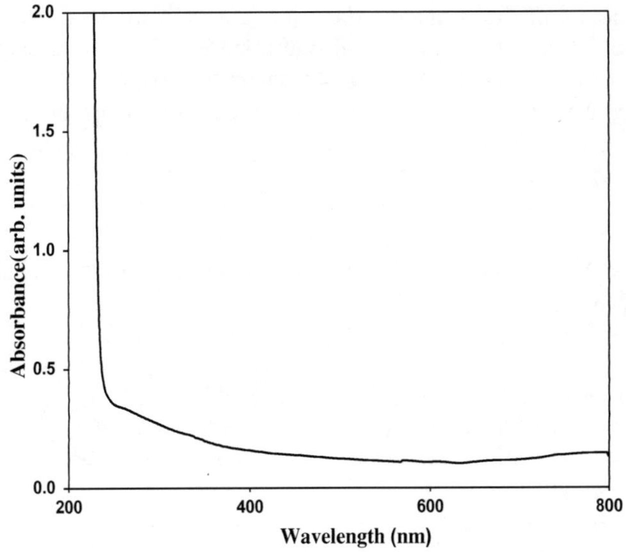

Figure 4. Optical absorption spectrum of LIDA crystal

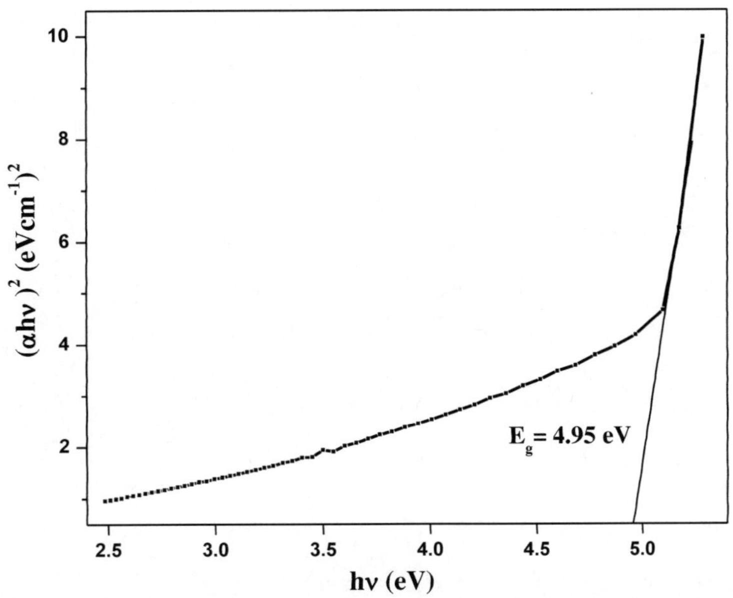

Figure 5. Optical band gap of LIDA crystal

IOP Conf. Series: Materials Science and Engineering **360** (2018) 012046 doi:10.1088/1757-899X/360/1/012046

4. Conclusion

Single crystal of L-Isoleucine D-Alanine (LIDA) was grown successfully by slow evaporation technique. Optimized structure of the isolated LIDA molecule obtained by DFT calculations give the minimum energy state. Existence of strong hydrogen bond in the donor accepter coupling was understood from bond lengths. First order hyperpolarizability of LIDA is calculated as $4.024860812 \times 10^{-30}$ esu and found useful in molecular designing. Theoretical and experimental spectroscopic studies exemplify the presence of various functional groups in the molecule. The UV-Vis spectrum reveals minimum absorption in the entire visible region.

Acknowledgment

The authors acknowledge University Grants Commission (UGC), India, for funding this research project-F.No.4-4/2015-16(MRP/UGC-SERO) - 2914.

REFERENCES

[1] Badan.J, Hierie.R, Perigaud.A, Zyes.J, Symp.A.C.S.ser 233 (1983) 81.
[2] Bellamy.L.J, *the infrared and Raman Spectra of Poly atomic molecules Wiley*: New York (1975).
[3] Silverstein R. M, Webster. F. X., *Spectrometric Identification of Organic Compounds*. John Wiley and sons, New York (2003).
[4] Socrates G., *Infrared and Raman Characteristic Group Frequency, third ed.*, Wiley, New York (2001).
[5] Dollish F.R., Fateley W.G., Bentley F.F., *Characteristic Raman Frequencies of Organic Compounds*, John Wiley & Sons, New York (1997).

Vibrational and Homo-Lumo Analysis of L - Asparарginium Tartrate by Density Functional Theory

E Kavitha[1], A Shiny Febena[2], R Subaranjani[2] and J Madhavan[2*]

[1]Department of Physics, Dr. M. G. R University, Chennai – 95

[2]Department of Physics, Loyola College, Chennai – 34

[*]**Corresponding Author: jmadhavang@yahoo.com**

Abstract. L - Asparарginium Tartrate (LAsT), a semi organic nonlinear optical material was synthesized by slow solvent evaporation technique. Fourier transform infrared (FT-IR) spectroscopic studies were performed for identifying different functional groups present in the compound and was compared with the theoretical data obtained from DFT studies. Non-linear optical (NLO) behavior of the LAsT was investigated by the determination of the dipole moment μ and the hyper polarizability β obtained by B3LYP/6-31G (d, p) method. The calculated highest occupied molecular orbital (HOMO) and lowest unoccupied molecular orbital (LUMO) with frontier orbital gap were presented.

1. INTRODUCTION

Nonlinear optical (NLO) crystals with high conversion efficiencies for second harmonic generation (SHG) and transparent in visible and ultraviolet ranges are required for numerous device applications. In the last few years, much progress has been made in the development of nonlinear optical (NLO) organic materials for second harmonic generation (SHG). However, most of the or Nonlinear optical (NLO) crystals with high conversion efficiencies for second harmonic generation (SHG) and transparent in visible and ultraviolet ranges are required for numerous device applications. In the last few years, much progress has been made in the development of nonlinear optical (NLO) organic materials for second harmonic generation (SHG). However, most of the organic NLO crystals are constituted by weak van der Waals and hydrogen bonds with conjugated π electrons [1]. In this respect, amino acids are interesting materials for NLO applications [2-3]. To the best of our knowledge, no density functional theory DFT wavenumber and structural parameters calculations of LAsT has been reported so far. The present work deals with the growth and characterization along with the detailed vibrational spectral investigation of LAsT.

2. EXPERIMENTAL PROCEDURE

L-Asparagine–L-tartaric acid (LAsT) was synthesized by dissolving one mole of L- Asparagine (Merck 99%) in double distilled water containing one mole of L-tartaric acid. The reaction is as follows

$$C_4H_8N_2O_3 + C_4H_6O_6 \longrightarrow C_8H_{14}N_2O_9$$

The synthesized salt was further purified by repeating the crystallization process at least thrice. The solubility of LAsT in deionized water (solvent) was determined by adding a solute in solvent till it is completely dissolved. The temperature dependence of solubility of LAsT is shown in Figure 1. Figure 2 shows the photograph of as grown crystal of LAsT.

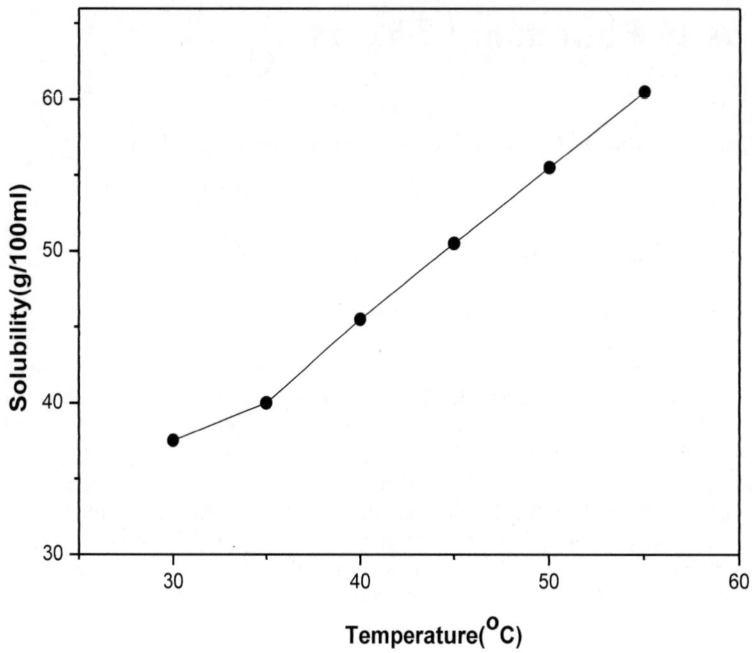

Figure 1.The solubility curve of LAsT crystal

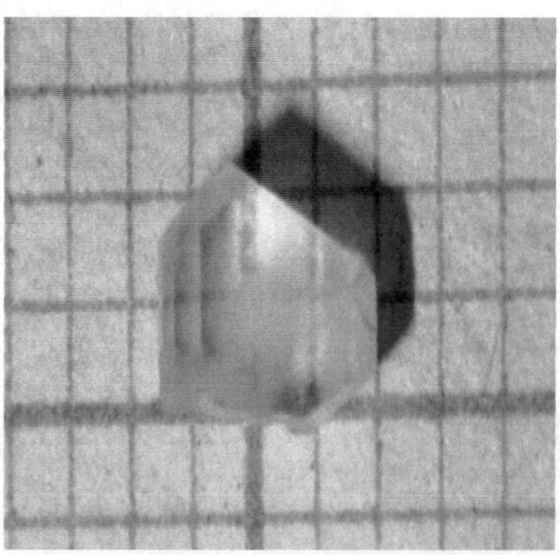

Figure 2.The grown LAsT crystal

3. RESULTS AND DISCUSSION

3.1 X-ray diffraction analysis

The selected well shaped, transparent, single crystal of LAsT was mounted on a BRUKER NONIUS CAD4/MAC4 X-ray diffractometer. Reflection data were measured at 20˚C using graphite monochromated Mo-Kα radiation and a ω-2θ scan mode. The structure was solved by direct methods and refined by full-matrix least-squares method using the SHELXS-97 software package [4]. From the single crystal analysis, it was observed that the crystal belongs to monoclinic crystal system having non-centrosymmetry with $P2_1$ space group. A summary of the key crystallographic information is given in Table 1.

Table 1. Crystal parameters of LAsT single crystal

Crystal data	
Emprical formula	$C_8H_{14}N_2O_9$
Crystal System	Monoclinic
Space Group	$P2_1$
a(Å)	5.086 Å
b(Å)	9.672 Å
c(Å)	11.834 Å
β(°)	95.31°

3.2 COMPUTATIONAL DETAILS

Quantum chemical density functional theory calculations were carried out with the 2003 version of the Gaussian program package using B3LYP functions [5] combined with the 6-31+G(d, p) basis set. The optimized geometrical parameters, fundamental vibrational frequencies, IR intensity, the atomic small charges, dipole moment, reduced mass and force constant were calculated.

VIBRATIONAL ASSIGNMENTS

The title molecule LAsT has 33 atoms. It has 93 (3N – 6) normal vibrational modes. The 93 normal modes of LAsT are distributed amongst the symmetry species as

$$\Gamma_{vib} = 63\ A' \text{ (in-plane)} + 30\ A'' \text{ (out-of-plane)}.$$

3.3 FT-IR ANALYSIS

FT-IR spectrum of the grown crystal was recorded in the range 500 cm^{-1} to 4000 cm^{-1}, using KBr pellet technique on BRUKKER IFS FT-IR Spectrometer. The experimental IR spectrum is compared with the results of B3LYP/6-31+ G (d, p) calculation carried out for the title compound. The experimental FT-IR is shown in the Figure. 3.

C-H Vibrations

Presence of band in the region 2700–3000cm^{-1} is the characteristic region for the identification of C–H stretching vibrations [6]. In this region the bands are slightly affected by the nature of the substituent. The IR vibrations are around 2900cm^{-1}. The experimental C–H vibrations are in good agreement with theoretical vibrations. For LAsT molecule prominent numbers of CH symmetric vibrations are obtained at frequencies 3092, 2977 and 2960cm^{-1}.

NH_2 Vibrations

The NH_2 asymmetric stretching vibrations [7] give rise to a strong band in the region 3390±60cm^{-1} and the symmetric NH_2 stretching is observed as weak band in the region 3210±60cm^{-1}. In our title molecule the theoretically computed NH_2 asymmetric and symmetric stretching vibration by B3LYP/6-311(d,p) method is at 3369 and 3344 cm^{-1}. But the recorded spectrum does not show any kind of peak in this region. In the present work, NH_2 symmetric stretching is assigned at 3320 cm^{-1} in IR region and also in 3280 cm^{-1}. The experimental values are well coincides with DFT values.

COO^- Vibration

Carboxyl group vibrations give rise to intense characteristic bands due to conjugation or formation of hydrogen bonds. These stretching and bending vibrations of acid group are generally expected in the region 1400–1200 cm^{-1} In the present work COO symmetric stretching is at 1420 cm^{-1} and 1553 cm^{-1} have asymmetric stretching in IR region.

O–H Vibrations

The O–H group gives rise to three vibrations (stretching, in-plane bending and out-of-plane bending vibrations). The O–H group vibrations are likely to be the most sensitive to the environment, so they show pronounced shifts in the spectra of the hydrogen bonded species. The hydroxyl stretching vibrations are generally observed in the region around 3500cm^{-1}[8] . In the title compound the O-H vibrations is present at 3450 cm^{-1} in IR region. The detailed vibrational assignments of LAsT for the experimental FT-IR along with the calculated frequencies for different basis sets are tabulated in Table 2.

Table 2. Vibrational Assignments of LAsT molecule

S.NO	FT-IR frequency		Vibrational Assignments
	Experimental	B3LYP	
1	3320	3327	υOH
2	3280	3308	υasym NH_3^+
3	1673	1678	βNH_2
4	1310	1307	βCH, υCC
5	1065	1048	γCO
6	690	689	Sym ring breath
7	510	510	φNCN

υ-stretching; **β**-bending; **δ**-in-planebending; **γ**-out-of-plane bending;

ω- wagging; φ-rocking;**τ**-torsion

3.4 Hyperpolarizability

The non-linear optical properties play an important role for the design of materials in modern communication technology, signal processing, optical switches and optical memory devices [9]. The non-linear optical properties of the organic molecules arise from delocalized π electrons that move along molecule. The increase of the conjugation on molecule leads to an increase in its nonlinear optical properties. One another way to increase non-linear optical properties is to add donor and acceptor groups. Acceptor group is opposite of donor group in the organic molecules containing donor and acceptor groups and π electron cloud moves from donor group to acceptor group. If the donor and acceptor groups are powerful, delocalization of π electron cloud on organic molecules increases and as a result of this the polarizability and first hyperpolarizability of organic molecules increase [10]. First-order molecular hyperpolarizability is a third rank tensor that can be described by 3 × 3 × 3 matrix. The 27 components of the 3D matrix can be reduced to 10 components due to the Kleinman symmetry . The calculated polarizability α and first hyperpolarizability β of LAsT is 1.765808 x 10^{-30} esu that is greater than that of urea (0.37289×10^{-30} esu), respectively.

3.5 HOMO-LUMO energy gap

Both the highest occupied molecular orbital (HOMO) and the lowest unoccupied molecular orbital (LUMO) are the main orbital take part in chemical stability. The HOMO represents the ability to donate an electron, LUMO as an electron acceptor represents the ability to obtain an electron. The HOMO and LUMO energy calculated by B3LYP/6-31G (d,p) method as shown in figure 4.

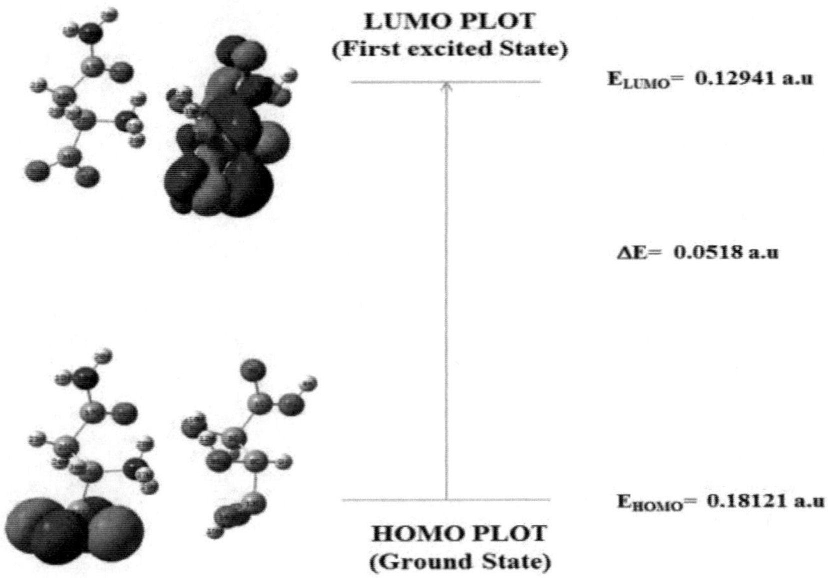

Figure 3. HOMO – LUMO plot of LAsT at B3LYP/6-31 G (d, p)

3.6 SHG efficiency studies

The second harmonic generation efficiency of the powdered material is measured using Kurtz and Perry method. A second harmonic signal of 161 mV is obtained for LAsT with reference to KDP (53 mV). Thus the SHG efficiency of LAsT is 3 times that of KDP.

Acknowledgment

The authors acknowledge University Grants Commission (UGC), India, for funding this research project-F.No.4-4/2015-16(MRP/UGC-SERO) - 2914.

4. CONCLUSION

Single Crystals of L-Asparagine–L-tartaric acid (LAsT) of dimension 5 x 3 x 8 mm^3 are conveniently grown by slow evaporation technique at room temperature. The XRD analysis proves that LAsT crystal belongs to monoclinic in structure with a non centrosymmetric space group $P2_1$. Fourier transform infrared (FT-IR) spectroscopic studies were performed for identifying different functional groups present in the compound. Non-linear optical (NLO) behavior of the LAsT was investigated by the determination of the dipole moment μ and the hyperpolarizability β obtained by B3LYP/6-31G (d, p) method. The calculated highest occupied molecular orbital (HOMO) and lowest unoccupied molecular orbital (LUMO) with frontier orbital gap were presented.

IOP Conf. Series: Materials Science and Engineering **360** (2018) 012047 doi:10.1088/1757-899X/360/1/012047

REFERENCES

[1] Xu D., Jiang M. and Tan Z. (1983), 'A new phase matchable nonlinear optical crystal L-arginine phosphate monohydrate', Acta Chem. Sinica, Vol. 41, pp. 570-573.

[2] Tapati Mallik and Tanusree Kar, (2005), 'Optical, thermal and structural characterization of an NLO crystal: L-arginine perchlorate', *Journal of Crystal Growth*, Vol. 274, pp.251-

[3] Tapati Mallik and Tanusree Kar, (2005) 'Growth and characterization of nonlinear optical L-arginine dihydrate single crystals', *J. Crystal growth*, Vol. 285, pp. 178-182.

[4] Kechen, Wu, Jaap, G, Snijders, & Chensheng, L 2002 'Theoretical studies of Nonlinear optical crystals in metal cluster compounds', *J. Phys. Chem B*, vol.106, pp.8954-8958.

[5] Becke, AD 1993 'Density functional thermochemistry. III. The role of exact exchange', *J. Chem. Phys*, vol. 98, pp. 5648–5652.

[6] Shanmugam, Sathyanarayana,1984 'Raman & polarized infrared spectra of pyridine-2- thione', *Spectrochimica Acta*, vol. 40A, no.8, pp. 757-761

[7] Roeges, NGP, 1994 '*A Guide to the Complete Interpretation of the Infrared Spectra of Organic Structures*', Wiley, NewYork

[8] Sajan, D, Hubert Joe,I, Jayakumar, VS, Zaleski, J 2006, 'Structural & electronic contributions to hyperpolarizability in methyl p-hydroxy benzoate' *J. Mol. Struct*.,vol. 785, pp. 43-53

[9] Thanthiriwatte K.S., Nalin de Silva K.M., Non-linear optical properties of novel fluorenyl derivatives—ab initio quantum chemical calculations *J. Mol. Struct*. (Theochem) 617 (2002)169–175.

[10]Gunasekaran, S, Balaji, RA, Kumeresan, S, Anand ,G, Srinivasan, S 2008 'Experimental & theoretical investigations of spectroscopic properties of N- acetyl-5- methoxytryptamine', Can. *J. Anal. Sci. Spectrosc* ,vol.53, pp.149-160

XRD, FE-SEM, FT-IR and ESR studies of cation deficient potassium tris(oxalato)ferrate(III)trihydrate

M. Narsimhulu and K.A. Hussain

Department of Physics, Kakatiya University, Warangal-506009, India

Corresponding author Email: narsimhulu257@gmail.com

Abstract. Cation deficient potassium tris(oxalato)ferrate(III)trihydrate $\{K_{2.72}[Fe(C_2O_4)_3]\cdot 3.17H_2O\}$ (CDPIOX) crystals were characterized by X-ray powder diffractogram, Fourier Transform Infrared Spectroscopy (FT-IR), quantitative elemental analysis of EDAX, Field Emission Scanning Electron Microscope (FE-SEM), Electron Spin Resonance Spectroscopy (ESR) and Thermogravimetry and Diffrential thermal analysis (TG-DTA). This compound crystallizes in the monoclinic system with $P2_1/c$ space group and shows three-dimensional (3D) chain structure. The elements present in the grown crystal were confirmed by the energy dispersive X-ray analysis (EDAX). The microstructural features on the surface of the single crystals were determined using FE-SEM. Powder X-ray diffraction results indicate the polycrystalline nature of the material. Infrared spectroscopy was used to study the vibrational modes of the compound. The room temperature crystalline ESR spectra indicate the high-spin ($S = 5/2$) of Fe(III) ion. Thermal behaviour of the compound has been studied using thermogravimetry and diffrential thermal analysis.

Keywords: Inorganic material, morphology, elemental analysis, FT-IR spectroscopy, ESR.

1. Introduction

The salts of metal oxalate complexes are interesting materials, which have technological applications as precursors to nanocrystalline metallic oxides [1, 2] and to molecular-based magnetic materials [3–6]. The interest in using metals with oxalate ions is due to the flexible bonding mode of oxalate ions with metal ions. The negative charge, planar shape and good donor capability due to the existence of four oxygen donors, make oxalate ligand very suitable to build coordination polymers in its interaction with metal ions [7]. The tris(oxalato)ferrate anion is an interesting "building block" due to its magnetic properties and possibility of different dimensional network formations [8].

IOP Conf. Series: Materials Science and Engineering **360** (2018) 012048 doi:10.1088/1757-899X/360/1/012048

The structural versatility of metal oxalate complexes is potential to engineering of multifunctional materials mainly in the field of molecular based magnetism and photophysics [9]. In this regard here, we report the results on XRD, FE-SEM, FT-IR, ESR and TG-DTA studies of CDPIOX.

2. Experimental Details

Energy dispersive analysis of X-rays was obtained from X-ray energy analyzer OXFORD INSTRUMENTES (MODEL: Inca Penta FET x3) attached to Carl Zeiss SEM (MODEL: EVO MA 15). The FE-SEM images of the compound were recorded using Carl Zeiss Ultra 55 model. The powder X-ray diffraction patterns were recorded using JEOL-JDX-8P X-ray diffractometer fitted with a scintillation counter and nickel filtered CuK_α radiation (λ=1.5406 Å). The infrared spectrum was recorded at room temperature using Nicolet NEXUS Spectrometer in the range of 400–4000 cm^{-1} using KBr matrix. ESR measurements were performed with a JEOL JES-FA200 spectrometer on a single crystal at room temperature. The TG-DTA measurements were performed using a Mettler Toledo instrument in the temperature range of room temperature to 1000 ^{o}C at a heating rate of 10 ^{o}C/min.

3. Results and discussion

All the chemicals were of reagent grade and used without further purification. CDPIOX single crystals were grown by slow evaporation of mixure of solutions obtained from adding potassium oxalate monohydrate ($K_2C_2O_4.H_2O$), ammonium oxalate monohydrate [$(NH_4)_2C_2O_4.H_2O$] and ferrous sulphate heptahydrate ($FeSO_4.7H_2O$) in 100 ml distilled water in equimolar ratio with constant stirring. This aqueous solution was bioled about 30 minutes and filtered at cold condition to remove any unreacted residue. This green colour solution was kept for crystallization. The as grown good quality single crystals are shown in Figure 1. Single crystal X-ray diffraction data reveals that the crystals are $K_{2.72}[Fe(C_2O_4)_3]\cdot 3.17H_2O$, crystallizes in the monoclinic system with $P2_1/c$ space group. The unit cell dimensions are a = 7.7573(4) Å, b = 19.8655(3) Å, c = 10.3489(6) Å, $\alpha = \gamma = 90^o$, β = 107.946 (6)^o and Z = 4. This compound is isostructutral with its parent potassium tris(oxalato)ferrate(III)trihydrate {$K_3 [Fe(C_2O_4)_3].3H_2O$} [10]

IOP Conf. Series: Materials Science and Engineering **360** (2018) 012048 doi:10.1088/1757-899X/360/1/012048

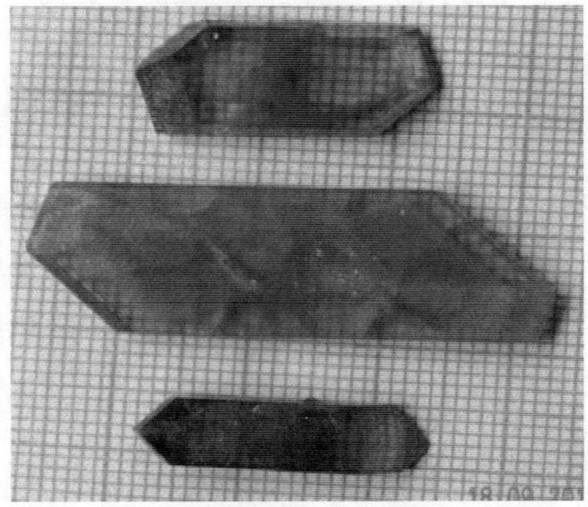

Figure 1. Photograph of CDPIOX crystals.

The elemetal analysis of CDPIOX shows the compound contains 14.95% (calc.14.90%) of carbon and 1.283% (calc.1.284%) of oxygen and it confirmed the presence of hydrogen in the crystal lattice. In order to confirm the presence of iron, potassium and oxygen, quantitative elemental analysis were performed on the application of EDAX. The EDAX spectrum (Figure 2) reveals the presence of chemical elements Fe, K and O in the crystal.

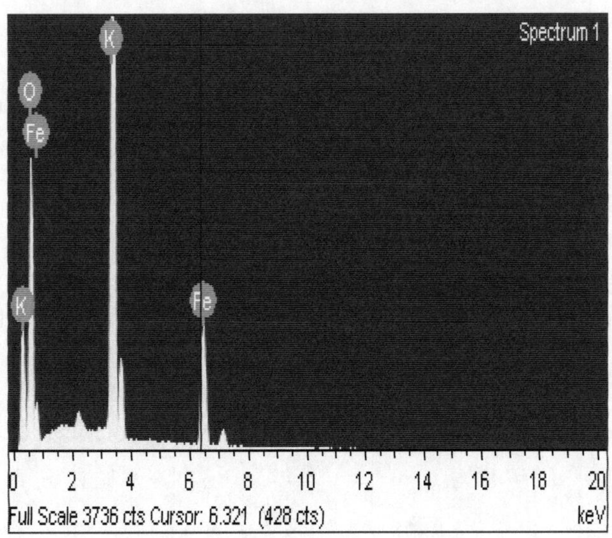

Figure 2. EDAX analysis for CDPIOX.

Field emission scanning electron micrographs of CDPIOX are presented in Figure 3. The FE-SEM images indicate that the compound has layered like and rod like morphology with irregular shapes.

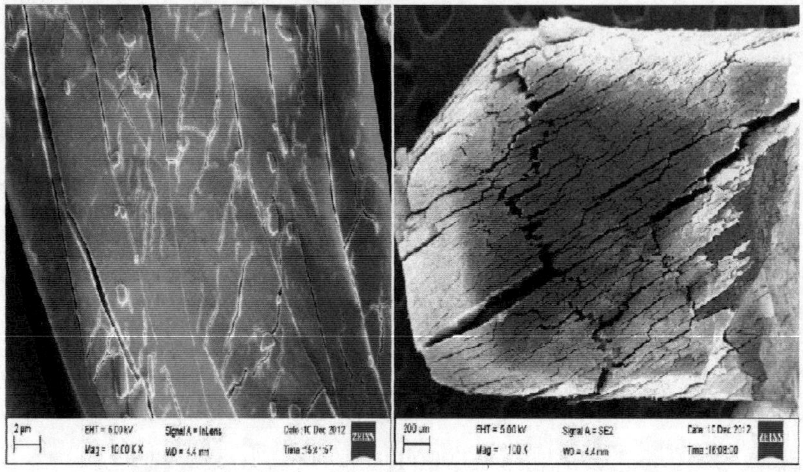

Figure 3. FE-SEM images of CDPIOX.

Figure 4 shows the powder X-ray diffraction pattern of CDPIOX. The diffraction peaks were indexed using powderX program [11]. The experimental and calculated interplanar spacings (d) and corresponding miller indices (h k l) are shown in Table 1. The average particle size of powder crystallites (t) of the compound has been calculated by Debye-Scherrer equation:

$$t = \frac{K\lambda}{\beta \cos\theta}$$

Where β is the integrated breadth of reflections located at 2θ (in radians), θ is the diffraction angle in radians and λ is the wave length of X-rays (1.5406Å for Cu Kα). Crystallite size of powder for the compound has been estimated as 37 nm.

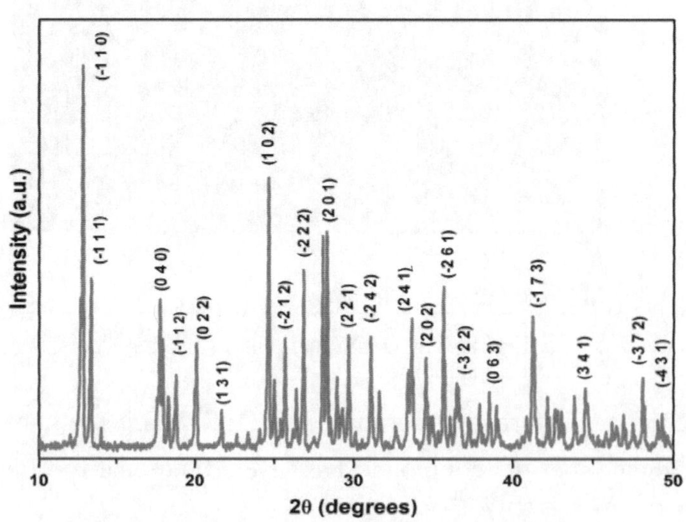

Figure 4. Powder XRD pattern for CDPIOX.

IOP Conf. Series: Materials Science and Engineering **360** (2018) 012048 doi:10.1088/1757-899X/360/1/012048

Table 1: d-spacing's and 2θ values for CDPIOX.

h k l	$2\theta_{exp}$	$2\theta_{cal}$	$2\theta_{diff}$	d_{exp}	d_{cal}
-1 1 0	12.784	12.786	0.002	6.918	6.917
-1 1 1	13.311	13.338	0.027	6.646	6.632
0 4 0	17.823	17.846	0.023	4.972	4.966
-1 1 2	18.814	18.851	0.037	4.712	4.703
0 2 2	20.088	20.116	0.028	4.416	4.410
1 3 1	21.721	21.746	0.024	4.088	4.083
1 0 2	24.635	24.618	0.018	3.610	3.613
-2 1 2	25.688	25.694	0.006	3.465	3.464
-2 2 2	26.853	26.861	0.009	3.317	3.316
2 0 1	28.347	28.301	0.046	3.145	3.150
2 2 1	29.699	29.722	0.023	3.005	3.003
-2 4 2	31.062	31.128	0.066	2.876	2.870
2 4 1	33.621	33.659	0.037	2.677	2.680
2 0 2	34.536	34.552	0.016	2.594	2.593
-2 6 1	35.622	35.662	0.040	2.518	2.515
-3 2 2	36.577	36.555	0.022	2.454	2.456
0 6 3	38.532	38.597	0.064	2.334	2.330
-1 7 3	41.241	41.334	0.093	2.187	2.182
3 4 1	44.534	44.493	0.041	2.032	2.034
-4 3 1	49.303	49.283	0.020	1.846	1.847

The FT-IR spectrum of CDPIOX is shown in Figure 5. The band assignments are made based on earlier IR studies of similar metal oxalate complexes [12–14]. The bands observed at 498 and 530 cm^{-1} are assigned to ν(Fe-O). The (C-C) symmetric stretching mode of oxalate ion and (O-C-O) deformational mode are observed at 889 cm^{-1} and 806 cm^{-1} respectively. The stretching oxalate modes were exhibited at 1250, 1270, 1390 and 1682 cm^{-1} respectively. The ν(O-H) mode at 3430 cm^{-1} shows the presence of lattice water molecules.

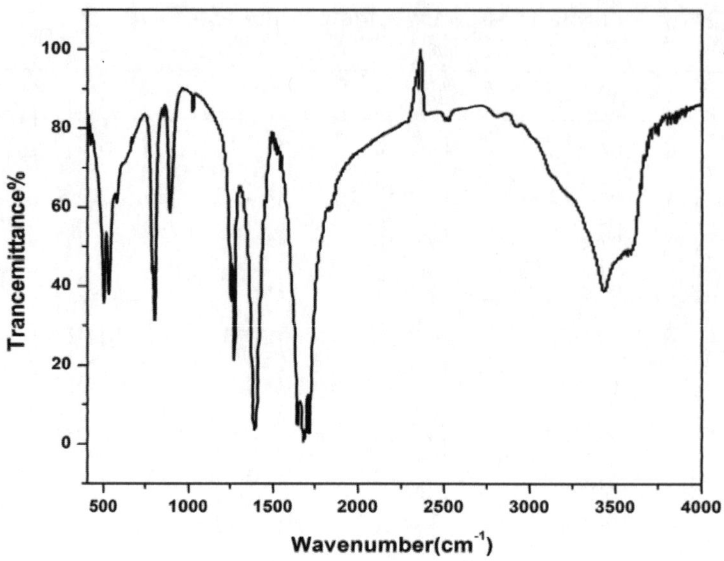

Figure 5. FT-IR spectrum of CDPIOX.

Figure 6 shows the room temperature X-band (9.45 GHz) ESR spectra of CDPIOX. It shows one broad intense resonance signal with g = 6.06 with both peak to peak line width of 51.89 mT, and two weak resonance signals with g = 2.07 and g = 1.13 respectively. The features of the ESR spectrum are very similar to those reported by Collision and Powell [15].

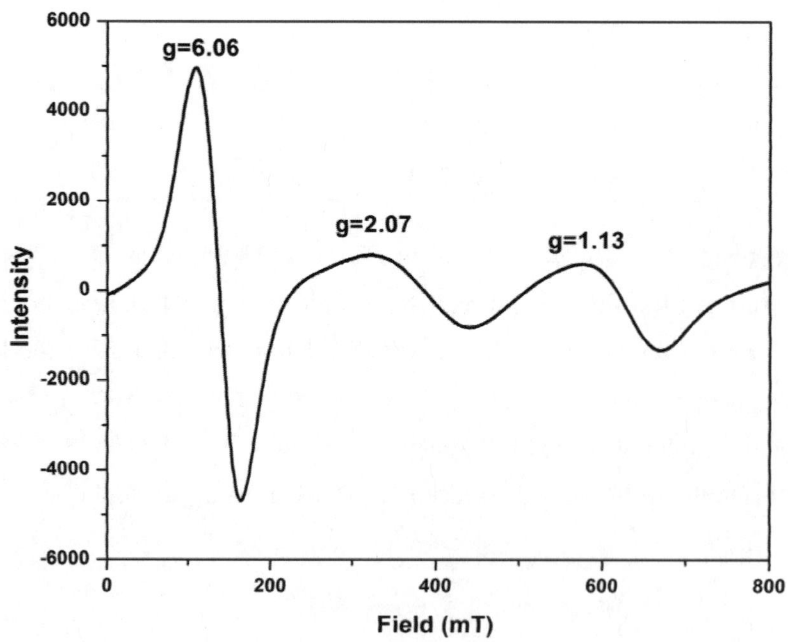

Figure 6. X-band ESR spectrum of CDPIOX.

Figure 7 shows the TG-DTA curves for CDPIOX. TG-DTA curves show five separate decomposition stages. An initial weight loss of 10% (calc. 10.6%) around 104 °C indicates the loss of three water molecules. The resulting anhydrous compound undergoes continuous decompositions as shown by the TG curve from 110 °C to 590 °C. The compound is stable up to 220 °C, above this temperature it starts to decompose in the temperature range 220-250 °C (exp. 4.4%, calc. 4.6%) with the release of half-a-mole of CO_2, resulting formation of mixture of potassium iron oxalate and potassium oxalate. The exothermic peak at 228 °C in the DTA curve shows this process.

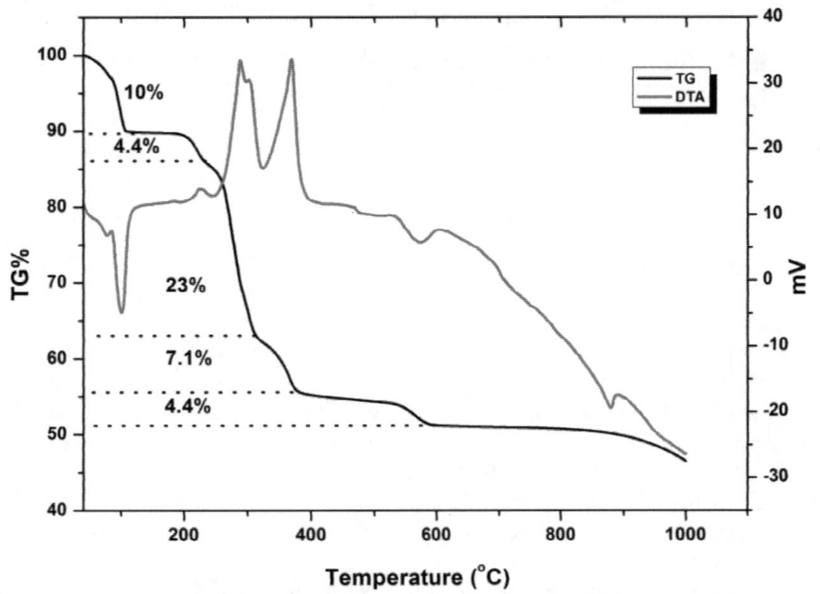

Figure7. TG-DTA curves for CDPIOX.

In the temperature range of 270-295 °C, the loss of one mole of CO_2 and two moles of CO corresponds (exp. 23%, calc. 23.97%) to the decomposition of potassium iron oxalate to potassium oxalate and iron oxide and the corresponding exothermic peak in DTA curve occurred at around 287 °C. The exothermic peak at 370 °C corresponds to formation of mixture of iron oxides with the loss of half-a-mole of CO_2 and half-a-mole of CO. The peak at 570 °C in DTA curve corresponds to decomposition of potassium oxalate to potassium carbonate. The endothermic peak at around 325 °C does not involve any weight loss and corresponds to the characteristic phase transition of $K_2C_2O_4$ [16]. A small endothermic peak at 887 °C corresponds to the melting point of K_2CO_3 formed. These observations indicate that this compound could be a potential for precursor to nano crystalline iron oxide.

4. Conclusion

The synthesized single crystals of CDPIOX were grown from solutions on slow evaporation. The crystallographic data indicates the compound crystallizes in the monoclinic system with space group $P2_1/c$. FE-SEM studies reveal that the crystals have layered and rod like morphology. FT-IR spectrum confirms the functional groups associated with the oxalate ions. ESR spectra confirm the state of Fe(III) ion in the compound is high-spin (S=5/2). The thermal analysis results show that the compound first decomposes to anhydrous potassium iron oxalate and then to the final mixture of oxides through formation of potassium carbonate and iron oxide as intermediates. Thermal decomposition of this material suggests that it may have potential application as a precursor for synthesis of iron oxide nano particles.

Acknowledgments

The authors are thankful to University Grants Commission, New Delhi, India for providing financial support under DRS-SAP programme to carry out this work.

References

[1] Louer D, Boultif A, Gotor F J and Criado J M 2000 *J. Solid. State Chem.* **150** 81
[2] Saritha A, Raju B, Narayana Rao D, Roy Chowdhury A, Das D and Hussain K A 2015 *Advanced Powder Technology* **26** 349
[3] Mathoniere C, Nutall C J, Carling S G and Day P 1996 *Inorg. Chem.* **35** 1201
[4] Coronado E, Galan-Mascaros J R and Gomez-Garia C J 2000 *J. Chem. Soc., Dalton Trans.* **2** 205
[5] Tamaki H, Zhong Z J, Matsumoto N, Kida S, Koikawa M, Achiwa N, Hashimoto Y and Okawa H 1992 *J. Am. Chem. Soc.* **114** 6974
[6] Larionova J, Mombelli B, Sanchez J and Khan O 1998 *Inorg. Chem.* **37** 679
[7] Carranza J, Sletten J, Lloret F and Julve M 2011 *Inorg. Chim. Acta.* **371** 13
[8] Muziol T M, Wrzeszcz G and Chrzaszcz L 2011 *Polyhedron*, **30** 169
[9] Decurtins S and Pellaux R 1998 *Comments on Inorganic Chemistry: A Journal of Critical Discussion of the Current Literature*. **20** 143
[10] Junk P C 2005 *J. Coord. Chem.* **58** 355
[11] Dong C 1998 PowderX: windows-95-based program for powder X-ray data processing, *J. Appl. Cryst.* **32** 838
[12] Fujiota J, Martell A E and Nakamoto K 1962 *J. Chem. Phys.* **36** 324
[13] Nakamoto K 1963 Infrared spectra of inorganic and coordination compounds, *John Wiley and Sons, Inc., New York*

[14] Bhattacharjee M N, Chaudhuri M K and Purkayastha R N D 1985 *Inorg. Chem.* **24** 447

[15] Collision D and Powell A K 1990 *Inorg. Chem.* **29** 4735

[16] Carlino S and Hudson M 1998 *Solid State Ionics* **110** 153

IOP Conf. Series: Materials Science and Engineering **360** (2018) 012049 doi:10.1088/1757-899X/360/1/012049

Origin of the high dielectric constant in $Sm_{2/3}Cu_3Ti_4O_{12}$ ceramics

A.K. Thomas[1], Kevin Abraham[1], Jini Thomas[2], K.V. Saban[3*]

[1] SMART LAB, Department of Physics, St. Berchmans College, Changanacherry, 686101, Kerala, India
[2] Department of Physics, Kuriakose Elias College, Mannanam, Kottayam, Kerala, India
[3] Principal, St.Aloysius College, Edathua, Alappuzha, Kerala,India

Email: kvsaban@gmail.com.

Abstract. The microstructure, crystalline structure, dielectric and electrical properties of polycrystalline $Sm_{2/3}Cu_3Ti_4O_{12}$ ceramics, prepared via conventional solid-state reaction were investigated. The structural evolution of these powders was analyzed by X-ray diffraction (XRD). Crystal structure investigated by Rietveld refinement was found to be cubic with space group *Im3*. It has been found that $Sm_{2/3}Cu_3Ti_4O_{12}$ ceramic shows giant dielectric permittivity values with low frequency (1 kHz) larger than 30,000 at room temperature. In the temperature domain, a new dielectric relaxation was clearly observed beyond 200 K, in addition to the well-investigated dielectric relaxation close to 100 K. This Maxwell–Wagner type of relaxation was found to be originating from the formation of external depletion layers at the electrode-sample interface. The dielectric relaxation at high frequencies in the dielectric dispersion spectra of $Sm_{2/3}Cu_3Ti_4O_{12}$ ceramics is caused by an IBLC effect associated with the insulating grain boundaries and the other one at low frequencies originates from an electrode polarisation effect. The activation energy of semi conducting grain is found to be similar to that of CCTO (~0.06 eV). The observed giant value of the dielectric constant in the $Sm_{2/3}Cu_3Ti_4O_{12}$ ceramics originates due to polarisation at the electrode- sample interface and at the insulating grain boundary interface. The results suggest that the IBLC effect mechanism that was formerly proposed for CCTO ceramics is also valid in explaining the high dielectric constant in the compositionally and structurally CCTO-like, $Sm_{2/3}Cu_3Ti_4O_{12}$ ceramics.

1. Introduction

Materials that exhibit colossal dielectric constant (CDC) have attracted significant attention because of their potential applications in electronic devices, such as high dielectric capacitors, capacitor sensors, and random access memories. Miniaturization as well as the production of low cost, highly efficient electronic components can be achieved using these materials. The recent discovery [1, 2] of "colossal" values of the dielectric constant, ε′, up to about 10^5 in $CaCu_3Ti_4O_{12}$ (CCTO) has aroused tremendous interest. Recent research revealed that many compounds $ACu_3Ti_4O_{12}$ (A = $La_{2/3}$, $Y_{2/3}$,

$Na_{1/2}Bi_{1/2}$, $Na_{1/2}La_{1/2}$) is isostructural to CCTO also shows giant dielectric constant accompanied by low dielectric loss at room temperature, and good temperature stability[3-6] . But so far, there are only few literatures reporting the dielectric properties of the compound $Sm_{2/3}Cu_3Ti_4O_{12}$ (SCTO) which belongs to the $ACu_3Ti_4O_{12}$ family. Impedance spectroscopic studies demonstrated CCTO ceramics are electrically heterogeneous and large dielectric constant was attributed to grain boundary barrier layer capacitance effects [7, 8]. Polarization effects at insulating grain boundaries accompanied by a strong Maxwell-Wagner relaxation mode are the primary cause for non intrinsic colossal values of CCTO. There are some reports on the observation of two extrinsic relaxations in the dielectric spectrum of both single crystals and ceramics [9-12]. These extrinsic relaxations were explained on the basis of a surface barrier layer capacitance (SBLC) effect due to the formation of Schottky barriers at the sample - electrode interfaces, and the giant value of permittivity was accounted for [9, 11]. Schottky barriers are interpreted to be formed due to the difference in the Fermi levels between the semi conducting ceramics and the metal electrode. The room temperature dielectric constant of SCTO ceramics reported by Subramanian *et al.* was only 1665 (measured at 100 kHz) [1].It is reported that processing conditions such as calcinations temperatures [13], sintering times [14] and sintering temperatures [15] have great influence on the dielectric properties of CCTO ceramics. For higher sintering temperatures and/ or longer sintering periods, the resistance of the grain boundary is found to decreases significantly to a level comparable to or smaller than the electrode. In this case electrode effects start to dominate and this results in two extrinsic responses and higher permittivity [16].

The present work is an attempt to enhance the dielectric properties of SCTO ceramics by varying the processing condition. SCTO ceramics were prepared by the solid-state reaction method. The phase structure, microstructure, and the dielectric properties of the prepared ceramics were analyzed. The complex impedance spectroscopy method is used to find the origin of giant dielectric constants of SCTO. The values of activation energy were obtained to reveal the differences of electric behaviors in SCTO ceramics.

2. Experimental section

Polycrystalline samples of $Sm_{2/3}Cu_3Ti_4O_{12}$ were prepared by solid- state reaction process using stoichiometric quantities of high purity Sm_2O_3, CuO, and TiO_2 (>99.9% purity). These reagents were mixed thoroughly using an agate mortar and pestle in an acetone medium. The calcinations of the mixed powder were done at 950 °C in a muffle furnace on Pt foil in successive cycles of 10 h. Cycles of regrinding and heat treatment were repeated until the phase compositions observed by XRD get stabilised. The calcined powder was reground and mixed uniformly with a few drops of 5 wt% polyvinyl alcohol as binder. The mixture was then uniaxially pressed under 5 tonnes and pellets of 13 mm diameter were yielded. These pellets were sintered for 24 hours in air at 1100 °C using heating and cooling rates 4 °C/min. The powder XRD data corresponding to 2θ values ranging from 10 to 120 ° at equal steps of 0.02 ° was collected using Cu-Kα radiation (λ=1.5418 Å) on a Bruker D8 Advance X-ray diffractometer. The Rietveld refinement of the data was done using FULLPROF program, assuming pseudo-Voigt profile for the peaks. The microstructure of the polished and thermally etched pellets was studied using scanning electron microscopy. The densities of the sintered pellets were determined by mass and volume measurements.

The pellet surfaces were fine polished with emery paper to make their faces flat and parallel. The low temperature dielectric measurements of the silver coated pellets were done using a NOVOCONTROL Alpha high-resolution dielectric analyzer (Alpha-S) in the temperature range 80 -300 K and the frequency

IOP Conf. Series: Materials Science and Engineering **360** (2018) 012049 doi:10.1088/1757-899X/360/1/012049

range 1–10^7 Hz. The complex impedance Z* of the sample was then converted in to complex permittivity using the equation,

$$\varepsilon^* = \varepsilon' - \varepsilon'' = \frac{1}{i\omega C_0 z^*} \quad (1)$$

In this equation ε',ε'' and ω have their usual meaning and $C_0 = \frac{\epsilon_0 A}{d}$ is the empty cell capacitance, where *A* is the sample area and *d* is the sample thickness.

3. Results and discussion

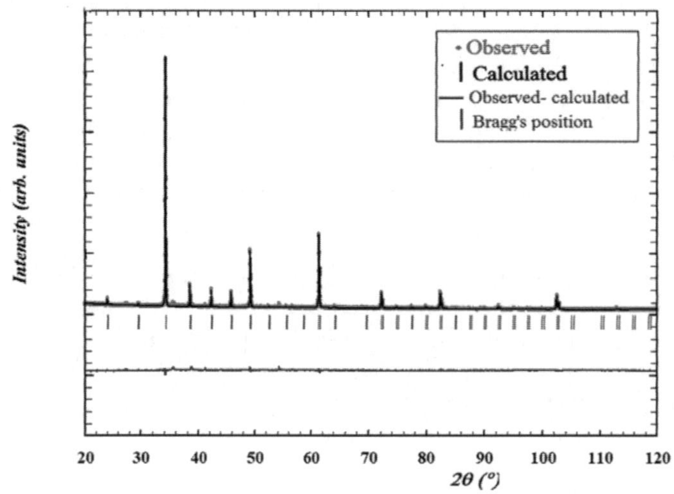

Figure 1. X- Rietveld refined XRD profiles of $Sm_{2/3}Cu_3Ti_4O_{12}$

Structural refinement of the XRD data was carried out using the Rietveld's refinement program ''Full Prof'' and the refined profile is shown in Figure 1. Refined profile shows that the sample is mono-phased and the difference between the observed and calculated intensities is negligible. From structural refinement it is confirmed that SCTO ceramics belongs to cubic structure with space group *Im3* at room temperature. The refinement resulted satisfactory agreement factors and lattice parameters which are listed in Table 1. Refined lattice parameter, 7.3923 Å, is in conformity with the value, 7.394 Å, reported earlier [17].

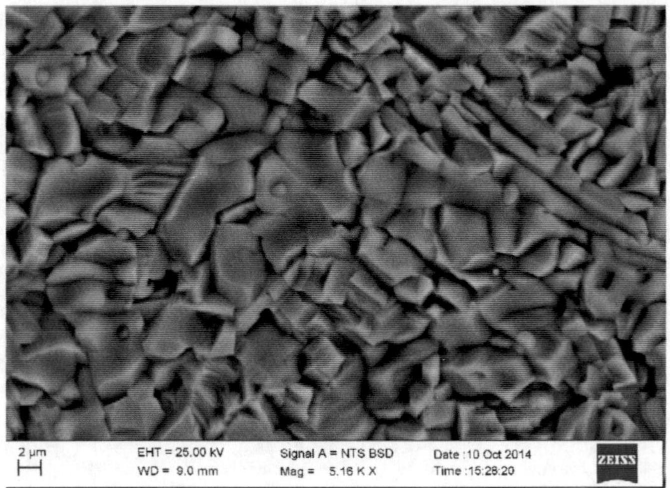

Figure 2.SEM micrograph of the polished surface of $Sm_{2/3}Cu_3Ti_4O_{12}$ ceramic.

Figure 2 shows the surface morphology of the polished and thermally etched SCTO sample. The sample is characterized by a dense microstructure of small and uniformly distributed grains with average grain size of 2-4 µm. A significant decrease in the grain size is observed on substituting A- site by Sm than that of pure CCTO sintered at 1100 °C (50-60 µm) [3] . Lower porosity and more uniform grain resulted in the formation of a dense ceramics with relative density value calculated to be within 94.9% of the theoretical value.

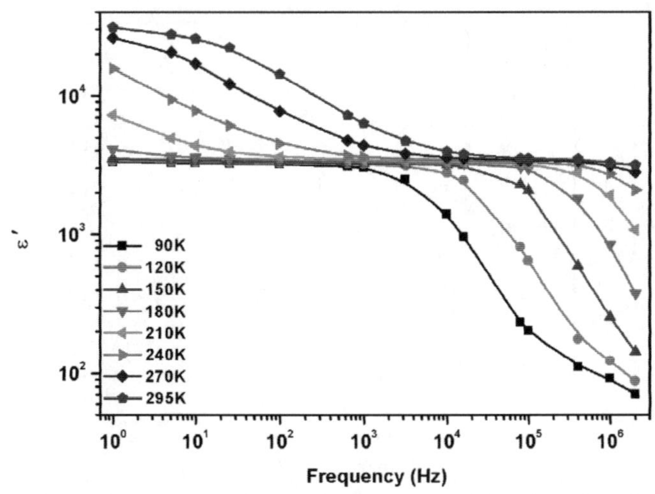

Figure 3. Frequency dependence of (a) dielectric constants ε′ for $Sm_{2/3}Cu_3Ti_4O_{12}$ ceramic in the temperature range (90-295 K)

Variation of real part of dielectric constant ε' as a function of frequency for SCTO samples at different temperature (90-290 K) is shown in Figure 3. It is clear from Figure. 3 that below 200 K the dielectric constant is considerably high and its variation is strictly in accordance with the typical Maxwell-Wagner relaxation process reported in $ACu_3Ti_4O_{12}$ ceramics [2]. It has been found that $Sm_{2/3}Cu_3Ti_4O_{12}$ ceramic shows giant dielectric permittivity values with low frequency larger than 30,000 at room temperature, which is much higher than that reported by Subramanian *et al* [1]. It is clear from Figure 3, that an additional relaxation is observed in the low frequency region (below 100 kHz) when the temperature is increased beyond 200 K. The magnitude of the low frequency plateau of ($\sim 10^5$) is approximately one order greater than that of the intermediate frequency plateau ($\sim 10^4$). This type of an additional low frequency plateau, reported in CCTO ceramics with relatively low (grain boundary resistivity) values had been attributed to electrode polarisation [18].

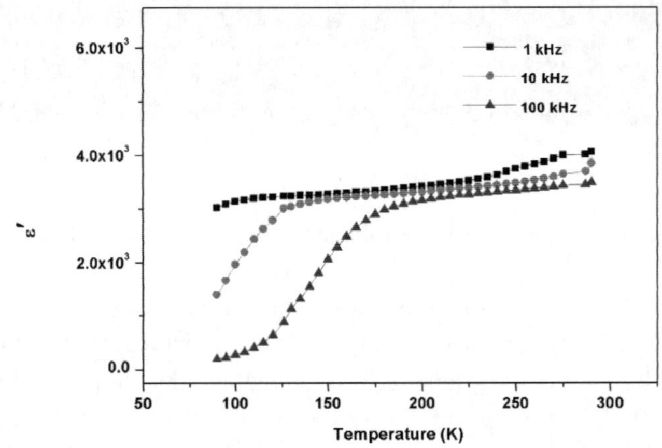

Figure 4.Temperature dependence of ε' at different selected frequencies.

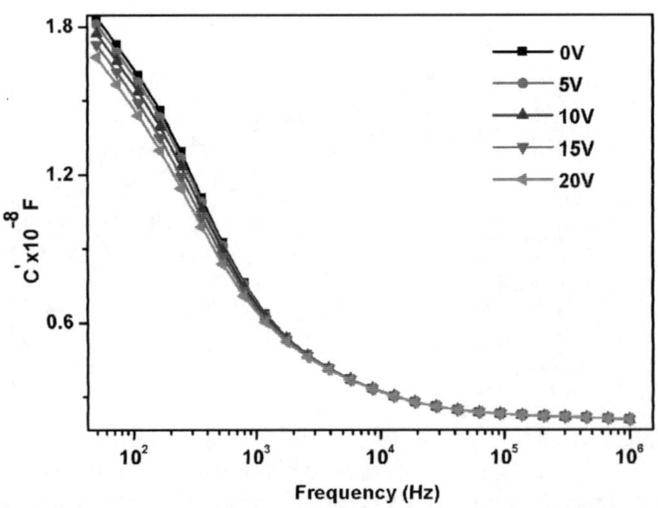

Figure 5. Effect of dc bias on the C'spectroscopic plot of $Sm_{2/3}Cu_3Ti_4O_{12}$ ceramic at 300 K

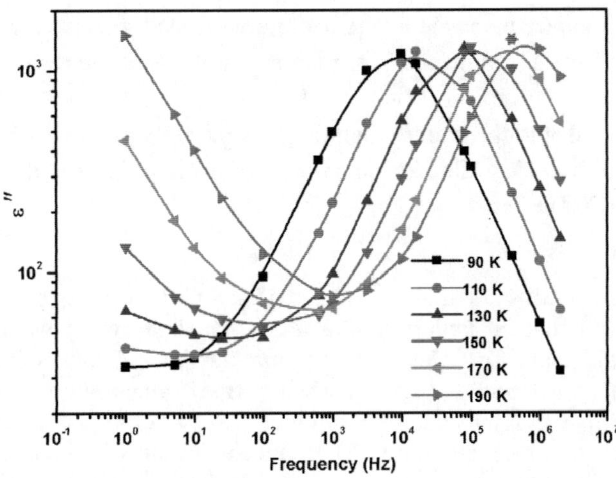

Figure 6. Variation of ε′ with frequency in the temperature range (90-295 K)

The temperature dependence of dielectric constant at selected frequency values is shown in Figure 4, which shows the bulk contribution to dielectric constant at low temperature and the grain boundary contribution at higher temperature, which is in accordance with the IBLC model. To check the polarisation at the sample –electrode interface, a dc voltage is superimposed with the applied ac signal. The effect of dc bias on the dielectric properties of the sample at room temperature is shown graphically in Figure 5. With increasing dc bias the capacitance value steadily decreases in the low frequency region and remains almost unaltered in the intermediate and high frequency regions. SinceR_e >> R_{gb}, dc bias drops almost totally across the electrode/sample interface resulting in the decrease of R_e (resistivity of the electrode sample interface) with increasing voltage. This result confirms the existence of a Schottky barrier at the sample electrode interfaces and an electrode effect in the low frequency response.

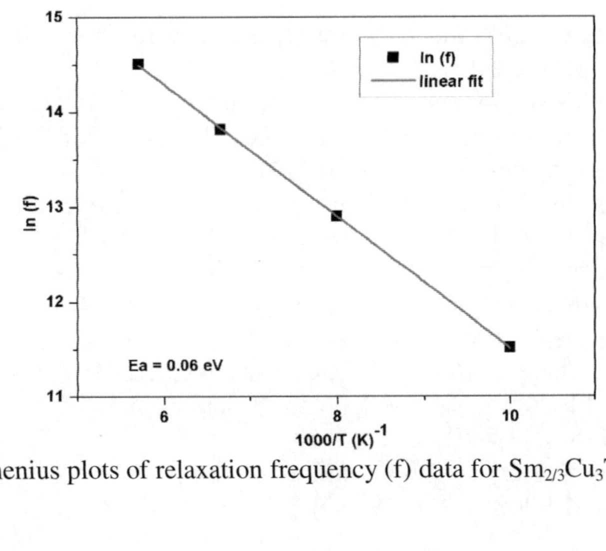

Figure 7. Arrhenius plots of relaxation frequency (f) data for $Sm_{2/3}Cu_3Ti_4O_{12}$ ceramics

The values of f ,the relaxation frequency at any temperature obeys Arrhenius law $f = f_0 \exp(-E_a / k_B T)$, where τ_0 is the relaxation frequency at infinitely high temperature, k_B is the Boltzmann constant, T is the absolute temperature, and E_a is the activation energy. The values of f corresponding to the relaxation peak frequencies in Figure 6 due to the grains when fitted using Arrhenius law (Figure. 7) resulted in an activation energy, Ea = 0.06 eV. This value is comparable to reported value of the activation energy for relaxation due to grains in CCTO.

4. Conclusion

A compositionally and structurally $CaCu_3Ti_4O_{12}$-like oxide, $Sm_{2/3}Cu_3Ti_4O_{12}$ ceramics were prepared and investigated to obtain a systematical understanding about the dielectric behaviours and the underlying related mechanism. A single phase $Sm_{2/3}Cu_3Ti_4O_{12}$ ceramics with a cubic crystal structure in space group *Im3* is obtained by solid-state reaction. A high density and fine grained microstructure is obtained for the $Sm_{2/3}Cu_3Ti_4O_{12}$ ceramic. $Sm_{2/3}Cu_3Ti_4O_{12}$ ceramics sintered at 1100°C for 24 h exhibited a giant dielectric constant (30,000) around room temperature. The colossal dielectric constant $Sm_{2/3}Cu_3Ti_4O_{12}$ was explained based on Maxwell-Wagner type contribution of depletion layers between the sample –electrode interface and at the grain boundaries.

5. Acknowledgments

Financial assistance from the Kerala State Council for Science Technology and Environment (KSCSTE) under its SRS scheme (File no.001/SRSPS/2010/CSTE) and a teacher fellowship under the Faculty Improvement Program of University Grants Commission (UGC – Govt of India) (FIP/XII Plan/KLKE-053/ TF-13) are gratefully acknowledged.

REFERENCE

[1] Subramanian M A, Li D, Duan N, Reisner B A and Sleight A W 2000 *J. Sol. St. Chem.* 151 323
[2] Homes C C, Vogt T, Shapiro S M, Wakimoto S and Ramirez A P 2001 *Science* 293 673
[3] Shri P B and Varma K B R 2008 *Phys. B* 403 2246
[4] Liang P F, Yang Z P, Chao X L and Liu Z H 2012 *J.Am. Ceram. Soc.* 95 2218
[5] Ren H M, Liang P P and Yang Z P 2010 *Mate. Res.Bull,* 45 1608
[6] Thongbai P, Yamwong T and Maensiri S 2012 *Mater. Res. Bull.* 47 432
[7] Sinclair D C, Adams T B, Morrison F D and West A R 2002 *Appl.Phys. Lett.* 80 2153
[8] Adams T B, Sinclair D C and West A R 2002 *Adv. Mater.* 14 1321
[9] Krohns S, Lunkenheimer P, Ebbinghaus S G and Loidl A 2007 *Appl. Phys. Lett.* 91 022910
[10] Mua C H, Liu P, Ying H, Zhoub J P and Zhang H W 2009 *J. Alloys Compd.* 471 137
[11] Krohns S, Lunkenheimer P, Ebbinghaus S G and Loidl A 2008 *J. Appl. Phys.* 103 084107
[12] Shao S F, Zhang J L, Zheng P, Zhong W L, and Wang C L 2006 *J. Appl. Phys.* 99 084106
[13] Cao L, Liu P, Zhou J P, Wang Y J ,Su L N, Liu C and Zhang H W 2010 *J. Ceram. Process. Res.* 11 453
[14] Patel P K and Yadav K L 2015 *Ceram. Intl.* 41 12386
[15] Bender B A and Pan M J 2005 *Mater. Sci. Eng.* B 117 339

[16] Li M, Shen Z J, Nygren M, Feteira A, Sinclair D C and West A R **2009** *J. Appl. Phys.* 106 104106
[17] Subramanian M A and Sleight A W **2002***Solid State Sci.* 4 347
[18] Li W and Schwartz RW **2007** *Phys. Rev. B* **75** 012104

IOP Conf. Series: Materials Science and Engineering **360** (2018) 012050 doi:10.1088/1757-899X/360/1/012050

Effect of Cadmium Dopant on Structure and Optical Properties of ZnO Thin Films Prepared by Spray Pyrolysis Technique

Sumanth Joishy, H S Sindhu, B V Rajendra
Department of Physics, Manipal Institute of Technology, Manipal University, Manipal- 576104
Corresponding Author: sumanthjoishy915@gmail.com

Abstract: $Cd_xZn_{1-x}O$ (x = 0.00, 0.10 and 0.20) thin films were deposited on glass substrate by the chemical spray pyrolysis technique. The precursor solutions were obtained by varying the concentration of zinc acetate and cadmium acetate in double distilled water. Substrate temperature was fixed at 723K. The structure, surface morphology and optical properties of deposited films were studied using X-ray diffraction, Scanning electron microscope and UV–Visible spectrophotometer. All the films were polycrystalline in nature. The energy band gap of the films was decreased with increase in cadmium content. The influence of cadmium concentration in the films on the structure and optical properties were studied in this study.
Keywords: Spray pyrolysis, Semiconductors, Transmittance, Energy band gap.

1. Introduction

Semiconducting nano films have attracted substantial attention because of their wide potential applications in the field of transparent electrodes, solar cells, gas sensors and other opto-electronic devices [1, 2]. Among these materials, Zinc oxide (ZnO) is an important material because of the wide band gap (3.37 eV), high optical transmittance and large free exciton binding energy (60 meV) at room temperature and in addition the excellent chemical stability, high melting point and small electron affinity [3, 4]. The energy band gap of ZnO films can be tailored by doping suitable elements. Cadmium oxide (CdO) has a cubic structure and a narrow direct band gap of approximately 2.34 eV. Hence the incorporation of Cd in the ZnO helps absorption edge to red-shift into the blue and green spectral range. Moreover, in view of the similarities in their ionic radii and other basic properties, a proper addition of Cd into ZnO is very useful for the fabrication of ZnO/$Cd_xZn_{1-x}O$ hetero-junctions or super-lattices, which are the key elements in ZnO-based emitters and detectors [5].

$Cd_xZn_{1-x}O$ thin films have been prepared by a variety of methods including sputtering [6], chemical bath deposition [7], sol-gel [8], chemical vapour deposition [9] and pulsed laser

IOP Conf. Series: Materials Science and Engineering **360** (2018) 012050 doi:10.1088/1757-899X/360/1/012050

deposition [10]. Out of these methods, spray pyrolysis is an attractive method to obtain thin films of $Cd_xZn_{1-x}O$ which is well known for its simplicity and inexpensive system that can be scaled up comparatively easily.

A few reports are available on synthesis and characterization of Cd doped ZnO films prepared by spray pyrolysis method. The present report provides the influence of doping of cadmium on structure and optical properties of a ZnO films with taking 0.0125M precursor of zinc acetate and cadmium acetate concentration using a spray pyrolysis method.

2. Experimental Details

$Cd_xZn_{1-x}O$ (x=0, 10 and 20 at. %) thin films were prepared by using a homemade spray pyrolysis experimental set up. Glass slides have used as substrates which were chemically cleaned. The spray solution consists of zinc acetate [Zn $(CH_3COO)_2$. H_2O] and cadmium acetate [Cd $(CH_3COO)_2$. H_2O] (supplied by Sigma Aldrich, >99% purity) of 0.0125 M concentration. The solutions were stirred continuously to get a clear and homogeneous solution. The spray rate was maintained at 2 ml/min. During the process, the spray pressure and nozzle to substrate distance were fixed at 5kg (f)/cm^2 and 28cm respectively. The substrate temperature was fixed at 723K and the process was carried out at atmospheric pressure.

Thickness of the films was in the range of 550-600nm calculated by gravimetric method. To examine the crystallinity, crystal phases of the deposits were performed using a Rigaku–Minflex 600, X-Ray Diffractrometer with CuKα radiation (λ = 1.54 Å). The optical transmission spectra of deposited thin films were measured by using Shimadzu -1800 UV-VIS spectrophotometer operated at a resolution of 1 nm from 350nm to 800nm.

3. Results and Discussion

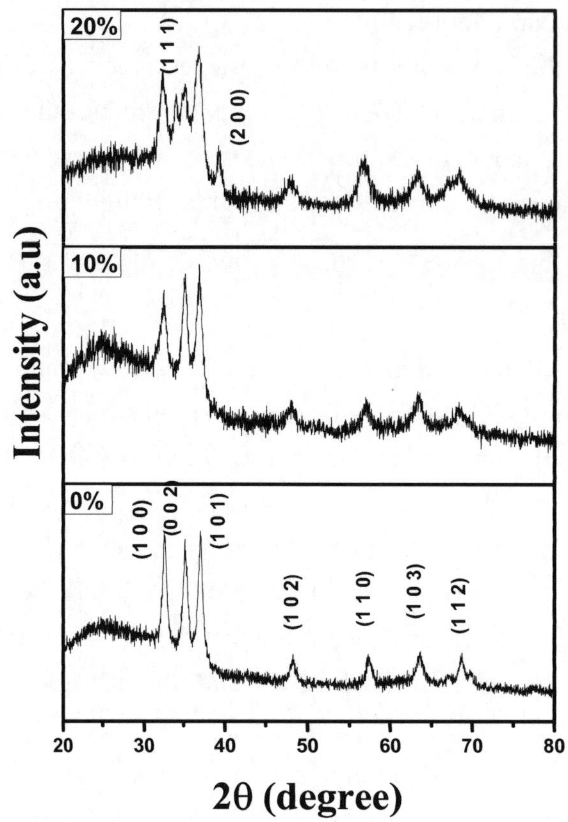

Figure 1. XRD patterns of $Cd_xZn_{1-x}O$ thin films(x=0.00-0.20) deposited at 723K.

Figure 1. Shows the XRD result of $Cd_xZn_{1-x}O$ films. All films were polycrystalline in nature and had the strongest XRD intensity of the (1 0 1) peak. Along with (1 0 1) plane other peaks corresponds to (1 0 0), (0 0 2), (1 0 2), (1 0 3), (1 1 2) and (0 0 4) planes to c-axis orientation with lesser intensity was noticed. The films of pure and 10% cadmium doped ZnO had hexagonal wurtzite-type structure and in these films the peaks corresponding to cadmium oxide were not discernible [11]. Cadmium ions can easily substitute into zinc ions on the corresponding crystallographic site. The replacement of Zn by Cd makes the lattice parameter of the $Cd_xZn_{1-x}O$ alloy thin films larger. Therefore, it can be certain that the Cd^{2+} ions have uniformly substitute into the Zn^{2+} sites in ZnO lattice. Two additional peaks of (1 1 1) and (2 0 0) belonging to cubic phase were found for 20% Cd dopant. With increasing doping concentration the crystalline quality of the films has weakened. This decrease of peak

IOP Conf. Series: Materials Science and Engineering **360** (2018) 012050 doi:10.1088/1757-899X/360/1/012050

intensity with Cd dopant may be affected by the saturation of new nucleating centres [12, 13]. The crystallite size was calculated by the Scherrer's formula [14],

$$D = 0.94\lambda / \beta\cos\theta$$

Where λ is the wavelength of X-ray, β is full width half maxima and θ is the Bragg's angle. Crystallite size decreased from 22-10nm with increasing cadmium concentration.

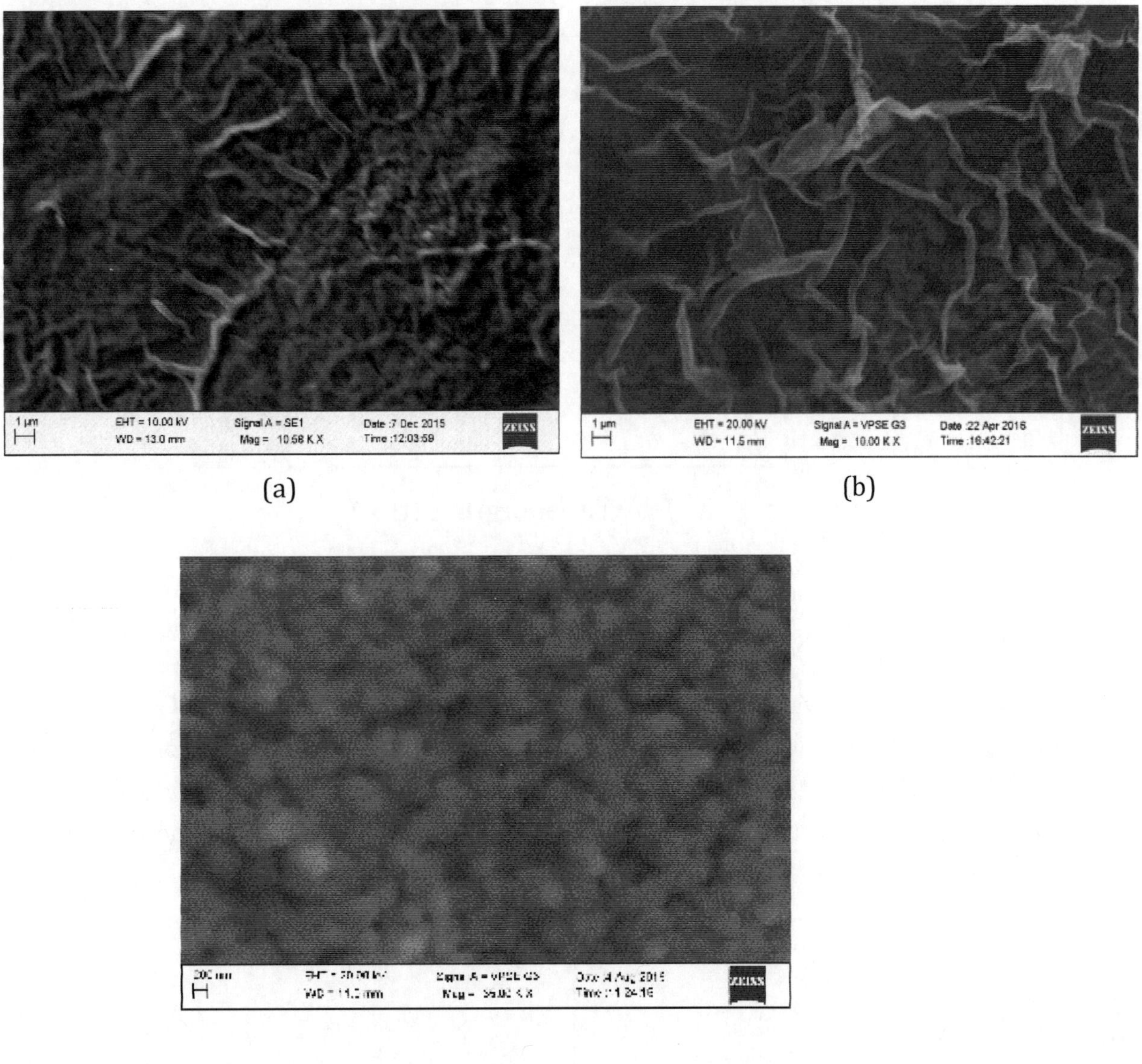

Figure 2. SEM images of $Cd_xZn_{1-x}O$ films for x = (a) 0.00, (b) 0.10, (c) 0.20 deposited at 723K.

The surface morphology studies of the $Cd_xZn_{1-x}O$ films deposited at 723K substrate temperature was studied by using scanning electron microscopy and results are displayed in Figures 2a, 2b and 2c. The surface morphology of the $Cd_xZn_{1-x}O$ films depends on the percentage of compositions of ZnO and CdO. The pure and 10% Cd doped ZnO films showed fibrous structure, while the film of 20% Cd dopant has shown grain structure [15]. The grain size of deposited films was decreased with increase in cadmium content as confirmed by XRD results [16].

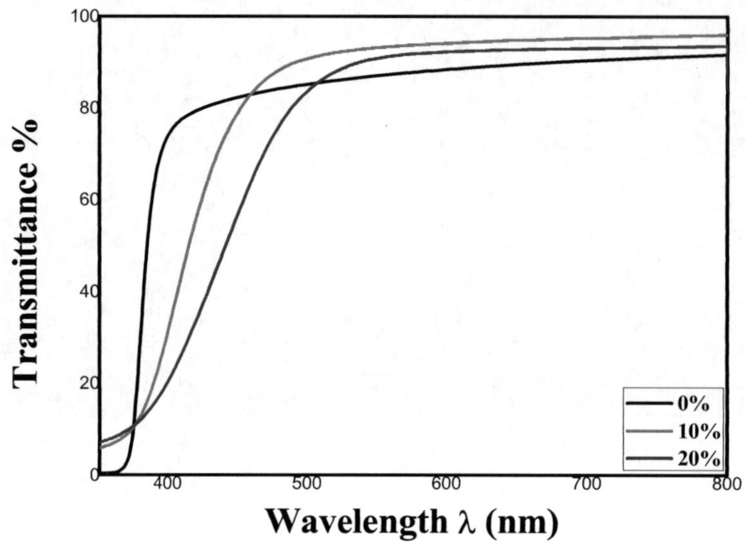

Figure 3. Variation of transmittance versus wavelength of Cd_xZn_{1-x} O thin films(x=0.00-0.20) deposited at 723K.

Figure 3. shows the variation of optical transmittance with wavelength for $Cd_xZn_{1-x}O$ films. All films had above 80% transmittance in the visible region. The transmittance was low in ultraviolet region. For pure ZnO, an absorption edge was rising sharply about 380nm and the absorption edge was red shifted for Cd doped ZnO films and have shown substantial improvement of light absorption [17].

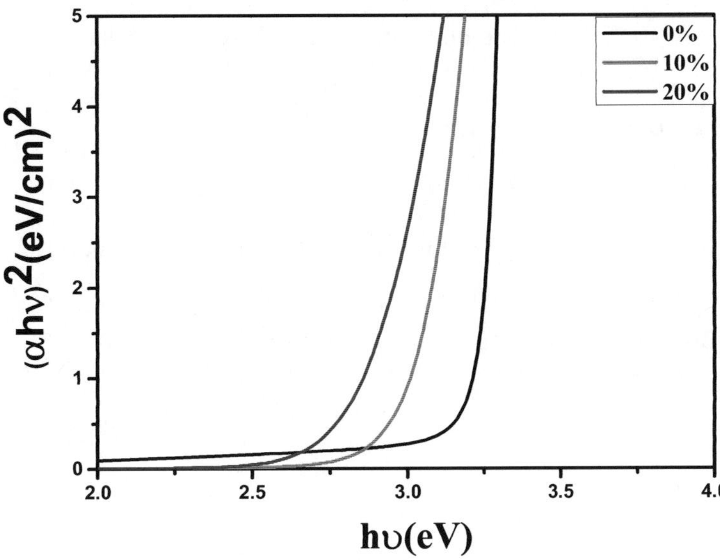

Figure 4. Variation of $(\alpha h\upsilon)^2$ versus photon energy(eV) of Cd_xZn_{1-x} O thin films(x=0.00-0.20) deposited at 723K.

The optical band-gap was determined from the absorption spectra measured in the range 350nm-800nm using the relation:

$$\alpha h\nu = A(h\nu - E_g)^n$$

Where E_g is the energy band gap, n is a constant and hν is the photon energy. Extrapolation of the straight-line portion of the $(\alpha h\nu)^2$ versus hν plot to zero absorption coefficient ($\alpha = 0$) gives the band gap energy (E_g) values. Figure 4. shows plot of $(\alpha h\nu)^2$ versus (hν). Energy band gap was decreased from 3.24 -2.87eV with increase in the Cd concentration [18]. The main reason for decrement in the band gap values is the fact that band gap of CdO (2.23 eV) is less than that of ZnO (3.37 eV). Also the effect of Cd doping concentration leads to red shift of the absorption edge which results in narrowing of the band-gap. This results in increase of carrier concentration with increasing Cd concentration

4. Conclusions

$Cd_xZn_{1-x}O$ thin films were prepared on glass substrate at 723K by spray pyrolysis technique. The effect of Cd-doping concentration on structure, morphological and optical properties was studied. All the films were polycrystalline in nature. Pure and 10% Cd doped ZnO films had hexagonal structure with (1 0 1) preferred orientation. While the film of 20% Cd dopant has shown mixed phase of hexagonal and cubic structure. The pure and 10% Cd doped ZnO films showed fibrous structure, while the film of 20% Cd dopant shown grains structure. Energy band gap was decreased from 3.24 to 2.87eV with the increase in Cd content from x=0.00 to 0.20.

Acknowledgment

Authors are grateful to Manipal Institute of Technology, Manipal for providing instrumentation facility and financial support to carryout research work. Authors thank UGC-DAE Consortium for Scientific Research, Mumbai Center India for sanctioning research projects (UDCSR/MUM/AO/CRS-M-212/2015/511).

References

1. J.W. Mares, M. Falanga, A.V. Thompson, A. Osinsky, J.Q. Xie, B. Hertog, A. Dabiran, P.P. Chow, S. Karpov and W.V. Schoenfeld, Journal of Applied Physics 104,093107-5 (2008).
2. A. Osinsky, J. Dong, B. Hertog, A. Dabiran, P. Chow, W. Schoenfeld, S. Pearton, D.C. Look and A. Cartwright, ECS Transactions 3, 323-331 (2006).
3. K. Matsubara, P. Fons, K. Iwata, A. Yamada, K. Sakurai and H. Tampo, Thin Solid Films 431,369-372 (2003).
4. C.H. Liu, J.A. Zapien, Y. Yao, X.M. Meng, C.S. Lee, S.S. Fan, Y. Lifshitz and S.T. Lee, Advanced Materials 15, 838-841 (2003).
5. T. Gruber, C. Kirchner, R. Kling, F. Reuss, A. Waag, F. Bertram, D. Forster, J. Christen, M. Schreck, Applied Physics Letters 83, 3290-3 (2003).
6. D.C. Look, G.M. Renlund, R.H. Burgener and J.R. Sizelove, Applied Physics Letters 85, 5269-5271 (2004).
7. L.R.L. Gutierrez, J.J.C. Romero, J.M.P. Tapia, E.B. Calva, J.C.M. Flores and M.O. Lopez, Materials Letters 60, 3866-3870 (2006).
8. J.S. Cruz, G.T. Delgado, R.C. Perez, C.I.Z. Romero and O.Z. Angel, Thin Solid Films. 515, 5381-5385 (2007).
9. S.T. Tan, X.W. Sun, Z.G. Yu, P. Wu, G.Q. Lo, and D.L. Kwong, Applied Physics Letters 91, 72101-3 (2007).
10. R.K. Gupta, K. Ghosh, R. Patel, S.R. Mishra and P.K. Kahol, Materials Letters 62, 4103-4105 (2008).

IOP Conf. Series: Materials Science and Engineering **360** (2018) 012050 doi:10.1088/1757-899X/360/1/012050

11. N.L. Tarwal, A.R. Patil, N.S. Harale, A.V. Rajgure, S.S. Suryavanshi, W.R. Bae, P.S. Patil, J.H. Kim and J.H. Jang, Journal of Alloys and Compounds. 598, 282-288 (2014).
12. M. Oztas and M. Bedir, Thin Solid Films. 516, 1703-1709 (2008).
13. G. Li, X.B. Zhu, H.C. Lei, W.H. Song, Z.R. Yang, J.M. Dai, Y.P. Sun, X. Pan and S.Y. Dai, Journal of Alloys and Compounds 505, 434-442 (2010).
14. B.D. Cullity, Element of X-ray Diffraction, 2nd ed. (London: Addition Wisely, 1978).
15. Mujdat Caglar, Saliha Ilican and Yasemin Caglar, Trakya University Journal of Sciences 7(2), 153-159 (2006).
16. J.A. Venables, G.D.T. Spille and M. Hanvucken, Reports on Progress in Physics 99,47 (1984).
17. A.D. Acharya, Shweta Moghe, Richa Panda, S.B. Shrivastava, Mohan Gangrade, T. Shripathi, D.M. Phasen and V. Ganesan, Thin Solid Films. 525, 49-55 (2012).
18. R.K.Gupta, M.Cavas and F.Yakuphanoglu, Molecular and Biomolecular Spectroscopy. 95, 107-113(2012).

IOP Conf. Series: Materials Science and Engineering **360** (2018) 012051 doi:10.1088/1757-899X/360/1/012051

Influence of Molar Concentration on the Structure and Optical Properties of ZnO Films Grown by Spray Pyrolysis Method

H S Sindhu[1], Sumanth Joishy, B V Rajendra and P.D. Babu[2]
[1]Thin film lab, Dept. Of Physics, Manipal Institute of Technology, Manipal-576104
[2]UGC-DAE CSR Mumbai Trombay-India-400086
Corresponding Author: sindhualve9@gmail.com

Abstract: ZnO films were deposited on glass substrates using precursor solution of 0.05M and 0.025M at 623K through spray pyrolysis technique. The comprehensive work has been carried out to analyze the change in the structure and optical properties of deposited films with respect to various precursor solutions concentrations. The XRD pattern shows that, deposited thin films are hexagonal wurzite structure. The deposited film transmittance and optical energy band gap increased with decrease in molar concentration of precursor solution.
Keywords: ZnO, glass substrate, spray pyrolysis, optical, structural

1. Introduction

n-type ZnO semiconductor which comes under the category of II –IV group with very important application in the field of optoelectronic and electronics [1]. It has a direct band gap of about 3.37 eV at room temperature. It has large exciton binding energy and highly sensitive to toxic and combustible gases [2]. A low conductivity and highly oriented nature of polycrystalline hexagonal structure ZnO thin films make use of these films in ultrahigh frequency electro acoustic transducers owing to their piezoelectric properties [3] and also employed in transparent and antireflective coating for the fabrication of solar cell [4]. To harness the complete advantage of the solar spectrum requires optimization of smoothness and thickness of the antireflection coating. Many techniques has been utilized for thin film fabrication viz.,RF magnetron sputtering [5], CVD [6] , pulsed laser deposition [7] ,molecular beam epitaxy [8] and spray pyrolysis [9]. From the various deposition technique, spray pyrolysis have been choosen for preparation of ZnO films. This method has peculiar advantages such as better homogeneity and stoichiometry. In general, this is a chemical deposition method in which some parameters are considered to be important i.e., the solution concentration, nozzle to substrate distance, substrate temperature, carrier gas and spray rates [10]. Herein, we report the preparation of transparent ZnO films deposited on glass substrates using spray pyrolysis technique and the effect of change in precursor solution concentration on structure and optical properties of the deposited films has been addressed.

2. Experimental Details

Films were prepared by spraying aqueous solution of Zinc Acetate dehydrate of different concentration of 0.025 and 0.05M at 623K temperature. The chemically cleaned glass slides were used as substrate. When zinc acetate solution was sprayed over hot substrates, fine droplets of the solution are formed. These droplets decompose thermally after falling over the hot surface of the substrates. This results in formation of zinc oxide films. The spray rate of 2ml/min was maintained using air as a carrier gas. The deposited samples were characterized by using Shimadsu -1800 UV-VIS spectrophotometer operated at a resolution of 1 nm from 400nm to 900nm and absorption spectra as well as transmittance spectra is taken. To examine the crystallinity crystal phases and grain size of ZnO films were performed using a Rigaku –Minflex 600 X-Ray Diffractometer operating at a voltage of 40kV maintained at a current of 15 mA.

3. Results and Discussion

XRD patterns of all deposited thin films with different concentration as shown in figure1.It reveals that increase in the precursor solution concentration facilitates the improvement of crystalline nature of the ZnO thin films. It was observed that for higher concentration crystallinity was found to be increased. The prominent peaks corresponding to (1 0 1) planes were observed in the spectra in addition to other peaks with less intensities were also found i.e (1 0 0), (0 0 2), (1 1 0) and (1 1 2). All the peaks in the diffraction pattern were polycrystalline in nature and hexagonal wurzite structure of ZnO [11]. There were no extra peaks of impurities were detected which indicates the good purity of prepared films. As the concentration increases, the number of Zn^{2+} ions increases which is the reason for higher intensity of (1 0 1) plane.

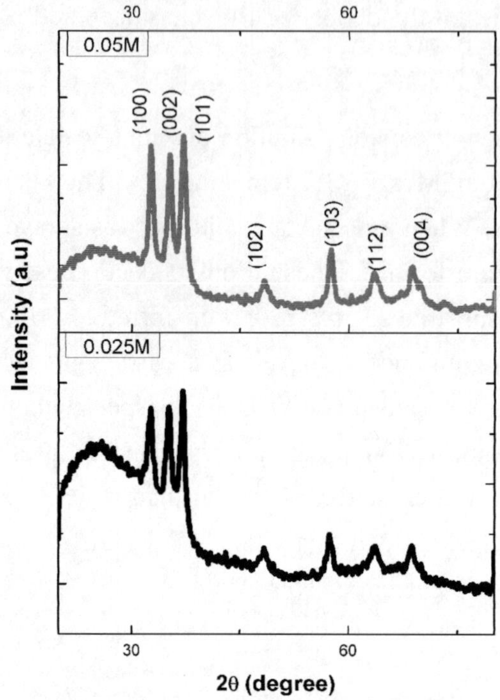

Figure 1. X-ray diffraction of pure ZnO thin films deposited at 673K substrate temperature using 0.05M and 0.025M concentration.

Generally, Debey-Scherrer [12] formula has been employed to calculate the average crystallite size (D). The formula is given by $D = k\lambda/B\cos\theta$, in which k is the shape factor i.e 0.94, λ is the wavelength of the X-ray (1.5406Å), θ is the Bragg's angle and B is the corrected FWHM for instrumental broadening. The crytallite size obtained for the films deposited using precursor solution 0.025 and 0.05M concentration to corresponding (1 0 1) peak was varied from 11.16 to 15.67nm.

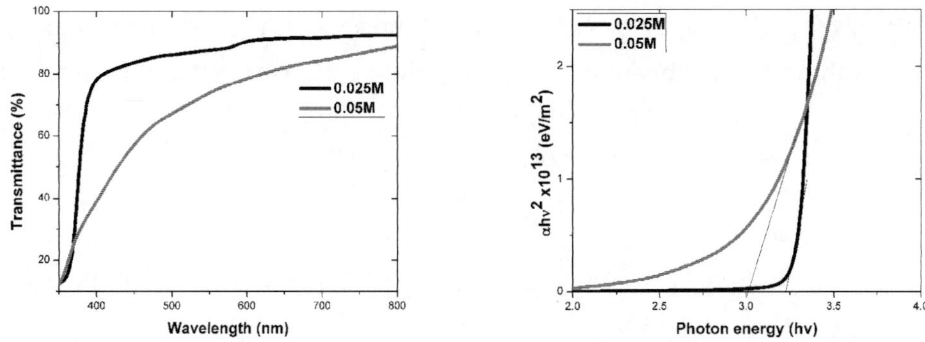

Figure 2. Optical properties of ZnO thin films (a) Wavelength versus transmittance (b) hv versus αhv2 for ZnO films deposited using of 0.025M and 0.05M precursor concentration.

Figure 2(a), depicts that the optical transmittance versus wavelength of ZnO thin films deposited at 623K temperature by utilizing two different concentration of precursor solution such as 0.025M and 0.05M. As the molar concentration of precursor solution increases from 80% 70%, it has been observed that transmittance of the films is decreased. It has also viewed from the spectra of optical transmittance of both the samples exhibits an enhanced absorption at 380nm which can be attributed to the strong photoluminescence emission [13].

Figure 2(b), shows the Optical energy band gap Eg of the film is derived from the absorption spectra by plotting $(\alpha hv)^2$ on y axis versus photon energy on x axis and it can calculated by using the Tauc relationship as follows [14] $\alpha hv = A(hv - E_g)^n$, in which α is absorption co-efficient, h is planks constant, A is constant, v is photon frequency the value of n is selected to be 1/2 , since ZnO is a direct band gap semiconductor. A slight decrease in the optical energy band gap from 3.23 to 3.05eV was observed in figure 2b. The decrement of the value of band gap of samples might be attributed to the variation of concentration [15].

4. Conclusions

In conclusion, ZnO thin films were obtained on glass substrate at 623K using different molar concentration 0.025 and 0.05M zinc acetate solution through spray pyrolysis technique. It exhibits polycrystalline nature and transparent in visible region. All the deposited films shows preferred orientation in (1 0 1) crystalline plan of hexagonal wurtzite structure and crystallite size increased with concentration. The films showed transparent 80-65% in visible region and optical energy band gap decreased.

Acknowledgment

Sindhu H S thanks to UGC-DAE Consortium for Scientific Research, Mumbai Centre, for financial support through Collaborative Research Scheme project entitled "Magnetic and Optical properties of pure and Mn doped ZnO nano films grown by spray pyrolysis method" sanctioned letter UDCSR/MUM/AO/CRS-M-212/2015.

References

1. M. Yuste, R. E. Galindo, I. Caretti, R. Torres, O. Sanchez, Journal of Physics D: Applied Physics 45, 025303 (2012).
2. S. Youssef, P. Combette, J. Podlecki, R.Al Asmar, A.S. Foucaran, Crystal Growth and Design 9,1088 (2009).
3. E. Sarica, V. Bilgin, Surface and Coatings Technology 286,1-8 (2016).
4. J.X. Wang, X.W. Sun, Y. Yang, H. Huang, Y.C. Lee, O.K. Tan, L. Vayssieres, Nanotechnology 17, 4995 (2006).
5. P. Nunes, D. costa, E. Fortunato, Vacuum 64, 293 (2002).
6. Y. Kashiwaba, F. Katahira, K. Haga, Journal of Crystal Growth 221, 431 (2000).
7. J. L. Zhao, X. M. Li, J. M. Bian, W. D. Yu, X. D. Gao, Journal of Crystal Growth 276, 507 (2005).
8. N. Kiyoshi, S. Tatsuya, K. Hee-Bog, Journal of Applied Physics 39, 534 (2000).
9. D. F. Paraguay, L. W. Estrada, D. R. Acosta, M. E. Andrade, MiKi Yoshida, Thin Solid Films 350, 192 (1999).
10. P. M. Ratheesh Kumar, K. P. Vijayakumar, C. SudhaKartha, Journal of Materials Science
    42, 2598-2602(2007).
11. V. Rakhesh, V. K.Vaidyan, Journal of optoelectronics and biomedical materials 1, 281-290(2009).
12. S S Shinde, P S Shinde, C H Bhosale, K Y Rajpure, Journal of Physics D: Applied Physics 41, 105109(2008).
13. J. L Zhuao,X-M Li, S Zhang,C Yang,X-D Gao, WD Yu, Journal of materials research 21,2185(2006).
14. S Shammon, B Jamil, Journal of Engineering & Technology 28, 6061(2010).
15. D.Mishra, K.C.Dubey, R.K.Shukla, Anchal, Atul Srivastava, Sensors & Transducers 105, 119 (2009).

IOP Conf. Series: Materials Science and Engineering **360** (2018) 012052 doi:10.1088/1757-899X/360/1/012052

Numerical Study of 1D Wave Propagation through Double Negative Photonic Crystal

Anusha N P and Alok Sharan*

Department of Physics, Pondicherry University, Puducherry, India– 605014

*aloksharan@gmail.com

Abstract. Photonic crystals are having many applications in different fields like wave guides, high efficiency antenna etc. The conventional photonic crystal contains alternating layers of Double Positive media having high and low positive refractive index values. Recently another kind of unconventional photonic crystal is proposed which consists of a Double Negative meta-material and a Double Positive medium. While in conventional photonic crystals the band gaps are formed due to Bragg scattering of fields, in unconventional photonic crystals the band gaps are formed due to both Bragg scattering and volume averaged zero refractive index of the unit cell. Here we propose a new kind of photonic crystal which consists of two Double Negative meta-materials having different refractive indices. The computational study of the propagation of electromagnetic wave through one dimensional Double Negative photonic crystal is reported. One dimensional Finite Difference Time Domain method is used for the study. Drude model is used in this paper to define the dispersion property of Double Negative media.

1. Introduction

A conventional photonic crystal consists of alternate layers of materials having different dielectric constants [1,2]. These periodic arrangements of material affect the propagation of electromagnetic wave by blocking or reflecting a particular range of frequencies. The frequencies which are forbidden by this structure is called photonic band gap. Here the refractive indices of the constituent materials are positive. Another kind of photonic crystal which consists of a Double Negative Material (DNG) and a Double Positive medium (DPS) is proposed by Jensen Li et.al [3]. In this DNG-DPS photonic crystal, volume averaged zero refractive index of the unit cell is responsible for the band gap [3,4]. A lot of research has been done on different combination of materials to form photonic crystals for different applications [5-12]. In this report we propose a photonic crystal in which both the layers are made up of Double Negative Materials. The values of permittivity and permeability of a DNG medium are negative [13-15]. The region of interest of our study lies in Mid Wave Infrared (MWIR) frequency range which is mainly used in guided missile technology applications. The possibility of enhancing the width of band gap in MWIR region using DNG-DNG photonic crystal is studied. If it is possible to increase the width of band gap, it will be useful for masking the object which is being traced with MWIR frequency for a wide range of frequencies. One dimensional Finite Difference Time Domain (FDTD) method is used in this paper to study the band gap.

2. Theory

For a DNG medium the energy density can be defined as [16]

$$W = \frac{\partial(\varepsilon\omega)}{\partial\omega}E^2 + \frac{\partial(\mu\omega)}{\partial\omega}H^2 \quad (1)$$

where E and H are the amplitudes of electric field and magnetic field, ω is the angular frequency and ε and μ are frequency dependent permittivity and permeability.

For modeling the frequency dependent permittivity and permeability, Drude dispersion model is used in this paper. For an x- polarized electric field and y-polarized magnetic field propagating in z direction, the one dimensional Maxwell's equation [16] can be written as

$$\begin{aligned} \partial_t E_x &= \frac{1}{\varepsilon_0}\left(-\partial_z H_y - J_x\right) \\ \partial_t H_y &= -\frac{1}{\mu_0}\left(\partial_z E_x + K_y\right) \end{aligned} \quad (2)$$

where J_x and K_y are electric current density and magnetic current density.
We use Drude model to define the current densities as [16]

$$\begin{aligned} \partial_t J_x + \Gamma J_x &= \varepsilon_0 \omega_{pe}^2 E_x \\ \partial_t K_y + \Gamma K_y &= \mu_0 \omega_{pm}^2 H_y \end{aligned} \quad (3)$$

where ω_{pe} and ω_{pm} are electric and magnetic plasma frequencies and Γ is the dissipation factor.

FDTD method [17,18] is used to solve these equations. The electric field, magnetic field and current densities are discretized in time and space using central difference formula. Standard Yee algorithm is used to implement FDTD method where electric field and magnetic fields are located in leap frog arrangement. The electric field is located at the cell edges for integer time step and the magnetic field is located at the cell center for half integer time step. Electric and magnetic current densities are located at the cell centers [16]. In this paper total field- scattered field (TF-SF) formalism is used to introduce a Gaussian pulse in the total field region [17,18]. One dimensional Mur Absorbing Boundary condition is used to terminate the FDTD lattice [17,18].

3. Observations and Calculations

A structure consists of two DNG slabs arranged in periodic order with 15 periods is modeled. The thicknesses of the slabs are chosen as $\lambda_0/4n$, where λ_0 is the target wavelength and n is the refractive index of each slab at the target wavelength. We fix the region of interest of our study in MWIR region (3μm - 8μm). A Gaussian pulse of bandwidth 100 THz is allowed to be incident on the periodic structure at normal incidence. The frequency dependent permittivity and permeability according to Drude model is [16]

$$\begin{aligned} \varepsilon(\omega) &= \varepsilon_0[1 - \frac{\omega_{pe}^2}{\omega(\omega + i\Gamma)}] \\ \mu(\omega) &= \mu_0[1 - \frac{\omega_{pm}^2}{\omega(\omega + i\Gamma)}] \end{aligned} \quad (4)$$

The electric plasma frequency and magnetic plasma frequency are properly chosen such that for the entire frequency range of interest the permittivity and permeability remain negative. For ω_{pe} = 17.36e14 rad/s and ω_{pm} = 6.14e14 rad/s, ε=-15 and μ=-2 and for ω_{pe} = 6.14e14 rad/s and ω_{pm} = 9.7e14 rad/s, ε=-1 and μ=-3 at the target frequency. With these values of permittivity and permeability two DNG slabs of refractive indices -3.873 and -2 are arranged periodically. The periodic structure starts at the 21st cell in the FDTD lattice. First 20 cells and last 20 cells in the FDTD lattice are filled with air. The reflection spectrum is recorded at the entrance of periodic structure and transmission spectrum is recorded at the exit of periodic structure. The reflection and transmission coefficients of DNG-DNG photonic crystal are calculated and plotted against frequency as shown in figure 1. The result is compared with DPS-DPS photonic crystal having refractive index +3.873 and +2 which is shown in figure 2. It is observed that the band gap is a little wider in the case of DNG- DNG photonic crystal.

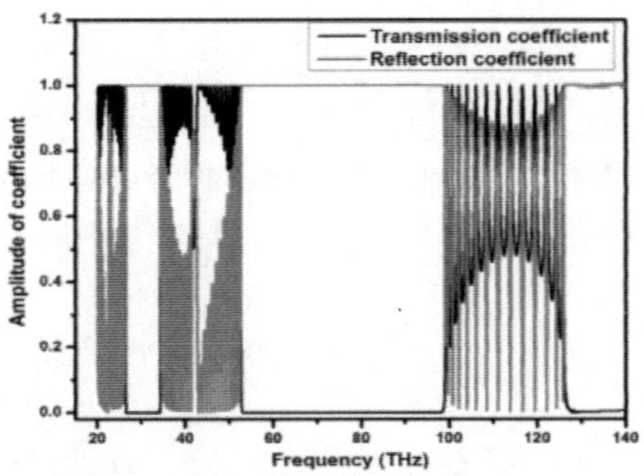

Figure 1. Band gap formed in DNG-DNG photonic crystal. Band gap is from 53.02THz to 98.65THz

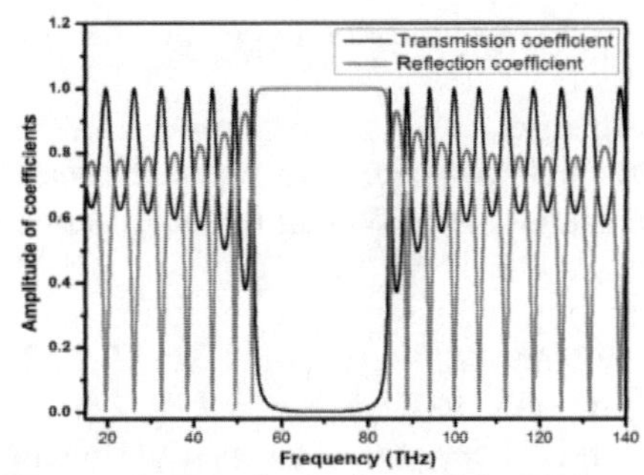

Figure 2. Band gap formed in DPS-DPS photonic crystal. Band gap is from 54.45THz to 83.26THz.

Here in figure 1, the band gap is formed between 53.02THz and 98.65THz (45.63THz wide) in MWIR region. From the band gap diagram of dielectric photonic crystal for identical geometry having the magnitude of refractive indices same as the DNG-DNG photonic crystal, it is seen that the band gap is smaller (28.81THz wide from 54.45THz to 83.26THz).

In dielectric photonic crystal, the band gap is formed due to Bragg scattering of electromagnetic waves incident on the periodic structure. In the same way, in DNG-DNG photonic crystal also, the band gap is formed due to Bragg scattering. Here the plasma frequencies are tuned to different values

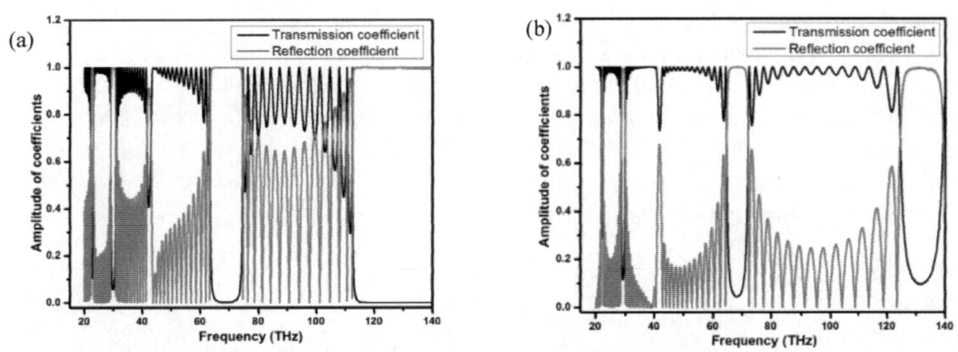

Figure 3. Transmission spectra of DNG-DNG photonic crystal for different combination of plasma frequencies. (a) ω_{pe1}=8.68e14rad/s, ω_{pm1}=10.63e14rad/s and ω_{pe2}=6.14e14rad/s, ω_{pm2}=9.7e14rad/s. (b) ω_{pe1}=8.68e14rad/s, ω_{pm1}=10.63e14rad/s and $\omega_{pe2}=\omega_{pm2}$=7.52e14rad/s

for the same refractive index and the band gaps are studied. The band gap study for different combination of electric and magnetic plasma frequencies are shown in figure 3a and figure 3b. It is found that when the difference between electric plasma frequency and magnetic plasma frequency increases, width of the band gap for the desired frequency range also increases.

4. Conclusion

The propagation of electromagnetic wave in the MWIR region through DNG-DNG photonic crystal structure is studied using a home built FDTD code. In a dielectric photonic crystal, band gaps are formed when electromagnetic wave passes through it due to Bragg scattering. In a similar way, it is found in DNG-DNG photonic crystal also a photonic band gap is formed in MWIR region. By varying the plasma frequencies which is used to define the permittivity and permeability of DNG material using Drude model, the band gaps are studied. It is found that as the separation increases the width of

IOP Conf. Series: Materials Science and Engineering **360** (2018) 012052 doi:10.1088/1757-899X/360/1/012052

band gap increases. For maximum separation of plasma frequencies, the band width is wider than that in the case of DPS-DPS photonic crystal having same refractive index.

Acknowledgement
The authors acknowledge Mr. Ashok Kumar M for the discussions during this paper work. ANP thanks UGC-BSR for providing financial support to carry out the work.

References
[1] Eli Yablonovitch 1987 *Phys. Rev. Lett.* **20** 58
[2] Sajeev John 1987 *Phys. Rev. Lett.* **23** 58
[3] Jensen Li, Lei Zhou, Chan C T and Sheng P 2003 *Phys. Rev. Lett.* **8** 90
[4] Jiang HT, Chen H, Li H, Zhang Y and Zhu S 2003 *Appl. Phys. Lett.* **83** 26
[5] Makhan M, Ramchurn S K, 2007 *J. Opt. Soc. Am. B* **24** 12 3040-47
[6] Tao Bo, Fu-Li Li and Yu Feng 2009 Modern *Physics Letters B* **23** 25 2943-53
[7] Li-Gang Wang, Hong Chen and Shi-Yao Zhu 2004 *Phys. Rev. B* **70** 245102
[8] Xiang Y, Dai X, Wen S and Fan D 2007 *J. Opt. Soc. Am. A* **24** 10
[9] Tang T, Chen F and Sun B 2011 *Optics & Laser Technology* **43** 237-41
[10] Zhang L, Wang J, Li He, Qiao W, Chen L, Wang Q and Zhao Y 2013 *Physica B* **431** 127-31
[11] Joseph S and Hafiz A K 2014 *Optik* **125** 2734-38
[12] Upadhyay M, Mehta A, Aswathi S K, Srivastava S K, Shukla S N and Ojha S P 2014 *J. Intense Pulsed Lasers and Applications in Advanced Physics* **4** 3 45-54
[13] Veselago V G 1968 *Soviet Physics. Uspekhi* **10** 4 509-14
[14] Anantha Ramakrishna S 2005 *Reports on Progress in physics* **68** 2
[15] John B Pendry and David R Smith 2004 *Physics Today* **57** 6
[16] Richard W Ziolkowski and Ehud Heyman 2001 *Physical Review E* **64** 056625
[17] Dennis M Sullivan 2000 *Electromagnetic Simulation using the FDTD method* (NewYork: IEEE Press)
[18] Allen Taflove and Susan C Hagness 2005 *Computational Electrodynamics Finite Difference Time Domain Method* (London: Artech House, INC).

IOP Conf. Series: Materials Science and Engineering **360** (2018) 012053 doi:10.1088/1757-899X/360/1/012053

Magnetic Properties of Magnesium Substituted Y-Type Hexaferrite

Jasvinderpal Singh Virdi

Veer SurendraSai University of Technology,
Sambalpur, Orissa-768018, India
Corresponding Author: jpsvirdi@gmail.com

Abstract- In the present work, we have synthesized Y-type hexaferrite $Ba_2Mg_2Al_{x/2}Cr_{x/2}Fe_{12-x}O_{22}$ (x= 0.0, 0.5, 1.0) using solution gel method. The structural and magnetic properties have been investigated using XRD, FTIR and VSM. X-ray study shows that grain size lies in nano range. Presence of two prominent peaks near 550-600 cm^{-1} to 500-600cm^{-1} in FTIR indicates the formation of hexaferrite. From VSM analysis, It is found that with increasing the amount of co-dopants ($Al-Cr$) the saturation magnetization, retentivity and increases for $x=0$ and $x=0.5$ it was observed that coercivity is zero which indicates that material is super-paramagnetic in nature.

1. INTRODUCTION

Development of ferrites materials is one of the major advances in the field of the nanotechnology. The Magnesium substituted hexaferrite has great importance for the modern society because they are component of a variety of electronic devices used in means of medical equipment, communication, distribution and generation of electric energy, automotive and magnetic recording [1]. Y-type hexagonal ferrite is a frontier material for these applications and has attractive for researchers for many years. This type of ferrite is composed of Y-type Magnesium hexagonal, maintaining a hexagonal structure, with the easy direction of magnetization along the c-axis. However, Co_2Y, Mg Y (where Co and Mg are present as divalent cations) are known as ferroxplana ferrites because they have an easy plane (basal plane) of magnetization perpendicular to the c-axis [2].

To synthesize Nano-ferrites there are lot of methods like mechanical grinding [3], glass crystallization [4], hydrothermal reaction sintering [5], micro- emulsion technique [6], route etc. In addition, many synthetic techniques such as combustion method [7], spray pyrolysis [8], ammonium nitrate melt [9], co-precipitation [10], Sol–gel [11], oxidation in nitric acid [12] have been exploited extensively. The sol–gel method is one of the most widely used methods for the formation of pure hexagonal ferrites. Using this method, it is possible to obtain to order of submicron particles for attaining useful properties for nano materials.

In the present work, we have investigated the effects of Mg substitution for Ba_2^+in barium based Y-type hexagonal ferrites ($Ba_2Mg_2Al_{x/2}Cr_{x/2}Fe_{12-x}O_{22}$) synthesized by sol–gel auto combustion method. The structural and magnetic properties have been investigated using X-ray diffraction (XRD) Fourier Transform Infrared Spectroscopy (FTIR) and Vibrating sample magnetometer (VSM).

2. CHARACTERIZATION TECHNUIQES

Y-type hexagonal ferrites ($Ba_2Mg_2Al_{x/2}Cr_{x/2}Fe_{12-x}O_{22}$) powders have been synthesized via sol gel combustion method using AR grade Barium Nitrate ($Ba(No_3)_2$), Chromium Nitrate Nonahydrate ($Cr(No_3)_3.9H_2O$), Aluminium Nitrate Nonahydrate ($Al(No_3)_3.9H_2O$), Magnesium Nitrate Hexahydrate ($Mg(No_3)_2.6H_2O$, Ferricnitrate Nonahydrate ($Fe(No_3)_3.9H_2O$ and Monohydrate Citric Acid ($C_6H_8O_7.H_2O$) citric acid. Aqueous solution of iron and metal salts are prepared separately in stoichiometric proportions by dissolving them in distilled water. Aqueous solution of citric acid is added to the salt solution with cations to citric acid molar ratio of 1:1.5. Ammonium hydroxide solution is further added drop by drop to attain the 7 pH value. Then the solution is heated at 80C–85C for 4–6 hours with continuous stirring using magnetic stirrer. After evaporation of water, the liquid gets converted into a homogenous brown colored gel. The viscous solution is dried over hot plate at 280–300 C for 3 hours to form the precursor material. Presintering is done at 500 C for 2 h. Then the precursor material is calcined at 700 C, 900 C, 1100 C and 1200 C for 5 h. Following devices were used to know the properties of synthesized Y- type hexaferrites.

3. RESULTS AND DISCUSSION

3.1 Structural properties

3.1.1 FTIR Analysis

Fourier Transform Infrared Spectroscopy is a technique that helps to know the chemical, structural and phase changes present in the sample during combustion and sintered process. Nearly 2 mg of ferrite powered is mixed with powdered KBr in the ratio 1:100 by weighing to ensure uniform dispersion. The infrared spectra of prepared sample $Ba_2Mg_2Al_{x/2}Cr_{x/2}Fe_{12-x}O_{22}$ was recorded in the range of 4000 cm^{-1} to 400 cm^{-1}.

For the sample $Ba_2Mg_2Fe_{12}O_{22}$ spectra is shown in the Fig 1.1.The band is in the range of 4000-400 cm^{-1}.They have dominant absorption bands at range of 400-600 cm^{-1}, 1250-1600 cm^{-1} and 3600-4000 cm^{-1}. The bands in the range of 400-600 cm^{-1} assigned to be for broadening of metal oxygen bonds which is the characteristics of hexaferrites. Further, the band in the range of 1250-2000 cm^{-1} indicates the presence of carboxylic acid (citric acid) and nitrile. The downward peak at 2400 cm^{-1} is due to presence of OH^-. The band from range 3400-4000 cm^{-1} exhibit the stretching of hydroxyl group which indicates the presence of free or absorbed moisture in the sample.

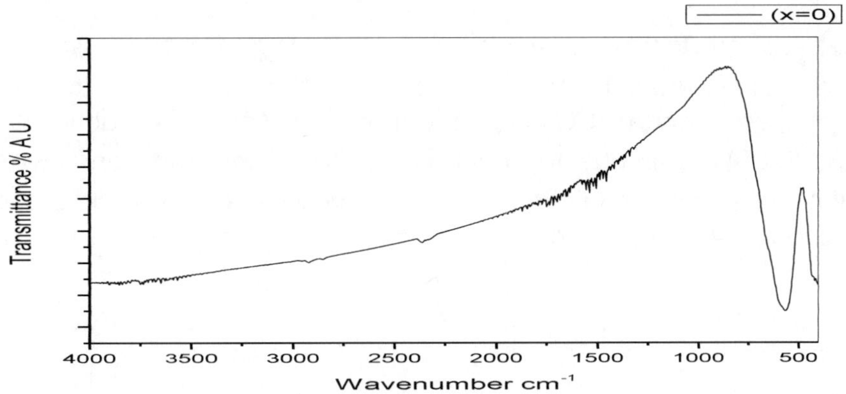

Fig 1.1: FT-IR Spectra of $Ba_2Mg_2Fe_{12}O_{22}$ at pH=7

Similarly Spectra of sample $Ba_2Mg_2Al_{0.25}Cr_{0.25}Fe_{11.5}O_{22}$ is shown in Fig 1.2. The band is in the range of 4000-400cm^{-1}.It have dominant absorption band at range of 400-600cm^{-1}, 1250-1600 cm^{-1},and 3500-3900cm^{-1}. The bands are in the range of 400-600 cm^{-1} is assigned to be for broadening of metal oxygen bonds which is the characteristics of hexaferrites. Further, the band in the range of 1250-1700cm^{-1} indicates the presence of carboxylic acid (citric acid) and nitrile.The downward peak at 2400cm^{-1} and 2800cm^{-1} is due to presence of OH^-.The band from range 3400-4000cm^{-1} exhibit the stretching of hydroxyl group which indicates the presence of free or absorbed moisture in the sample.

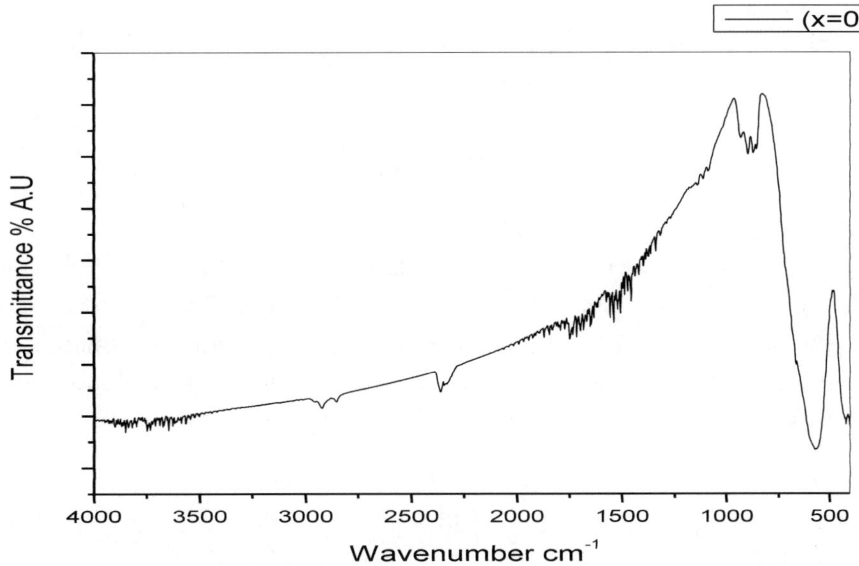

Fig 1.2: FT-IR Spectra of $Ba_2Mg_2Al_{0.25}Cr_{0.25}Fe_{11.5}O_{22}$ at pH=7

Similarly, Spectra of sample $Ba_2Mg_2Al_{0.5}Cr_{0.5}Fe_{11}O_{22}$ is shown in Fig 1.3. The band is in the range of 4000-400cm^{-1}.It have dominant absorption band at range of 400-600cm^{-1},1250-1600 cm^{-1},and 3500-3900cm^{-1}. The bands in the range of 400-600

cm^{-1} is assigned to be for broadening of metal oxygen bonds which is the characteristics of hexaferrites. Further, the band in the range of 1250-1800 cm^{-1} indicates the presence of carboxylic acid(citric acid) and nitrile. The downward peak at 2400 cm^{-1} and 2900 cm^{-1} is due to presence of OH^-.The band from range 3600-4000 cm^{-1} exhibit the stretching of hydroxyl group which indicates the presence of free or absorbed moisture in the sample.

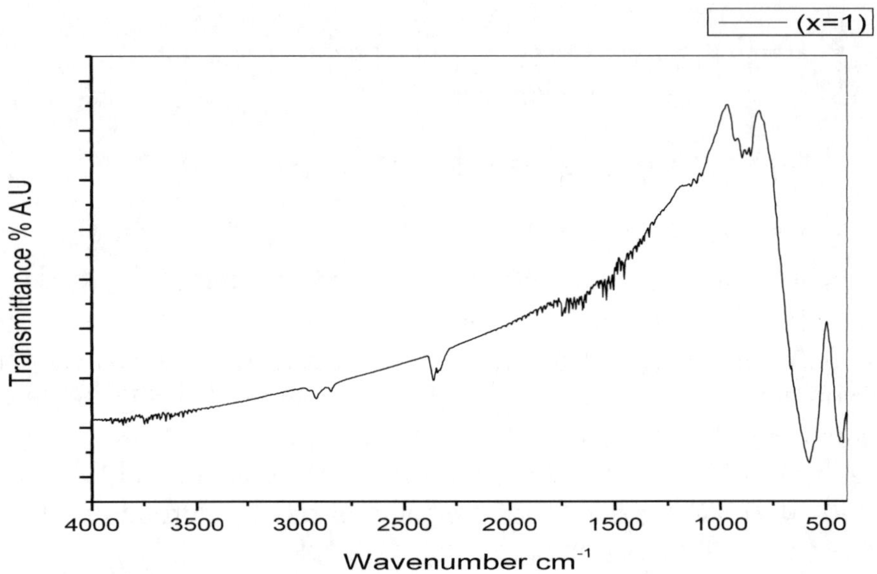

Fig 1.3: FT-IR Spectra of $Ba_2Mg_2Al_{0.5}Cr_{0.5}Fe_{11}O_{22}$ at pH=7

3.1.2 XRD Analysis

Fig 1.4 shows the X-ray diffraction pattern of samples $Ba_2Mg_2Al_{x/2}Cr_{x/2}Fe_{12-x}O_{22}$ $(x = 0,0.5,1.0)$ nanoparticles. Pure hexaferrites phase was the formation at a temperature of 1000°c for 6 hrs. The average crystallite size of the Nano-crystalline hexaferrite synthesized at different molar ratios of substitution for the peaks was calculated from the XRD data using the Scherrer formula.

$$D = \frac{k\lambda}{\beta \cos\theta}$$

Where D is the crystallite size. The shape factor of K can be 0.62–2.08 and is usually taken as about 0.89 for hexagonal systems. λ is the X-ray wavelength (1.54Å).It was observed and calculated and was found out to be 44.9434nm for $x = 0$, 57.9684nm for $x = 0.5$ and 38.379nm for $x = 1$.

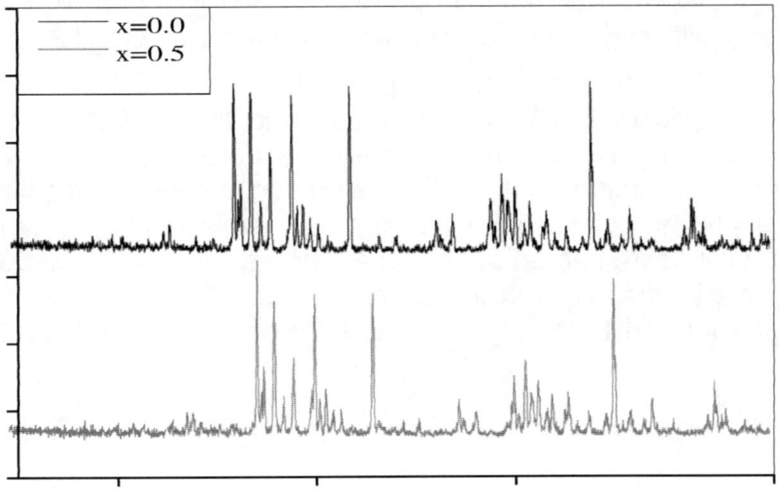

Fig 1.4: XRD Patterns of hexaferrite $Ba_2Mg_2Al_{x/2}Cr_{x/2}Fe_{12-x}O_{22}$ at different composition

The lattice parameter (a and c) and cell volume (V) was also calculated using formula given below.

$$\frac{1}{d_{hkl}^2} = \frac{4}{3}\left[\frac{h^2 + hk + k^2}{a^2}\right] + \frac{l^2}{c^2}$$

$$V = 0.8666a^2c$$

With usual parameter. The lattice parameters of all the samples synthesized at different concentrations were calculated and is shown in the Table1.1. It was observed that lattice parameter 'c' decreases and lattice parameter 'a' increases with increase in dopants.

Table 1.1: Value of different parameter calculated for hexaferrite

S.No.	Sample	Lattice constant a (Å)	Lattice constant c (Å)	Cell Volume (Å)
1	$x=0$	4.40	47.86	802.956
2	$x=0.5$	5.85	44.432	1317.12
3	$x=1$	5.87	39.76	1187.24

IOP Conf. Series: Materials Science and Engineering **360** (2018) 012053 doi:10.1088/1757-899X/360/1/012053

3.1.3 VSM Analysis

Figure 1.5 to 1.7 shows the room temperature hysteresis loop plots for all of the Nano size samples with the chemical formula of $Ba_2Mg_2Al_{x/2}Cr_{x/2}Fe_{12-x}O_{22}$ sintered at 1000°C for 6hrs. The values of saturation magnetization (Ms) and coercivity (Hc) for all samples were accomplished from the loops, and variation of these parameters as a function of aluminum and chromium contents. According to Figure it is clear that by an increase in substitution content, the saturation magnetization was gradually increased. For $x = 0$ and $x = 0.5$ the value of coercivity is zero, which says nanomaterial is super-paramagnetic in nature whereas for $x = 1$ it is 838.7025.The retentivity also gets increasing with increase in substitution contents. The value of Retentivity, Coercivity and Saturation Magnetization is shown in Table1.2 for each sample.

Table 1.2: The value of Retentivity, Coercivity, and saturation magnetization for (x=0, 0.5, and 1)

S. No	Samples (x)	Retentivity (Mr) emu/g)	Coercivity (Hc) (Oe)	Saturation Magnetization (Ms) (emu/g)
1	$x = 0$	4609.39	0	19441.4095
2	$x = 0.5$	5183.01011	0	19796.5037
3	$x = 1$	9628.5168	838.7025	21510.5163

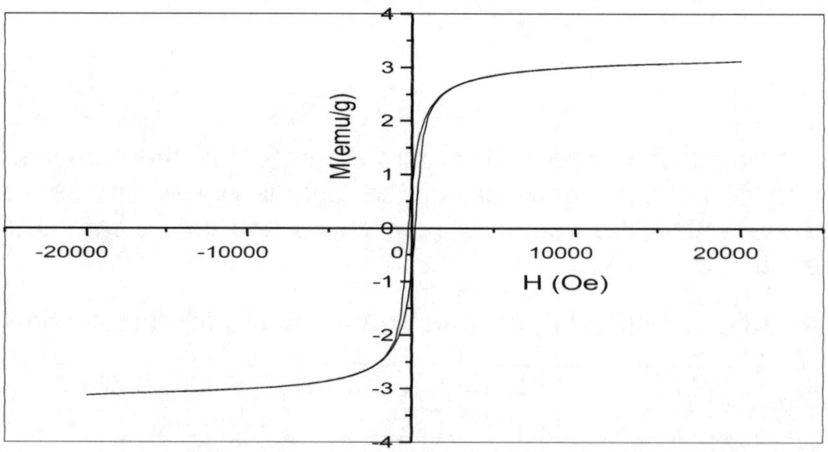

Fig.1.5:Magnetic-Hysteresis loop for x **=0**

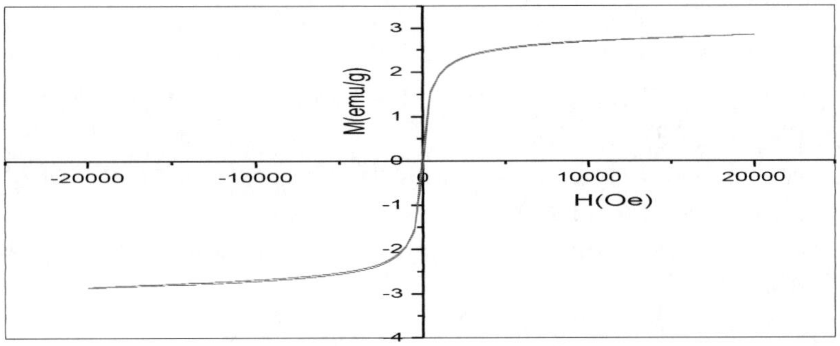

Fig 1.6:Magnetic-Hysteresis loop for x =0.5

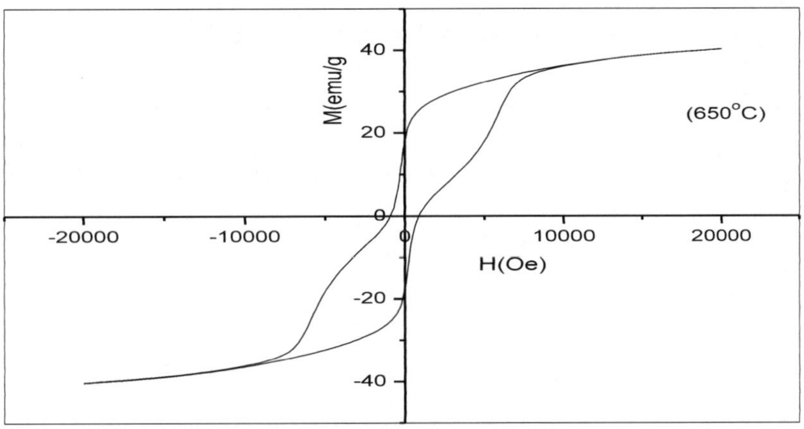

Fig 1.7: Magnetic-Hysteresis loop for x =1

5. CONCLUSIONS

The Y-type Nano-hexaferrites have been synthesized using sol-gel technique and was characterized using FT-IR, XRD and VSM. X-ray study shows that grain size lies in Nano range. Presence of two prominent peaks near 550-600 cm^{-1} to 400-600 cm^{-1} in FTIR indicates the formation of hexaferrite. From VSM analysis, It is found that with increasing the amount of co-dopants $(Al-cr)$ the saturation magnetization, retentivity and increases for and it was observed that coercivity is 0 for $x=0$ and $x=0.5$ which indicates that material is super-paramagnetic in nature.

REFERENCES

1. R. B. Jotania, H.S. Virk, *Solid State Phenomena.* 189 209**(2012)**.
2. Z. Haijun, Yao Xi, Zhang Liangying, *J. Eur. Ceram. Soc.* 22 835 **(2002).**

IOP Conf. Series: Materials Science and Engineering **360** (2018) 012053 doi:10.1088/1757-899X/360/1/012053

3. Z. Jin, W. Tang, J. Zhang, H. Lin, Y. Du, *Journal of Magnetism and Magnetic Materials.*182 231, **(1998).**
4. L. Rezlescu, E. Rezlescu, P.D. Popar, N. Rezlescu, *Journal of Magnetism and Magnetic Materials.* 193 288 **(1993)**.
5. X. Liu, J. Wang, L.M. Gan, S.C. Ng, *Journal of Magnetism and Magnetic Materials.*195 452 **(1999).**
6. X. Liu, J. Wang, L.M. Gan, S.C. Ng, J. Ding, *Journal of Magnetism and MagneticMaterials,*184 344 **(1998).**
7. L. Junliang, Z. Wei, G. Cuijing, Z. Yanwei, *Journal of Alloys and Compounds.* 479 863**(2009).**
8. P. Ren, J. Guan, X. Cheng, *Materials Chemistry and Physics,* 98 90 **(2006).**
9. H. Sözeri, I. Kucuk, H. Ozcan, *Journal of Magnetism and Magnetic Materials.* 323 1799 **(2011)**.
10. M. Radwan, M.M. Rashad, M.M. Hessien, *Journal of Materials Processing Technology,* 181 106**(2007).**
11. T. Kaur, B. Kaur, B. H. Bhat, S. Kumar, A.K. Srivastva,*Physica B*, 456 206 **(2015).**
12. H. Sözeri, *Journal of Magnetism and Magnetic Materials,* 321 2717 **(2009).**

Synthesis and Characterization of Li^{+}, $Ag^{2+}$and Cu^{2+} metals doped MgO nanostructured in Gel Method

V. Rajendran[1], B. Deepa[1*] and R. Mekala[1]

[1]PG Research Department of Physics, Presidency college, Chennai 600 005, Tamilnadu, India.

* Email: b.deepasenthil@gmail.com

Abstract The MgO nano particles doped with alkali, Nobel and transition metals (Li^{+}, Ag^{2+}, Cu^{2+}) were synthesized by gel method. The structures, morphologies and size of synthesized samples were characterized and confirmed by Powder X-ray diffraction (PXRD), Scanning Electron Microscope (SEM) and High Resolution Transmission Electron Microscope (HR-TEM). The antibacterial activity for pure and doped MgO nano particles were studied against S.auerus. Among the results, the Li^{+} and Cu^{2+} doped MgO nano particles are best inhibition against the bacterial strains due to the presence of active oxygen species.

Keywords: Nanoparticles, MgO, Strains, PXRD.

1. Introduction

The metal oxides have attracted attention of many applications such as in sensors, optoelectronic, and plasma display panel and as insulating layers[1-3]. Among these material MgO has good inorganic antibacterial agent with cubic periclase structure also high oxygen affinity and low temperature of Mg causes oxide compound easy to synthesis[2]. However, many studies have been done to improve the antibacterial activities of various metal oxides such as ZnO, AgO and ZrO_2 by reducing the grain size or by doping transition metal elements[4, 5]. Doping of some metal ions into oxide material is effective way for modifying the physical and chemical properties such as high defect concentrations[6]. These defect concentrations (active oxygen species AOS) played a significant role towards the antibacterial activities of nano crystal[7, 8]. Therefore, we synthesis the bare MgO and various metal doped (Li^{+}, Ag^{2+}, Cu^{2+}) MgO via cost effective sol-gel method. In that Li ions elements has a similar ionic radius (0.76Å) to Mg (0.72Å). These Li ions easily accommodate within the crystal lattice and enhance the antibacterial activity[9]. Second, the Nobel metal Ag has bigger ionic radius (1.26Å) than Mg. So that, Ag ions spreading on the surface grain then reduce to Ag^{0} in higher temperature[10], it's inhibit the crystal growth and increases the surface area and generation of hydrogen peroxide[5]. The Cu elements has same ionic radius of Mg. The similarity of ionic radius promotes the substitution incorporation of the dopants[11]. So, these metal ions (Li^{+}, Ag^{2+}, Cu^{2+}) dopants are attributes to the generation and stabilization of AOS. The aim of this work is to clarify the effect of the dopants on antibacterial activity, crystalline size and structure morphology.

2. Experimental

All the chemicals used here were analytical grade and directly used as received without further purification. 0.1 M Magnesium ethonate tetra hydrate was initially dissolved in 50ml of the methanol solvent under constant stirring. The pH of the solution was adjusted to 5 using 1M of BDC. The mixture was continuously stirred until a thick white gel is formed and it is left overnight for further gelation process before been dried in an oven at 100°C for 24 h. The fluffy dried material was grounded to fine powder precursors. The resultant synthesized material was annealed at 600°C for 6 h to form MgO nanostructures. Doped MgO was also prepared by same procedure with addition of dopands. The doping concentration is 2.5mol% for each dopand. In the sol-gel reaction, the magnesium ethonate Mg $(CH_3COO)_2.4H_2O$ and benzene -1,2 Dicarboxylic acid (BCD) $C_8H_6O_4$ are hydrolyzed in the methanol to form the mixture of magnesium (Mg^{2+}) cation and phthalate $(C_8H_4O_4)^{-}$ anion, respectively. When the reaction medium reaches a pH of about 5, these ions form complexes and are polymerized to give a thick white gel. The gel formed is identified as magnesium phthalate. Besides that, the side products such as acetic acid and water were evaporated during drying at 100 °C for 24 h[12]. Then the Mg $(C_8H_4O_4)$ is decomposed into MgO nanostructures.

2.1 Characterization Techniques

The synthesized samples were subjected into various characterizations. The thermal profiles of the precursors were studied using Thermo gravimetric analysis (TGA Q500 V20.10 Build 36). The Powder X-ray diffraction (PXRD) patterns of the periclase were obtained by the XRD Analytical X'Pert Pro MPD with CuKα radiation. The morphology and size of the periclase crystallites were determined using a

IOP Conf. Series: Materials Science and Engineering **360** (2018) 012054 doi:10.1088/1757-899X/360/1/012054

scanning electron microscope (SEM; JEOL JSM-7600 F,) and a High resolution transmission electron microscope (HRTEM; JEOL JEM-2100 F).

2.2. Antibacterial studies

The antibacterial activity of the sample was studied by using disc diffusion method. Pure culture of ***Staphylococcus aureus*** species of bacteria was used for the testing antimicrobial activity of MgO nanoparticles. Nutrient agar medium plates were prepared by suspending 28 g of nutrient agar powder in one liter of distilled water then to sterilize the medium through autoclave at 121°C for 20-25 minutes. Pour 25-35 ml of medium into Petri plate under sterile conditions via laminar air flow chamber. Medium allowed to solidification under room temperature and then the bacterial cultures were swabbed on plates. The sterile discs were dipped as prepared nanoparticles solution at different concentrations (25, 50, 75 and 100 of μg/ml) and kept for incubation at 37°C for 24 hours. Zones of inhibition were measured with ***Hi Media ruler***.

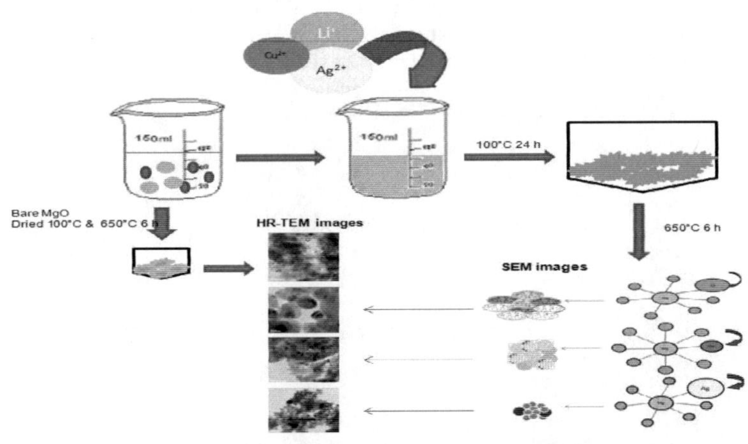

Fig. 1 Schematic diagram of Bare and doped MgO samples.

3. Result and discussion

Fig (2) shows a XRD diffraction pattern of as prepared samples. The observed pattern for bare and doped MgO sample can be indexed to the cubic phase with periclase structure. As the presence of dopands, one observes that the intensity of the peaks varied and also the diffraction peaks shifted slightly in the (200) plane position, indicating a lattice distortion of the doped catalyst[13] shown in **Table. 1**. The higher intensity values observed in Li doped MgO sample and lower intensity values for Ag which is due to ionic radius of dopants. The ionic radius is much bigger, smaller and similar than that of Mg^{2+} ions, the dopants substituting for MgO crystal lattice distortion[14]. Moreover, the crystallite of doped samples it appears that later decreased from Li to Ag, which indicates changes in crystal strains. The crystalline size of bare and doped MgO nano powder was calculated by Scherer equation. Compare to bare MgO (16.033nm) Li - doped MgO nano structured has largest size and Nobel metal Ag^{2+} has the smallest crystallite size. Then Cu^{2+} doped MgO has nearly similar to the bare MgO. In XRD diffraction pattern, Ag doped peak has some metallic silver elements were detected for exceeding the silver due to its solubility limit of Ag in MgO[15].

Samples	2θ(°)	d-spacing(nm)	FWHM	Crystalline Size (nm)	Lattice Constant (Å)
MgO	42.8167	0.210951	0.5270	16.033	4.2190
Li-Doped	42.7897	0.211080	0.3905	21.635	4.2216
Cu-Doped	42.7892	0.210536	0.5632	15.00	4.2107
Ag-Doped	42.8187	0.210942	0.7859	10.751	4.2188

Table 1. XRD parameters for Bare and Doped MgO samples.

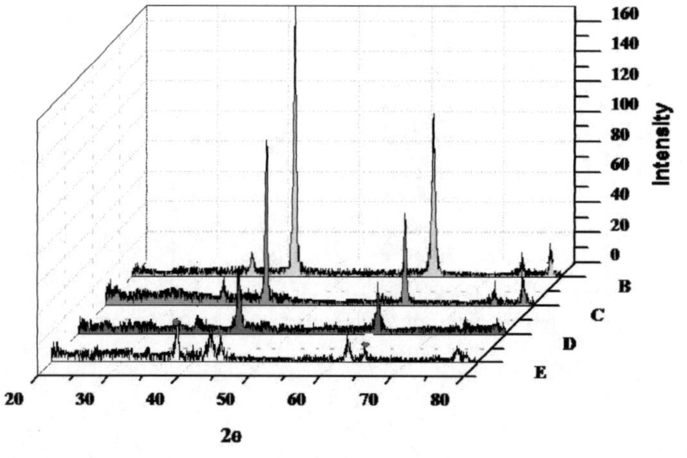

Fig.2 XRD Diffraction pattern (B) Bare MgO (C) Li-doped (D) Cu-doped (E) Ag-doped.

Typical SEM micrographs of different samples were investigated in Fig (3) while considering the doped MgO. The morphological structure was converted from homogeneous agglomerated particle for bare MgO into disc like shapes and ball like shapes besides the dispersed particle for doped MgO sample. The high resolution TEM image of the as synthesized MgO nano crystalline shown in Fig (4). It clearly shows the lattice fringes, indicating the established crystalline of MgO nano powder. Since a crystallite can be defined by studying the orientation of the lattice fringes, one can see the average grain size in the synthesized MgO nano particle is about 10-20 nm in diameter, which is good agreement with calculated values from XRD analysis. The spacing of a set of fringes is proportional to lattice spacing when the corresponding lattice planes meet the Bragg condition[16]. The distance between the lattice fringes was found to be 0.21 nm, which was good agreement with the lattice spacing of (200) plane in cubic phase of MgO.

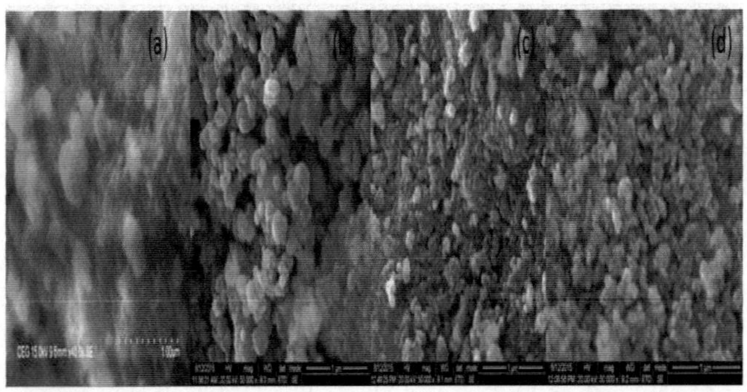

Fig. 3 SEM for (a) Bare MgO (b) Li-doped (c) Cu-doped (d) Ag-doped.

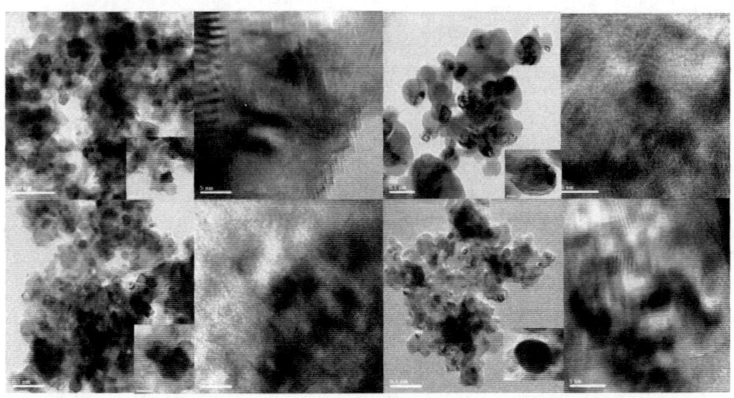

Fig. 4 HR-TEM images of (a) (c) (e)(g) Bare MgO, Li, Cu, Ag-doped samples (b) (d) (f) (h) lattice fringes for Bare, Li, Cu, Ag-doped samples.

The antibacterial activities of Li^{+}, Ag^{2+},and Cu^{2+} doped MgO nano crystallites are investigated using MBC test and anti bactericidal rate against S.auerus. The MBC values of the as synthesized nano particles towards S.auerus are listed in **Table 2**. Clearly Li^{+} and Cu^{2+} doped MgO has a lowest concentration (75µg/ml) observed the best anti bacterial activity followed by standard antibiotic ciprofloxacin. The Li^{+} ions have low valance state and occupy the MgO

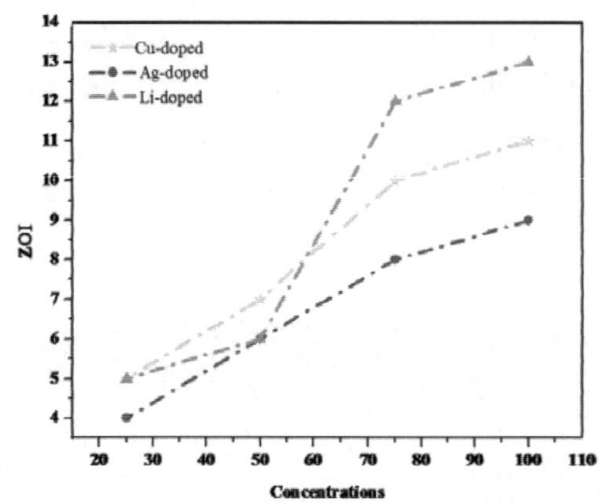

Fig. 5 Comparative graph of ZOI and Concentrations of Li, Ag and Cu dopants.

crystal lattices which generate the oxygen vacancy to inhibit the bacterial growth[17]. In the same, Cu^{2+} ion has similar ionic radius of Mg^{2+}, the similarity of ionic radius promotes substitution incorporation of the dopand, it's has to increase the number of oxygen vacancies[18]. However, Ag- doped MgO have relatively less antibacterial activity because of lower susceptibility of gram positive bacteria. The lower susceptibility of gram positive bacteria to silver doped nanoparticles has been attributed to difference in the cell wall structure. Thus, gram positive bacteria may allow less Ag^{+} ions to reach to cytoplasmic membranes of the bacterial strains. Compared with dopants the bare MgO has a highest antibacterial activity due to high AOS generation in which inhibits the growth of S.auerus.

Table 2. Treatment- ZOI of Bare and doped MgO samples.

Microorganism *Staphylococcus aureus*	Concentration of MgO Nps (µg/ml)	Li-MgOSamples ZOI (mm)	Cu-MgO Samples ZOI (mm)	Ag-MgO Samples ZOI (mm)
Standard **Ciprofloxin** ZOI (11mm)	25	5	5	4
	50	7	6	6
	75	12	10	8
	100	13	11	9

4. Conclusion

In conclusion, the bare and alkali, Nobel and transition metal (Li^{+}, Ag^{2+}, Cu^{2+}) doped nanostructured MgO were synthesized by gel method. The Powder XRD, SEM and HRTEM analysis observed that the Periclase cubic structure, disc like and ball like morphologies. The metal dopands obviously influenced the crystallization, various structures and antibacterial activity of nanostructured MgO. The Li and Cu-doped MgO exhibit the good antibacterial activity against S.auerus due to an increase in defect concentration (AOS). However, Ag-doped samples exhibit less antibacterial activity than the other dopands. All these results indicate the ionic radius of dopands is important key factor for the structural, morphological and antibacterial activities of nanostructured MgO.

Acknowledgement

The author is grateful to the Department of Science and Technology (DST) for extending financial assistance to carry out this work.

Reference

[1] T. Abukawa, S. Sato, Y. Matsuoka, 2010 *Surface science,* **164**, 604.
[2] S. Benedetti, H.M. Benia. N. Nilius, S.Valeri, H.J.Freund, 2006 *Chemical physics letters* **430**, 330.
[3] D.S.E.Atanassova, 2002 *Microelectronic Reliability* **42**, 1171.
[4] V.K.Sharma, R.A. Yngard, Y.Lin, 2009 *Advanced colloid interface science* **145**, 83-96.
[5] Neha sharma, J.Kumar, S.Thakur, S.Sharma, V. Shrivastava, 2013 *Drug invention today*, **5** 50-54.
[6] J.H.He, I.Ichinose, S. Fujikawa, T. Kunitake, A. Nakao, 2002 *Chemical Matter* **14,** 3493-3500.
[7] L. Huang, D.Q.Li, Y.J.Lin, M.Wei, D.G.Evans, X. Duan, 2005 *Journal of Inorganic Biochemical* , **99**, 986-993.
[8] F.Al-Hazmi, F. Alnowaiser, A.A. Al-Ghamdi, M.M. Aly, R.M. Al- Tuwirqi, F. El-Tantaway, 2012 *Superlattices Microstructures* **52**, 200-209.
[9] Y. Rao, W.Wang, F. Tan, Y. Cai, J. Lu, X. Qiao, 2014 *Creamics International,* **40** 14397-14403.
[10] H. E. Chao, Y.U.Yun, H.U. Xingfang, A. Larbot, 2003 *Journal of European Creamic Society* **23**, 1457-1464.
[11] W. Li, S. Ismatshah, C.P. Huang, O.Jung, C. Ni, 2002 *Material Science Engineering B*, **96**, 247.

[12] M. Mastuil, N. Kamarulzaman, M. Nawawi, A. Mahat, R. Rusdi, N. Kamarudin, 2014 *Nano. Rsch. Lett,* **9**, 134.

[13] J.Yuan, M.X. Chen, J.W. Shi, W.F. shangguan, 2006 *International Journal of Hydrogen energy* **31**,1326-1331.

[14] W. Kallel, 2006 *Journal of non-crystalline solids,* **352**, 3970-3978.

[15] H. S. Park, D. H. Kim, S. J. Kim, K. S. Lee, 2006 *Journal of Alloys and Compounds,* **415**, 51-55.

[16] S. Qiu, S. J. Kalita, 2006 *Materials Science and Engineering A* **435-436**, 327-332.

[17] I. Balint, K. Aika, 2001 *Applied Surface Science* **173**, 296-306.

[18] S. Egger, R.P. Lehmann, M.J. Height, M.J Loessner, M.Schuppler, 2009 *Applied Enviromentel Microbiology*, **75**, 2973.

Wavelength Dependent Ammonia Sensing Characteristics of SnO_2 based Fiber Optic Sensor

S Narasimman[1], L Balakrishnan[2,*], S R Meher[3], R Sivacoumar[4], Elizabeth Rufus[4], and Z C Alex[4]

[1]School of Electronics Engineering, VIT University, Vellore 632 014, India.
[2]Department of Physics, Government College of Technology, Coimbatore 641 013, India.
[3]Department of Physics, School of Advanced Sciences, VIT University, Vellore 632 014, India.
[4]Department of Sensor and Biomedical Technology, School of Electronics Engineering, VIT University, Vellore 632 014, India.

*Email: bslv85@gmail.com

Abstract. SnO_2 nanoparticles were synthesized by co-precipitation technique. The synthesized SnO_2 nanoparticles were characterized by X-ray diffraction (XRD), scanning electron microscope (SEM) and UV-visible diffused reflectance spectrometer (DRS). The XRD results addressed that the SnO_2 nanoparticles was crystallized in rutile tetragonal structure. The SEM analysis confirms that the prepared nanopowder is composed of nanoparticles. Analysis of the DRS UV-Vis spectrum showed that the band gap of the synthesized SnO_2 is to be ~3.4 eV. The small cladding portion of the optical fiber was replaced with the synthesized nanoparticles. The ammonia sensing characteristics of the prepared SnO_2 nanoparticles were analyzed for different wavelength ranges (red, yellow and blue) using polymethyl methacrylate fiber. It has been found that the synthesized SnO_2 nanoparticles shows high sensitivity (~23) in yellow wavelength range compared with red and blue wavelength ranges at room temperature.

Keywords: SnO_2, Nanopowder, Fiber optic sensor, Ammonia sensing, Wavelength dependent sensing.

1. Introduction

Ammonia (NH_3) has widespread use in chemical processes, living environment, medical treatments and industrial applications. Ammonia is an indispensable for many biological processes and serves as a precursor for amino acid and nucleotide synthesis. The amount of ammonia manufactured every year by human is almost equal to the amount produced by nature every year. Threshold limit of ammonia concentration in air is only 25 ppm for human beings. Generally, ammonia has been found in air, soil and water samples at hazardous wastes. About 80% of the manufactured ammonia is used as fertilizer in agriculture and also used to manufacture synthetic fibers, plastics and explosives. Still, ammonia is a toxic gas and severe irritant to the respiratory system. Generally, the minimum explosive limit for ammonia is 15% [1]. Depending on those conditions and sensibilities, the critical requirement of flexible and effective ammonia monitoring everywhere is hurried. Metal oxides based ammonia gas sensors are being developed extensively as they are highly sensitive to NH_3 and also easy to materialize [2]. These sensors are traditionally of the electrical resistive whose resistance varies when they are exposed to the detecting gas. However, these sensors exhibit enhanced gas sensitivity only at high operating temperatures (above 200 °C) and also respond to many gases (CO, NH_3 and methanol) [3,4]. In the recent decades, fiber optic gas sensors based on metal oxides as the sensing medium have been reported for improving gas sensitivity and selectivity for room temperature operation. These sensors are immune to electromagnetic interference, low cost and could be used in the hostile environments [5,6]. The gas sensing technique is based on cladding transition technology, in which a centre portion of clad in the optical fiber in denuded and coated with the gas sensing material. The intensity of the light passing through the fiber is very sensitive to the change in the refractive index of the sensing medium which gets changed due to gas species interaction [7].

In this article, the fiber-optic sensing surface is prepared by modifying the original cladding material with a chemical sensitive material, SnO_2 on a certain portion of optical fiber. We have studied the wavelength dependent ammonia sensing characteristics of SnO_2 nanoparticles at different gas concentrations.

2. Experimental Details

SnO_2 nanopowders were synthesized by co-precipitation technique. Tin (II) chloride ($SnCl_2$) was used as a precursor, tri-ethanolamine ($C_6H_{15}NO_3$) act as a solvent and ammonia is used as precipitating agent respectively. For the synthesis of SnO_2 nanopowders, 25 ml of 0.2 M tin (II) chloride and 25 ml of 0.5 M tri-ethanolamine solutions were prepared. Initially, tin (II) chloride and tri-ethanolamine were mixed and stirred continuously for 5 hr. Then, the aqueous ammonia solution was added to the mixed solution drop wise for 1 hr. After adding a particular amount of ammonia, the precipitation starts and the concentrations of precipitates increases as the addition was increased. Till the p^H of the solution reaches 8.0, ammonia was added. Further, the solution was stirred for another 30 min and then the particles were washed with double distilled water to remove the residual ions then filtered with filter paper. Finally, the powder was calcined at 600 °C for 4 hr.

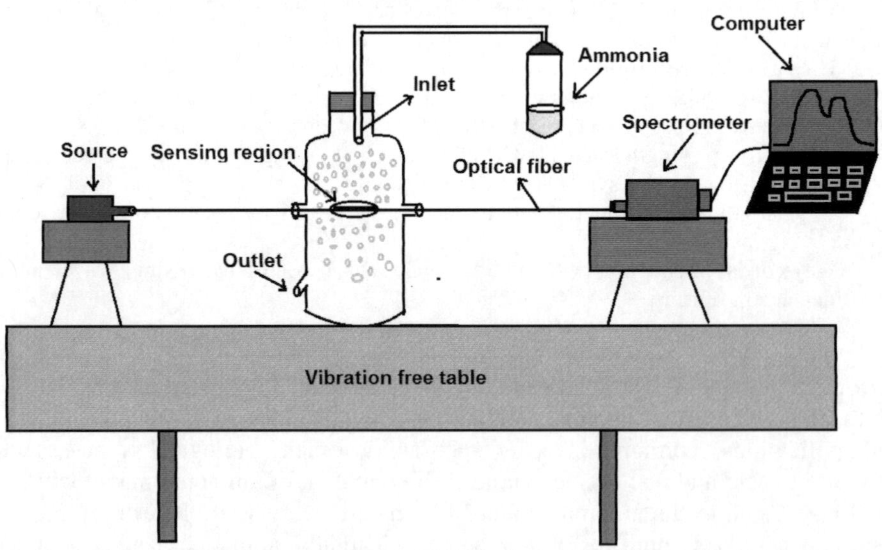

Figure 1. Fiber optic gas sensing setup.

Figure 1 shows the scheme of fiber optic gas sensing setup used for the present study. The sensor consists of multimode step index plastic fiber having a length of 42 cm, diameter of 750 μm and numerical aperture of 0.51. The refractive index of the core and cladding are 1.492 and 1.402, respectively [8,9]. The fiber is made up of polymethyl methacrylate and is cleaved at both ends, and coupled with light source and spectrometer. The sensor setup contains a broadband light source

(Halogen light source-SLS201/M) with the wavelength range from 300 to 2600 nm and fiber optic spectrometer (CCS200/M) having a spectral response of 200 to 1000 nm. In a middle portion of an optical fiber, the sensing region was formed by abolishing mechanically the original cladding of the fiber up to the core for a length of 3 cm. Initially, the buffer was cleaved. Then the original cladding surface was etched away using acetone solution. The etched surface was cleaned and coated with sensing material to serve as a new cladding. The SnO_2 nanopowder was mixed with double distilled water to form a paste and coated in the clad removed region by dip coating method as new cladding [10]. At different wavelength ranges for different concentrations of ammonia the intensity variations were recorded using spectrometer interfaced with computer. The sensor studies were carried out under normal atmosphere pressure and room temperature.

3. Results and Discussion

The X-ray diffraction pattern of SnO_2 nanopowders are shown in the Figure 2. The XRD pattern of the synthesized SnO_2 nanopowders could be addressed to rutile tetragonal structure (PCPDF card no-88-0287). The sharp and narrow diffraction peaks corresponds to (110), (101), (111) and (211) different planes confirms the formation of rutile structured nanopowder [11]. There is no impurity phases was detected within the XRD detection limits.

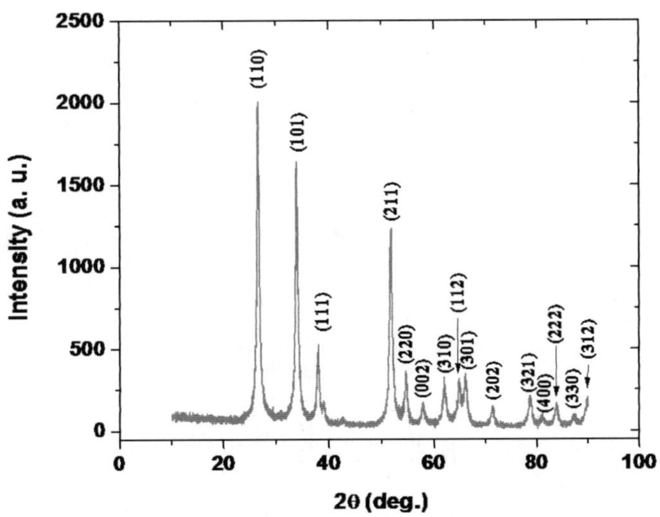

Figure 2. XRD pattern of synthesized SnO_2 nanopowders.

The average crystallite size of the nanopowders are estimated as ~29 nm from Williamson and Hall (W-H) equation for Cauchy-Lorentizian is given by [12],

$$\beta\cos\theta = \frac{C\lambda}{D} + 4\varepsilon\sin\theta \quad (1)$$

where, D is the crystallite size, C is the shape factor usually taken to be 0.89, β is full width at half maximum (FWHM), θ is Bragg diffraction angle and λ is the wavelength of radiation (λ=1.54 Å). Figure 3a and 3b shows the surface morphology of the SnO_2 nanopowders at different magnification. It exhibits the formation of spherical shaped nanoparticles having an average size of 275 nm which are uniformly distributed.

Figure 3. SEM micrographs of SnO_2 nanopowders at two different magnification (a) 500 nm and (b) 300 nm.

The optical absorption of the synthesized SnO_2 nanopowders is given in Figure 4a. It shows strong absorption in the visible region (285 nm). The energy band gap of the sample was examined by Tauc's relation. According to this relation, the absorption co-efficient is given by,

$$\alpha(h\nu) = C(h\nu - E_g)^m \tag{2}$$

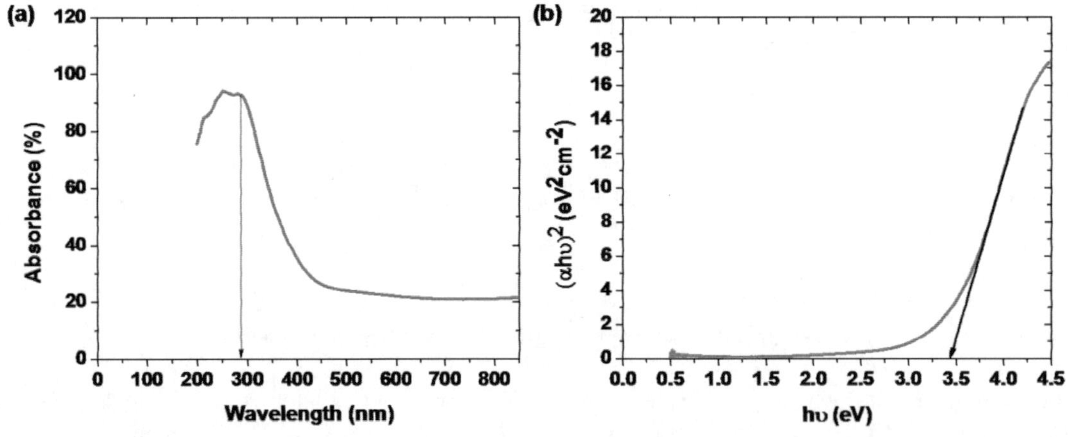

Figure 4. (a) Absorption spectra of SnO_2 nanopowders and (b) Tauc's plot.

where, C is a constant, E_g is the energy gap, hν is the photon energy and m is an exponent which assumes the value depends on the electronic transition. In order to determine the optical energy gap of the SnO_2 sample, the graph of hν versus (αhν) has been plotted by replacing n=1/m where m takes the value of 2 for allowed direct transition [13]. From Figure 4b, the value of band gap for the prepared SnO_2 nanopowder was found as ~3.4 eV.

The optical fiber works on the principle of total internal reflection between core and cladding. Generally, the light intensity is not fully reflected but a small amount of light intensity pierces into the cladding region and its intensity decays exponentially away from the interface (it is called evanescent field) [14]. The cladding region which is having refraction index of 1.492 was replaced with the sensing material having refractive index of 2.0. In such condition the fiber will operate in leaky mode. The change in output characteristics may be attributed to the change in refractive index of new cladding region which may increase or decrease from 2.0.

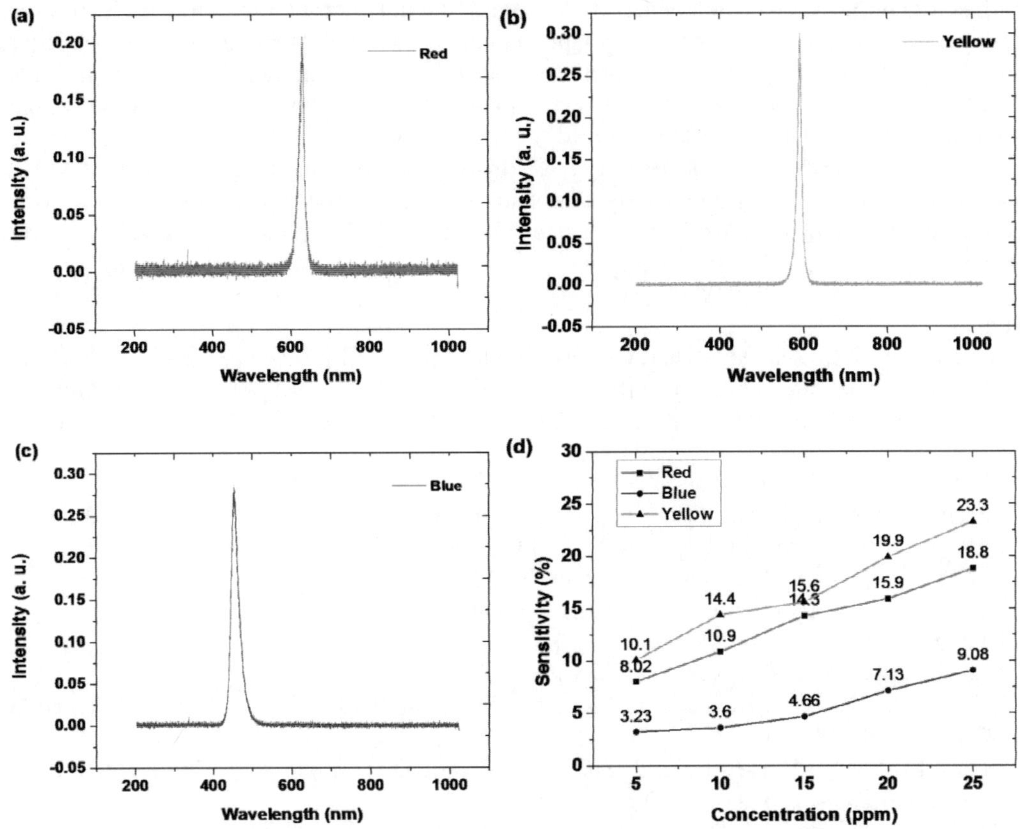

Figure 5. Spectral response and sensitivity of SnO_2 nanopowders for various concentration of ammonia at different wavelength ranges (a) Red (b) Yellow and (c) Blue. (d) Variation of sensitivity at different concentration of ammonia and at different wavelengths.

Figure 5a, 5b and 5c shows the output spectral characteristics of SnO_2 nanopowders upon exposure to different concentration (0 to 250 ppm) of ammonia at room temperature. The spectra exhibits sharp peak for different LEDs of red, yellow and blue around 620 nm, 588 nm and 450 nm, respectively. This peak intensity expresses large variation with the concentration of the gas.

The sensitivity was calculated using the given formula [15],

$$S(\%) = \frac{\left| R_{gas} - R_{air} \right|}{R_{air}} \quad (3)$$

where, R_{air} is the intensity in air ambience and R_{gas} is the intensity in presence of gas. The ammonia sensing properties of the clad modified fiber was studied at different wavelength of 620 nm, 580 nm and 450 nm for red, yellow and blue LEDs, respectively. From the Fig. 5d, it is seen that for all the gases sensitivity increases with the increase in concentration and shows highest sensitivity (~23) in yellow wavelength range compared with red and blue wavelength ranges at room temperature.

4. Conclusion

The rutile structured SnO_2 nanopowders were synthesized by co-precipitation method. The SEM image showed the formation of spherical shaped nanoparticles with the average size of 275 nm. The spectral characteristics of fiber optic sensor for the detection of ammonia in different wavelength ranges (red, yellow and blue) were studied at room temperature. It shows linear increase in sensitivity for all the wavelengths with increase in ammonia concentration. It has been found that the synthesized SnO_2 nanoparticles shows highest sensitivity (~23) in yellow wavelength range compared with red and blue wavelength ranges. This study clearly demonstrates that the fiber optic sensor works successfully for the detection of ammonia using SnO_2 as the modified cladding material in yellow wavelength region.

References

[1] Agency for Toxic substances and Disease Registry (ATSDR) 2004 Toxiological profile for ammonia. Atlanta GA:U.S. Department of Health and Human Services Public Health Service
[2] Pourfayaz F, Khodadadi A, Mortazavi Y, and Mohajerzadez S S 2005 *Sens. Actuators, B* **108** 172
[3] Zhang M, Yuan Z, Song J, and Zheng C 2010 *Sens. Actuators*, B **148** 87
[4] Ahmed F, Arshi N, Anwar M S, Danish R, and Koo B H 2013 *Curr. Appl. Phys.* **13** 64
[5] Fernandez Vallejo M, and Lopez Amo M 2012 *Sensors* **12** 3929
[6] Cao W and Duan Y 2005 *Sens. Actuators, B* **110** 252
[7] Xu J, Han J, Zhang Y, Sun Y A, and Xie B 2008 *Sens. Actuators, B* **132** 334
[8] Huang J, and Wan Q 2009 *Sensors* **9** 9903
[9] Yuan J, and El-Sherif M A 2003 *IEEE Sens. J.* **3** 05
[10] Vijayan A, Fuke M, Hawaldar R, Kulkarni M, Amalnerkar D, and Aiyer R C 2008 *Sens. Actuators, B* **129** 106
[11] Tian Z M, Yuan S L, He J H, Li P, Zhang S Q, Wang C H, Wang Y Q, Yin S Y, and Liu L 2008 *J. Alloys Compd.* **466** 26
[12] Zak K, Abd. Majid W H, Abrishami M E, and Yosefi R 2011 *Solid State Sci.* **13** 251
[13] Yang J, Wang Y, Kong J, Yu M, and Jin H 2016 *J. Alloys Compd.* **657** 261
[14] Morisawa M, Amemiya Y, Kohzu H, Liang C X, and Muto S 2001 *Meas. Sci. Technol.* **12** 877
[15] Liu X, Chen N, Han B, Xiao X, Chen G, Djerdz I, and Wang Y 2015 *Nanoscale* **7** 14872

Acknowledgments

The authors express their sincere thanks to Department of Science and Technology (DST), New Delhi, India for providing the financial support through FIST (Fund for Improvement of S&T Infrastructure in Higher Education Institution) project [SR/FST/ETI-015/2011].

Synthesis and UV detection characteristics of TiO_2 thin film prepared through sol gel route

Parthasarathy P

Research Scholar
School of Electrical Engineering, VIT University, Vellore 632 014, India.

Email: arjunsarathii@gmail.com

Abstract. In this work, TiO_2 thin films were synthesized by sol-gel route initiating titanium isopropoxide and ethanol mixture as a precursor solution. Crystallization and Phase transitions of thin films were studied by X-ray diffraction (XRD). The XRD spectrum shows polycrystalline anatase phase without any considerable impurity phase. UV-VIS spectroscopy of the films shows 70-80% of transmission in the visible region. Optical band gap was calculated from the Tauc's plot and was found to be around 3.3 eV. Scanning electron microscope was used to study the morphological characteristics of TiO_2 thin films. The AFM images indicate that the surface morphology of the film is found to be uniform. Above the film, the aluminum interdigitated electrodes were deposited by thermal evaporation technique. The photo response of the film was measured at different incident powers. The film showed a good photo response with linear behavior for the incident light intensity and applied bias. The better performance of the presented TiO_2 detector suggests that it will have promising potential for UV detector applications.
Keywords: TiO_2, thin film, spin coating, UV detector.

.

1. Introduction

Detection of ultraviolet (UV) radiation is becoming increasingly important in the field of solar UV monitoring, biological research, and flame alarm. UV detectors based on wide band gap semiconductors such as SiC [13], GaN [14], and ZnO [15] show excellent wavelength selectivity and possibility of room temperature operation. These detectors exhibit good detection performance and operation. The complexity of the fabrication and the high cost of single wafer raise the cost of products and reduce their commercial applications. Tio2 is also a well-known wide band gap semiconductor. It possess an excellent physical, chemical and optical properties and has been employed extensively in the field of photo catalysis, gas sensors, dye sensitized solar cells, and so on. Furthermore, the wide band gap (3.2 eV for anatase and 3.0 eV for rutile structure) make it simultaneously very suitable for UV detection applications. Recently, Tio_2 based detectors, especially the metal-semiconductor-metal (MSM) structure detectors [10, 11], have drawn much attention due to the simplicity in structural design, ease of optoelectronic integration, and low cost of the wafers. Thus, they are predominantly used in the commercial applications. High UV responsivity is achieved in these devices, while large dark current and undesired persistent photoconductive effect are also produced due to the small crystallinity and the sub stoichiometry in the growth of the TiO_2 films [10]. Actually, achievement of low dark current is crucial to produce UV detectors with large signal-to-noise ratio and high operation reliability, and reduction of the photoconductive gain is helpful to keep the device stability in operation. To improve the combined device performances, it's necessary to further improve the crystallinity of the TiO_2 films and reduce the oxygen vacancies at the film surface during the material growth and device fabrication process. Sol gel synthesized spin coating technique and optimized annealing conditions are employed to improve the quality of the TiO_2 films in this paper. Sol-gel synthesized TiO_2 thin film based photo detectors have drawn much attention due to the simplicity in structural design, ease of optoelectronic integration, and low cost of wafers. Good quality of TiO_2 thin films can be easily obtained by sol gel spin coating method. Therefore, high photo electric characteristics would be possible when UV detectors based on TiO_2 are realized. TiO_2 films are known to have three crystalline polymorphs, namely: anatase, rutile, and brookite. Anatase TiO_2 films are proven to have wider band gap, larger electron mobility, and lower defect concentration at the film surface, and hence are preferred in the fabrication of the UV detectors. In this work, TiO_2 thin films were grown on glass substrate and TiO_2 detector with aluminum interdigitated electrodes were fabricated. The TiO_2 photo detector exhibit a good photo response in the UV region and sensitivity was obtained.

2. Experimental Details

Titanium dioxide (TiO_2) sol were synthesized through a sol-gel method using titanium isopropoxide (TTIP, $Ti[OCH(CH_3)_2]_4$) as precursor and ethanol (CH_3CH_2OH) acts as a solvent. Briefly, the titanium isopropoxide of 4.2 ml was dissolved in 25 ml of ethanol. The solution was vigorously stirred for 30 min in order to form sols. Nitric acid was used to adjust the pH and for restrain the hydrolysis process of the solution. The mixture of these solutions was stirred for (6-8) hours at ambient temperature. After aging for 24 hours, the sols were transformed into gels. TiO_2 thin films were deposited on glass substrate using spin coating technique. First the substrate is cleaned thoroughly with HNO_3, then in acetone and ethanol for several minutes with ultrasonic agitation. Then the substrates were spin coated in the solution at a constant arte of 3000 rpm. After each coating, the films pre-heated for 10 min at 200 °C. Subsequently, the thin films were heat treated in the temperature range 350-450 °C for 2 hours in the furnace.

Interdigitated Al contacts were developed over the TiO_2 film with diverse finger spacing of 1 mm. The fingers of those contacts were 1 mm width, 11.5 mm length, and 20 mm height. The device was fabricated by PVD method. Under the UV irradiation with the different incident powers of 2W, 4W, 6W, 8W, 10W high energy photons will be absorbed by the TiO_2 deposited layer. With an appropriate bias, photon generated carriers will drift towards the contact electrodes and photo current will be measured. Current-voltage (I-V) measurements were performed to investigate the electrical properties of TiO_2 thin film. I-V characteristics of the TiO_2 detector were measured by applying bias voltage from 0 to 5V range using the instrument model B2901A, precision source/measure unit (Agilent). Fig. 6 shows that the sensitivity of the synthesized TiO_2 thin films was increases with an increase in different incident power from 2W to 10W. Dark current measurements were done at room temperature when the samples were kept inside a closed box.

IOP Conf. Series: Materials Science and Engineering **360** (2018) 012056 doi:10.1088/1757-899X/360/1/012056

3. Results and Discussion

XRD analysis was carried out to study the crystal structure of TiO_2 thin films. Fig.1. shows that XRD pattern of synthesized TiO_2 thin films are tetragonal in structure with anatase phase (JCPDS card no-89-4921), and a preferred grain orientation in the (101) direction. The crystallite size was determined to be around 7.9 nm by applying the Scherrer's equation to the major peaks of the diffraction pattern:

$$D = \frac{K\lambda}{\beta cos\theta} \quad (1)$$

where D is average crystallite size in angstroms (A), K is shape factor taken as 0.9, λ is the wavelength of X-ray radiation (λ = 1.54 A), β is full width at half maximum after making approximate baseline correction, θ is the diffraction angle at the position of highest peak.

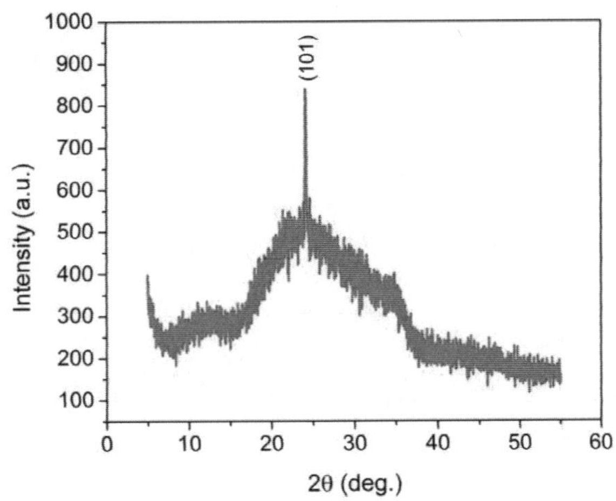

Figure 1. XRD Spectrum of TiO_2 thin films annealed at 400 °C

The absorption was recorded at a range of 100 – 1200 nm. The spectrum (Fig.2) shows sharp absorption around 370 nm. To measure the energy band gap from absorption spectra, a graph $(\alpha h\upsilon)^{1/2}$ vs photon energy (hυ) was plotted and the band gap of TiO_2 was determined to be around 3.3 eV which is shown in the fig.3. As an indirect band gap semiconductor, TiO_2 has an absorption coefficient α obeying the following relation,

$$\alpha(h\upsilon) = C\,(h\upsilon\text{-}Eg)^{1/2} \quad (2)$$

where, C is constant, Eg is band gap energy, and υ is the photon frequency.

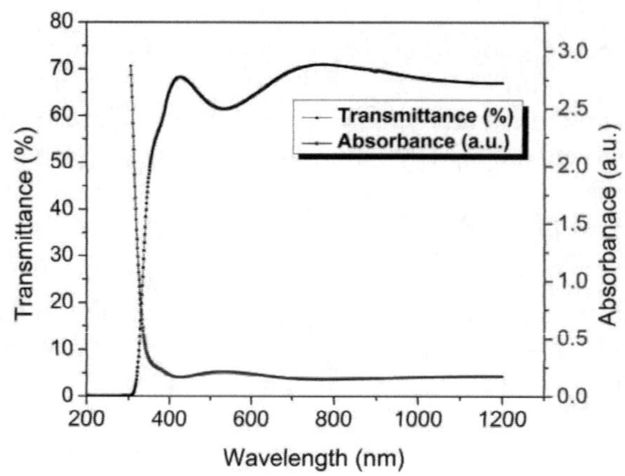

Figure 2. Optical Transmittance and Absorption spectrum of TiO_2 thin film

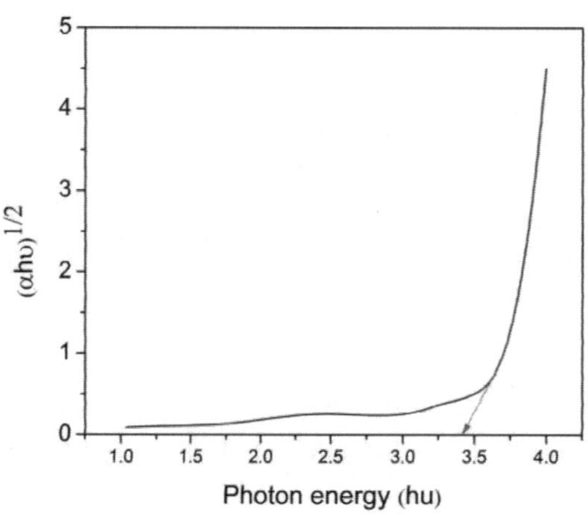

Figure 3. Tauc's Plot of TiO_2

Fig.4. shows the SEM micrographs of synthesized TiO_2 thin film. As a result, the average grain size of the film was measured around 120 nm.

Figure 4. SEM micrographs of synthesized TiO2 film used in the fabrication of UV photo detector

Atomic force microscopy (AFM) is a convenient technique to study the surface morphology of thin films. The surface roughness was examined by recording AFM images in non-contact mode with a micro fabricated cantilever. The roughness was determined around 54 nm and 14 nm for two different scan rates using the commercial software (Nanosurf Easyscan 2). Fig5. Shows the three-dimensional AFM images (a – 1 µm) & (b - 183 nm) of the surface of TiO_2 thin film on glass substrate. As to the analysis, crack free and homogeneous films were synthesized.

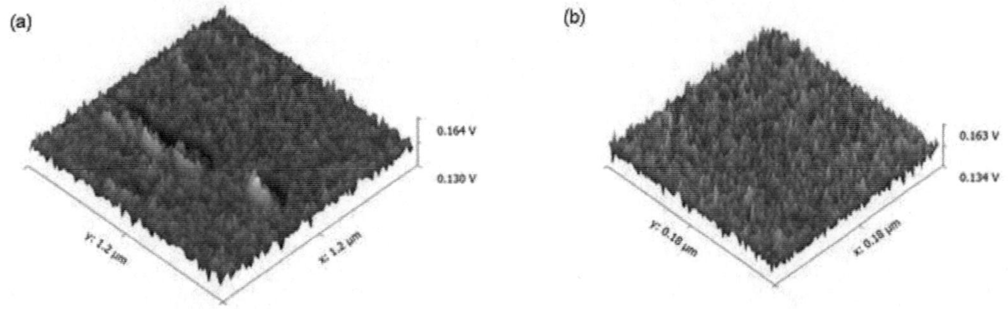

Figure 5. Three-dimensional AFM images (a – 1 µm) & (b - 183 nm) of the surface of TiO_2 thin film on glass substrate

In the absence of light (i.e. in dark), an intrinsic semiconductor or pure single crystal may exhibit negligible electrical conductivity because the energy of the incident photon is very small as compared to the energy gap (i.e. $h\upsilon << E_g$) between the top of the valence band and bottom of the conduction band. When UV radiation is incident there is a considerable increase in electrical conductivity were observed. Thus the photoconductivity is the increase in the electrical conductivity of certain insulating crystalline solids, as a result of exposure to UV radiation. The increase in the electrical conductivity is due to the production of electron-hole pairs by the absorbed photons. When light impinges onto the TiO2 layer, the photons were absorbed by the TiO2 deposited layer. With an appropriate bias, photo generated carriers drift towards the contact electrodes and the photo current was measured.

The sensitivity was determined using the following formula,

$$S = \frac{R_d \sim R_l}{R_d} * 100$$

where, R_d is the resistance in the dark and R_l is the resistance under UV illumination. In the present work, the photosensitivity of the TiO_2 thin films was measured. Fig. 6 shows that the sensitivity of TiO_2 film with different incident powers.

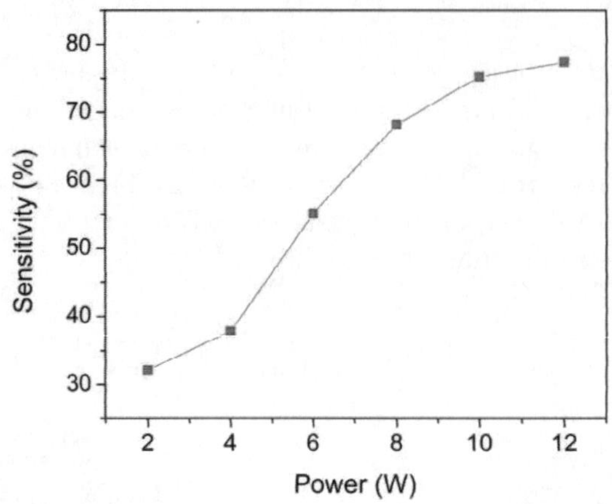

Figure 6. Sensitivity plot of TiO2 thin film under the UV illumination of different incident powers.

4. Conclusion

UV photodetectors were fabricated using TiO2 thin films, and their properties were investigated. TiO_2 thin films were synthesized by sol gel spin coating method. It addresses good crystal structure and anatase phase of TiO_2 with tetragonal structure. The optical transmittances of synthesized TiO_2 films were determined around 70 % and the optical absorption spectrum indicates a sufficient UV photon absorption of the films. The photoresponsivity characteristics implies that the TiO2 thin films are highly sensitive towards UV illumination and the sensitivity plot shows the linear in both dark and UV light. Considering the combined characteristics of the synthesized TiO_2 based detectors, ease of fabrication process, and low cost, the detectors are very promising for various commercial and military applications.

5. Acknowledgments

The authors express their sincere thanks to Department of Science and Technology (DST), New Delhi, India for providing the financial support through FIST (Fund for Improvement of S&T Infrastructure in Higher Education Institution) project [SR/FST/ETI-015/2011].

References

[1] Huolin Huang, Yannan Xie,Zifeng Zhang, feng Zhang, Qiang XU and ZhenyumWu 2014 Applied Surface science **293** 248
[2] Mazhar Ali Abbasi, Zafer Hussain Ibupoto, Azam Khan, Omer Nur, Magnus 2013 Material Letters, **108** 149
[3] Y.Choi, S.Yamamoto, T.Umebayashi, M.yoshikawa 2004 Solid State ionics, **172** 105
[4] 4. Liang-Wenji, Yu-JenHsiao, Sheng joue young, Walter Water, Shiau-Meilin, Wei Shun Shih 2015 IEEE sensors Journal, **15** 2
[5] R.Vijayalakshmi and V.Rajendran 2012 Archives of Applied Science research, **4(2)** 1183
[6] C.Garzella, E.Comini, E.Tempesti, C.Frigeri, G.Sberveglieri 2000 Sensors and Actuators B, **68** 189
[7] D.Basak, G.Amin, B.Mallik, G.K. Paul, S.K.Sen 2003 Journal of Crystal growth, **256** 73
[8] Jin Hyung Jun, Hujun Seong, Kyoungah Cho, Byung-Moo Moon, Sangsig Kim 2009 Ceramics International, **35** 2797

[9] W.c.Lein, D.S.Tsai, D.H.Lein, D.G. Senesky, A.P.pisano 2012 IEEE Electron devices letter, **33** 1586

[10] 10. J.L.Hou, S.J.Chang, M.C.Chen, C.H. Liu, T.J.Hsueh, J.K.Sheu 2013 IEEE Trans, Electron devices, **60** 1178

[11] 11. Young Ran Park and Kwang Joo Kim 2005 Thin solid films, **34-38** 484

[12] Sakae Tanemura, LeiMiao, Wilfried Wunderlich, Masaki Tanemura, Yukimasa mori, ShoichiToh and Kenji Kaneko 2005 Science and Technology of Advanced Materials,**6** 11-17

[13] R.Pandeeswari, R.K.Karan and B.G.jeyaprakash 2014 Sensors and Actuators B, **194** 420-477

[14] M.Miki-Yoshida, M.Rojas villabos, M.Garcia-Guaderrvrama and J.A.Aguilar-Martinez 2015 Journal of alloys and compound, **647**

[15] Liu X, Chen N, Han B, Xiao X, Chen G, Djerdz I, and Wang Y 2015 Nanoscale **7** 14872.

IOP Conf. Series: Materials Science and Engineering **360** (2018) 012057 doi:10.1088/1757-899X/360/1/012057

Crystal growth and optical characterization of an organic single crystal for frequency conversion applications

M. Divya Bharathi[a], G. Ahila[a], J. Mohana[a], R. Bhuvaneswari[a], G. Anbalagan[b*]

[a]Department of Physics, Presidency College, Chennai-600005, Tamil Nadu, India

[b]Department of Nuclear Physics, University of Madras, Chennai-600025, Tamil Nadu, India

[*]anbu24663@yahoo.co.in

Abstract. Organic nonlinear optical 2-methylquinolinium L-malate single crystals have been grown by slow evaporation solution growth technique from a mixture of 2-methylqiunoline and L-malic acid in ethanol solution at ambient temperature. Single crystal X-ray diffraction analysis reveals that grown crystal in monoclinic system with non-centrosymmetric space group $P2_1$ and the lattice parameters are a = 7.35 Å, b = 26.51 Å, c =10.83 Å, $\alpha = \gamma = 90°$, $\beta = 102.95°$ and V = 2057.4 $Å^3$. UV-vis spectrum indicates that the crystal is transparent (75%) in the entire visible region with a cut-off wavelength of 437 nm and optical energy band gap E_g is found to be 2.71 eV. Microhardness measurement reveals the mechanical strength of the grown crystal. The photoluminescence spectrum shows the blue emission of the crystal. Laser damage threshold studies was carried out to ascertain the suitability of grown crystal for laser applications.The relative second harmonic generation efficiency of 2-methlquinolinium L-malate crystal was found to be two times greater than that of KDP.

Keywords: Crystal growth, Photoluminescence, Optical band gap, Laser damage threshold, SHG studies.

1. Introduction

Nonlinear optics (NLO) has emerged in a grand manner with the evolution of material science in the growth of optimistic organic and inorganic single crystals [1]. Second-order NLO materials have evoked much attention because of their application in frequency shifting, optical modulation, signal processing and telecommunications [2]. Organic molecular materials are appealing in controlling a light beam by an electric field because they offer many possibilities to tune their optical properties and obtain large electro-optic effects [3]. Currently, much attention has been given to hybrid organic–inorganic crystals because of their large hyperpolarizabilities and high laser damage resistance compared to inorganic materials [4, 5]. The main advantage of organic materials is that their structures can be easily altered to get the craved optical properties [6, 7]. The hydroxyl groups of carboxylic acids and heterocyclic nitrogen atoms has been proved to be a useful and powerful organizing force for the formation of supramolecules due to its strong hydrogen bonding. It is evident that these structures are normally held together by hydrogen-bonding, and in this contemplate, the most frequently used moieties with hydrogen bonding capability are pyridyl and carboxyl. As a pyridyl derivative, besides the methyl group, 2-methylquinoline bears more aromatic π-electrons which can be a better group in creating aromatic stacking interactions. Many of the dicarboxylic salts are delineated to be active in second-harmonic generation (SHG) and it may be useful to review complexes with carboxylic acids and their properties. The intramolecular hydrogen bond in malic acid is very strong, also malic acid forms crystalline malate of various organic molecules through hydrogen bonding and π–π interactions. It is popular that malic acid acts not only as an acceptor to form various π-stacking complexes with other aromatic molecules but also as an acidic ligand to form salts through determined electrostatic or hydrogen bond interactions. 2-methylquinolinium L-malate (MLM) is one such π donor–acceptor molecular compound in which L-malic acid fetches one of its proton to the 2-methylquinoline, hence the asymmetric unit consists of 2-methylquinoline molecules in protonated form and a L-malic acid in monoionized state. The structure of 2-methylquinolinium L-malate crystal was illustrated by Jing Zhang [8]. In the contemporary work, we report the crystal growth and optical characterization of an organic single crystal for frequency conversion applications.

2. Experimental Procedure

2.1. Material Synthesis

The 2-methylquinolinium L-malate has been synthesized from a solution of ethanol containing 2-methylquinoline (SRL) and L- malic acid (Merck) in 2:1 molar ratio by slow evaporation technique at room temperature. The reaction scheme is shown in Fig.1. The calculated amount of 2-methylquinoline was dissolved separately in the ethanol. L-malic acid was added slowly to the 2-methylquinoline solution with continuous stirring. The prepared mixture was stirred well for 5 h and clear solution was obtained. The solution was filtered off to remove insoluble impurities. Then the solution was allowed to evaporate at room temperature. Optically good quality single crystals were harvested from the mother solution in a growth period of 20 days and the photograph of the grown crystal is given in Fig. 2.

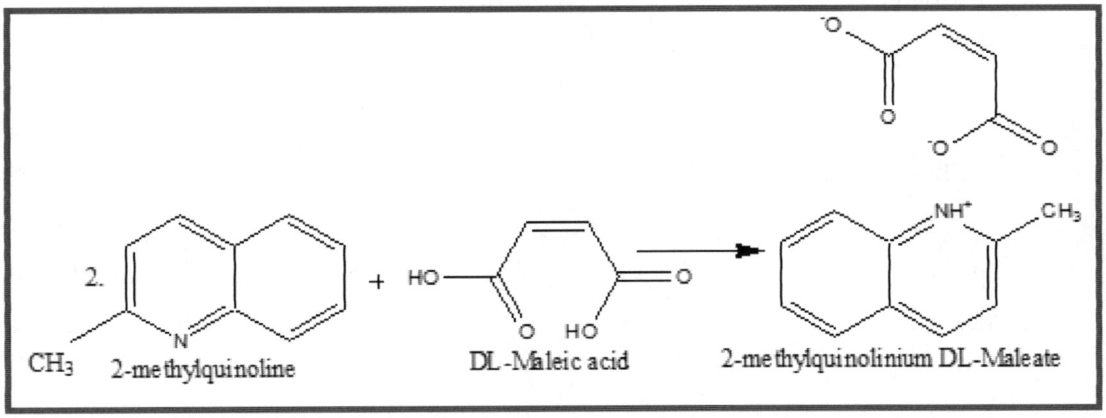

Figure 1. Reaction scheme of MLM crystal

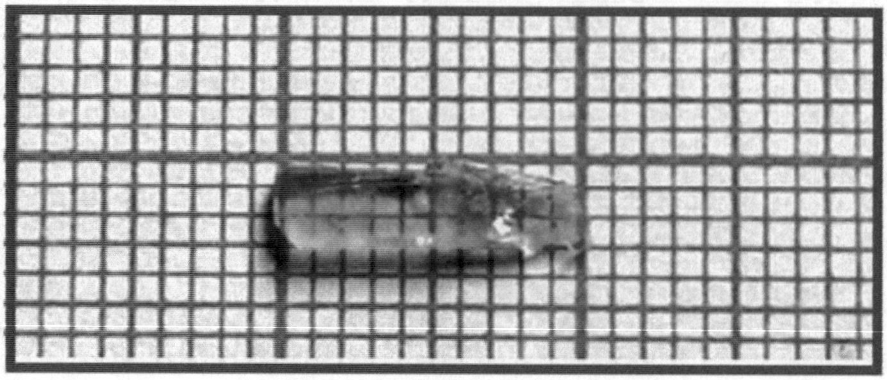

Figure 2. As grown MLM crystal

3. Result and Discussion

3.1. Single Crystal X-Ray Diffraction Analysis

Single crystal X-ray diffraction was analyzed using Bruker AXS Kappa Apex II CCD diffractometer with MoKα radiation (λ=0.71073 Å; graphite monochromator). Single crystal X-ray diffraction analysis reveals that MLM crystal belongs to the monoclinic system with non-centrosymmetric space group $P2_1$. The obtained lattice parameters are presented in Table 1. These values are in good agreement with the literature [8].

Table 1. Crystallographic data of MLM

Parameter	Presented study	Reported values [1]
a	7.3530 Å	7.3530(6) Å
b	26.5102 Å	26.5100(2) Å
c	10.8307 Å	10.8301 (9)Å
α	90 °	90 °
β	102.950°	102.951(2) °
γ	90 °	90 °
V	2057.4 $Å^3$	2057.4(3)$Å^3$
System	Monoclinic	Monoclinic
Space group	$P2_1$	$P2_1$

3.2. Uv- Vis Spectral Studies

UV-vis transmission spectrum of MLM crystal was recorded in the range of 190-900 nm using T-90+ UV-vis Spectrophotometer with sample thickness 1.2 mm. The dependence of optical absorption coefficient with the photon energy helps to understand the band structure and the type of transitions of electrons [9-11]. Fig. 3(a) shows the transmittance spectrum of MLM crystal. It has good transparency of about 75% with cut-off wavelength of 437 nm. This crystal is active in the entire visible region and hence MLM material could be viable alternative for optical materials in the entire visible region and there is no significant absorption between 438 - 900 nm, which enables it to be a potential candidate for optoelectronic and frequency conversion applications.

$$\alpha = \frac{2.303 \log\left(\frac{1}{T}\right)}{t} \tag{1}$$

where T is the transmittance, t is the thickness of the crystal. The quantity α can be displayed in a number of ways as described by the Tauc's relation [12, 13],

$$(\alpha h\upsilon)^2 = A(E_g - h\upsilon) \tag{2}$$

where α is the absorption coefficient, h is Planck constant, υ is the frequency of the incident radiation, A is a constant and E_g represents the optical band gap. A plot of $(\alpha h\upsilon)^2$ versus hυ was drawn by extrapolating the linear part of the plot to the axis (hυ) as shown in Fig.3(b) and the optical band gap was found to be 2.71 eV. As a significance of a wide band gap, the grown crystal has a large transmittance in the visible region.

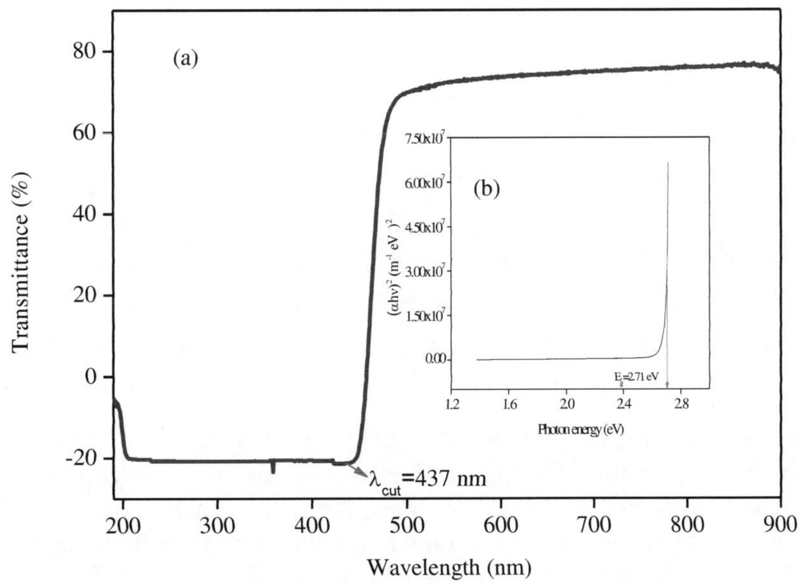

Figure 3(a). UV-Vis transmittance spectrum **(b)** Tauc's Plot of MLM crystal

3.3. Photoluminescence Studies

Photoluminescence (PL) spectroscopy is one of the effective tools to provide relatively direct information about the physical properties of materials at the molecular level, including shallow and deep level defects and band gap states. The photoluminescence (PL) spectrum of MLM recorded in the range between 400-650 nm with excitation wavelength of 437 nm at room temperature is shown in Fig.4. The broad emission spectrum showed a single peak centered at about 2.62 eV (473 nm) and no other visible emission was observed. The enhanced PL emission is due the presence of electron donating group COOH and electron-withdrawing group NH. It also revealed that MLM crystals had blue fluorescence emission.

The maximum intensity peak at 473 nm is due to the protonation of carboxyl group from L-malic acid to 2-methylquinoline of nitrogen group. As the energy of emission peak is lower than the band gap energy, the observed PL is not related to a direct electronic transition between the valence and conduction bands and hence it could be associated with the radiative recombination involving trapped electrons and holes which occur between localized states situated in the band gap [14].The intensity and broadness of peak reveal the crystalline nature.

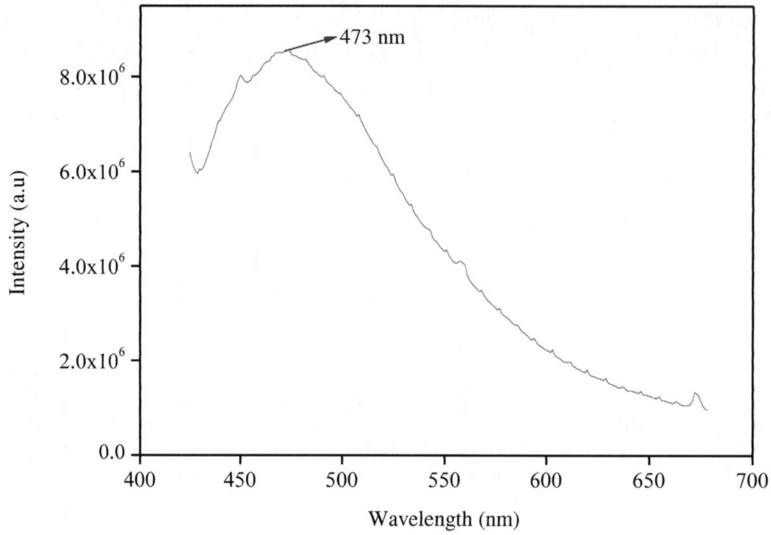

Figure 4. PL emission spectrum of MLM crystal excited at 437 nm

3.4. Mechanical Studies

Hardness is a measure of the resistance offered by a material to the localized plastic deformation caused by scratching [15]. A microhardness study was carried out on the smooth and flat surface of the MLM crystal for various loads ranging from 10 – 100 g with a constant time of indentation (10 s) using Leitz - Wetzlar Vickers microhardness tester fitted with diamond pyramidal indenter attached to an optical microscope. The Vickers microhardness number H_v was calculated using the relation

$$H_V = 1.8544\left(\frac{P}{d^2}\right) kg/mm^2$$

where 'P' is applied load in g and 'd' is the diagonal length of impression in μm. For an applied load above 100 g, the crystal gets crack initiation and materials starts breaking. The variation of microhardness (H_V) with applied load (P) is shown in Fig.5. From the plot it is observed that the hardness number increases with increase of applied load, which reveals that the MLM crystal exhibits reverse indentation size effect (RISE). For load above 100 g crack started developing around the impression mark, which may be due to the release of internal stress [16].

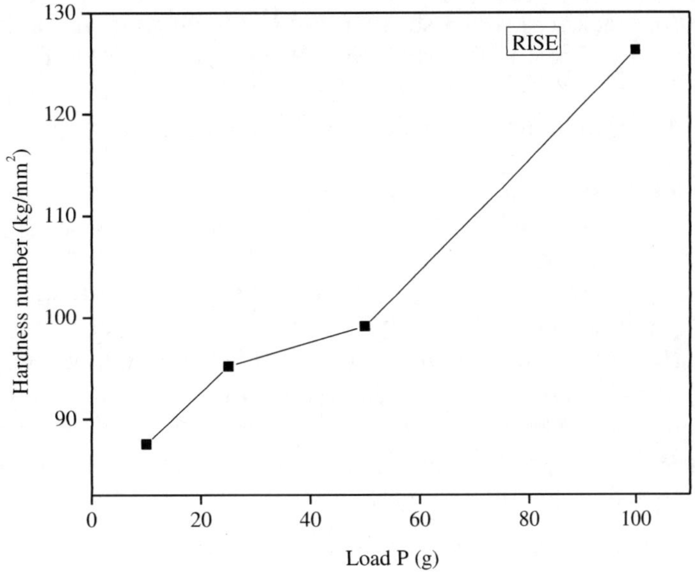

Figure 5. Variation of microhardness number with applied load for MLM crystal

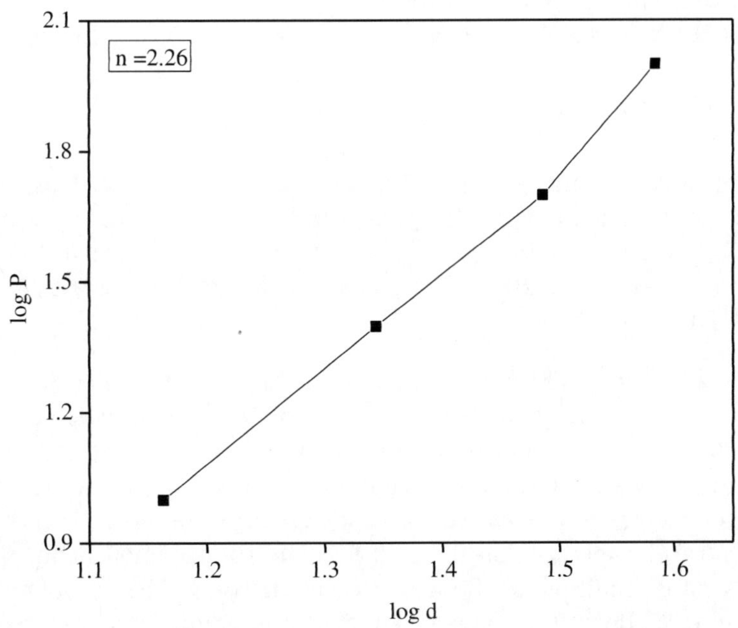

Figure 6. log P with log d of MLM crystal

The size of indentation and load are related through Meyer's law [17]. $P = k_1 d^n$ where 'k_1' is standard hardness constant and 'n' is Meyer's index or work hardening coefficient. A plot of log P versus log d before cracking gives a straight line Fig.6 which is in good agreement with Meyer's law. The slope of the plot is found to be 2.26. According to Onitsch and Hanneman 'n' should lie between 1 and 1.6 for harder materials and above 1.6 softer materials [18, 19]. Hence the grown MLM crystal belongs to the soft material category.

3.5. Laser Damage Threshold Studies

LDT depends on many external factors like repetition rate, spot size, pulse duration, beam size, location of beam, irradiance and experimental geometry. A well polished good quality MLM crystal surface was treated with high power laser and its laser damage resistance values was calculated. In the present study, an actively Q-switched high energy Nd: YAG laser input pulse width of 6 ns at 1064 nm in the repetition rate of 10 Hz. The output intensity of the laser was controlled with a variable attenuator and delivered to the test sample located at the near focus of the converging lens. The energy density of the laser beam was recorded using a power meter which the crystal gets damaged. The surface damage threshold of the MLM crystal was calculated by using the following relation,

$$\text{Power density P (d)} = \frac{E}{\tau \pi r^2}$$

where E is the input energy (mJ), τ is the pulse width (ns) and r is the radius of the circular spot size (cm^2). The measured multiple shot laser damage threshold value is 5.58 GW/cm^2 which is higher than that of KDP and Urea [20]. The surface damage of the crystal using high power laser limits the performance of the NLO applications of the materials.

3.6. Second Harmonic Generation Measurement

The polarizability of MLM and KDP were calculated using Clausius –Mosotti relation:

$$\alpha = \frac{3M}{4\pi\rho N_A}\frac{\varepsilon_r - 1}{\varepsilon_r + 2}$$

where M is the molecular weight, N_A is Avogadro number, ρ is the density and ε_r is the dielectric constant. The calculated values for polarizability for both MLM and KDP are given as 5.0591×10^{-23} C m^{-3} and 2.1499×10^{-23} C m^{-3} respectively. Since the polarizability of MLM is found to be higher than that KDP, the SHG of MLM is theoretically interpreted as higher than that of KDP.

The second harmonic generation efficiency of MLM crystal has been measured by Kurtz–Perry technique [22]. The grown single crystal of MLM is powdered with a uniform particle size and then packed in a micro-capillary tube of uniform bore and exposed to laser radiation. A high intense beam from the Q-switched Nd: YAG laser with the wavelength of about 1064 nm with an input power of 1.2 mJ, and repetition rate of 10 Hz with a pulse of width 8 ns is allowed to be incident on the sample. The fundamental beam was filtered by using an IR filter. A photo-multiplier tube was used as detector. The output from the sample was monochromator to collect the intensity of 532 nm component and to eliminate the fundamental radiation. The efficiency of energy (frequency) conversion is confirmed by the emission of green light from the powder sample. The SHG output for MLM and KDP samples were found to be 200 mV and 100 mV, respectively. Thus, it is observed that SHG efficiency of MLM crystal was found to be two times higher than that of KDP. Hence, the experimental

IOP Conf. Series: Materials Science and Engineering **360** (2018) 012057 doi:10.1088/1757-899X/360/1/012057

result is found to be good agreement with the theoretically predicted value from the polarizability values.

4. Conclusion

Good quality of single crystals 2-methylquinolinium L-malate were grown by the slow evaporation solution growth technique in ethanol as solvent at room temperature. Grown crystals were characterized by single crystal XRD and confirmed that the crystal belongs to monoclinic system with space group $P2_1$. UV-vis study revealed that the crystal possesses good optical transmittance (75%) in the visible region. The optical cut-off wavelength and band gap energy of MLM crystal were found to be 437 nm and 2.71 eV, respectively. The photoluminescence spectrum shows the blue emission for the crystal. Vickers microhardness studies showed that MLM crystal belongs to soft material (n= 2.26) category. The laser damage threshold value for MLM crystal was found to be 5.58 GW/cm^2 and is higher than that of KDP. The relative second harmonic generation efficiency of the MLM crystal was found to be two times greater than that of KDP. Thus, MLM crystal seems to be a promising material for frequency conversion applications.

Acknowledgement

The author acknowledges B.S. Abdur Rahman University, Chennai for providing the laser facility to take laser damage threshold studies.

REFERENCES

[1] Harsh Yadav, Nidhi Sinha, Nidhi Tyagi, Binay Kumar, Cryst. Growth. Des. 15(2015) 4908.

[2] Weigo Zhang, Xutang Tao, Chenqian Zhang, Huaijian Zhang, Minhua Jiang, Cryst. Growth. Des. 9(2009) 2633.

[3] A. Stanculescu, S. Antohe, H.V. Alexandru, L. Tugulea, F. Stanculescu, M. Socol, Synth. Met. 147(2004) 215.

[4] T.L. Fonseca, H.C.B. de Oliveira, O.A.V. Amaral, M.A. Castro, Chem. Phys. Lett. 413 (2005) 356.

[5] S. Manivannan, S. Dhanuskodi, K. Kirschbaum, S.K. Tiwari, Cryst. Growth. Des. 5 (2005) 1463.

[6] J. Zyss, J.F. Nicoud, M. Coquillay, J. Chem. Phys. 81 (1984) 4160.

[7] I. Ledoux, J. Badan, J. Zyss, A. Migus, D. Hulin, J. Etchepare, G. Grillon, A. Antonetti, J. Opt. Soc. Am. B 4(1987) 987.

[8] Jing Zhang, Shouwen Jin, Lin Tao, Bin Liu, Daqi Wang, Journal of Molecular Structure 208 (2014) 1072.

[9] V.G. Drimitriev, G.G. Gurzadyan, D.N. Nikoyosyan, Hand Book of Nonlinear Optical Crystals third ed., Springer, Berlin (1999).

[10] S. Krishnan, C. Justin Raj, S. Dinakaran, S. Jerome Das, Crystal Research and Technology 43(2008) 670.

[11] M. Tariq Bhatti, M. Ishag Raza, A.M. Rana, J. Ros, Journal of Research in Science 15(2004) 369.

[12] J. Tauc, Amorphous and Liquid Semoconductors, Plenum, London, (1994).

[13] J. Tauc, R. Grigovici, A. Vencu, Phys. Status solidi B 15(1996) 627.

[14] G. Shanmugam, S. Brahadeeswaran, Spectrochim. Acta A 95(2012) 177.

[15] P. Rajesh, P. Ramasamy, Binaykumar, G. Bhagavannarayana, Physica B 405(2010) 2401.
[16] P.M. Ushasree, R. Jayavel, C. Subramanian, P. Ramasamy, Bull. Electrochem. 14 (1998) 407.
[17] E. Meyer, Z. Ver, Dtsch, Ing 52(1908) 645.
[18] E.M. Onitsch, Mikroscopia 2(1947) 131.
[19] M. Hanneman, Metall, Manch. 23(1941) 135.
[20] G.C Bhar, A.K. Chaudhury, P. Khumbhakar, Appl. Surf. Sci. 161(2005) 155.
[21] N.L. Boling, M.D. Crisp, G. Dube, Appl. Opt. 12(1973) 650.
[22] S.K. Kurtz, T.T. Perry, J. Appl. Phys. 39(1968) 3798.

IOP Conf. Series: Materials Science and Engineering **360** (2018) 012058 doi:10.1088/1757-899X/360/1/012058

Structural and Optical Properties of ZnS Thin Films by SILAR Technique obtained by acetate Precursor

Ashith V K, Gowrish Rao K
Department of Physics, Manipal Institute of Technology, Manipal University, Udupi, Karnataka 576104, India
Corresponding author: ashithvk@gmail.com

Abstract: In the group of II-VI compound semiconductor, zinc sulphide (ZnS) has numerous potential applications in optoelectronic devices. In this effort, we have prepared ZnS thin films on glass substrates by Successive Ion Layer Adsorption and Reaction (SILAR) method. Zinc acetate and sodium sulphide were used as cationic and anionic precursors for the films. The crystal structure and surface morphology of the films were studied by X-ray diffractometer (XRD) and Scanning electron microscope (SEM). The deposited ZnS thin films showed polycrystalline with cubic phase. The increment in grain size is an effect of the increase in thickness of the films in accordance with the immersion cycles. The SEM images of the films confirmed that films were uniformly distributed in substrates and pin hole free. The band-gap of the films was estimated.
Keywords: Semiconductor, Thin films, SILAR, Band gap.

1. Introduction

In the current development, the wide band gap semiconductor of zinc sulphide (ZnS) thin films are auspicious for optoelectronic device applications, such as electroluminescent devices and solar system [1]. Among the various sulphides, zinc sulphide (ZnS) have been broadly studied from both experimental and theoretical points of view [2, 3]. Epitomizing the results from the literature, it is found that ZnS thin films have been deposited by physical methods like thermal evaporation [4], pulsed laser deposition [5], sputtering [6], and chemical methods like Metal organic chemical vapour deposition [7], electrodeposition [8], sol gel [9], spray pyrolysis [10], chemical bath deposition [11, 12], SILAR [13] etc.
Among these techniques, SILAR (Successive Ionic Layer Adsorption and Reaction) technique is simple and economical because of the utilisation of environment production conditions. SILAR technique, introduced by Nicolau [14], is a unique method in which thin films of compound semiconductors can be deposited by alternate immersing of a substrate into the solutions containing ions of each precursors. This method is improved version of Chemical Bath Deposition. But in this method, the precursor solutions are kept in different beakers. Hence it is easy to control the growth of the film. From the growth cycles, the

controlled thickness can be possible [15, 16]. By tuning the experimental parameters, i.e., deposition cycles, dip duration, etc., the growth rate can be controlled. SILAR offers a quick operation to deposited thin films from an aqueous solution of precursors at room temperature. The tedious dipping process has motivated the development of many automatic deposition apparatus [17].

In the current work, the effects of immersion cycles on the structural and optical properties of SILAR deposited ZnS have been analysed.

2. Experimental details

ZnS thin films were prepared by SILAR method using glass substrates. Before deposition substrates were cleaned in dilute hydrochloric acid, and then in acetone. Later they were rinsed with double distilled water.

For this deposition, cationic precursor used as zinc acetate ($Zn\ (CH_3COO)_2$) and anionic precursor as sodium sulphide (Na_2S). Well cleaned glass substrates were dipped into aqueous solution of zinc acetate. The surface of the substrate was absorbed by zinc ions. Then substrate was rinsed by double distilled water to expulsion of loosely bound Zn^{2+} ions. Next to this, the substrates were dipped into aqueous solution of sodium sulphide. The sulphide ions reacted with adsorbed zinc ions. Finally, to avoid precipitation the substrates were rinsed with distilled water. This is the growth cycle of SILAR method. To obtain desired thickness of the films, these growth cycles have been replicated.

The reactions involved were:

$$Zn\ (CH_3COO)_2\ (\text{Aqueous}) \rightarrow Zn^{2+}\ (\text{Adsorbed}) + 2CH_3COO^-\ (\text{In solution})$$

$$Na_2S\ (\text{Solid}) + H_2O \rightarrow 2Na^+ + SH^-\ \text{-----------}\ (1)$$

$$OH^-\ (\text{In solution}) + SH^-\ (\text{In solution}) \rightarrow S^{2-}\ (\text{In solution}) + H_2O$$

$$Zn^{2+}\ (\text{In solution}) + S^{2-}\ (\text{In solution}) \rightarrow ZnS\ (\text{Adsorbed})\ \text{---------}\ (2)$$

In this work, the prepared ZnS thin films were characterized to study structural and optical properties with the effect of immersion cycles. Powder X-Ray diffraction analysis studied the crystal structure of the films. The surface feature of the films was analysed by a scanning electron microscope. The UV - VIS spectrophotometer was used to measure the absorbance in the wavelength range 300 - 600nm and from these measurements, the band gap of the films was calculated.

3. Results and discussion

ZnS thin film deposited by varying the number of immersion cycles. The thickness of the film was estimated by gravimetric method and verified by cross-sectional SEM. The thickness of the film increased from 210 nm to 400 nm with number of immersion cycles.

The XRD patterns of the ZnS films are shown in Figure 1. which influenced by immersion cycles. The patterns show polycrystalline with cubic phase and peaks corresponds to (111), (220) and (311) planes.

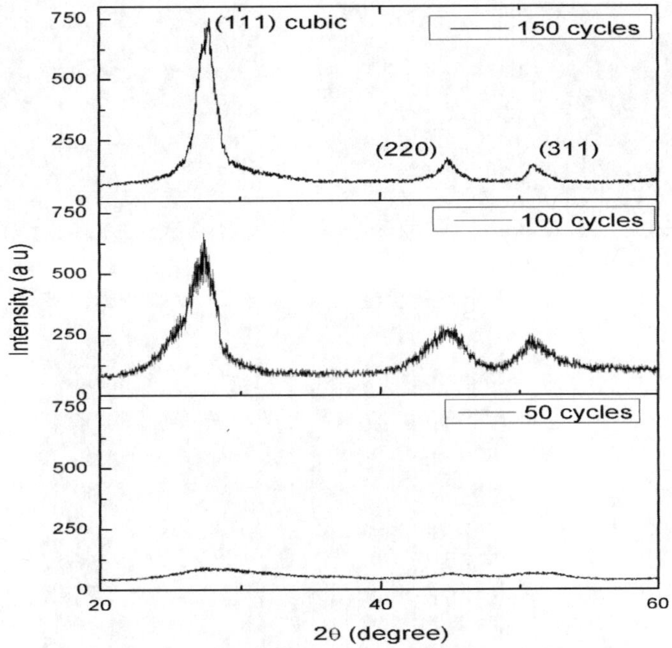

Figure 1. XRD patterns of ZnS thin films with different number of immersion cycles

From Scherrer formula, the grain size 'D' of the films was calculated were shown in Table 1.

$$D = \frac{0.9\lambda}{\beta\cos\theta}$$

Where λ is the wavelength of the x-rays used, β is the full-width-at-half maximum (FWHM) at Bragg angle θ.

Table 1. Grain size of the films with respect to the number of immersion cycles

Immersion cycles	Grain Size (Å)
50	15
100	36
150	49

The deposited films were attributed to the significant improvement in the crystallinity of the films. For lower immersion cycles 50 and 100, the growth rate of ZnS has been slightly lower.

IOP Conf. Series: Materials Science and Engineering **360** (2018) 012058 doi:10.1088/1757-899X/360/1/012058

So, broad peak observed for 100 cycles. For 150 immersion cycles, there was sharp peak observed with increase of crystallinity.

The deposited films were uniform substrate coverage and no void, pin hole free clearly showed in Figure 2.

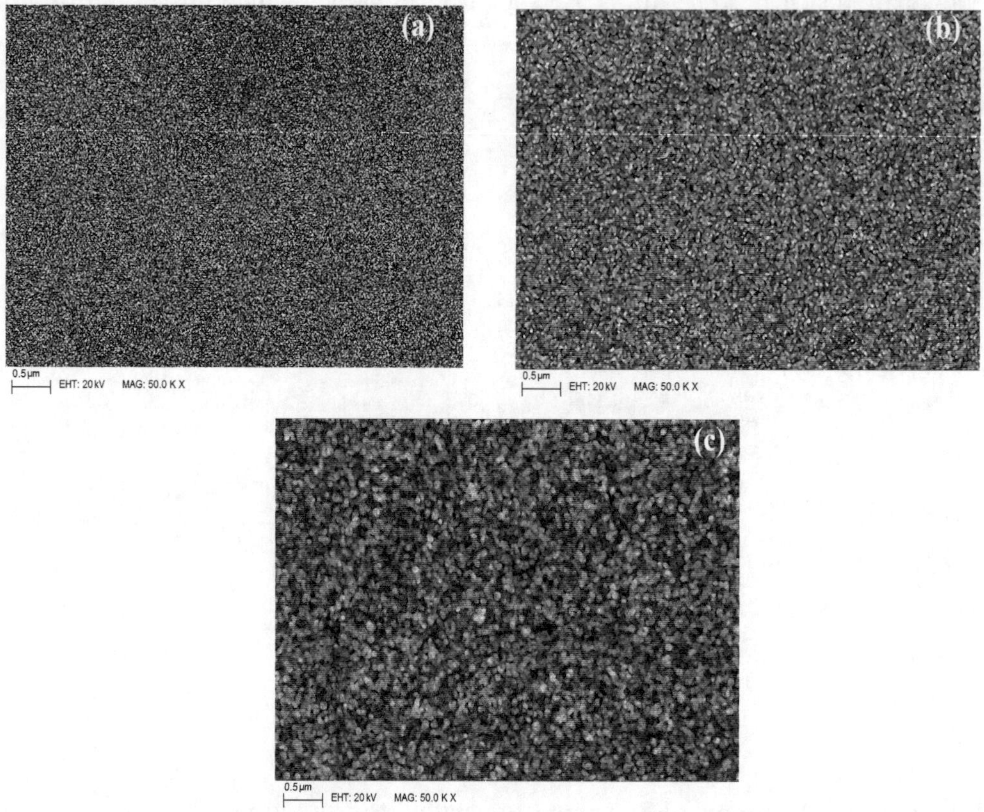

Figure 2. SEM images of ZnS thin films with different number of immersion cycles (a) 50 (b) 100 (c) 150 cycles.

The growth in the grains are visible in the SEM images. The optical properties of the ZnS films were estimated from the wavelength measurements in the range 300–600 nm shown in Figure 3.

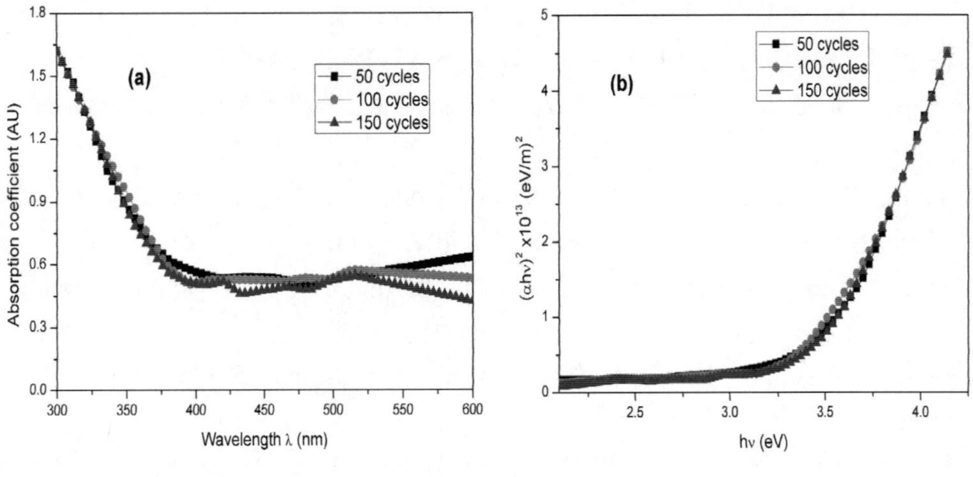

Figure 3. The optical properties of ZnS films deposited for different numbers of immersion cycles: (a) absorbance (b) Tauc plots

The films show high absorption for wavelengths below visible region and band gap of the films observed closer to 3.5 eV.

4. Conclusion

ZnS films were deposited by SILAR technique showed uniform substrate coverage. The thickness of the films can be controlled by the number of immersion cycles. The deposited films showed uniform substrate coverage. Increase in the grain size was observed with the increase in deposition cycles. The bandgap of the films obtained 3.5 eV.

Acknowledgment

The research work presented here is fully funded and MU scholarship by Manipal Institute of Technology and Manipal University, Manipal, India. Part of the characterization work was done in Central Instrumentation Facility, MIT Innovation Centre, Manipal University.

References

[1] S. Yamaga, A. Yoshokawa, H. Kasain, Journal of Crystal Growth, 86, 252, 1998.

[2] S. W. Shin, R. R. Kang, J. H. Yun, A. V. Moholkar, J. H. Moon, J. Y. Lee, J. H. Kim, Solar Energy Materials and Solar Cells, 95, 856, 2011.

[3] N. K. Allouche, T. B. Nasr, N. T. Kamoun, C. Gausch, Materials Chemistry and Physics, 123, 620, 2011.

[4] W. Xiaochun, L. Fachun, L. Limei, Jing Lv, Binping Zhuang, Qu Yan, Zhigao Huang, Applied Surface Science, 254, 6455, 2008.

IOP Conf. Series: Materials Science and Engineering **360** (2018) 012058 doi:10.1088/1757-899X/360/1/012058

[5] M. McLaughlin, H. F. Sakeek, P. Maguire, W. G. Graham, J. Molloy, T. Morrow, S. Laverty, J. Anderson, Applied Physics Letters, 63, 1865, 1993.

[6] Dongjun Yoo, Moon Suk Choi, Seung Chan Heo, Chulwon Chung, Dohyung Kim, Changhwan Choi, Metals and Materials International, 19, 1309, 2013.

[7] J Fang, P H Holloway, J E Yu, K S Jones, B Pathangey, E Brettschneider T J Anderson, Applied Surface Science, 70, 701, 1993.

[8] C. D. Lokhande, M. S. Jadhav, S. H. Pawar, Journal of Electrochemical Society, 136, 2756, 1989.

[9] B. Bhattacharjee, D. Ganguli, K. Iakoubovskii, A. Stesmans, S. Chaudhuri, Bulletin of Materials Science, 25, 175, 2002.

[10] M. C. Lopez, J. P. Espinos, F. Martin, D. Leinen, J. R. Ramos-Barrado, Journal of Crystal Growth, 285, 66, 2005.

[11] Huan Ke, Shuwang Duo, Tingzhi Liu, Qi Sun, Chengxiang Ruan, Xiaoyan Fei, Jilin Tan, Sheng Zhan, Materials Science in Semiconductor Processing, 18, 28, 2014.

[12] Aixiang Wei, Jun Liu, Mixue Zhuang, Yu Zhao, Materials Science in Semiconductor Processing, 16, 1478, 2013.

[13] G. Laukaitis, S. Lindroos, S. Tamulevicius, M. Leskela and M. Rackaitis, Materials Science and Engineering A, 288, 223, 2000.

[14] Y. F. Nicolau, Applied Surface Science, 22, 1061, 1985.

[15] S. Lindroos, T. Kanniainen T, M. Leskela, Applied Surface Science, 75, 70, 1994.

[16] S. Lindroos and M. Leskela, in Solution Processing of Inorganic Materials, Edited D. B. Mitzi, Chap. 8, John Wiley & Sons, New York, 2009.

[17] K. Gowrish Rao, V. K. Ashith, Journal of Physics and Chemistry of Solids, 77, 14, 2015.

IOP Conf. Series: Materials Science and Engineering **360** (2018) 012059 doi:10.1088/1757-899X/360/1/012059

Optical non-linearities by Z-scan measurements, thermal characterization of single crystal: 8-hydroxyquinolinium 3-carboxy-4-hydroxy benzene sulfonate monohydrate

R. Bhuvanesawri[a], M. Divya Bharathi[a], K. Sakthi Murugesan[a*]

[1*] *Departmantof Physics, Presidency College, Chennai-600005, Tamil nadu, India*

[*]ksakthimurugesan2492@gmail.com,

Abstract. Single crystals of 8-hydroxyquinolinium 3-carboxy-4-hydroxy benzene sulfonate monohydrate were grown by slow evaporation technique at room temperature by using ethanol as solvent. Single crystal X-ray diffraction study elucidated that the crystal structure of8-hydroxyquinolinium 3-carboxy-4-hydroxy benzene sulfonate monohydrate belongs to monoclinic crystal system with space group $P2_1/n$. The solid state physical parameters have been also determined for the grown crystal from the single crystal data. The cut-off wavelength and optical band gap energy of grown crystal was found to be 420 nm and 2.67 eV respectively.Thermogravimetric and differential thermal analysis reveals that the 8-hydroxyquinolinium 3-carboxy-4-hydroxy benzene sulfonate monohydrate single crystal is thermally stable up to 144°C without any weight loss. The mechanical properties of the grown crystal were studied by Vickers microhardness technique and it is found that 8-hydroxyquinolinium 3-carboxy-4-hydroxybenzene sulfonate monohydrate belongs to hard material category. The third order nonlinear refractive index (n_2), nonlinear absorption coefficient (β) and nonlinear optical susceptibility (χ^3) of the grown crystal were also measured by Z-scan studies.

Key words: Solution growth, Optical properties, Mechanical studies, Thermal properties

1. Introduction

The potential applications such as optical electronics, laser technology, bio imaging, telecommunication and data processing are relay upon the organic nonlinear optical crystals, hence the researchers have taken keen interest in these fields [1-4]. Organic nonlinear materials are highly active than inorganic materials due to the presence of weak Van der Waals forces and hydrogen bonds. Hence, it provides a high degree of delocalization for charge transfer [5]. Large optical laser damage threshold, low frequency dispersion and response time of nonlinear optical crystals of second and third order polarizability stimulated the investigation towards the synthesis of novel π-conjugated organic molecules [6-8]. 8-hydroxyquinoline is a versatile compound having extraordinary antifungal and antimicrobial activities where the complexes are formed due to variety of charge transfer chromophores [9-12]. The optical nonlinearities of 8-hydroxyquinoline compounds are improved due to increasing acceptor character of pyridine or by increasing the donor character of benzene ring. So, we can ascertain that 8HQ acts both as acceptor as well as donor [13]. The grown crystals were characterized by single crystal X-ray diffraction, UV–Visible spectroscopy, thermal, mechanical,Z-scanstudies and obtained results were discussed.

2. Experimental Procedure:

2.1. Material Synthesis and Crystal Growth

The 8HQ3C4S compound was synthesized using 8-hydroxyquinoline (Merck) 5-sulfosalicyclic acid (Merck) raw materials in 1:1 molar ratio. The calculated amount of

IOP Conf. Series: Materials Science and Engineering **360** (2018) 012059 doi:10.1088/1757-899X/360/1/012059

5-sulfosalicyclic acid and8-hydroxyquinoline were dissolved in ethanol solvent. The solution continuously stirred using a motorized magnetic stirrer to ensure homogenous solution and concentration throughout the volume of the solution. The chemical synthesis scheme of the reaction is shown in Fig.1.The resultant saturated solution kept in a dust free atmosphere at room temperature, allowed for slow evaporation towards bulk growth. A good quality 8HQ4C3S crystal harvested having dimension of 23 × 4 × 4 mm^3 over the period of 15 days. The photograph of the grown crystal is shown in Fig.2.

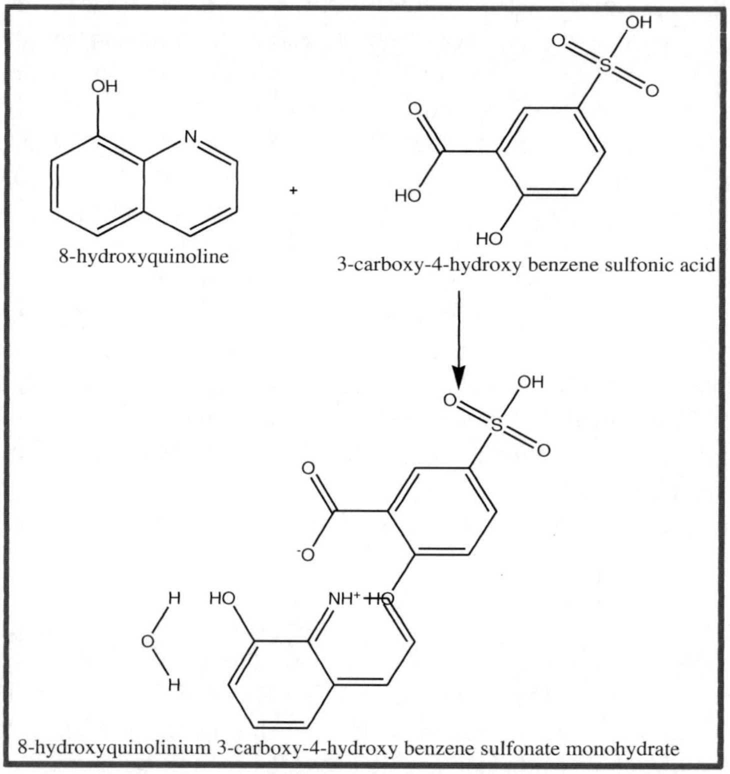

Figure 1. Reaction scheme of 8HQ3C4S crystal

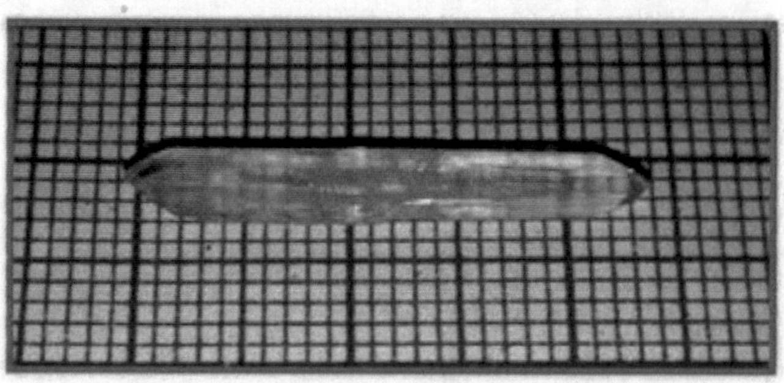

Figure 2. As-grown 8HQ3C4S crystal

3. Results and Discussion

3.1. X-Ray Diffraction Studies

The crystal structure was studied from single-crystal X-ray diffraction analysis using Bruker AXS Kappa Apex II CCD diffractometer with Mo-Kα radiation (λ=0.71073 Å, graphite monochromator). The XRD data reveal that the 8HQ3C4S crystal belongs to the monoclinic system with the space group of $P2_1/n$. The calculated cell parameter values are good agreement with reported values a = 13.24 Å, b = 10.62 Å, c = 13.56 Å, α=γ= 90°, β=119.11° and volume V = 1665 $Å^3$ [14]. The density of grown crystal was found to be ρ = 1.518 mg/m^3 and the dielectric constant at 5 MHz is ε_∞= 46.86. The molecular weight of the 8HQ3C4S crystal is M= 381.35 and the total number of valance electron Z = 198. The valance electron plasma energy, $\hbar\omega_p$ is given by the relation

$$\hbar\omega_p = 28.8\left(\frac{Z\rho}{M}\right)^{1/2} \quad (1)$$

where, Z is the total number of valance electrons, ρ is the density and M is the molecular weight of 8HQ3C4S crystal. The fermi energy in terms of plasma energy is given by the relation [15],

$$E_P = \frac{\hbar\omega_P}{(\varepsilon_\infty - 1)^{1/2}} \quad (2)$$

and

$$E_F = 0.2948\,(\hbar\omega_P)^{4/3} \quad (3)$$

Polarizability α is obtained using the relation [16]

$$\alpha = \left[\frac{(\hbar\omega_p)^2 S_o}{(\hbar\omega_p)^2 S_o + 3E_P{}^2}\right] \times \frac{M}{\rho} \times 0.396 \times 10^{-24} cm^{-3} \quad (4)$$

where, S_0 is a constant for a particular material and is given by

$$S_0 = 1 - \left[\frac{E_P}{4E_F}\right] + \frac{1}{3}\left[\frac{E_P}{4E_F}\right]^2 \quad (5)$$

The value of α so obtained agrees well with that of clausius - Mossotti equation given by,

$$\alpha = \frac{3M}{4\pi N_A \rho}\frac{\varepsilon_\infty - 1}{\varepsilon_\infty + 2} \quad (6)$$

where, N_A is Avogadro number (6.6022 x 10^{-23}) and the calculated fundamental data on the grown crystal of 8HQ3C4S are given in Table.1.

Table.1.Theoretical data for 8HQ3C4S crystal.

Parameters	Values
Plasma energy (eV)	26.07
Penn gap (eV)	1.79
Fermi gap (eV)	22.78
Polarizability (cm^{-3}) using Penn analysis	9.241x 10^{-23}
Polarizability (cm^{-3}) using Clausius-Mossotti equation	9.257x 10^{-23}

3.2. Uv-Visible Spectral Studies

The optical properties of solid provide an important tool for studying energy band structure, impurity levels, lattice vibrations, localized defects, excitons and certain magnetic excitations. The band gap energy and refractive index are the two fundamental physical quantities which characterize the electronic and optical properties of materials. The different arrangement of ions and free charges in a crystal gives raise to the energy band-gap. These lattice arrangements given by the different elements combined (binary, tertiary, etc,) and doping result in different refractive indices, thus light propagates differently in every crystal [17]. The UV-vis transmission spectrum of an 8HQ3C4S crystal was recorded in the range 250-800 nm using a PerkinElmer Lambda 650 instrument spectrophotometer. The crystal has a sufficient transmission in the visible region as shown in Fig.3(a). The lower cut-off wavelength of 8HQ3C4S was found to be 420 nm.

3.2.1. Determination of Optical Band Gap

The measured transmittance (T) was used to calculate the absorption coefficient (α) using the relation,

$$\alpha = \frac{2.3026 \log\left(\frac{1}{T}\right)}{t} \quad (7)$$

The optical band gap (E_g) was evaluated from the transmissions spectrum and the optical absorption coefficient (α) near the absorption edge is given by [18]

$$h\nu\alpha = A\left(h\nu - E_g\right)^{1/2} \quad (8)$$

where $E_g$the optical band gap, A is a constant of the material, h is a Planck's constant and υ the frequency of the incident photons. The optical band gap was calculated from the plot between hυ and $(\alpha h\upsilon)^{1/2}$depicted inFig.3(b) and the band gap energy of the crystal was found to be 2.69 eV.

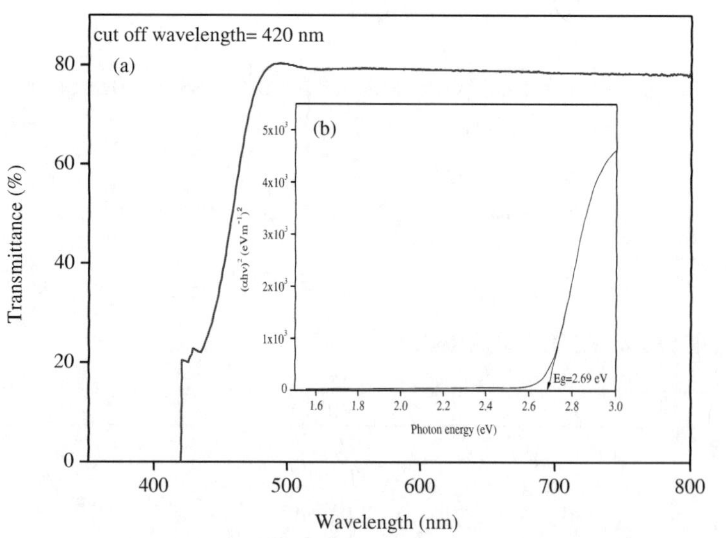

Figure 3(a). UV-vis spectrum (b) Plot of $(\alpha h\upsilon)^2$ vs. photon energy of 8HQS crystal.

IOP Conf. Series: Materials Science and Engineering **360** (2018) 012059 doi:10.1088/1757-899X/360/1/012059

3.2.2. Refractive Index

The refractive index of the material can be measured from the reflectance (R) in terms of absorption coefficient can be obtained from the following relation [19]

$$R = \frac{\exp(-\alpha t) \pm \sqrt{\exp(-\alpha t) - \exp(3\alpha t)T + \exp(-2\alpha t)T^2}}{\exp(-\alpha t) + \exp(-2\alpha t)T} \quad (9)$$

The refractive index (n) can be determined from the reflectance data using

$$n = \frac{-(R+1) \pm 2\sqrt{R}}{(R-1)} \quad (10)$$

The refractive index 1.64 obtained the 8HQ3C4S crystal is used to calculate third order nonlinear successbility.

3.3. Thermal Studies

Thermal stability was analysed for the grown 8HQS crystal by thermogravimetric analysis using the model NETZSCH STA 449F3 in a nitrogen atmosphere at a heating rate 20 K/ min in the range from room temperature to 500°C in the alumina crucible. The thermogravimetric (TG) and differential thermal analysis (DTA) are the important characterization technique to study the thermal stability, melting point, phase transition and different stages of decompositions of the sample. The thermogram TGA and DTA of the compound is shown in Fig.4. From TGA trace, The TG curve shows a three-stage weight loss pattern when the material was heated from 30°C to 500°C. The initial weight loss up to 144°C reveals the presence of 1 mole of water in the lattice. Removal of water is immediately followed by decomposition of 8HQ3C4S and this decomposition extends up to 283°C**.** In DTA curve shows the initial exothermic peak at 128°C is due to desorption of water. The sharp exothermic peak at 223°C is due to the decomposition of the compound. From the thermal analysis the 8HQ3C4S crystal is thermally stable up to 144°C without any weight loss and can be useful for NLO applications.

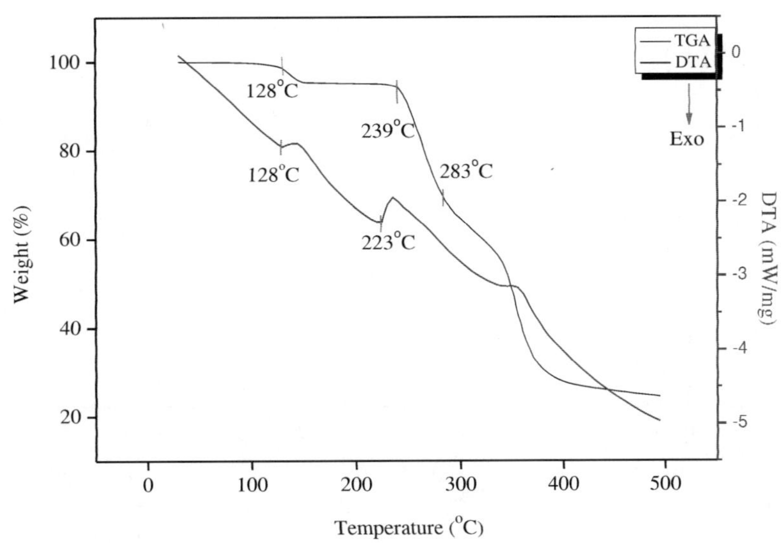

Figure 4. TG-DTA curve of 8HQS compound

IOP Conf. Series: Materials Science and Engineering **360** (2018) 012059 doi:10.1088/1757-899X/360/1/012059

3.4. Mechanical Studies

Hardness of a crystal sample takes over enormous information such as the strength, molecular bindings and elastic constant hence it plays a major role in device fabrication [20-22]. Mechanical behaviour of single crystal is related mainly to the molecular structure and composition of crystalline solids. Mechanical strength of 8HQ3C4S crystal is tested using Vicker's microhardness study. It was carried out on a crack free smooth polished surface of the grown crystal. The indentations were taken for different loads ranging from 10g to 100g with a constant indentation time of 10s at room temperature in all of each case. The standard equation for finding out the Vicker's Microhardness number (H_v) is given by,

$$H_v = 1.8544(P/d^2)\ (kg/\ mm^2) \quad (11)$$

The parameters 'H_v' is the hardness number in kg/mm, 'P' is the applied load in gm and 'd' is the mean diagonal length of indentation impression in mm. The variation of hardness number (H_v) with respect to the applied load (P) is represented in the Fig .5. From the figure it is evident that the microhardness of 8HQ3C4S sample gradually increases with increase in applied loads from 10g to 100g, and on further applying load after 100g the smooth crystal surface breaks due to the external stress created by the indentation [23]. The relation connecting the applied load (P) and the mean diagonal length (d) of the indenter is given by the Meyer's Law as shown below:

$$P = K_1 d^n (12)$$

where, 'k' is the material constant and 'n' is the Meyer's index. This relation is used to plot the graph between log (d) and log (P) as drawn in Fig.6.for the 8HQ3C4S crystal. The slope of this straight line gives the Meyer's index number 'n'. According to the concept of Onitsch 'n' should lie between 1 and 1.6 for harder material and above 1.6 for softer materials [24,25]. The Meyer's index number of 8HQ3C4S sample was calculated to be 1.15, thus the grown 8HQ3C4S crystal belongs to harder material category.

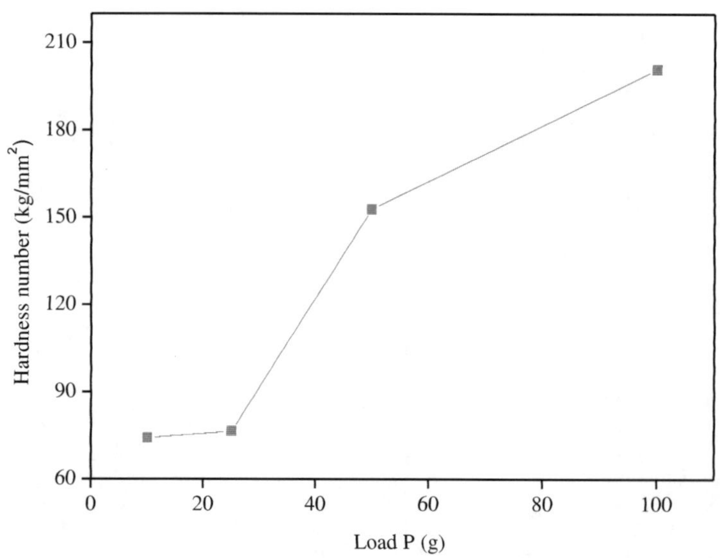

Figure 5. Variation of hardness number (Hv) with respect to the applied load (P).

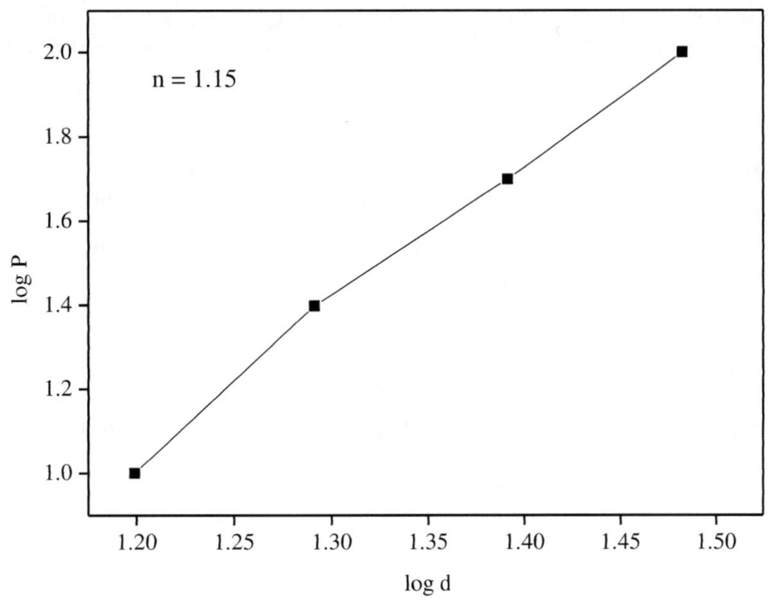

Figure 6. Plot of log P with log d of 8HQS crystal

3.5. Z-Scan Technique

The third order nonlinearities of 8HQS were investigated with Z-scan method by using A CW He-Ne laser of wavelength of 632.8nm. The laser beam was focused to a waist of 45μm with the help of a lens of focal length 12cm to give the intensity of 6.25×10^7 W/m^2 at the focus. The sample is moved along the optic axis (z-direction) through the focus of the lens. The energy transmitted through an aperture is recorded as a function of the sample position. Generally, typical z-scan data with fully open aperture is insensitive to nonlinear refraction [26]. Therefore, the data are expected to be symmetric with respect to focus. For materials with two – photon absorption, there is a minimum transmittance in focus (valley) and for saturable absorber samples there is a maximum transmittance in the focus (peak). The saturation of absorption enhances the peak and suppresses the valley, while saturation produces the opposite effect [27]. It is a successful technique to study the third order NLO properties of the grown crystal, namely, nonlinear refractive index (n_2), nonlinear absorption coefficient (β) and the third order susceptibility($\chi^{(3)}$). In the z-scan measurements, it is assumed that the sample thickness is much less than Rayleigh's range (z_0). It is written as,

$$Zo = \frac{k\omega_o^2}{2} \qquad (13)$$

where,k is the wave vector and ω_0 is the beam waist radius given by

$$\omega_o = \frac{f\lambda}{D} \qquad (14)$$

IOP Conf. Series: Materials Science and Engineering **360** (2018) 012059 doi:10.1088/1757-899X/360/1/012059

Here, D is the beam radius at the lens used, $\lambda$is the wavelength of the source and f is the focal length of the lens.Fig.7. shows the normalized transmission is symmetric for the open curvature (OA). The transmission is symmetric with respect to the focus (Z=0), where it has a minimum transmission. The phase shift at focus (Z=0) is calculated using the relation,

$$\Delta\phi = \frac{BI_o L_{eff}}{Z} \qquad (15)$$

where,L_{eff} is the effective thickness of the sample, I_o is the intensity of the beam focused at the focus, Z is the propagation depth of the light and β is the non-linear absorption coefficient. B and L_{eff} can be calculated using the following relation,

$$\beta = \frac{2\sqrt{2}\Delta T}{I_o L_{eff}} \qquad (16)$$

where ΔT is the value at the open curvature Z-scan curve. L is the linear absorption co-efficient at 632.8 nm and d is the thickness of the sample. The imaginary part of the third order nonlinear is given by

$$\chi_I^{(3)} = \frac{10^{-2}\varepsilon_o c^2 n_o^2 \lambda\beta}{4\pi^2} \qquad (17)$$

where n_o is the linear refractive index (n_o= 1.661 at λ = 632.8 nm), ε_o is the permittivity of free space and c is the velocity of light. The non-linear refractive index of the solid crystal can be calculated using the relation,

$$n_2 = \frac{\Delta\Phi_o}{KI_o L_{eff}} \qquad (18)$$

The real part of the third order non-linear susceptibility is given by the relation,

$$\chi_R^{(3)} = \frac{10^{-4}\left(\varepsilon_o c^2 n_o^2 n_2\right)}{\pi} \qquad (19)$$

The normalized transmittance for closed aperture is shown in Fig.8. The importance of this curve is to understand the sign of the nonlinear refractive index and to calculate the nonlinear susceptibilities. In Fig.8, the peak is followed by valley. This suggests that the nonlinear refractive index and nonlinear susceptibilities are positive exhibiting defocusing effect. The third order nonlinear coefficientis -1.380 $\times 10^{-6}$esu. Theabsolute value of $\chi^{(3)}$ is given as,

$$\chi^{(3)} = \sqrt{\left(\chi_I^{(3)}\right)^2 + \left(\chi_R^{(3)}\right)^2} \qquad (20)$$

The negative sign of the nonlinear absorption coefficient is attributed to the saturation absorption process in 8HQ3C4S material [28]. The third-order nonlinearity of a material is dominated by nonlinear absorption and nonlinear refraction, which lead to the strong optical limiting of the laser. The Z-scan measurements indicate that 8HQ3C4S exhibits negative nonlinear optical properties. It is worth noting that the value of nonlinear susceptibility ($\chi^{(3)}$) for 8HQ3C4S is larger than those of some representative third-order nonlinear optical materials [29-31].

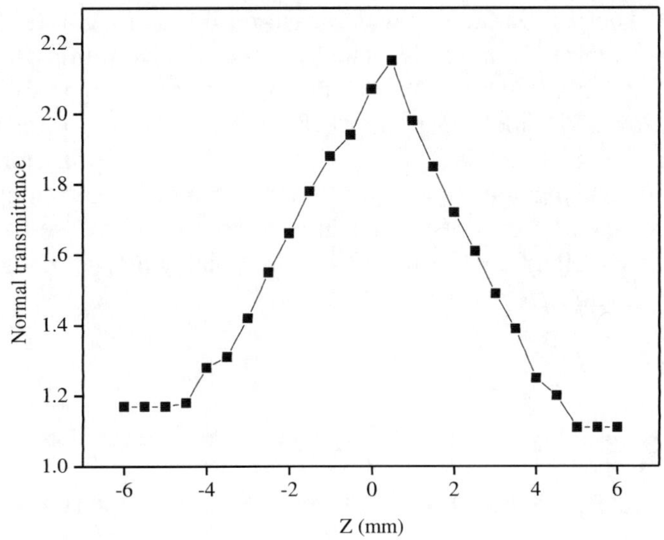

Figure 7. Normalized Transmission for Open aperture (OA)

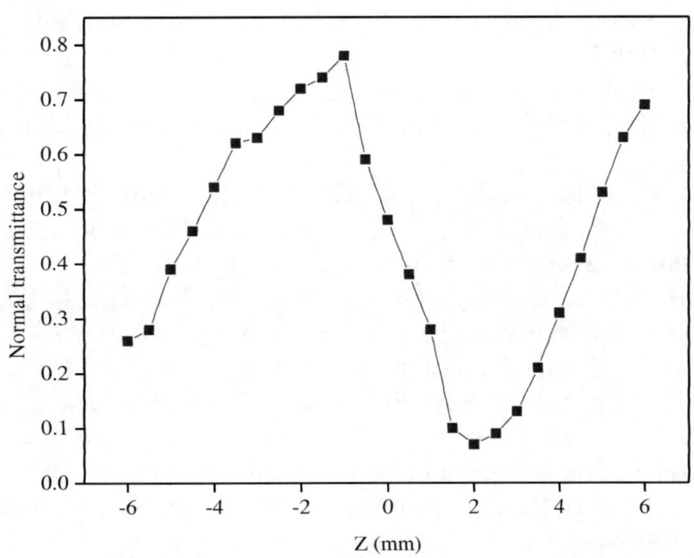

Figure 8. Normalized Transmission for Closed aperture (CA)

IOP Conf. Series: Materials Science and Engineering **360** (2018) 012059 doi:10.1088/1757-899X/360/1/012059

4. Conclusion

Good quality organic NLO crystals of 8HQ3C4S have been successfully grown by slow evaporation technique. The cell parameter was measured by single crystal XRD method and confirms the crystal system as monoclinic. The absence of absorption in the entire visible region in UV–Vis spectrum of 8HQ3C4S makes the crystal in to a suitable material for optoelectronic applications. The melting point of the compound was found to be 223°C from the thermal study. The Vickers microhardness numbers H_v for the crystals were calculated and it increases with increasing load.The third order nonlinear coefficients were estimated using z-scan technique and these coefficients are found to exhibit self-defocusing effect. Hence 8HQ3C4S is an excellent third order NLO material which can find applications in optoelectronic, photonics and holographic applications.

Reference

[1] A.M. Petrosoyan, R.P. Sukiyasan, H.A. Karpetyan, S.S.Tenzya, R.S. Feigeison, J. Cryst. Growth 213, (2000) 103

[2] S.B. Monaco, L.E. Davis, S.P. Velsko, F.T. Wong, D. Eimeri, J. Cryst. Growth. 85, (1987)252

[3] B.B. Bozena, A.V. Bree, B.O. Patrick, J.R. Scheffer, J. Trotter, can. J. Chem. 76 (1998)1616

[4] J.E. Reeve, H.L Anderson, K. Clays, Phys, chem., chem., phys, 12, (2010)13484

[5] R. Santhakumari, K. Ramamurthi, G. Vasuki, B.M. Yamin, G. Bhagvanarayana, Spectrochim. Acta Part Amol. Biomol. Spectrosc, 76, (2010) 369

[6] J. Zyss, J.F. Nicoud, M. Coquillay, J. Chem. Phys, 81, (1984)4160

[7] I. Ledoux, J.Badan, J.Zyss, A. Migus, D. Hulin, J. Etchepare, G. Grillon, A. Antonelli, J. Opt. Soc. Am B4, (1987) 987

[8] R. Gunaseelan, S. Selvakumar, P. Sagayaraj, ScienciaActaXaveriana 3, (2012)45

[9] R. E. Bambury, in: M.E, Wolff (Ed.), Burger's Medicinal chemistry, Part II, John Wiley, New York, (1979)

[10] S. Yamaguchi, M. Goto, H. Takayanagi, H. Ogura, Bull. Chem. Soc. Jpn. 61, (1998)1026

[11] V. Krishna Kumar, R. Nagalakshmi, Spectrochim. Acta Part A 66, (2007)924

[12] I. M. Khan, A. Ahmad, Spectrochim. Acta Part A 63, (2009)966

[13] R. Santhakumari, K. Ramamurthi, G. Vasuki, B.M. Yamin, G. Bhagavannarayana, Spectrochim. Acta Part A: Molecular and Biomolecular Spectroscopy. 76, (2010)369

[14] Graham Smith Urs, D. Wermuth and Jonathan M.WhiteActaCryst. C60, (2004)o575

[15] N.M. Ravindra, R.P. Bharadwaj, K.Sunil Kumar and V.K. Srivastava, Infrared Phys 21, (1981)369

[16] S. Suresh, D. Arivuoli, Journal of Optoelectronics and Biomedical Materials3, (2011)63

[17] V. Sangeetha, K. Gayathri, P. Krishanan, N. Sivakumar, N. Kanagathara, G. Anbalagan, J. Crys. Growth. 389, (2014)30

[18] A. Ashour, N. El. Kadry, S. A. Mahmoud, Thin Solid Flims 269, (1995)117

[19] M. A. Kaid, A. Ashour, Appl.Surf, Sci. 253, (2007) 3029

[20] K. Li. X. Wang, D. Xue, Mater. Focus 1, (2012) 142

[21] K.Li, P, Yang, L, Niu, D. Xue, J. Phys. Chem. A 116, (2012) 6911

[22] K.Li, X. Wang, F.Zhang, D.Xue, Phys.Rev.Lett.100, (2008)235504

[23] T. Umadevi, N. Lawrence, R. Rameshbabu, K. Ramamurthi, J. Cryst. Growth 310, (2008)116

IOP Conf. Series: Materials Science and Engineering **360** (2018) 012059 doi:10.1088/1757-899X/360/1/012059

[24] E.M. Onitsch, Mikroskopie 95 (1998)12.
[25] S. Karan, S.P.S. Gupta, Mater. Sci. Eng. A 398, (2005)198
[26] M. Sheik-Bahae, A. A. Said, T.H. Wei, D.J. Hagan, E.W. Van stryland, J. Quantum Elect.26, (1990)760
[27] G. Peramaiyan, P. Pandi, V. Jayaramakrishnan, R. Subhasis Das, M. Mohan Kumar, Opt.Mater.35, (2012)307
[28] T.C. SabariGirisun, S. Dhanuskodi, Cryst. Res. Technol. 44, (2009)1297
[29] H.J. Ravindra, A. John Kiran, K. Chandrasekaran, H.D. Shashikala, S.M. Dharma prakash, Appl. Phys. B 88, (2007)105
[30] Z. Li, Z.-H. Jin, K. Kasatani, H. Okamoto, S. Takenaka, Jpn. J. Appl. Phys. 44, (2005)4956
[31] P.V. Dhanaraj, N.P. Rajesh, J. Cryst. Growth 318, (2011)974

IOP Conf. Series: Materials Science and Engineering **360** (2018) 012060 doi:10.1088/1757-899X/360/1/012060

Study of effect of temperature on phase transition pressure of GaAs and GaSb

Pooja Pawar[1], Abdul Shukoor V[1], Sadhna Singh[1]
[1]High Pressure Research Lab, Department of Physics, Barkatullah University Bhopal, MP 462026, India

Corresponding author Email: poojapawarphy@gmail.com

Abstract. We have presented the pressure induced phase transition of GaAs and GaSb from zinc blende (ZB) to rocksalt (NaCl) structure associated by using realistic interaction potential model (RIPM), which is modified by taking effect of temperature. This model consists of coulomb interaction, three body interaction and short range overlap repulsive interaction up to second neighbor. Phase transition pressures are associated with a sudden collapse in volume showing the occurrence of first order phase transition. The phase transition pressure (P_t) and volume collapse obtained from this model show a reasonable agreement with experimental and theoretical results.

1. Introduction

During the last generation, III-V semiconductors have been broadly served as substances for electronic and optoelectronic devices. These applications make them as vital elements in an advanced information society [1]. The III–V compounds generally crystallize in zinc-blend (ZB) and wurtzite (W) structure. In this group GaAs has wide applications due to its direct band gap and having a higher saturated electron velocity as well as higher electron mobility, especially where the reliability is a major concern. [2]. Among III–V compounds, the narrow band gap antimonide based compound semiconductors (ABCS) are broadly regarded as the first nominee materials for the manufacturing of the third generation infrared photon detectors and ICs with ultra-high speed and ultra-low power consumption. Their distinctive physical aspects and band gap structures provide a large space to design varied unique devices [3]. They differ structurally only in their third-nearest-neighbour atomic arrangement. Even the difference in total energy between these two phases is very small. The study of materials at high pressure by high pressure technique like diamond anvil technique [4] exhibits new crystal phases and novel behaviour under pressure.

Among these materials, GaAs shows complicated phases under high pressure. As the pressure increased, the experimentally observed phase transition sequence in GaAs is GaAs-I, the ZB structure (space group F-43m), to GaAs- II, the orthorhombic phase (space group Pmm2) [5], which can be seen as a distortion of the rock-salt (RS) structure (space group Fm-3m) [6–8]. First-principles studies had predicted the existence of a field of stability for an sc16 phase (binary analog of the bc8 phase observed in Si and Ge), which was, however, close to the numerical precision of the calculations (9-10).

2. Methodology

The three body interaction effect [11] arising from the deformation of the overlapping electron shells of the adjacent ions in solids is important. This effect is introduced in the expression of Gibbs free energy (G = U+PV–TS), in order to obtain the stability condition for a crystal structure. Here U is the internal energy, which at T= 0K is equivalent to the lattice energy, S is the vibrational entropy at absolute temperature T. Since the theoretical calculation are done at T= 0K, hence the Gibbs free energy is equivalent to enthalpy (H). This is not a realistic approach because experiments are carried out at room temperature and not at T=0K. This fact causes discrepancy in comparability of theoretical results with experimental data. To obtain better comparability with experimental results we have taken account of the room temperature (300K) in pressure induced theoretical calculations.

The Gibbs free energies for ZB (B3) and rock salt (B1) structures at room temperature T=300K are given by:

$$G_{B3}(r) = U_{B3}(r) + PV_{B3} - TS_1 \quad (1)$$

$$G_{B1}(r) = U_{B1}(r) + PV_{B1} - TS_2 \quad (2)$$

With V_{B3} (=3.08r^3) and V_{B1} (=2.00r^3) and as the unit cell volume S1 and S2 are the entropies for B3 and B1 phases, respectively. In fact, the condition for a transition is that the difference in free energy between two phases approaches to zero.

$$\Delta G = \Delta H - T\Delta S \quad (3)$$

The first term in energies of eq. (1) and (2) are lattice energies for B3 and B1 structures. As an isolated phase is stable only if its free energy is at minimum, we have adopted the method of minimization of free energies for both the phases. The pressure at which $\Delta G \rightarrow 0$ is called the transition pressure, shows a sudden collapse in volume.

The first term in eq (1) and eq (2) are lattice energies for B_3 and B_1 structures and they are expressed as,

$$U_{B_3}(r) = -\frac{\alpha_m e^2 Z^2}{r} - \frac{4\alpha_m e^2 Zf(r)}{r} + 4b\beta_{ij}\exp\left[\frac{r_i + r_j - r_{ij}}{\rho}\right] + 6b\beta_{ii}\exp\left[\frac{2r_i - 1.63r_{ij}}{\rho}\right] - 6b\beta_{jj}\exp\left[\frac{2r_j - 1.63r_{ij}}{\rho}\right] \quad (4)$$

$$U_{B_1}(r) = -\frac{\alpha'_M e^2 Z^2}{r'} - \frac{6\alpha'_M e^2 Zf(r')}{r'} + 6b\beta_{ij}\exp\left[\frac{r_i + r_j - r'_{ij}}{\rho}\right] + 6b\beta_{ii}\exp\left[\frac{2r_i - 1.414r'_{ij}}{\rho}\right] + 6b\beta_{jj}\exp\left[\frac{2r_j - 1.414r'_{ij}}{\rho}\right] \quad (5)$$

Here β_{ij} is the Pauling coefficient, r_i (r_j) is the ionic radii of i(j) ions, ρ (b) are the range (hardness) parameters.

3. Table

Table 1. Input parameters and Model parameters of GaAs and GaSb.

Input parameters		Model parameters		
r	B(GPa)	b($\times 10^{-12}$ergs)	ρ(Å)	f(r)
2.45[a]	72.4[b]	3.8012	0.748	0.007256
2.65[a]	55.4[b]	1.740	0.837	-0.0878

[a]Ref.[12], [b]Ref.[13]

Table 2. Calculated transition pressure and volume collapse of GaAs and GaSb.

Compounds	Transition Pressure (GPa)	Volume Collapse (%)
GaAs	11.8	17.34
	12±1.5[a] (exp.)	17.4[b] (exp.)
	11.4[c] (without temp.)	17.3[c] (without temp.)
GaSb	7.6	18.59
	6-8[b] (exp.)	18.6[b] (exp.)
	8[c] (without temp.)	18.5[c] (without temp.)

[a]Ref. [14] , [b]Ref. [15] ,[c]Ref. [13]

IOP Conf. Series: Materials Science and Engineering **360** (2018) 012060 doi:10.1088/1757-899X/360/1/012060

5. Figure

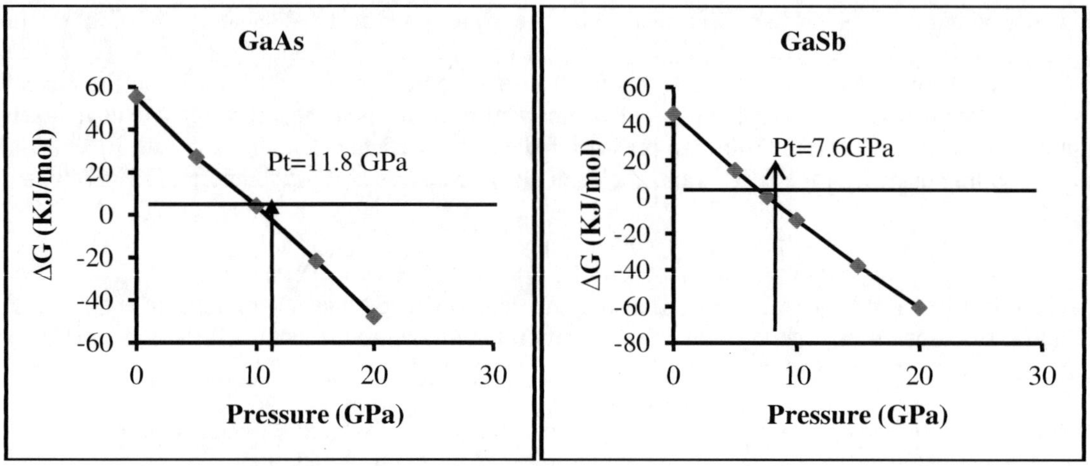

Figure 1.Variation of Gibb's free energy with pressure for GaAs and GaSb.

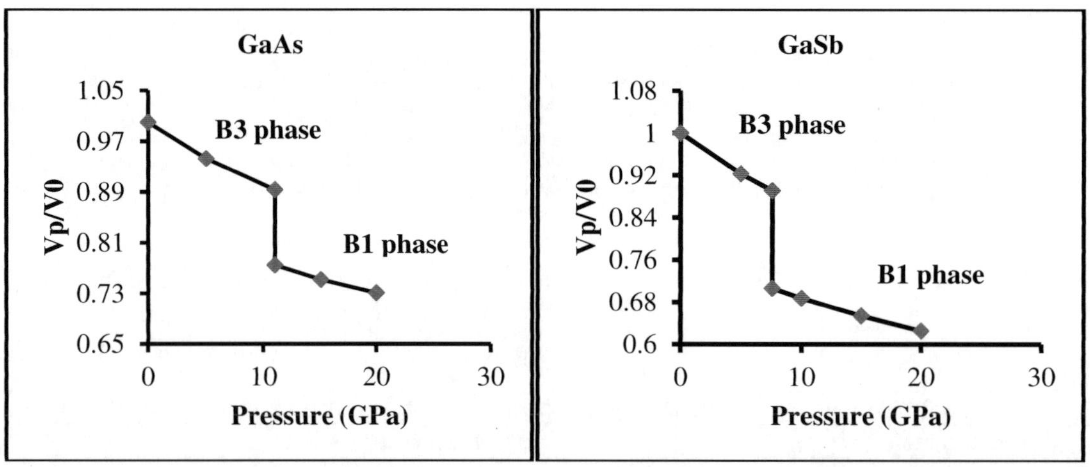

Figure 2. Variation of volume change V_p/V_0 with pressure for GaAs and GaSb.

4. Results and Discussion

The stable phase is associated with minima of energy. The structural phase-transition pressures were calculated by minimizing the lattice energies given in Eqs. (4) and (5) at different pressures in order to obtain the inter-ionic separations r and r' corresponding to B_3 and B_1 phases. Now we have evaluated the corresponding Gibbs free energies $GB_3(r)$ and $GB_1(r')$ and their respective differences $\Delta G = (GB_3(r) - GB_1(r'))$. As the pressure increases this value of ΔG decreases and approaches zero at the transition pressure (Pt). Beyond this pressure, ΔG becomes negative as the phase B_1 becomes stable. This difference has been plotted against pressure (P) as shown in Figure 1 and is indicated by arrows for GaAs and GaSb. It is clear from Figure 1 and Table 1 that phase transition B3 →B1 occurs at about 11.8 GPa for GaAs and 7.6 GPa for GaSb, which is close to experimental results. Besides, the values of relative volume change (V(P, T)/V(0, T)) associated with various compressions have been obtained and plotted against pressure in Figure 2 the values corresponding to phase transition are well listed in Table 2. It is clerc from Table 2 and Figure 2

IOP Conf. Series: Materials Science and Engineering **360** (2018) 012060 doi:10.1088/1757-899X/360/1/012060

that our calculated volume collapses from realistic interaction potential model are 17.34% for GaAs and 18.59% for GaSb, which are close to experimental values and reported by other works (without temperature).

5. Conclusion

Finally, from the present theoretical investigation we have noticed that the crystallographic pressure-induced phase transition from B3 to B1 structure analysis should be done at room temperature rather than at T = 0K. In order to meet the above defined condition, we obtained good comparative results regarding the phase transition pressure and the volume discontinuity in pressure–volume phase diagram with the experimental data using our realistic model.

References

[1] Adachi S 2004 *Handbook on Physical Properties of Semiconductors* (Boston : Kluwer Academic Publishers)
[2] Ferhat M and Zaoui A 2006 *Appl. Phys. Lett.* **88** 161902
[3] Liu C, Li Y and Zeng Y 2010 *Engineering* **2** 617–624
[4] Schilling J S and Shelton R N 1981 *Physics of Solids under Pressure* (New York:North-Holland)
[5] Weir S T, Vohra Y K, Vandenborgh C A and Ruoff A L 1989 *Phys. Rev. B* **39** 1280
[6] Mujica A, Rubio A, Munoz A and Nedds R J 2003 *Rev. Mod. Phys,* **75** 863
[7] Garcia A and Cohen M L 1993 *Phys. Rev. B* **47** 6751
[8] Rino J P, Chatterjee A, Ebbsjo I, Kalia R K, Nakano A, Shimojo F and Vashishta P 2002 *Phys. Rev. B* **65** 195206
[9] Crain J, Piltz P O, Ackland G J, Clark S J, Payne M C, Milman V, Lin J S, Hatton P D and Nam Y H 1994 *Phys. Rev. B* **50** 8389
[10] Mujica A, Needs R J and Munoz A 1995 *Phys. Rev. B* **52** 8881
[11] Singh R K 1982 *Phys. Rep. (Neth.)* **85** 259
[12] Verma A S, Singh R K and Rathi S K 2009 *J. Alloys Comp.***489** 795
[13] Sarwan M and Singh S 2012 *Computational Materials Science* **58** 167–174
[14] Cai J, Chen N and Wang H 2007 *J. Phys. Chem. Solids* **68** 445
[15] Zhang S B and Cohen M L 1987 *Phys. Rev. B* **35** 14

IOP Conf. Series: Materials Science and Engineering **360** (2018) 012061 doi:10.1088/1757-899X/360/1/012061

Isoelectronic Sn-Ge substitution effect on electronic and thermoelectric properties of ZrNiSn compounds: a DFT Study

Sukhwinder Singh

Department of Physics, Panjab University, Chandigarh-160014

*Corresponding author Email: Sukhwinderapsk@gmail.com

Abstract. In this work, we have studied the Isoelectronic Sn-Ge substitution effect on electronic and thermoelectric properties of ZrNiSn compounds using the density functional theory and Boltzmann transport equations. The calculation shows that band gap decreases with doping and exhibit semiconductor nature. The negative value of Seebeck coefficient of all systems indicates that conduction is due to electrons. With increase in concentration of Ge, Seebeck coefficient decreases, while electrical conductivity and electronic thermal conductivity increases. At x=0.125, the maximum value of power is attained at 1200 K, which is equal to $9x10^{11}V^2SK^{-2}ms^{-1}$.

1. Introduction

Intensive attention has recently been paid to energy conversion using thermoelectric principles, which can directly convert both waste heat and solar energy into electricity. At absolute temperature T, the performance of thermoelectric materials is characterized by dimensionless figure of merit [1], which is defined as

$$ZT = \frac{S^2\sigma T}{k} \qquad (1)$$

where S,σ,T,k are Seebeck coefficient, the electrical conductivity, absolute temperature, total thermal conductivity respectively.

Efficient thermoelectric materials have high seebeck coefficient, high electrical conductivity and low lattice thermal conductivity. In present time, discovering new classes of materials are great challenge in the field of thermoelectric research. Half-Heusler [HH] compounds are intermetallic alloys [2,3] and interesting in thermoelectric field due to high seebeck coefficient and good chemical stability. At room temperature, the value of seebeck coefficient of MNiSn (M=Zr , Hf ,Ti) alloys lies in -200 to -400 μV/K (n-type) range[4,5]. The performance of these materials improves by doping. According to the literature, doping on Sn sites tunes the carrier concentration [6] and on M sites gives mass fluctuations which reduce the lattice thermal conductivity [7]. Thus enhance their properties. For pure ZrNiSn compounds, experimentally it was found that the maximum dimensionless figure of merit about 0.3 achieved at 900K [7]. In $Zr_{0.5}Hf_{0.5}NiSn$ compound, maximum ZT value was achieved about 0.5 at 700K [8].Recently, R.Akram et al [9] reported that ZT =0.53 achieved at 923 K for compound $La_{0.01}Zr_{0.99}NiSn$. At 800k, ZT~0.6 was achieved in $Zr_{0.5}Hf_{0.5}NiSn$ while Doping of 20% Pd in this compound enhance the ZT~0.7 at same temperature [10]. Yuanteng liu et al was found that in $Ti_{0.1}Zr_{0.9}NiSn$ compound doping of 2.5% Ge gives 70% reduction in lattice thermal conductivity [11]. At 693k, the maximum dimensionless figure of merit 1.5 was achieved in $Ti_{0.5}(Zr_{0.5}Hf_{0.5})_{0.5}NiSn_{0.998}Sb_{0.002}$ by N.

IOP Conf. Series: Materials Science and Engineering **360** (2018) 012061 doi:10.1088/1757-899X/360/1/012061

Shutoh et al [8]. From previously study, it is confirmed that that ZrNiSn promising acts as good thermoelectric materials like SiGe, PbTe, Bi2Te3 (ZT_{max}~1). On the theoretical side, using ab initio calculations and the Boltzmann calculations, Yang et al [12] was studied electronic transport properties of ZrNiSn compounds. D. F. Zou [13] et al was investigated the electronic structure and thermoelectric properties of $Zr_{0.5}Hf_{0.5}NiSn$ compound. Theoretical research on half heusler materials is very rare.

To the best of our knowledge, Theoretical calculations as well as experimental investigation for this doped system are not available in the literature. This study might be provide guidance to future researchers.

2. Computational Methodology:

ZrNiSn system has the cubic MgAgAs-type structure, corresponding to (group no. 216) space group. To simulate the ZrNiSn1-x Gex compound, we have constructed 2*1*1 super cell which contains 24 atoms (8 Zr atoms, 8 Ni atoms and 8 Sn atoms).

In this study, we adopted same methodology which we have used in our earlier work [14]. The electronic structure calculations were performed using density functional theory (DFT) based on plane wave pseudo potential method as implemented in the Quantum Espresso package [15]. The Local Density approximation (LDA) of Perdew-Wang [16] was used for exchange –correlation functional. The cutoff for the kinetic energy was set to 40 Ry for the plane-wave expansion of the electronic wave functions. The charge-density cutoff was kept at 400Ry and the marzari-vanderbilt cold smearing size was fixed at 0.003 Ry. The Brillouin zone integration was performed using Monkhorst–Pack scheme [17] with 4 × 8 × 8 meshes. The lattice constant of ZrNiSn1-xGex (x=0, 0.125, 0.25) system was optimized until the total energy converged to at least 10-6 Ry, and the forces between atoms became smaller than 10-4 Ry/bohr. For the density of state (DOS) calculation, we used tetrahedron method with 16x16x16 denser k- point mesh.

In order to calculate the thermoelectric transport properties, we have used BoltzTraP code [18] which is based on semi classical theory Boltzmann theory.

3. Results and Discussion

3.1. Electronic Structures

The calculated lattice constant of pure system is 6.15Å and this is in good agreement with the experimental work reported by W. Jeitschko et al [19].The calculated band gap of pure and doped ZrNiSn systems are summarized in Table 1. The calculated band gap of pure ZrNiSn compound is 0.622eV which is comparable with the previous theoretical work [20]. In the experimental work, the band gap of ZrNiSn is 0.18eV [21] which is very small as compared to the calculated value. We calculated the band gap for perfect crystal, that's why our reported value is larger than the experimental value [20]. According to the literature, during the experimental work, the band gap shrink due to structural imperfections such as anti-site defects which are inherent feature of HH compounds [22, 23]. The calculated band structure of $ZrNiSn_{1-x}Ge_x$ (x=0, 0.125, 0.25) systems are depicted in figures 1. In $ZrNiSn_{1-x}Ge_x$ systems, the highest occupied valance band (HOVB) and the lowest unoccupied conduction band (LUCB) are lies at Γ high symmetry point. So that, pure and doped systems are a direct band gap semiconductors. Here occurrence of direct band gap for all systems is due to the band folding because in our work we used supercell. As the concentration of Ge ($0 \leq x \leq 0.250$) increases, the Fermi level shifts from the bottom of Conduction band to the top of valence band as shown in the fig. 2(d) The valence electron of Ge and Sn are equal in number, therefore it neither acts as a donor and acceptor. Due to the lower electronegativity of Sn than the Ge, it loses the electron from the outer shell and thus the carrier concentration increases.

IOP Conf. Series: Materials Science and Engineering **360** (2018) 012061 doi:10.1088/1757-899X/360/1/012061

Table no. 1 Calculated electronic band gaps of $ZrNiSn_{1-x}Ge_x$ in comparison with previous results

System	This work(eV)	Previous calculations(eV)	Experimental values (eV)
ZrNiSn	0.622	0.5[20]	0.18[21]
$ZrNiSn_{0.875}Ge_{0.125}$	0.619	-----	-------
$ZrNiSn_{0.75}Ge_{0.250}$	0.618	------	--------

To have an explicit understanding of the interaction between the different orbitals of atoms, the total DOS and projected density of states (PDOS) has been calculated and depicted in the Fig. 2 and 3. It's clear from PDOS that at x=0, 0.125 the bands below 2eV are due Sn s orbitals but at x=0.25 the bands are also due to the Sn *s*- and Ge *s*- orbitals and in this range no contribution of all other atoms. At x=0, the bands spans below the Fermi level are due to Zr-d, Ni-d and Sn-p orbitals, in this range Ni-d orbitals give high contribution as compared to other atoms and there are negligible contribution of Zr-s ,Ni-s-p and Sn –s orbitals. Above the Fermi level, at x=0, the bands are due the Zr d ,Sn s and p orbitals but there is negligible contribution due to the orbitals of Ni atom. At x=0.125, the bands spans below Fermi level due to Zr-d,Ni-d,Sn-p and Ge-p and at x=0.25 the band spans due to Zr-d ,Ni-d, Sn-p, Ge-p and all other orbitals contribute negligible. At,x=0.125, above Fermi level the bands are due to the, Zr-s, d, Sn-s, p and Ge-p orbitals and at x=0.250 these are due to the Zr-s, d, Sn-s, p , Ge-p, s orbitals.

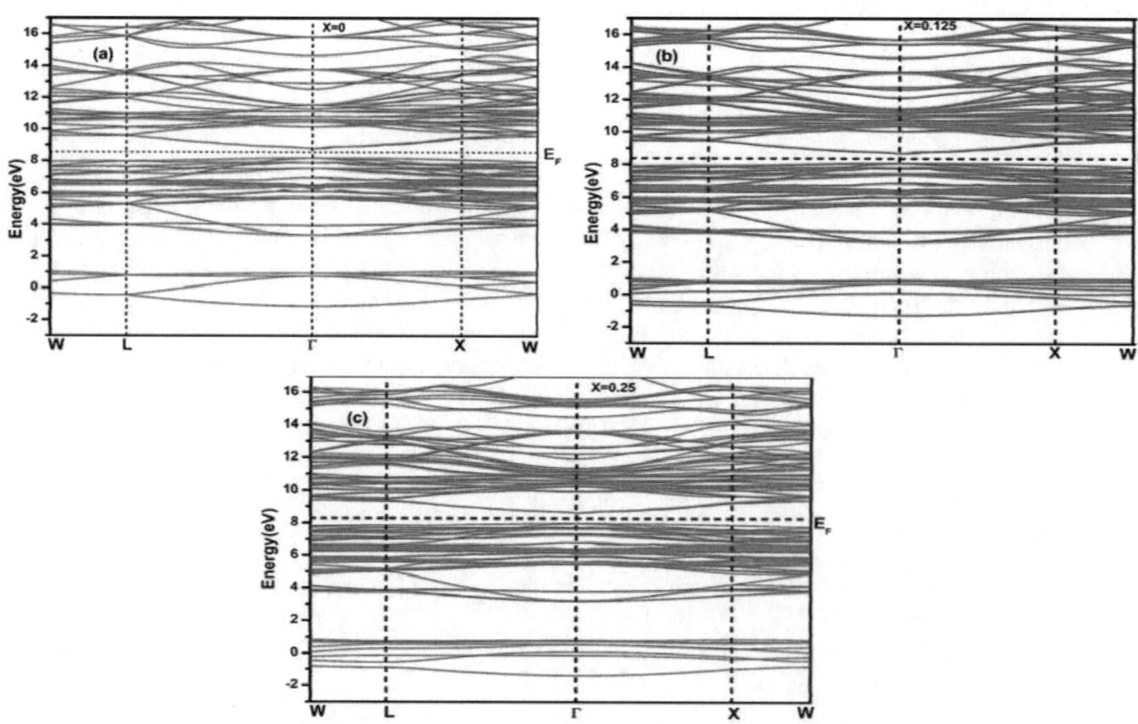

Fig.1.Calculated Band structure of ZrNiSn1-xGex compound ($0 \leq x \leq 0.25$)

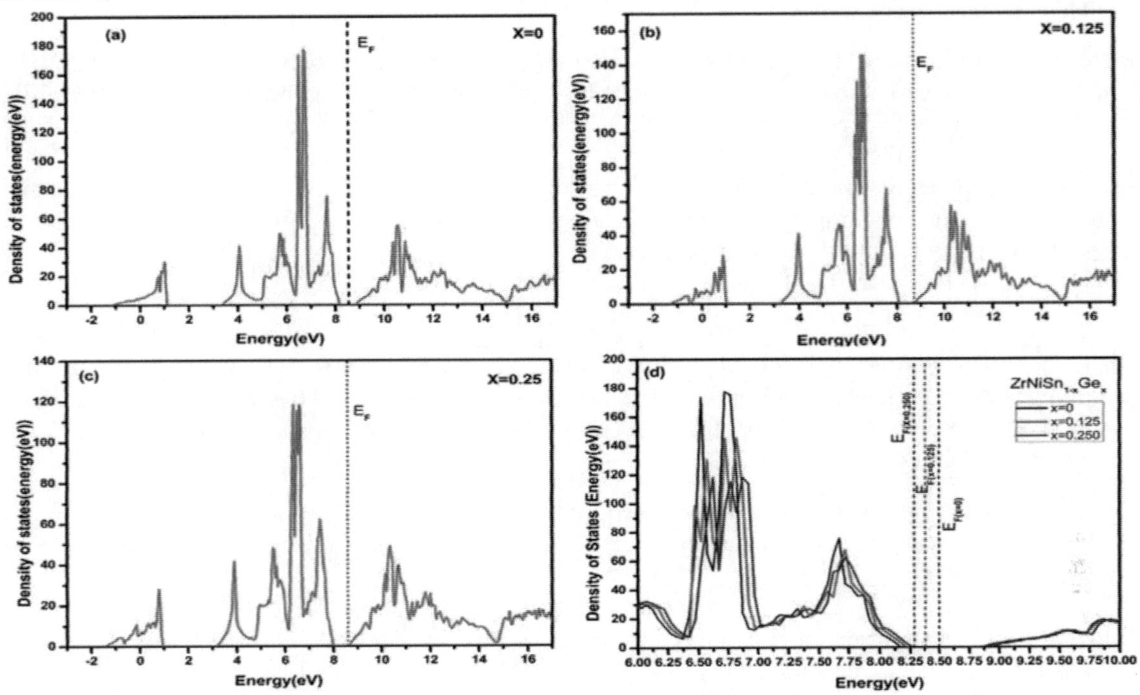

Fig.2. Calculated Density of states of $ZrNiSn_{1-x}Ge_x$ compound ($0 \leq x \leq 0.25$)

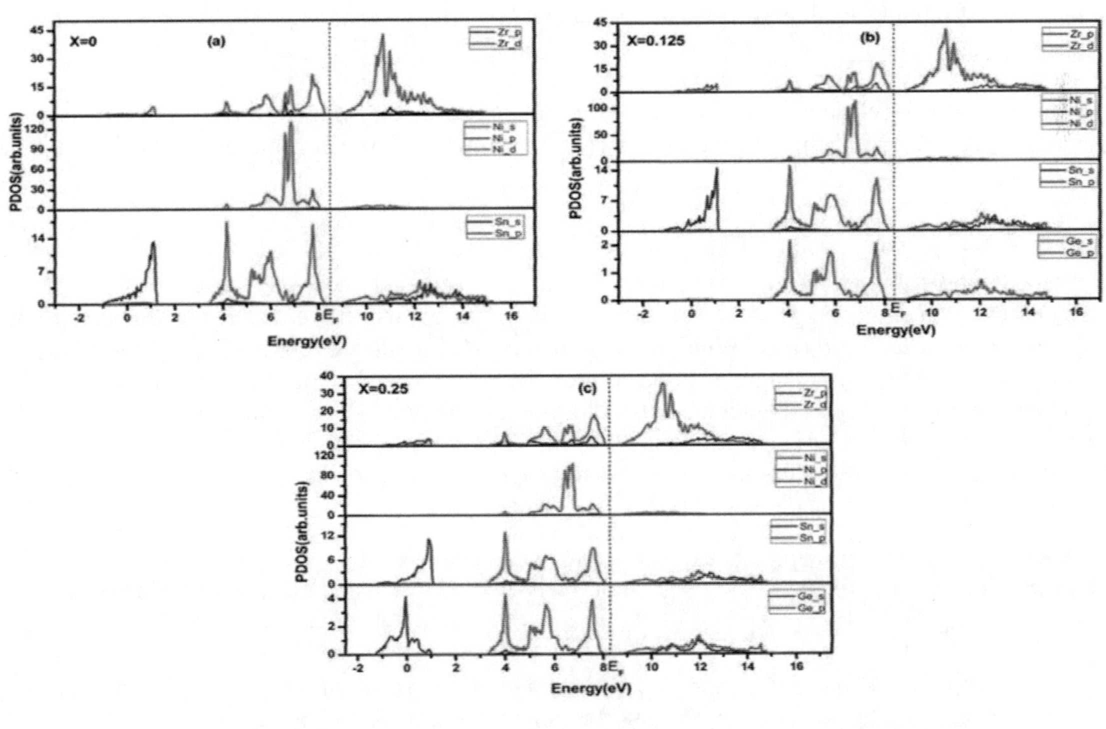

Fig.3.Calculated Projected Density of states of $ZrNiSn_{1-x}Ge_x$ compound ($0 \leq x \leq 0.25$)

3.2. Thermoelectric properties:

In this subsection, a quantitative investigation on thermoelectric properties of pure and Ge doped ZrNiSn is undertaken. The variation of Seebeck coefficient(S) as the function of temperature is depicted in Fig 4(a). For pure and doped system, the sign of Seebeck coefficient is negative, which indicates that the conduction is due to the electrons. For pure system, the value of Seebeck coefficient increases with increase in temperature, but at 1000K it decreases due to presence of minority carriers and our this result is in good agreement with the work reported by H. Mutta et al [7]. The Seebeck coefficient in terms of carrier concentration (n) is expressed as:

$$S = \frac{8\pi^2 k_b^2}{3eh^2} m^* T \left[\frac{\pi}{3n}\right]^{\frac{2}{3}} \qquad (2)$$

Where k_b is Boltzmann constant, m^* is effective mass and n denotes the carrier concentration. According to the equation (2), Seebeck Coefficient is inversely proportional to the carrier concentration. Therefore, the absolute value of Seebeck coefficient of doped systems decreases with increase in the concentration of Ge.

Fig. 4(b) show the electrical conductivity scaled by relaxation time (σ/τ) of pure and doped system as the function temperature. The electrical conductivity in terms of carrier concentration (n) and mobility(μ) is given as:

$$\sigma = ne\mu \qquad (3)$$

Where e is electronic charge and μ is mobility. Therefore according to equation (3), with increase in concentration, the electrical conductivity increases.

But at high temperature range, there is a minor decrease in the electrical conductivity for pure and doped system. This is due to decrease in the mobility ($\mu \alpha T^{-3/2}$) [24, 25]. According to equation (3), electrical conductivity is not only a function of carrier concentration but also function of mobility. Therefore, decrease in mobility results into decrease in electrical conductivity for all systems. For pure system, our calculated electrical conductivity follows the same trend reported by D. F. Zou et al [13].

Fig 4 (c) shows the electronic thermal conductivity of $ZrNiSn_{1-x}Ge_x$ ($0 \leq x \leq 0.250$) as the function of temperature. The electronic thermal conductivity(k_e) in terms of carrier concentration (n) expressed as:

$$k_e = \frac{n\,\pi^2 T k_b^2}{3m^* v_e} \qquad (4)$$

In previous subsection, (figure 2(d)) we discuss that with increase in concentration of Ge ($0 \leq x \leq 0.250$) Fermi level and band gap also decreases (table1). The availability of states near the fermi level with increase in concentration results into more thermal excitation of electrons at all temperature ranges. Therefore, at all temperature range, addition of more thermal excited charge carriers increases charge carriers concentration in the conduction band. According to equation (4), electronic thermal conductivity directly depends upon carrier concentration (n), therefore electronic thermal conductivity increase with increase in the concentration. In our work, the calculated value of the electronic thermal conductivity for pure system follows same trend as reported by Liu et al [26].

To finalized, our study, we calculate the power factor scaled by relaxation time (S^2 σ/τ) of $ZrNiSn_{1-x}Ge_x$ system which is equal to the product of square of Seebeck coefficient and electrical conductivity:

$$P = S^2\sigma \qquad (5)$$

The calculated power factor of pure and doped system as the function of temperature is shown in Fig. 4(d). For pure system, the power factor follows the same trend as reported by Liu et al [26]. At x=0.125, the maximum power factor is attained. In doped systems, the enhancement in the power factor is due to increase in the electrical conductivity. In this work, we conclude that electrical conductivity plays a major role to enhance the power factor.

IOP Conf. Series: Materials Science and Engineering **360** (2018) 012061 doi:10.1088/1757-899X/360/1/012061

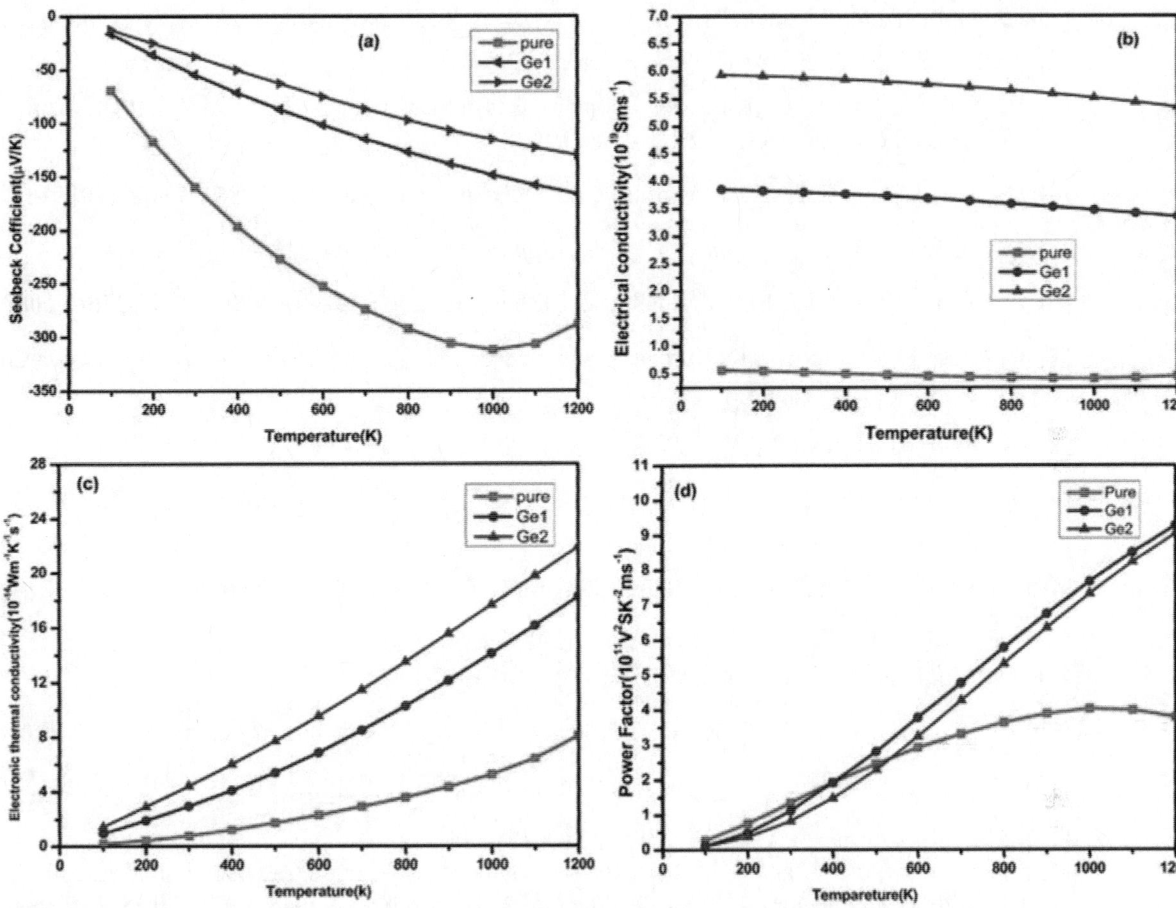

Fig.4: Variation of (a) Seebeck coefficient (S) (b) Electrical conductivity (σ/τ) (c) Electronic Thermal conductivity(k_e) (d) Power factor ($S^2\sigma/\tau$) with temperature.

4. Conclusion:

In this work, an attempt is made to employ first-principle calculations in order to predict the effect of Ge substitution on Sn sites in ZrNiSn half heusler system. As concentration of Ge ($0 \leq x \leq 0.250$) increases, Seebeck coefficient and band gap decreases. But the electrical conductivity and the electronic thermal conductivity increases. The increase in the carrier concentration is due to lower electronegativity of Sn which increase the electrical conductivity and electronic thermal conductivity. The negative sign of Seebeck coefficient of all systems indicates that the conduction is due to electrons. The maximum power factor is achieved at x=0.125.

Acknowledgment
One of the authors, Sukhwinder Singh, thanks the SHE DST Inspire under no. 792/2012 for providing scholarship.

References

[1] F.J. Disalvo, Science 285 703–706 **(1999).**

[2] F.G. Aliev,V.V. Moshchalkov.,V.V. Kozyrkov., M.K. Zalyalyutdinov, V.V. Pryadun.,R.V. Scolozdra J. Mag. Magne. Mater. 76 -77 **(1988).**

[3] W. Kafer, K. Fess, C. Kloc, K. Friemelt, E. Bucher, January In *CONFERENCE SERIES-INSTITUTE OF PHYSICS* 152,185-188 **(1998)**.

[4] C. Uher, J Yang, S. Hu, D.T. Morelli, G.P. Meisner, Phys. Rev. B, 59, 8615–8621**(1999).**

[5] M.Onoue, F. Ishii, T. *J. Oguchi, Phy. Soc. Japan*, *77*(5), 054706 **(2008)**.

[6] C. Yu, T. J. Zhu, R. Z. Shi, Y. Zhang, X. B.Zhao, J.He *Acta Mater. 57*(9), 2757-2764 (**2009).**

[7] H. Muta, T. Yamaguchi, K. Kurosaki, S. *Yamanaka, ICT 2005. 24th International Conference on* 351-354). IEEE **(2005).**

[8] S. Sakurada, N. Shutoh, *Appl. Phys. Letters*, *86*(8), 082105 **(2005).**

[9] R. Akram, Q. Zhang, D. Yang, Y. Zheng, Y.Yan, X. Su, X. *J. Tang, Elect. Mater*, *44*(10), 3563-3570 **(2015).**

[10] Q. Shen, L. Chen, T. Goto, T. Hirai, J. Yang,G. P. Meisner,C. Uher, *Appl. Phy. Lett.*, *79*(25), 4165-4167 **(2001).**

[11] Y. Liu, P. F. Poudeu, J. Mater. Chem A 23**(2015)**.

[12] J. Yang, H. Li, T. Wu, W. Zhang, L. Chen, and J. Yang, Adv. Funct. Mater. 18, 2880 **(2008).**

[13] D. F. Zou, S. H. Xie, Y. Y. Liu, J. G. Lin, J. Y. Li,. J. Appl. Phy., 113(19), 193705 **(2013).**

[14] S.Singh ,R.kumar, J.Alloy.Compd. 722 **(2017)** 544-548.

[15] P. Giannozzi, S. Baroni, N. Bonini, M. Calandra, R. Car, C.Cavazzoni, D.Ceresoli, G.L. Chiarotti,M. Cococcioni, I. Dabo, A.D. Corso, S. de Gironcoli, S. Fabris, G. Fratesi, R. Gebauer, U. Gerstmann, C. Gougoussis, A. Kokalj, M. Lazzeri, L.Martin-Samos, N.Marzari, F.Mauri, R.Mazzarello, S.Paolini, A. Pasquarello, L. Paulatto, C. Sbraccia, S. Scandolo, G. Sclauzero, A.P. Seitsonen, A. Smogunov, P. Umari, and R.M. Wentzcovitch , J. Phys. Condens. Matt. 21 395502 **(2009)**.

[16] J.P.Perdue, Y.wang, phys. Rev B, 45 13244**(1992).**

[17] H.J.Monkhorst, and J.D Pack, Phys. Rev. B 13 5188 **(1976)**.

[18] K. H. Madsen and J. Singh, Comput. Phys. Communications 175 67-71 **(2006)**.

[19] W. Jeitschko, Metall. Trans. A 1, 3159 **(1970)**.

[20] D. F.Zou, S. H.Xie, Y. Y.Liu, J. G.Lin, J. Y.Li . *J. Appl. Phy.*, *113*(19), 193705 **(2013)**.

[21] F. G. Aliev, , N. B. Brandt, V. V. Moshchalkov,, Kozyrkov, V. V., R. V. Skolozdra, A. I. Belogorokhov, *Zeitschrift für Physik B Cond. Matt.*, *75*(2), 167-171 **(1989)**.

[22] L. F. Mattheiss, Phys. Rev. B **43**, 1863 **(1991).**

[23] J.W. Simonson and S. J. Poon, J. Phys.: Condens. Matter **20**, 255220 **(2008)**.

[24] R.J.LaBotz, D.R.Mason, D.F.Okwane J. Electrochem. Soc.110 127 **(1963)**.

[25] Y.Noda, H.Kon, Y. Furukawa, N. Otsuka, I. Nishida, K. Masumoto Mater. Trans. JIM33851 **(1992).**

[26] X. H.Liu, J.He, H. H.Xie, X. B.Zhao, T. J.Zhu *I. J. Smart and Nano Mater.*, *3*(1), 64-71 **(2012)**.

IOP Conf. Series: Materials Science and Engineering **360** (2018) 012062 doi:10.1088/1757-899X/360/1/012062

Ammonia, acetone and ethanol gas sensing characteristics of CuO thin films grown by sputtering

Anju Anna Jacob[1], L. Balakrishnan[2,*], S. R. Meher[3], R. Sivacoumar[4], K. Shambavi[5], and Z. C. Alex[4]

[1]School of Electronics Engineering, VIT University, Vellore 632 014, India.
[2]Department of Physics, Government College of Technology, Coimbatore 641 013, India.
[3]Department of Physics, School of Advanced Sciences, VIT University, Vellore 632 014, India.
[4]Department of Sensor and Biomedical Technology, School of Electronics Engineering, VIT University, Vellore 632 014, India
[5]Department of Communication, School of Electronics Engineering, VIT University, Vellore 632 014, India.

*Email: bslv85@gmail.com

Abstract. Here, we have reported the sensing characteristics of CuO thin films towards volatile gases like ammonia, acetone and ethanol. The CuO films were deposited on glass substrate by RF magnetron sputtering at 350 °C and 400 °C. The characteristics of CuO thin films grown at both temperatures were analyzed using X-ray diffractometer (XRD), UV-Vis spectrometer, atomic force microscope (AFM), scanning electron microscope (SEM) and Fourier transform infrared spectrometer (FTIR). The thin films showed mesoporous morphology with average crystallite size of 78 nm and 36 nm for films grown at 350 °C and 400 °C, respectively. As the film grown at 400 °C showed better properties, it has been preferred for the gas sensing analysis. Gas sensing properties of the film towards different concentration (50-250 ppm) of ammonia, acetone and ethanol were studied in chemi-resistive mode. As expected, an increase in resistance with concentration of gases was observed due to the p-type nature of CuO thin films. Further, the film showed comparatively better sensitivity towards ethanol than other gases.

Keywords: Thin film, CuO, sputtering, Gas sensor, Ethanol sensing.

1. Introduction

Novel gas sensors development are of great concern due to its widespread applications in automobiles, safety, environmental control, indoor air quality, food, medicine etc., [1–5]. For the past few decades, semiconducting metal oxide (SMO) based gas sensors are extensively studied and developed. SMOs strike as being best candidate for gas sensing materials, considering their high sensitivity along with easy fabrication, low cost, miniaturized dimension, reliability etc.,

A wide range of SMOs like SnO2, ZnO, CuO, TiO_2, WO_3, Fe_2O_3 etc., have been reported as gas sensing materials[6–11]. Cupric oxide (CuO) is one among the SMOs that exhibits p-type conductivity with a narrow band gap of 1.2 eV and has widespread applications in the field of gas sensors,

IOP Conf. Series: Materials Science and Engineering **360** (2018) 012062 doi:10.1088/1757-899X/360/1/012062

catalysis, electrochemical cells, field emission displays, etc. Considerably good electrical properties, high absorption coefficient, catalytic reactivity, stability in air and earth abundance make it a promising candidate for the fabrication of gas sensors[12–14]. Taking up the advantages as sensing material, p-type SMOs are prone to well exchange of lattice oxygen with target gas. Furthermore, the conductivity of p-type oxides are remarkably less dependent on temperature at higher ranges. Also, CuO is a prominent catalytic material. All these benefits for the betterment of sensitivity, stability and life time of the sensors[15–17].

CuO thin films for gas sensing can be grown using various methods like chemical vapor deposition, thermal oxidation, solvo-thermal technique, DC or RF magnetron sputtering etc.,[8,12,18–20]. Among these the film properties like thickness and uniformity can be effectively controlled using RF magnetron technique. Also, it has the advantages of better tuning of grain morphology, stoichiometry and crystallinity of the film[21,22].

In this work, we have reported the deposition of uniform and crystalline CuO thin films by RF magnetron sputtering technique and its sensing response towards acetone, ammonia and ethanol gases.

2. Materials and Methods

CuO thin films were deposited on glass substrate of size 2×2 cm^2 by RF reactive magnetron sputtering technique using metallic Cu target. Oxygen and argon were used as the reactive and sputtering gases with gas pressure 0.02 mbar and 0.01 mbar, respectively. The substrates were fixed on a substrate holder at a distance of 7 cm above the target. For deposition of CuO thin films sputtering was done for 20 minutes at 350 °C and 400 °C.

The properties of films were studied using X-Ray diffractometer (XRD), atomic force microscope (AFM), scanning electron microscope (SEM), Fourier transform infra-red (FTIR) spectrometer and UV-Vis spectrometer. For gas sensing characteristics, aluminum interdigitated electrodes were made on the films using metal evaporation technique. The electrical characteristics of the films for gas sensing analysis were studied using Agilent B2901A Precision Source/Measure Unit. The gas sensing characteristics of the film was studied in chemi-resistive mode. The change in resistivity due to the variation in concentration of free electrons through the reaction with the target gas can be monitored for the detection and quantification of gases[23].

3. Results and Discussion

The X-ray diffraction pattern of CuO thin films grown by sputtering at substrate temperature of 350 ° C and 400 ° C are shown in Fig. 1a and 1b respectively. The patterns confirm that the films deposited were polycrystalline in nature with monoclinic structure and the film coated at 400 °C shows better crystallinity. Both the films exhibited a single diffraction peak at $2\theta = 37.73$ corresponding to (111) orientation of CuO thin film which conveys the preferential orientation of the films along (111) plane. The grain size of the films (D) was evaluated from the full width half maximum (FWHM) intensity of X-ray diffraction peak of CuO using Scherrer's relation,

$$D = \frac{k\lambda}{\beta\cos\theta} \qquad (1)$$

where, k is shape factor generally it has been taken as 0.89 for spherical shape, β is the FWHM of the (111) peak measured in radians. The calculated grain sizes of the films are 78 nm and 36 nm for substrate temperatures 350 °C and 400 °C, respectively.

The UV-Vis transmission and absorption spectra of the cupric oxide thin films deposited on glass substrates by RF magnetron sputtering is shown in Fig. 2a and 2b. Both the films showed approximately 57% of transmittance in the visible region. Sharp fall in the optical transmittance at shorter wavelengths resulted from the excitation of charge carriers across the optical band gap.

To determine the optical band gap, absorption coefficient (α) of the film was calculated using the following relation,

$$\alpha(\nu) = 2.303\left(\frac{Abs}{t}\right) \tag{2}$$

where, Abs is the optical absorbance and t is the film thickness. The optical band gap of the films was estimated by using Tauc's equation,

$$\alpha h\nu = C\left(h\nu - E_g\right)^{1/2} \tag{3}$$

where, hν is the energy of the incidence photon, C is a constant and Eg is the optical band gap.

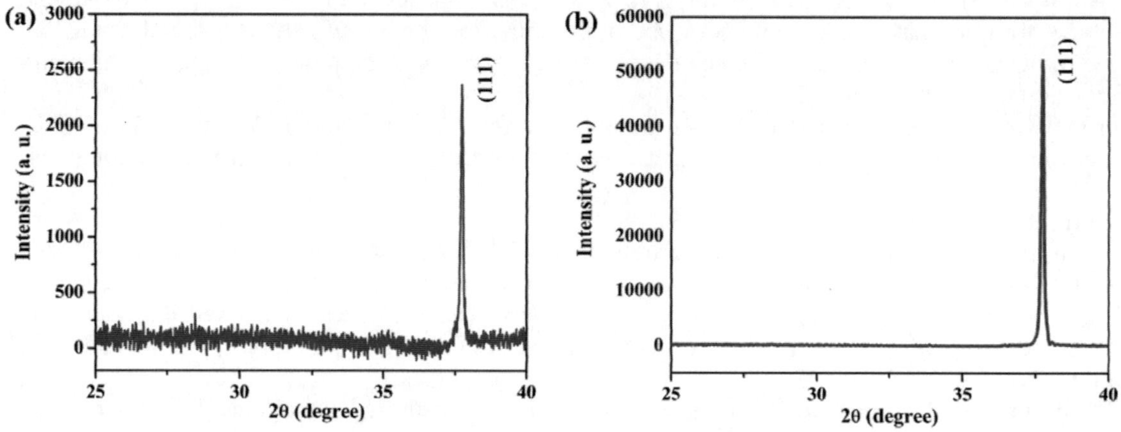

Figure 1. XRD pattern of CuO film at (a) 350 °C and (b) 400 °C.

The sample possesses a direct band gap of 2 eV for the films coated at 350 °C and 2.4 eV for the films coated at 400 °C. The change in band width is mostly due to increase in growth temperature, which results in improved crystallinity. Also, the deviation from bulk optical band gap may be attributed to the Burstein-Moss effect and quantum confinement[15,24].

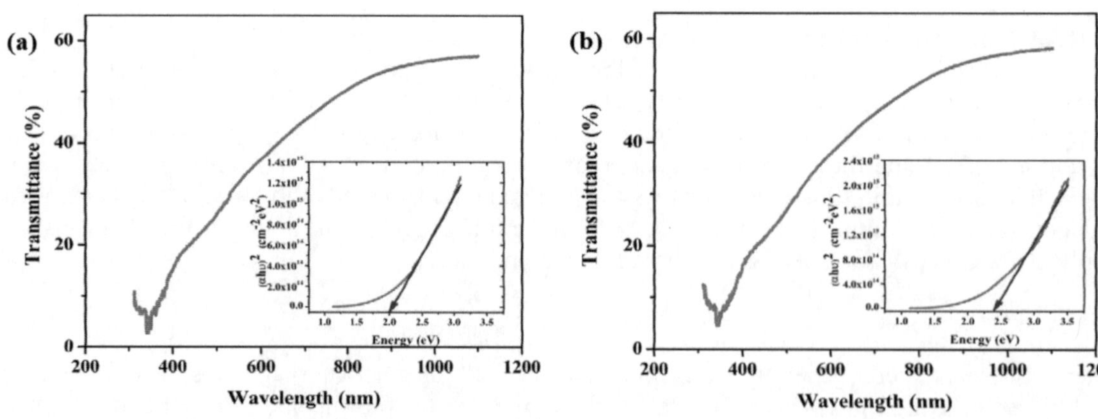

Figure 2.UV-Spectrum of CuO film at (a) 350 °C and (b) 400 °C.

Surface morphology of the CuO thin films were studied by atomic force microscopy (AFM). Figures 3a and 3b shows the three dimensional AFM morphology of CuO thin films deposited at different temperatures. The peak to valley roughness of the CuO thin films grown at 350 °C and 400 °C are 86.96 and 44.75 nm, respectively and the average roughness of the films are found as 6.41 and 3.83 nm. It was observed that films are polycrystalline and consisted of smaller grains. The results do exhibit apparent evolution in film grain morphologies with increasing substrate temperatures. The deposited CuO thin film at 400 °C exhibits much smoother surface morphology as compared to the film grown at 350 °C.

Figures 4a and 4b shows the SEM images of the films grown at 350 °C and 400 °C, respectively. The average particle sizes calculated from SEM images were 67 nm and 31.6 nm for the films grown at 350 °C and 400 °C respectively. The SEM image shows the spongy porous appearance and randomly disperses particles on the surface of the thin film. Both the films showed slight mesoporous surface morphology which can enable them to use efficiently for gas sensor applications.

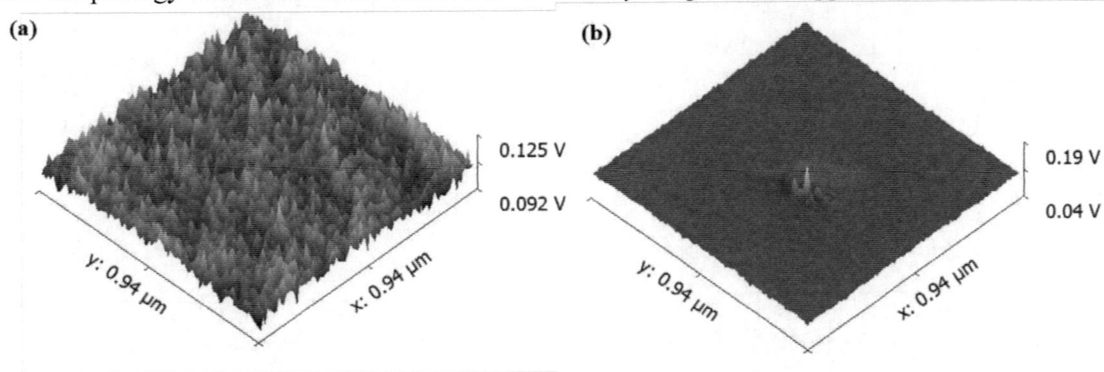

Figure 3. AFM images of CuO film at (a) 350 °C and (b) 400 °C.

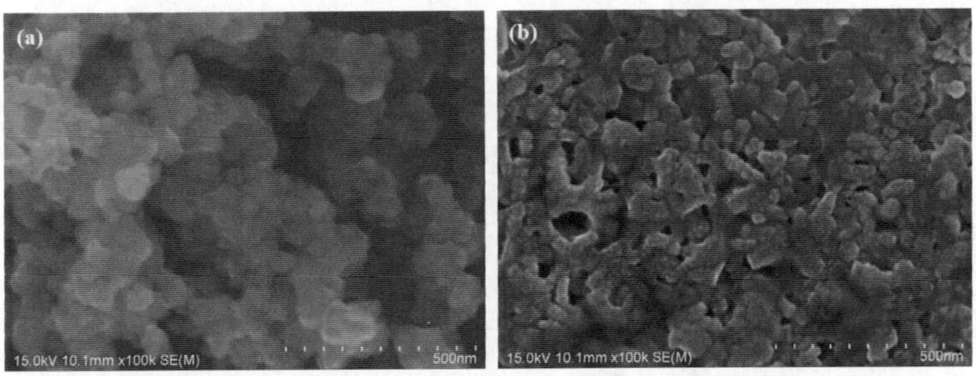

Figure 4. SEM images of CuO film at (a) 350 ° C and (b) 400 ° C.

Figures 5a and 5b shows the room temperature Fourier transform infrared (FTIR) spectra of CuO thin films on glass substrate deposited by sputtering. The FTIR spectrum of CuO thin films shows three characteristic strong peaks associated with the Cu-O vibrations of monoclinic CuO. At 350 °C the peaks are at 442, 467 and 560 cm^{-1}, whereas at 400 °C, the peaks are at 416, 442 and 558 cm^{-1}. The strong band located at about 558 cm^{-1} in both films is due to the Cu-O stretching along (111) direction.

From the XRD, AFM and SEM analysis it has been found that the film grown at 400 °C shows better properties compared to the film grown at 350 °C, hence it has been selected for further sensing studies. The sensitivity of the film towards ammonia, acetone and ethanol gases were analyzed in chemi-resistive mode and are shown in Fig. 6.

The adsorbed oxygen on the surface of the thin film traps electrons from p-type CuO film which thereby increases the hole concentration. When the gases ethanol, acetone and ammonia are introduced, which interacts with adsorbed oxygen or lattice oxygen, releases electrons which in turn decreases the hole concentration and hence reduces conductivity. This change is resistivity/conductivity of film, with varying sensing gas concentration is obtained from the I-V characteristics to measure the sensitivity. Sensitivity is calculated using the following relation,

$$S = \left[\frac{R_{gas} - R_{air}}{R_{air}} \right] \times 100 \tag{4}$$

where, R_{air} is the resistance of the film in air ambience and R_{gas} is the resistance of the film upon gas exposure.

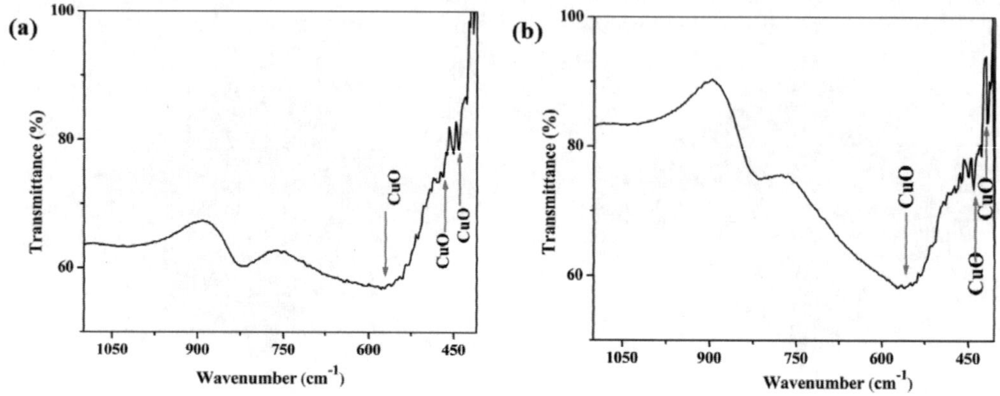

Figure 5. FTIR Spectrum of CuO film at (a) 350 °C and (b) 400 °C.

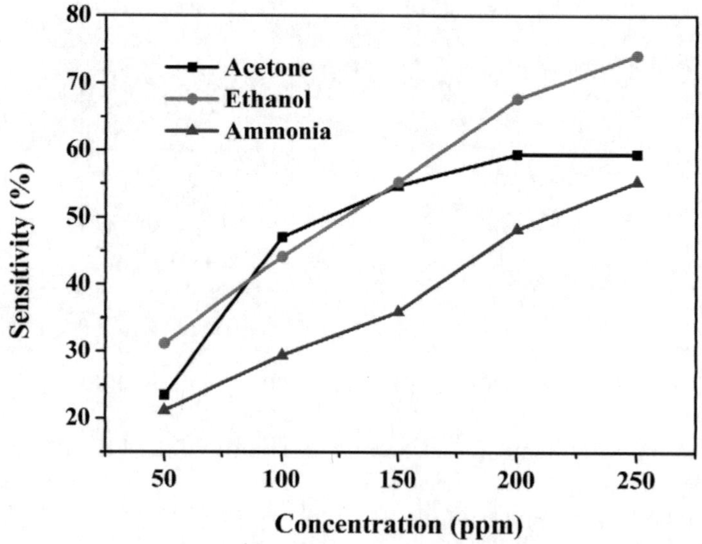

Figure 6. Sensitivity response of CuO thin film sensor towards different concentration of acetone, ethanol and ammonia gases.

Further, it has been found that the film showed better sensitivity towards ethanol compared to acetone and ammonia due to the enhanced catalytic activity towards ethanol. The good sensitivity of the film can be attributed to its porous morphology and also in the case of CuO film, both adsorbed oxygen and lattice oxygen will take part in sensing reaction.

4. Conclusion

Polycrystalline CuO thin films were deposited using RF magnetron sputtering technique at 350 °C and 400 °C growth temperatures. After detailed structural, morphological and compositional study the film grown at 400 °C which exhibited more advantageous properties was selected for further sensing

IOP Conf. Series: Materials Science and Engineering **360** (2018) 012062 doi:10.1088/1757-899X/360/1/012062

studies. The film showed good sensitivity towards ethanol compared to other gases. The enhanced catalytic activity towards ethanol and porous structure makes CuO thin films preferable candidates for the fabrication of ethanol gas sensors.

References

[1] Afzal A, Cioffi N, Sabbatini L and Torsi L 2012 *B Chem.* **171–172** 25–42
[2] Kanan S M, El-Kadri O M, Abu-Yousef I A and Kanan M C 2009 *Sensors* **9** 8158–96
[3] Kim Y S, Hwang I S, Kim S J, Lee C Y and Lee J H 2008 *Sensors Actuators, B Chem.* **135** 298–303
[4] Hu P, Du G, Zhou W, Cui J, Lin J, Liu H, Liu D, Wang J and Chen S 2010 *ACS Appl. Mater. Interfaces* **2** 3263–9
[5] Nam J M, Stoeva S I and Mirkin C A 2004 *J. Am. Chem. Soc.* **126** 5932–3
[6] Bhanjana G, Dilbaghi N, Kumar R, Umar A and Kumar S 2015 *Electrochim. Acta* **169** 97–102
[7] Rai P, Raj S, Ko K J, Park K K and Yu Y T 2013 *Sensors Actuators, B Chem.* **178** 107–12
[8] Yang C, Su X, Xiao F, Jian J and Wang J 2011 *Sensors Actuators, B Chem.* **158** 299–303
[9] Bayata F, Saruhan-Brings B and Ürgen M 2014 *B Chem.* **204** 109–18
[10] Hubner M, Simion C E, Haensch A, Barsan N and Weimar U 2010 *Sensors Actuators, B Chem.* **151** 103–6
[11] Tang H, Yan M, Zhang H, Li S, Ma X, Wang M and Yang D 2006 *Sensors Actuators B Chem.* **114** 910–5
[12] Parmar M and Rajanna K 2011 *Int. J. Smart Sens. Intell. Syst.* **4** 710–25
[13] Rakhshani A E 1986 *Solid State Electron.* **29** 7–17
[14] Liu X-W, Wang F-Y, Zhen F and Huang J-R 2012 *RSC Adv.* **2** 7647
[15] Korotcenkov G 2007 *Mater. Sci. Eng. B Solid-State Mater. Adv. Technol.* **139** 1–23
[16] Lu Y-M, Chen J-Y and Wey T-S 2004 *Mater. Res. Soc. Symp. Proc.* **822** 55–64
[17] Heiland G 1954 *Phys.* **138** 459–64
[18] Barreca D, Comini E, Gasparotto A, Maccato C, Sada C, Sberveglieri G and Tondello E 2009 *Sensors Actuators, B Chem.* **141** 270–5
[19] Raksa P, Gardchareon A, Chairuangsri T, Mangkorntong P, Mangkorntong N and Choopun S 2009 *Ceram. Int.* **35** 649–52
[20] Zoolfakar A S, Ahmad M Z, Rani R A, Ou J Z, Balendhran S, Zhuiykov S, Latham K, Wlodarski W and Kalantar-Zadeh K 2013 *Sensors Actuators, B Chem.* **185** 620–7
[21] Ogwu A A, Bouquerel E, Ademosu O, Moh S, Crossan E and Placido F 2005 *Acta Mater.* **53** 5151–9
[22] Pierson J F, Thobor-Keck A and Billard A 2003 *Appl. Surf. Sci.* **210** 359–67
[23] Wetchakun K, Samerjai T, Tamaekong N, Liewhiran C, Siriwong C, Kruefu V, Wisitsoraat A, Tuantranont A and Phanichphant S 2011 *Sensors Actuators, B Chem.* **160** 580–91
[24] Kamat P V, Dimitrijevic N M and Nozik A J 1989 *J. Phys. Chem.* **93** 2873–5

Acknowledgements

The authors would like to acknowledge Department of Science and technology (DST), New Delhi, India for providing the financial support through FIST (Fund for Improvement of S & T Infrastructure in Higher Educational Institutions) project [SR/FST/ETI-015/2011].

IOP Conf. Series: Materials Science and Engineering **360** (2018) 012063 doi:10.1088/1757-899X/360/1/012063

Effect of Dy substitution on the Structural and Gas sensing Properties of $BiFeO_3$

P.A. Ghadage [a], U. R. Ghodake [b], J.Y. Patil [a], and S.S. Suryavanshi [a*]

[a] Ferrite Material Laboratory, Department of Physics, Solapur University, Solapur (India) - 413255

[b] Department of Electronics, Shri. Shivaji Mahavidyalaya, Barshi - 413411, Dist-Solapur, Maharashtra, India

Abstract

Facile synthesis of Dy substituted $BiFeO_3$ were employed using Glycine combustion method. X-ray diffraction (XRD) and Scanning electron microscopy (SEM) were employed to characterize the structure and morphology of the material. The effect of Dy doping on the crystalline structure, morphologies and gas sensing properties of the final products have been investigated. The gas sensing properties of $Bi_{1-x}Dy_x FeO_3$ (x=0.00, 0.05) materials for semiconductor gas sensors are presented. For composition with x=0.05, sensor exhibited highest response (87%) towards acetone vapor at an operating temperature 250°C. The response and recovery time of the sensor for x=0.05 were quite low (5sec and 40sec) respectively.

Keywords: Multiferroic, Glycine combustion method, Selective acetone sensor.

***Corresponding Author:** Tel.: +91 217 2744771 Fax: +91 217 2744770
Email address: sssuryavanshi@rediffmail.com, ghadagepa@gmail.com
(Prof. Sharad Shivaji Suryavanshi)

1. Introduction

In recent years, a rhombohedrally distorted perovskite (ABO_3) type $BiFeO_3$(BFO) have been investigated for the multiferroic properties. Due to the interesting physical properties of BFO, it acts as a potential applicant for future technological applications[1-3]. For the formation of BFO nanoparticles we adopted the superficial glycine combustion method[4-5]. In this short communication, we discuss the effect of Dy substitution on the structural, morphological and gas sensing properties of BFO.

2. Experimental details

$BiFeO_3$(BFO) and $Bi_{0.95}Dy_{0.05}FeO_3$(BDFO) nanoparticles were synthesized by using metal nitrates and glycine as a starting materials. This mixture was dissolved in 100ml distilled water with vital magnetic stirring for 30min. which is maintained at constant 70^0C temperature. Whereas one more solution with 5% Dy doped BFO sample was also prepared. The precursor solution was kept on the heater, which undergoes a flameless combustion and the fluffy foam of our final product was obtained[4]. The grinded calcined powder was used for the pellet formation. The final pellet samples were sintered at 650^0C and silver contacts were drained from it for the measurement of gas sensing properties. In presence of reducing gases the P-type materials showed that the change in its electrical resistance. The gas sensing properties of the BFO and BDFO samples sintered at 650^0C were studied at various temperatures in the range of 200-450^0C. The resistance of the sample was measured by two-probe dc measurement technique and the sensitivity of sensor for reducing gases were calculated by following formula as[6]: $S(\%) = \left[\frac{Rg - Ra}{Ra}\right] \times 100$

IOP Conf. Series: Materials Science and Engineering **360** (2018) 012063 doi:10.1088/1757-899X/360/1/012063

Where, R_a is resistance of sensor in air and R_g resistance after the exposure of test gas.

3. Results and discussion

The X-ray diffraction (XRD) pattern showed that pure BFO sample have rhombohedral R3c Phase with secondary phases $Bi_2Fe_4O_9$, where as BDFO sample shows modification in the structure due to the lattice distortion is shown in figure 1. The XRD patterns show the structural transformation from rhombohedral R3c with (JCPDS card No. 71-2494) to orthorhombic Pbnm/Pbam (JCPDS card No. 73-1098) phase due to 5% Dy doping[7]. Also it confirms the values of lattice parameters are as (a=b=5.6204Å to 5.6185Å and c=13.6930Å to13.6925Å). The existence of $Bi_2Fe_4O_9$ and $Bi_{36}Fe_2O_{57}$ as secondary phases have been reported by several authers[8-9].

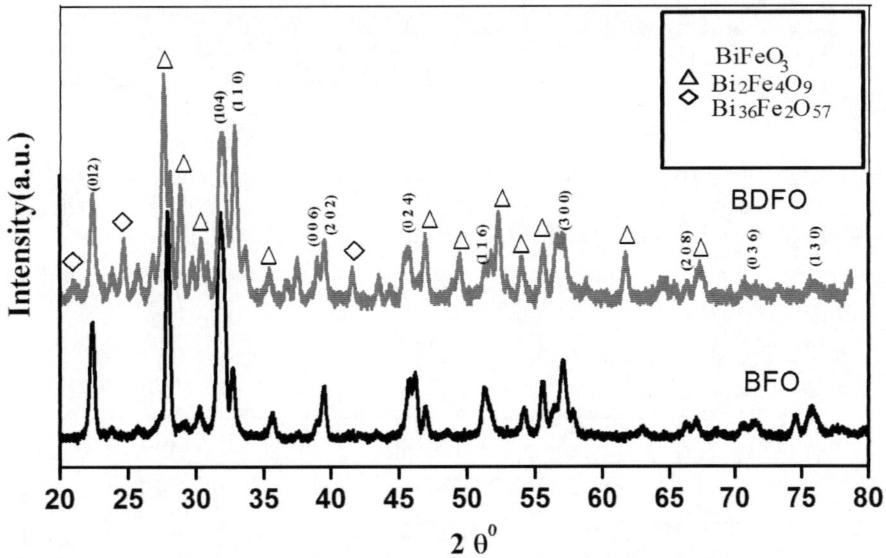

Figure 1. XRD pattern of BFO and BDFO sintered at 650^{0}C

Scanning electron microscope (SEM) micrographs of BFO and BDFO samples sintered at 650^{0}C show the agglomeration of little particles having unequal shape, size and absorbent structure, which forms the non homogeneous grains. The average grain size of the samples BFO and BDFO are 1.56μm to 0.635μm respectively. With addition of Dy content the average grain size decreases as compared to BFO. The decrease in grain size provides more surface area for gas adsorption which is an important requisite for gas sensing application and hence it exhibits improved gas sensing properties.

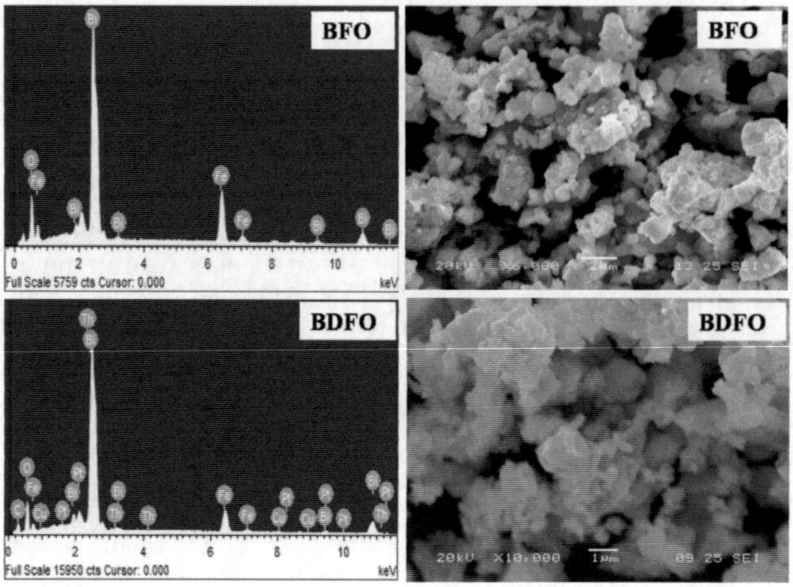

Figure 2. EDAX and SEM images

Electron diffraction spectrum (EDAX) shows the confirmation of elemental composition which is shown in figure 2. The reducing gases were tested on the BFO and BDFO samples sintered at 650^{0}C sample for gas sensing properties (figure 3).

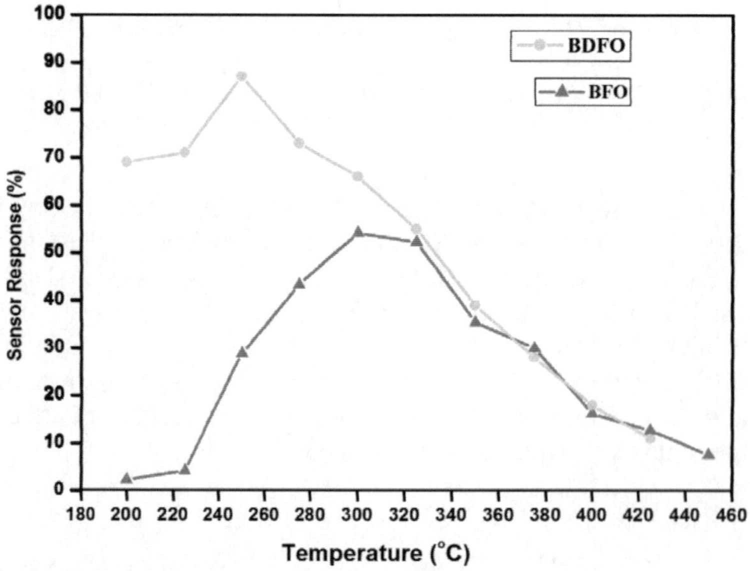

Figure 3. Gas sensing properties of BFO and BDFO samples sintered at 650^{0}C

The BDFO sample shows selective highest response (87%) for acetone gas at optimal operating temperature 250^{0}C as shown in figure 4. The response and recovery time observed in BDFO sample were quite low (5sec and 40sec) respectively which is shown in figure 5.

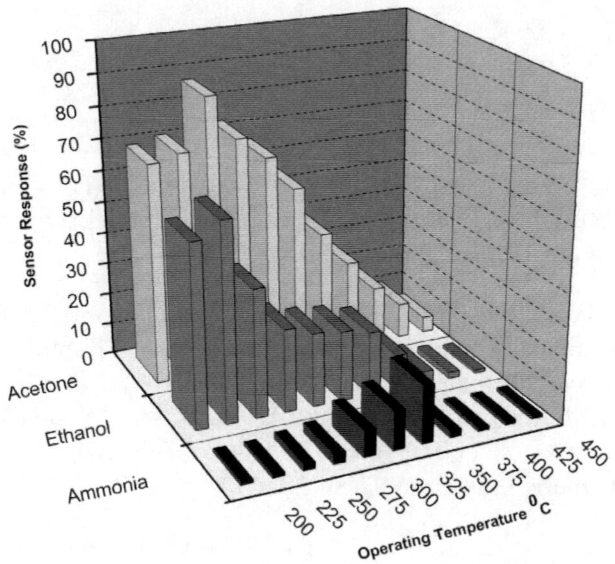

Figure 4. Selectivity of the BDFO sintered at 650^0C

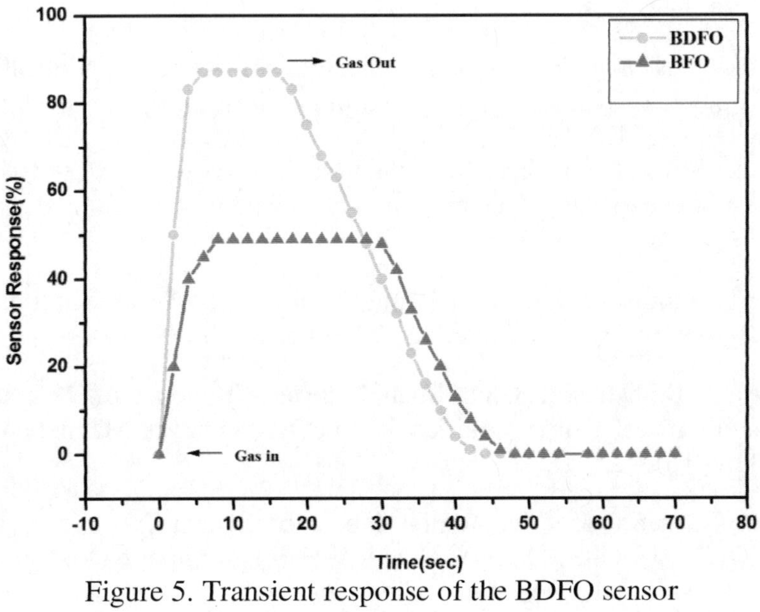

Figure 5. Transient response of the BDFO sensor

4. Conclusion

The BFO and BDFO nanoparticles were successfully synthesized by a facile glycine combustion method. XRD confirms the transformation of rhombohedral R3c phase to orthorhombhic Pbnm/Pbam phase of BFO. EDAX technique identifies the chemical composition of Bi, Fe and O. The BDFO sensor showed highest response towards acetone gas at optimal operating temperature which was mainly accredited to its absorbent structure.

IOP Conf. Series: Materials Science and Engineering **360** (2018) 012063 doi:10.1088/1757-899X/360/1/012063

References:

1] Prakash Chandra Sati, Manisha Arora, Sunil Chauhan, Manoj Kumar, Sandeep Chhoker, Effect of Dy substitution on structural, magnetic and optical properties of $BiFeO_3$ ceramics, Journal of Physics and Chemistry of Solids75,105(2014)

2] V. Koval , I. Skorvanek , M. Reece , L. Mitoseriu , H. Yan, Effect of dysprosium substitution on crystal structure and physical properties of multiferroic $BiFeO_3$ ceramics, Journal of the European Ceramic Society 34,641(2014)

3] M.J. Miah , M.N.I.Khan, A.K.M.Akther Hossain, Synthesis and enhancement of Multiferroic properties of $(x)Ba_{0.95}Sr_{0.05}TiO_3$–$(1\text{-}x)BiFe_{0.90}Dy_{0.10}O_3$ ceramics Journal of Magnetism and Magnetic Materials397,39(2016)

4] Roberto Köferstein, Synthesis, phase evolution and properties of phase-pure nanocrystalline
$BiFeO_3$ prepared by a starch-based combustion method, Journal of Alloys and Compounds 590, 324(2014)

[5] J. Yang, X. Li , J. Zhou, Y. Tang, Y. Zhang, Y. Li, Factors controlling pure-phase magnetic $BiFeO_3$ powders synthesized by solution combustion synthesis, Journal of Alloys and Compounds 509, 9271(2011)

6] M. Dziubaniuk, R. B. Koro´nska, J. Suchanicz, J. Wyrwa, M. R ̨ekas, Application of bismuth ferrite protonic conductor for ammonia gas detection, Sensors and Actuators B 188, 957(2013)

7] Shuxia Zhang, Wanju Luo, Dongliang Wang, Yanwei Ma, Phase evolution and magnetic property of $Bi_{1-x}Dy_xFeO_3$ ceramics, Materials Letters 63,1820(2009)

8] A.Sathiya Priya , I.B.Shameem Banu , Shahid Anwar, Influence of Dy and Cu doping on the room temperature multiferroic properties of $BiFeO_3$ Journal of Magnetism and Magnetic Materials 401,333(2016)

9] Bogdan Jurca , Carmen Paraschiv, Adelina Ianculescu, Oana Carp, Thermal behaviour of the system $Fe(NO_3)_3$ - $9H_2O$–$Bi_5O(OH)_9(NO_3)_4$ - $9H_2O$–glycine/urea and of their generated oxides ($BiFeO_3$)J Therm Anal Calorim 97,91(2009)

IOP Conf. Series: Materials Science and Engineering **360** (2018) 012064 doi:10.1088/1757-899X/360/1/012064

Ultrasonic Characterization of Silver Nano Fluid

A. Nanda, A. Tiadi, S. K. Mallik, R. Giri and G. Nath *
Department of Physics, Veer Surendra Sai University of Technology,
Burla, Sambalpur -768 018, Odisha, India

*Email: ganeswar.nath@gmail.com

Abstract. Nano fluids are stable suspensions of nano particles in a liquid. Metal oxide nanoparticles offer great merits over controlling rheological, thermal, chemical and physical properties of solutions. The effectiveness of a nanoparticle to modify the properties of a fluid depends on its diffusive properties with respect to the fluid. An important application of Nano fluids containing nano particles is as a coolant, since the addition of only a few volume percent of nano particles to a liquid coolant and significantly improves its thermal conductivity. A heat exchange device using nanofluid needs to operate at best nanoparticle loading to get the maximum heat transfer performance. Due to its antimicrobial properties, the future therapeutic directions may include nano-silver as an anti-inflammatory and as anti-platelet agents and antiviral drugs. Owing to the applications of nano silver particles, attempts has been made in the present study to synthesis silver nano particles and characterized with the help of ultrasonic technique. Different acoustic properties of $AgNO_3$ nanofluids are studied with the help of Ultrasonic technique. Silver nanofluids are synthesized by chemical reduction method. The chemical reduction method involves the reduction of $AgNO_3$ in aqueous solution by a reducing agent in the presence of a suitable stabilizer, which is necessary in protecting the growth of silver particles from aggregation. Ultrasonic velocity and density values are measured for different concentrations of $AgNO_3$ nanofluids. The acoustic parameters like adiabatic compressibility, intermolecular free length, and acoustic impedance are calculated from the experimental data. The variation of ultrasonic velocity and other parameters with particle concentration are discussed on the basis of interaction of ultrasonic wave with the nano sized silver particles.
Keywords: Nanofluid, silver nanofluid, ultrasonic velocity, acoustic parameter

1. Introduction

Nano science has blossomed its demand and dominancy from the last decade and the need for nanotechnology will increase as miniaturization becomes more and more essential in areas such as computing, sensors and biomedical applications. Nanofluids are novel suspensions of nano solid particle in base liquids. When a very small amount of guest nano particles, dispersed uniformly and suspended stably in base liquid, they provide dramatic improvements in the physico - chemical and thermal properties of base liquid [1]. One fascinating feature of nano fluids is that they have anomalously high

IOP Conf. Series: Materials Science and Engineering **360** (2018) 012064 doi:10.1088/1757-899X/360/1/012064

thermal conductivity [2-3], which make nano fluids strong material for the next generation of coolants for improving the design and performance of thermal management systems. The Brownian motion of nano particles in base liquid is proposed to be one of the major physical mechanisms of the thermal conduction of nano fluids [4-6].

Therefore; it is significant to investigate the movements of nano particles in nano fluids. Ultrasonication is an accepted technique for study of such Brownian motion and dispersing aggregated nanoparticles for the preparation of aqueous nano suspensions. Most of the studies [7] have focused on the effect of sonication time on the stability of nanofluids and all of them used continuous ultrasonic waves to sonication. The ultrasonic property of solid–liquid suspensions in micrometer size of particles causes a change in ultrasonic velocity and ultrasonic attenuation which affects the thermal conductivity of the nanofluids [8]. More recently there has been an increasing interest in the acoustical properties of suspensions for acoustic telemetry through drilling fluids as well as arising demand for ultrasonic particle size instrumentation. Several scientists have made the study of ultrasonic propagation behavior through the suspension of solid particles particularly in micrometer or millimeter size in a liquid aiming to identify the mechanism that enable useful information to be extracted from the behavior of ultrasonic properties, such as particle size, concentration and mechanical properties of the constituents. Because of their high surface to volume ratio, nanoparticles suspended in base liquid, significantly revealing some new properties which are not present in either of the pure materials. Therefore, the investigation of the influence of nanoparticles on the properties of a base liquid is necessary in order to be able to better predict the final properties of the complex fluids. In the present work we have made the study of the propagation of ultrasonic wave in a colloidal solution with dispersed Ag-metal nanoparticles .We prepared stable nanofluids containing silver metal nanoparticles suspended in the base fluid like water. Even though reports are available on the thermal conductivity and viscosity of nanofluid, very little work is reported on the acoustical parameters of nanofluid. Hence, an attempt is made in this work to prepare nanofluid of Ag and study their acoustical parameters. Ag nanofluid is prepared using one step chemical method with water as base fluid. Ultrasonic velocity measurements are carried out on Ag nanofluid and the results are discussed in detail.

2. Material and chemicals

Silver nitrate ($AgNO_3$, 99.9%), tannic acid ($C_{76}H_{52}O_{46}$, M_w = 1701), K_2CO_3 were all purchased from Sigma-Aldrich and used without further treatment. All of the solutions were freshly made for the synthesis of silver nanoparticles, especially the freshly made tannic acid aqueous solution, which was ice bathed before use. All solutions were prepared using distilled water.

The present work, describes a green protocol using tannic acid, a polyphenolic plant extract, as both the reducing and stabilizing agent. Tannic acid has been studied extensively for its antioxidant properties and as a chelating agent for several inorganic cations. The representative structure of tannic acid, corresponding to its average formula weight, is shown in Fig.1. It consists of a central core of glucose that is linked by ester bonds to polygalloyl ester chains. Tannic acid at its natural acidic pH is known to be a weak reducing agent that can only grow seeds into nanoparticles at room temperature.

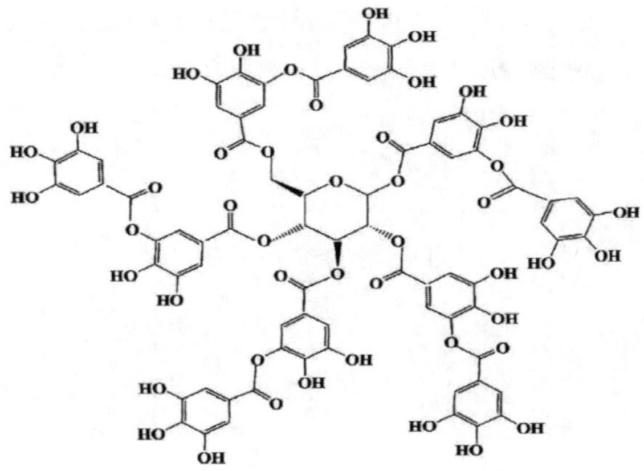

Fig 1. Molecular structure of tannic acid ($C_{76}H_{52}O_{46}$)

Tannic acid has a pH value between seven and eight, depending on its extent of dissociation and is known to partially hydrolyse under mild acidic/basic conditions into glucose and gallic acid units. Gallic acid at alkaline pH reduces silver nitrate into silver nanoparticles rapidly at room temperature, but the particles form aggregates in solution as gallic acid is a poor stabilizing agent. Glucose is a weak reducing agent at room temperature, but it is a good stabilizing agent at alkaline pH. These findings suggest that tannic acid could be an ideal reducing and stabilizing agent under alkaline conditions at room temperature.

3. Synthesis of silver nano particles

One-step chemical method is followed for synthesis of metallic nanoparticles, since the nanoparticles are dispersed in the base fluid as they are produced. This process helps in preventing oxidation of the particles and minimized the agglomeration of nanoparticle. Silver nano fluid was prepared by chemical reduction method using tannic acid as a reducing agent. 35mg of tannic acid was weighed and dissolved in 150 ml of water in a beaker. From this solution 20 ml was further diluted with 40 ml of distilled water which gives total 60 ml solution. After this 30 mg of K_2CO_3 powder was weighed and added to this 60 ml solution due to which there is slight change in colour i.e. pale yellow colour appeared and it was covered with a parafilm. Then 0.22g of $AgNO_3$ was exactly taken and transferred to 100ml volumetric flask to get 0.00132M standard $AgNO_3$ solution. When all this work was completed then the 60ml solution was kept on a magnetic stirrer and a magnetic bar is used for stirring inside the solution. Then while stirring 1ml of $AgNO_3$ solution was added slowly with the help of a siring. This solution was kept on a magnetic stirrer for 1hr. and then gradual darkening of the solution was formed as a function of the amount of silver nitrate added to the solution. Similarly, 2ml, 3ml ,4ml,and 5ml concentration of $AgNO_3$ solution was added to the more four same amount of 60ml solution and this solution was also kept on a magnetic stirrer for 1hr as shown in Fig.2(a) and (b).

IOP Conf. Series: Materials Science and Engineering **360** (2018) 012064 doi:10.1088/1757-899X/360/1/012064

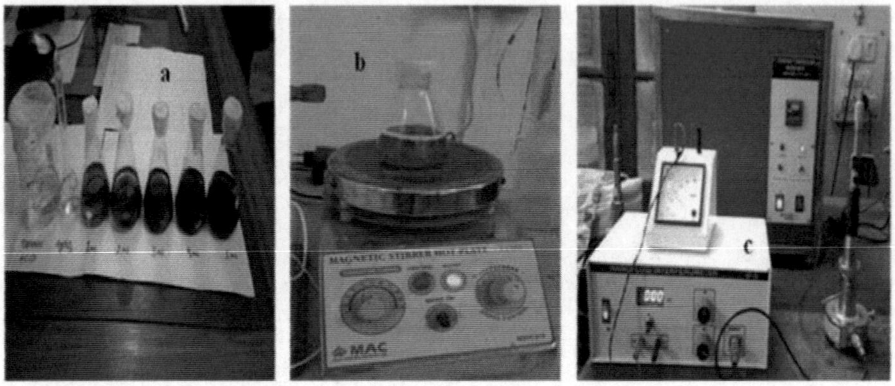

a. Different concentration of silver nano particle

b. Uniform dispersion of silver nano particle by magnetic stirrer

c. Ultrasonic velocity measurement

Fig 2. Synthesis of silver nanofluid and measurement of ultrasonic velocity

Thus the probable reaction mechanism for the formation of Ag nanoparticles by tannic acid reduction of $AgNO_3$ can be represented as below:

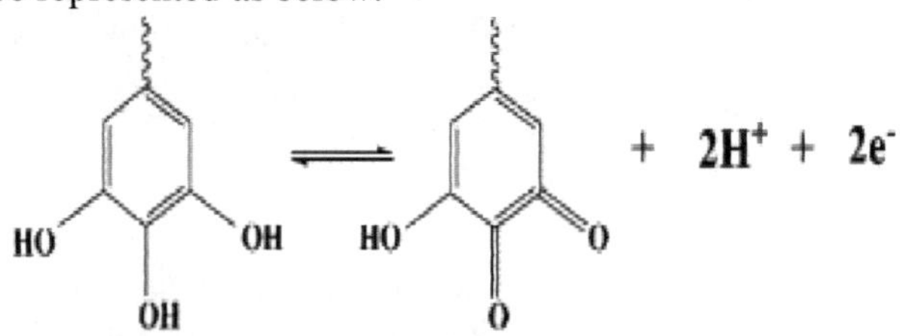

(Phenolic form) (Quinone form)

$$Ag^{+} + 2e^{-} \Longrightarrow Ag^{0}$$

In this model reaction, phenol took part in redox reactions to form quinones and donate electrons. The Ag^{+} converted to be Ag atoms followed by electrons, forming Ag clusters, and producing Ag-particles by cluster growth. Ag atoms are stabilized from oxidative reactions and aggregations when capping in tannin molecules.

4. UV- Vis spectroscopy measurements

The absorption spectrum of synthesized silver nanofluid for different concentration was observed by ultraviolet-visible spectrophotometer (HITACHI). Fig.3 shows the UV-Vis absorption spectra of silver nanofluid scanned in the wavelength range of 300 - 600 nm. The strong absorption peaks of UV-visible spectrum revealed the presence of silver nano particles. The UV-visible spectrum shows strong absorption peaks at 410 nm and 416 nm due to the plasmon oscillation modes of conduction electrons in the colloidal nanoparticles-liquid suspensions. This indicates that size of Ag nanoparticles increases with increase of concentration. The peak shifting towards the shorter wavelength region was observed indicating the decrease in the size of nanoparticles.

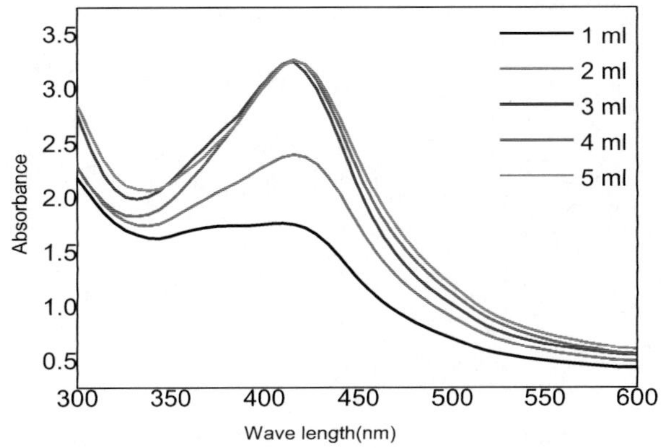

Fig 3. Absorption spectra of silver nanofluids

4. Ultrasonic velocity measurements

Ultrasonic velocity measurements have been made at 2 MHz of frequency with help of a variable path interferometer at temperature 30^0C. The temperature stability is maintained within 0.1K by circulating thermo stated water around the interferometer cell that contains the sample, with circulating pump.The velocity of ultrasonic waves in the liquid is calculated as

$$C = fx\lambda \quad \ldots (1)$$

where f is the frequency of the generator and λ is the wave length of ultrasonic wave measured from the interferometer. The measured ultrasonic velocity data and density of nano fluid are used to compute the acoustical parameter such as adiabatic compressibility (β), intermolecular free length (L_f) and acoustic impedance (Z) using relations [9]

Adiabatic compressibility, $\beta = (\rho\, C^2)^{-1}$ ... (2)

Intermolecular free length, $L_f = k\beta^{1/2}$ ... (3)

Acoustic impedance, $Z = \rho C$ ... (4)

The constant k is temperature dependent which is given as [93.875+ (0.375T)] $\times 10^{-8}$ and 'T' being the absolute temperature.

.

IOP Conf. Series: Materials Science and Engineering **360** (2018) 012064 doi:10.1088/1757-899X/360/1/012064

5. Result and discussion

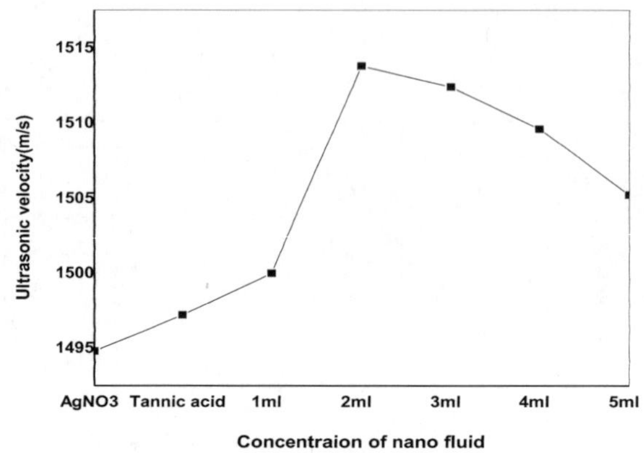

Fig 4. Variation of ultrasonic velocity with concentration of silver nano fluid

From Fig.4 it is found that ultrasonic velocity increases with increase in concentrations of nanofluids. But at certain concentrations it shows variation in the values. The increase in velocity with increase in mole fraction of particles is due to structural changes occurring in the liquid system which results in weakening of intermolecular forces. The gradual increase in velocity up to 2ml Ag nanofluid concentration is due to interaction between Ag^+ ions and water molecules. Increase in velocity in solution indicates that the interaction may be due to the surface effect because of hydrogen bonding between particles and water molecules. Hence, there might be particle fluid interaction which favors increase in the velocity values. The random movements of nanoparticles are increased with increase in concentration and when the ultrasonic vibration is propagated in nanofluid, Brownian motion stops the fluid particles in suspension, leading to decrease in velocity. Therefore, at higher concentrations it is found that there is deviation in ultrasonic velocity. Also agglomeration in particles leads to collisions between the suspended particles.

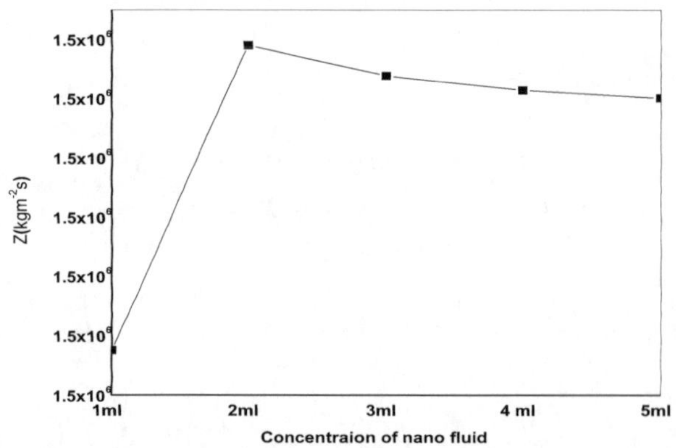

Fig 5. Variation of acoustic impedance with concentration of silver nano fluid

It is also observed that at 2ml concentration, there is a decrease in velocity value which suggests that there may be weak interactions between particles and water molecules. The structural changes affect the compressibility and hence there is a change in ultrasonic velocity. The decreased velocity value shows that there is decrease in nanoparticle-fluid interaction and increase of particle – particle interaction. Thus it may suggest that, the ultrasonic velocity is quite sensitive to the size, morphology and dispersion of the particles. The increase in density with concentration is due to interactions between particles and water molecules.

It is observed that acoustic impedance values increases with increase in concentration of particles and then it decreases with further increase in concentration. It implies that the Z-values show similar behavior to that of ultrasonic velocity values as shown in Fig.5. From the profile it is observed that acoustic impedance increases with increase in Ag^+ ion concentrations in water. But there is a decrease of Z- value at 2ml concentration which shows weak interactions similar to ultrasonic velocity. The higher values of acoustic impedance indicate that there is a significant interaction between the particle and base fluid molecules which may affect the structural arrangement. The interaction between particles and base fluid molecules increases the intermolecular distance between the molecules which in turn causes impedance in the propagation of ultrasonic waves.

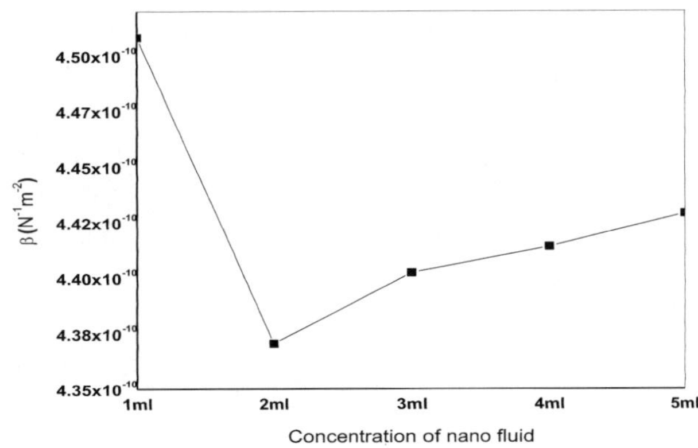

Fig 6. Variation of adiabatic compressibility with concentration of silver nano fluid

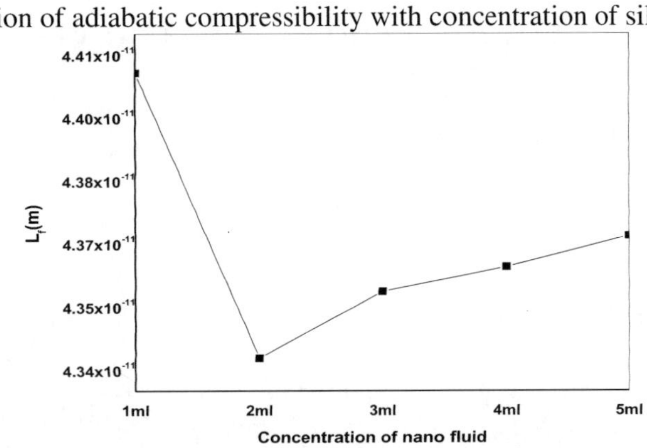

Fig 7. Variation of intermolecular free length with concentration of silver nano fluid

It is also found that the adiabatic compressibility and free length decreases with increasing concentration of particles. The decrease in adiabatic compressibility and free length shows (Fig.6 and Fig.7) the weaker force of interaction between particles and base fluid molecules. Compressibility decreases due to the fact that Ag^{+} ions form a core compact structure with the solvent molecules through hydrogen bonding. Deviations in the values of adiabatic compressibility and intermolecular free length exhibit in the system may be due to weak forces operating with some specific interaction between molecules in this. Decrease in free length value causes rise in sound velocity in the system which is an indication of clustering together of the molecules into agglomerates.

6. Conclusion

In the present paper silver nanoparticles are obtained using one step chemical method. From ultraviolet-visible spectrophotometer analysis, the nano size confirmed comparing with the literature data [10-11] and particle size of silver nano particle with tannic acid at room temperature is greater than 15nm. Homogeneous suspensions of silver nanofluids are prepared in different concentrations with the aid of magnetic starrier. Ultrasonic velocity studies were carried out for the prepared nanofluids at room temperature (303K). The acoustical parameters including adiabatic compressibility, intermolecular free length, and acoustic impedance are calculated for all the samples. Interaction between nanoparticles and water molecules was analyzed using acoustical parameters. The ultrasonic studies provide comprehensive investigations between Ag^{+} ion and water molecules. From the analysis, it is evident that the particle–fluid interaction in microfluids decreases with increasing particle loading [12]. But for the nanofluids the particle –fluid interaction increases with increase of concentration up to a critical concentration of 2ml above which the particle –fluid interaction weakens due to strong particle – particle interaction. It is observed that there is particle – fluid interaction which favors increase in velocity. Thus, particle–fluid interaction studies are helpful to understand the reasons behind anomalous enhancements in physical properties of nanofluids in nanoscale.

Acknowledgement

The authors are thankful to University Grant Commission, New Delhi for providing the financial support under the BSR Scheme, No.F.20-1(28)/2012(BSR) to carry out the research work.

References

1. Das S K et al 2008 *Nanofluids: Science and Technology* Hoboken, John Wiley & Sons, Inc. 1-36
2. Eastman J A et al 2001 *Appl. Phys. Lett.* **78** 718
3. Das S K et al 2003 *J. Heat Transfer* **125** 567
4. Prasher R et al 2005 *Phys. Rev. Lett.* **94** 025901
5. Chon C H and Kihm K D 2005 *Appl. Phys. Lett.* **87** 153107
6. Xuan Y and Li Q 2006 *J. Appl. Phys.* **100** 043507
7. Chung S J 2009 *Powder Technol.* **194** 75
8. Yadav R R 2008 *Ultrasonics* **48** 591
9. Nath G and Paikaray R 2009 *Indian J. of Phys.* **9** 763
10. Zhao S Y 2004 *Physica E* **23** 92
11. Bulut E and Oezacar M 2009 *Ind. Eng. Chem. Res.* **48** 5686
12. Hemalatha J 2011 *Microfluid Nanofluid* **10** 263

Crystal growth and physical characterization of bismaleato lead (IV) grown in silica gel

V Mahalakshmi, A Lincy, J Thomas and K V Saban*

Smart Materials Analytic Research and Technology (SMART), St. Berchmans College, Changanassery 686101, Kerala, India

*Corresponding author. Tel: +91 9847390370, Fax: +91 4812401472, E-mail address: smartlabindia@gmail.com.

Abstract. Single crystals of bismaleato lead(IV) [$Pb_2(C_4H_2O_4)_2$] are grown by the controlled diffusion of lead ions in gelated hydrosilica matrix impregnated with maleate ions. The various functional groups present in the compound is identified using FTIR spectrum. The structure of the crystal was elucidated using single crystal XRD studies. It crystallizes in the monoclinic system with space group $P2_1/c$. The TG- DTA analysis was employed l to explore the thermal stability and decomposition the pattern. the dielectric measurements were made over wide frequency and temperature ranges. The optical band gap was determined using diffuse reflectance spectroscopy.

1. Introduction

Metal coordinated dicarboxylates have wide applications in the regimes of ferroelectricity, piezoelectricity, photoconductivity and optical nonlinearity [1-7]. Among dicarboxylates, metal complexes of maleic acid have displayed optical, magnetic and electrical properties [8-10]. A detailed report only on the structural characterization of a lead maleate complex grown by the slow evaporation method [11] is available. As the solubility of metal dicarboxylates is very low in pure water, it is not so easy to grow a them by the slow cooling or slow evaporation methods. Strain free single crystals of good optical quality can be grown by the controlled diffusion technique in inert hydrosilica gel medium and employing this method we have succeeded in growing good crystals of a few maleates [12-15]. Here we present a concise report on the structural, thermal, dielectric and optical properties of lead maleate crystals grown by the above technique.

2. Experimental

Lead maleate crystals were grown in gel matrices taken in straight tubes by titrating maleic acid against sodium meta silicate solutions of different densities (1.03-1.07 g/cc, in steps of 0.01) and pH varying from 4 to 7 [16]. Concentration of the maleic acid was also varied. After ensuring firm gel setting, supernatant solution, lead nitrate, of different molarities (0.25 – 1M) were poured. The tubes were then hermitically sealed and kept undisturbed at room temperature.

The powdered sample was subjected to FTIR Spectroscopic investigations using a Thermo Nicolet Avtar 370 model FTIR Spectrophotometer, in the range 400-4000 cm^{-1} at a resolution of 4 cm^{-1}, using KBr pellet method. The single crystal X-ray diffraction studies were conducted on a good crystal of size 0.15 x 0.15 x 0.10 mm^3 using Bruker AXS Kappa Apex2 CCD Diffractometer with graphite monochromated Mo-Kα radiation. The software APEX2/SAINT [17] was used for the data collection, cell refinement and data reduction of the crystal.The crystal structure was solved using SIR92 [18] and was refined using SHELXL-97 [19-20]. The Full-matrix least squares refinement based on 931 reflections and 83 parameters along with 48 restraints, converged the residuals to R_1= 0.0490, wR_2= 0.1464. The thermal behaviour of the sample was explored by the Perkin Elmer made Pyris Diamond TG–DTA analyzer. For this the powdered sample was heated from 50 °C to 1030 °C at the rate of 20 °C/min. The dielectric properties of the grown crystal was carried out for wide frequency and temperature ranges using Hioki 3532-50 Hi tester. For this pellets were prepared by the application of 4 tonnes pressure for two minutes. The DRS analysis of the powdered sample was conducted using Jasco UV-Vis-NIR spectrophotometer in the range 200 to 2000 nm.

3. Results and discussion

3.1. Crystal Formation

Optically transparent single crystals with flat surfaces formed within the gel in a period of about 8 weeks. Excellent crystals developed in a gel of density 1.06 gm/cc with 0.75 M concentration of inner maleic acid at pH 6. The outer supernatant lead nitrate solution of strength 0.25 M was poured gently on the set gel. Crystals of dimensions up to 2 x 1 x 1 mm^3 were obtained (Figure 1).

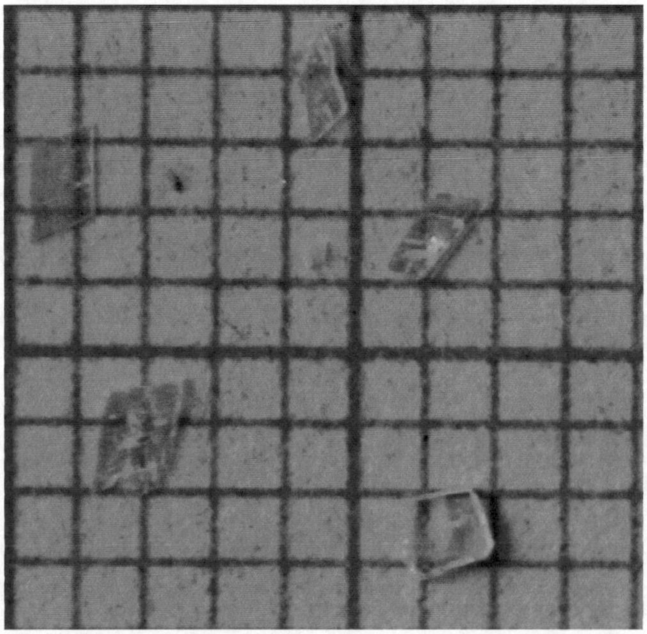

Figure 1. Photographs of the grown crystals.

3.2. FTIR Spectral Analysis

The recorded FTIR spectrum of the sample (Figure 2) is interpreted in the light of the established data of similar compounds [21-23] and found to bear the signatures of all the functional groups in the compound. The stretching mode of the C-H group is depicted with the band at 3008 cm^{-1}. The stretching of the covalent bond C=O is revealed by the peak at 1719 cm^{-1} and its asymmetric stretching is shown by the band at 1672 cm^{-1}. Asymmetric and symmetric stretching of the carboxylate groups occur at wave numbers 1534 cm^{-1} and 1382 cm^{-1} in order. The peaks at 1408 cm^{-1} and at 1303 cm^{-1} represents the deformation of C-H group. The peaks at 1495 cm^{-1}, 1174 cm^{-1} and 905 cm^{-1} is credited to the asymmetric and symmetric stretching of the CC bond. Rocking, twisting and vibration modes of the C-H group happen at 973 cm^{-1}, 834 cm^{-1} and 741 cm^{-1} in order. The metal oxygen stretching and its combination along with C-C stretching give rise to the respective vibrations at 696 cm^{-1} and 542 cm^{-1}. Vibration mode of the carboxylate group is assigned with the peak at 615 cm^{-1}.

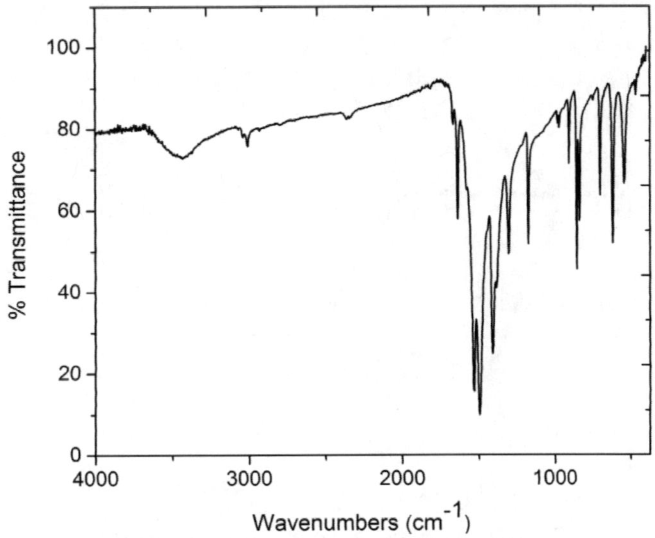

Figure 2. FTIR spectrum of lead maleate.

Table 1. Spectral assignment of IR peaks.

Wave number (cm^{-1})	*Assignment*	*Wave number (cm^{-1})*	*Assignment*
3008	υ C-H	1174	$υ_{as}$ C=C
1719	υ C=O	973	$ρ_r$ C-H
1672	$υ_{as}$ C=O	905	$υ_s$ C=C
1534	$υ_{as}$ COO	834	$ρ_t$ C-H
1495	$υ_{as}$ C=C	741	π C-H
1408	δ C-H	696	υ M-O
1382	$υ_s$ COO	615	π COO
1303	δ C-H	542	υ M-O + υ C-C

3.3. *Crystal Structure Analysis*

The single crystal X-ray diffraction studies reveal that the crystallized complex is bismaleatolead(IV) [$Pb_2(C_4H_2O_4)_2$]. It crystallizes in the monoclinic system with space group $P2_1/c$ and unit cell dimensions a = 9.8778(13) Å, b = 6.9501(9) Å, c = 8.2630(11) Å, $\alpha = \gamma = 90$ ° and β =111.191(7) °. This matches well the reported structural data [11]. The crystal structure and refinement is displayed in Table 2. The bond lengths and bond angles are listed in Tables 3 and 4. The ORTEP [24-26] of the molecule with thermal ellipsoids at 50% probability is shown in figure 3. The structure of the molecule reveals that each maleate anion acts as tetradendate ligand - bidendate through the oxygen atoms (O1 and O3) of the two carboxylate groups to one Pb^{2+} ion (Pb1) and two monodendate coordination by the other two oxygen atoms (O2 and O4) to adjacent Pb^{2+} ions (Pb2 and Pb3). Thus the compound gets chelated and heightens the thermodynamic stability. The intermolecular hydrogen bonds also increase the stability of the structure. The packing diagram shows that (Figure 4) each Pb^{2+} ion is heptacoordinated to oxygen atoms forming a stable distorted octahedron. Also each Pb^{2+} ion is found to be bonded with two maleate anions.

IOP Conf. Series: Materials Science and Engineering **360** (2018) 012065 doi:10.1088/1757-899X/360/1/012065

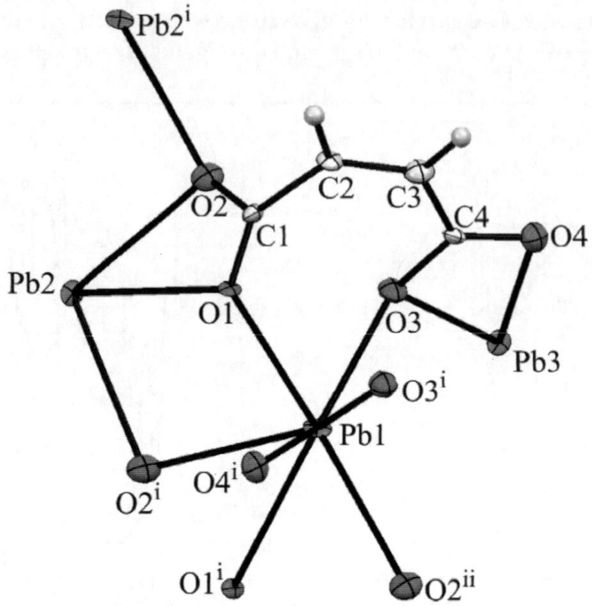

Figure 3. ORTEP of the molecule with 50 % probability.

Table 2. Crystal data and structure refinement.

Empirical formula	$C_8 H_4 O_8 Pb_2$
Formula weight	642.49
Temperature	293(2) K
Wavelength	0.71073 Å
Crystal system, space group	Monoclinic, $P2_1/c$
Unit cell dimensions	
	a = 9.8778(13) Å, α = 90 °
	b = 6.9501(9) Å, β = 111.191(7) °
Volume	c = 8.2630(11) Å, γ = 90 °
Z, Calculated density	528.91(12) Å^3
Absorption coefficient	2, 4.304 Mg/m^3
F(000)	31.822 mm^{-1}
Crystal size	560
θ range for data collection	0.15 mm x 0.15 mm x 0.10 mm
Limiting indices	2.21 ° to 25 °
Reflections collected / unique	$-11\leq h \leq 11$, $-7 \leq k \leq 8$, $-7\leq l \leq 9$
Completeness to θ = 25.00 °	4087 / 931 [R(int) = 0.0647]
Absorption correction	100.0 %
Max. and min. transmission	Semi-empirical from equivalents
Refinement method	0.1865 and 0.0845
Data / restraints / parameters	Full-matrix least-squares on F^2
Goodness-of-fit on F^2	931 / 48 / 83
Final R indices [I>2σ(I)]	1.120
R indices (all data)	R_1 = 0.0490, wR_2 = 0.1464
Extinction coefficient	R_1 = 0.0527, wR_2 = 0.1494
Largest diff. peak and hole	0.0031(9)
	3.728 and -3.235eÅ^{-3}

Table 3. Bond lengths in bismaleatolead (IV).

Bond	Lengths (Å)	Bond	Lengths (Å)
C(1)-O(2)	1.26(2)	O(1)-Pb(1)#2	2.708(10)
C(1)-O(1)	1.257(19)	O(2)-Pb(1)#2	2.622(12)
C(1)-C(2)	1.46(2)	O(3)-Pb(1)#1	2.422(11)
C(2)-C(3)	1.32(2)	O(3)-Pb(1)	2.660(11)
C(2)-H(2)	0.9300	O(4)-Pb(1)#1	2.523(12)
C(3)-C(4)	1.48(2)	Pb(1)-O(3)#3	2.422(11)
C(3)-H(3)	0.9300	Pb(1)-O(4)#3	2.523(12)
C(4)-O(4)	1.204(19)	Pb(1)-O(2)#4	2.622(12)
C(4)-O(3)	1.318(18)	Pb(1)-O(1)#4	2.708(10)
C(4)-Pb(1)#1	2.814(16)	Pb(1)-C(4)#3	2.814(16)
O(1)-Pb(1)	2.432(10)		

Table 4. Bond angles of bismaleatolead(IV).

Bond	Angles (°)	Bond	Angles (°)
O(2)-C(1)-O(1)	121.6(14)	C(4)-O(4)-Pb(1)#1	90.9(9)
O(2)-C(1)-C(2)	117.7(14)	O(3)#3-Pb(1)-O(1)	81.4(4)
O(1)-C(1)-C(2)	120.5(14)	O(3)#3-Pb(1)-O(4)#3	53.1(4)
C(3)-C(2)-C(1)	126.9(15)	O(1)-Pb(1)-O(4)#3	80.5(3)
C(3)-C(2)-H(2)	116.6	O(3)#3-Pb(1)-O(2)#4	72.8(4)
C(1)-C(2)-H(2)	116.6	O(1)-Pb(1)-O(2)#4	73.0(4)
C(2)-C(3)-C(4)	123.4(14)	O(4)#3-Pb(1)-O(2)#4	122.5(4)
C(2)-C(3)-H(3)	118.3	O(3)#3-Pb(1)-O(3)	142.2(4)
C(4)-C(3)-H(3)	118.3	O(1)-Pb(1)-O(3)	70.6(3)
O(4)-C(4)-O(3)	122.4(14)	O(4)#3-Pb(1)-O(3)	96.5(3)
O(4)-C(4)-C(3)	121.8(13)	O(2)#4-Pb(1)-O(3)	119.8(4)
O(3)-C(4)-C(3)	115.8(13)	O(3)#3-Pb(1)-O(1)#4	79.8(3)
O(4)-C(4)-Pb(1)#1	63.7(9)	O(1)-Pb(1)-O(1)#4	121.5(2)
O(3)-C(4)-Pb(1)#1	59.3(8)	O(4)#3-Pb(1)-O(1)#4	125.6(3)
C(3)-C(4)-Pb(1)#1	169.7(10)	O(2)#4-Pb(1)-O(1)#4	48.6(4)
C(1)-O(1)-Pb(1)	121.7(9)	O(3)-Pb(1)-O(1)#4	136.6(3)
C(1)-O(1)-Pb(1)#2	92.6(8)	O(3)#3-Pb(1)-C(4)#3	27.9(4)
Pb(1)-O(1)-Pb(1)#2	110.0(4)	O(1)-Pb(1)-C(4)#3	82.0(4)
C(1)-O(2)-Pb(1)#2	96.7(9)	O(4)#3-Pb(1)-C(4)#3	25.3(4)
C(4)-O(3)-Pb(1)#1	92.8(9)	O(2)#4-Pb(1)-C(4)#3	99.7(4)
C(4)-O(3)-Pb(1)	121.4(9)	O(3)-Pb(1)-C(4)#3	120.2(4)
Pb(1)#1-O(3)-Pb(1)	112.7(4)	O(1)#4-Pb(1)-C(4)#3	103.1(4)

Symmetry transformations used to generate equivalent atoms:
#1 x,-y+1/2,z-1/2 #2 -x+1,y+1/2,-z+1/2
#3 x,-y+1/2,z+1/2 #4 -x+1,y-1/2,-z+1/2

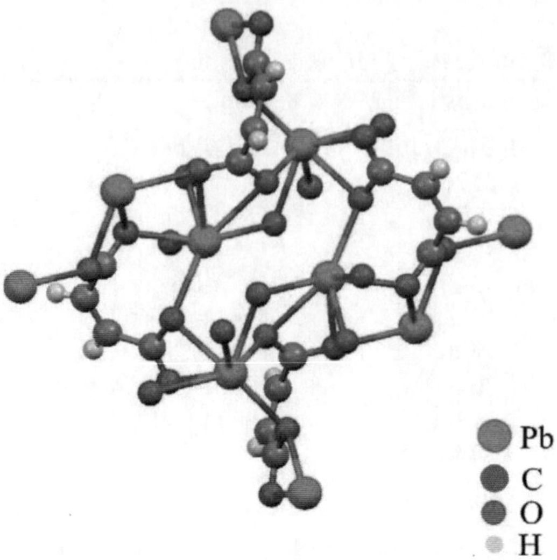

Figure 4. Polymeric structure of lead maleate

3.4. Thermal Studies

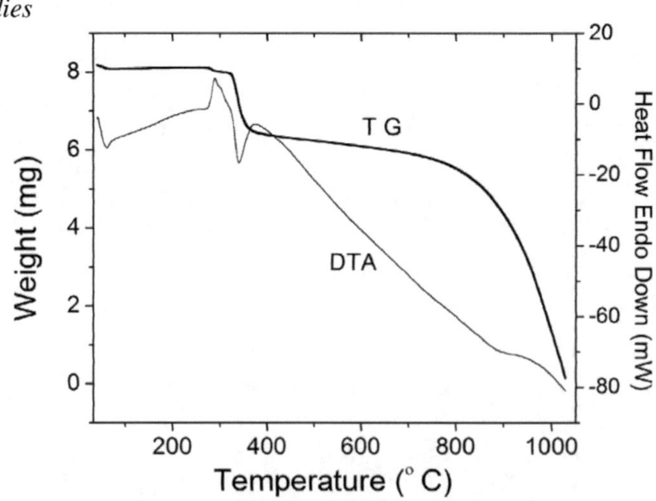

Figure 5. TG- DTA curves of bismaleatolead(IV).

The analysis of the TG-DTA curves (Figure 5) reveals the thermal stability and decomposition pattern of the grown sample. The TG curve indicates that the material is thermally stable up to 274 °C. It is also to be noted that the TG curve does not show any considerable weight loss till this temperature. Upon heating the sample to 338 °C, the compound turns to lead oxalate by the removal of acetylene molecules. The exothermic peak at 289 °C in DTA confirms this decomposition. During the next stage of heating to 355 °C, expulsion of carbon monoxide takes place leaving behind lead carbonate. The broad exothermic peak which starts immediately after 340 °C in the DTA is indicative of this stage. Further heating the material up to 799 °C, it gets transformed into lead oxide by the elimination of carbon dioxide molecules. This transformation is shown in the DTA by the endothermic peak which starts beyond 378 °C. The agreement between the observed and calculated mass loss matches well in each stage and is given in Table 5. The TG curves also displays that the weight loss continues even after 1000 °C, which may be due to reduction of lead oxide to the stable metallic lead.

The stoichiometry of the crystallized complex is evident from the thermal studies also and hence the chemical reaction leading to the formation of the title compound is:

$$2\ Pb(NO_3)_2 + 2\ C_4H_4O_4 \rightarrow Pb_2(C_4H_2O_4)_2 + 4\ HNO_3.$$

IOP Conf. Series: Materials Science and Engineering **360** (2018) 012065 doi:10.1088/1757-899X/360/1/012065

Table 5. The decomposition process of $Pb_2(C_4H_2O_4)_2$.

Stage	Decomposition temperature(°C)	Product after decomposition	Molecules evolved	Observed mass loss(%)	Calculated mass loss(%)
1	274-338	$Pb_2(C_2O_4)_2$	$2C_2H_2$	8.193	8.104
2	338-355	$Pb_2(CO_3)_2$	2CO	8.844	8.719
3	355-799	2PbO	$2CO_2$	14.207	13.699

3.5. *Dielectric Studies*

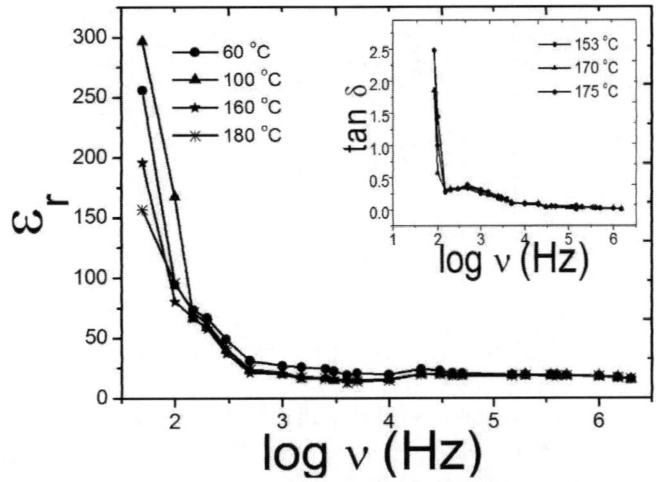

Figure 6. Frequency response of dielectric constant and dielectric loss.

Figure 6 shows that the value of dielectric constant and loss tangent is greater at low frequencies. The higher value of dielectric constant at low frequencies is mainly due to the presence of interfacial polarization and that of loss tangent is attributed to the space charge polarization [27-28]. The dielectric constant decreases with increasing frequency, depending on the fact that beyond certain frequency of the electric field, the dipole does not follow the applied alternating field [29-30]. The low dielectric loss at high frequency ensures that the grown crystals are of good optical quality.

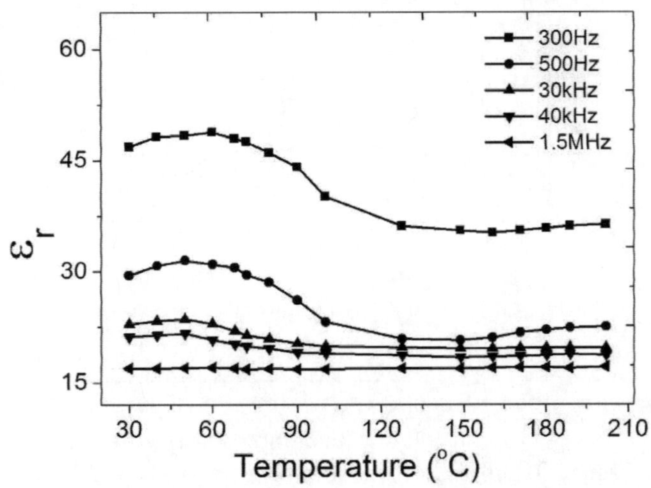

Figure 7. Thermal response of dielectric constant.

IOP Conf. Series: Materials Science and Engineering **360** (2018) 012065 doi:10.1088/1757-899X/360/1/012065

The variation of the dielectric constant and loss with temperature is depicted in figure 7. The slightly higher dielectric permittivity for low frequencies confirms the presence of interfacial polarization in the material. The dielectric constant displays a slight increase near 60 °C only at lower frequencies and for higher frequencies (40 kHz and 1.5 MHz), the dielectric constant is almost temperature independent, indicating the good chemical homogeneity [31] and structural stability of the grown crystals.
The ac conductivity of the material is estimated using the relation $\sigma_{ac} = \acute{\varepsilon}\ \varepsilon_0\ \omega \tan\delta$ and it shows almost similar behaviour both for high and low frequencies (Figure 8). It can also be seen that the conductivity displays usual behaviour of maleates with slightly elevated value for high frequencies and almost temperature independent behaviour.

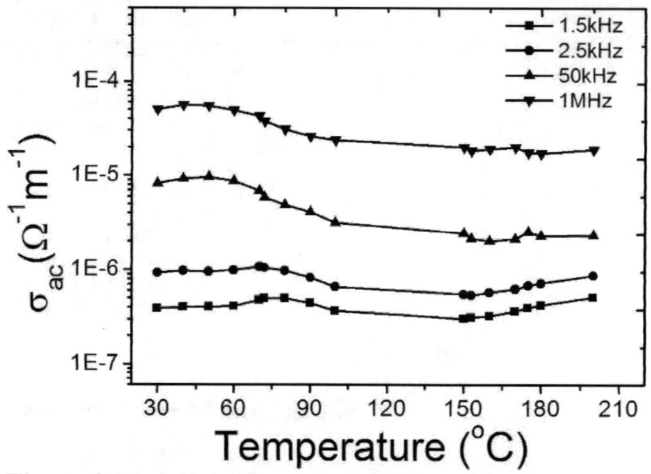

Figure 8. Variation of a.c. conductivity with temperature.

3.6. *Diffuse Reflectance Spectroscopy*

The diffuse reflectance spectrum of the title compound in the UV-Vis-NIR region is shown in figure 9. The band gap is determined using the Kubelka Munk Theory [32-33]. The Kubelka–Munk remission function F(R) at each wavelength λ is given by $F(R) = (1-R)^2/2R = k/s$, *R* being the reflectance of the sample, *k* is the absorption coefficient and *s* is the scattering coefficient.

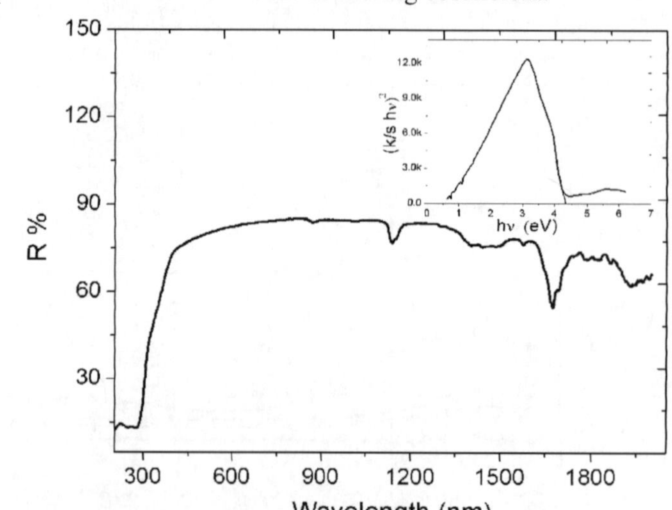

Figure 9. Diffuse reflectance spectrum of the compound.

The plot of $((k/s)h\nu)^2$ versus $h\nu$ (inset of figure 9) was drawn and the optical band gap (E_g) is evaluated from the linear fit of the curve on the energy axis is 4.32 eV.

IOP Conf. Series: Materials Science and Engineering **360** (2018) 012065 doi:10.1088/1757-899X/360/1/012065

4. Conclusions

Well faceted, transparent single crystals of bismaleato lead have been developed by the controlled diffusion of aqueous solutions of maleic acid and lead nitrate in silica gel. The functional groups were determined from FTIR spectroscopy. The single XRD studies revealed that the complex crystallized in the monoclinic system with space group $P2_1/c$ with unit cell dimensions $a = 9.8778(13)$ Å, $b = 6.9501(9)$ Å, $c = 8.2630(11)$ Å, $\alpha = \gamma = 90$ ° and $\beta = 111.191(7)$ °. The TG-DTA analyses showed that the compound is thermally stable up to 274 °C. The dielectric and ac conductivity studies show that the compound exhibits normal dielectric behavior. The optical band gap of the material explored using the DRS studies is 4.32 eV.

5. References

[1] Dalal P V and Saraf K B 2006 *Bull. Mater. Sci.* **29** 421

[2] Arora S K, Patel V, Patel R G, Amin B and Kothari A 2004 *J. Phys. Chem. Solids* **65** 965

[3] Jini T, Saban K V and Varghese G 2006 *Cryst. Res. Technol.* **41** 250

[4] Jini T, Saban K V, Varghese G, Naveen S, Sridhar M A and Prasad J S 2007 *J. Alloys Compd.* **433** 211

[5] Lincy A, Mahalakshmi V, Tinto A J, Thomas J and Saban K V 2010 *Physica B* **405** 4673

[6] Doreswamy B H, Mahendra M, Sridhar M A, Prasad J S, Varughese P A, George J and Varghese G 2005 *Mater. Lett.* **59** 1206

[7] Varghese M, Lizymol X, Mahadevan C K and Abraham K E 2011 *Phys. Scr.* **83** (3) 035801

[8] Bharthasarthi T, Arulmozhichelvan P and Murugakoothan P 2010 *Mater. Lett.* **64** 1506

[9] Natarajan S, Britto S A M and Ramachandran E 2006 *Cryst. Growth Des.* **6** (1) 137

[10] Oczko G, Legendziewicz J, Wickleder M S and Meyer G 2002 *J. Alloys Compd.* **341** 255

[11] Bonhomme F et al. 2005 *Inorg Chem.* **44** 7394

[12] Mahalakshmi V, Lincy A, Thomas J and Saban K V 2012 *J. Phys. Chem. Solids* **73** 584

[13] Mahalakshmi V, Lincy A, Thomas J and Saban K V 2013 *IOSR J. Appl. Phys.* **4** 67

[14] Mahalakshmi V, Lincy A, Thomas J and Saban K V 2014 *Optik – Int. J. Light Electron Opt.* **125** 191

[15] Mahalakshmi V, Lincy A, Thomas J and Saban K V 2015 *Optik – Int. J. Light Electron Opt.* **126** 2419

[16] Henisch H K 1970 *Crystal growth in Gels* (Pittsburg: Pennsylvania University Press) p 20

[17] Bruker 2004 *APEX2, SAINT and XPREP* Bruker AXS Inc. Madison, Wisconsin, USA

[18] Altomare A, Cascarano G, Giacovazzo C and Guagliardi A 1993 *J. Appl. Crystallogr.* **26** 343

[19] Sheldrick G M 1997 SHELXL-97 *Program for the Refinement of Crystal Structures* Germany: University of Göttingen

[20] Sheldrick G M 2008 *Acta Cryst.* A **64** 112

[21] Bellamy L J 1975 *The IR Spectra of Complex Molecules* vol 1 (London: Chapman and Hall) 3rd ed

[22] Nakamoto K 1963 *Infrared Spectra of Inorganic and Coordination Compounds* part III (New York: John Wiley and sons) 6th ed p 62

[23] Edsall J T 1937 *J. Chem. Phys.* **5** 508

[24] Spek A L 1990 *Acta Cryst.* A **46** C34

[25] Spek A L 1998 *PLATON* A Multipurpose Crystallographic Tool (The Netherlands: University of Utretcht)

[26] Farrugia L J 1997 *J. Appl. Cryst.* **30** 565

[27] Boaz B M, Selvaraj R S, Kumar K S and Das S J 2009 *Indian J. Phys.* **83** (12) 1647

[28] Raj C J, Krishnan S, Dinakaran S, Linet J M, Uthrakumar R, Robert R and Das S J 2009 *J. Mater. Sci. Technol.* **25** 745

[29] Sharma K K, Kotru P N, Tandon R P and Wanklyn B M 1998 *J. Phys. Cond. Matt.* **10** 5277

[30] Mazen S A, Metawe F and Mansour S F 1997 *J. Phys. D: Appl. Phys.* **30** 1799

[31] Babu G A, Sreedhar S, Rao S V and Ramasamy P 2010 *J. Cryst. Growth* **312** 1957

[32] Kubelka P and Munk F 1931 *Z. Tech. Phys.* (Leipzig) **12** 593

[33] Kubelka P 1948 *J. Opt. Soc. Am.* **38** 448